AGING: OXIDATIVE STRESS AND DIETARY ANTIOXIDANTS

AGING: OXIDATIVE STRESS AND DIETARY ANTIOXIDANTS

Edited by

VICTOR R. PREEDY

King's College London,
London, UK

ELSEVIER

AMSTERDAM • BOSTON • HEIDELBERG • LONDON
NEW YORK • OXFORD • PARIS • SAN DIEGO
SAN FRANCISCO • SINGAPORE • SYDNEY • TOKYO
Academic Press is an imprint of Elsevier

AP

Academic Press is an imprint of Elsevier
The Boulevard, Langford Lane, Kidlington, Oxford, OX5 1GB, UK
225 Wyman Street, Waltham, MA 02451, USA

First published 2014

British Library Cataloguing in Publication Data
A catalogue record for this book is available from the British Library

Library of Congress Cataloguing in Publication Data
A catalogue record for this book is available from the Library of Congress

ISBN: 978-0-12-405933-7

For information on all Academic Press publications
visit our website at store.elsevier.com

Printed and bound in the United States

14 15 16 17   10 9 8 7 6 5 4 3 2 1

Contents

8. Antioxidants, Vegetarian Diets and Aging

S. WACHTEL-GALOR, P.M. SIU, I.F.F. BENZIE

9. Enteral Nutrition to Increase Antioxidant Defenses in Elderly Patients

JOSÉ EDUARDO DE AGUILAR-NASCIMENTO

10. Herbs and Spices in Aging

SUHAILA MOHAMED

11. Coenzyme Q_{10} as an Antioxidant in the Elderly

ELENA M. YUBERO-SERRANO, ANTONIO GARCIA-RIOS, JAVIER DELGADO-LISTA, PABLO PÉREZ-MARTINEZ, ANTONIO CAMARGO, FRANCISCO PEREZ-JIMENEZ, JOSE LOPEZ-MIRANDA

12. Vitamin C and Physical Performance in the Elderly

KYOKO SAITO, ERIKA HOSOI, AKIHITO ISHIGAMI, TETSUJI YOKOYAMA

13. Tryptophan and Melatonin-Enriched Foodstuffs to Improve Antioxidant Status in Aging

M. GARRIDO, A.B. RODRÍGUEZ, M.P. TERRÓN

14. Protective Effects of Vitamin C on Age-Related Bone and Skin Phenotypes Caused by Intracellular Reactive Oxygen Species

SHUICHI SHIBUYA, HIDETOSHI NOJIRI, DAICHI MORIKAWA, HIROFUMI KOYAMA, TAKAHIKO SHIMIZU

15. S-Equol, an Antioxidant Metabolite of Soy Daidzein, and Oxidative Stress in Aging: A Focus on Skin and on the Cardiovascular System

RICHARD L. JACKSON, JEFFREY S. GREIWE, RICHARD J. SCHWEN

24. Lycopene and Other Antioxidants in the Prevention and Treatment of Osteoporosis in Postmenopausal Women

L.G. RAO, N.N. KANG, A.V. RAO

25. Zinc, Oxidative Stress in the Elderly and Implications for Inflammation

ANANDA S. PRASAD

26. Antioxidant Supplementation in the Elderly and Leukocytes

DAVID SIMAR, CORINNE CAILLAUD

27. Metabolic Mobilization Strategies to Enhance the Use of Plant-Based Dietary Antioxidants for the Management of Type 2 Diabetes

DIPAYAN SARKAR, KALIDAS SHETTY

Contributors

Shadwan Alsafwah Division of Cardiovascular Diseases, University of Tennessee Health Science Center, Memphis, TN, USA

Fawaz Alzaid Diabetes and Nutritional Sciences Division, School of Medicine, King's College London, Franklin-Wilkins Building, London, UK

B. Andallu Sri Sathya Sai Institute of Higher Learning, Anantapur, A.P., India

Raza Askari Division of Cardiovascular Diseases, University of Tennessee Health Science Center, Memphis, TN, USA

Sylvette Ayala-Peña Department of Pharmacology and Toxicology, University of Puerto Rico Medical Sciences Campus, San Juan, Puerto Rico

Mario Barbagallo Geriatric Unit, Department of Internal Medicine DIBIMIS, University of Palermo, Italy

I.F.F. Benzie Department of Health Technology & Informatics, The Hong Kong Polytechnic University, Hung Hom, Kowloon, Hong Kong

Syamal K. Bhattacharya Division of Cardiovascular Diseases, University of Tennessee Health Science Center, Memphis, TN, USA

Brunna Cristina Bremer Boaventura Department of Nutrition, Health Sciences Center, Federal University of Santa Catarina, Campus Trindade, Florianópolis/SC, Brazil

Corinne Caillaud Exercise Physiology and Nutrition, Faculty of Health Sciences, University of Sydney, Lidcombe NSW, Australia

Antonio Camargo Lipids and Atherosclerosis Unit, IMIBIC/Reina Sofia University Hospital/University of Cordoba, and CIBER Fisiopatologia Obesidad y Nutricion (CIBEROBN), Instituto de Salud Carlos III, Córdoba, Spain

José Eduardo de Aguilar-Nascimento Department of Surgery, Julio Muller University Hospital, Federal University of Mato Grosso, Cuiaba, Mato Grosso, Brazil

Javier Delgado-Lista Lipids and Atherosclerosis Unit, IMIBIC/Reina Sofia University Hospital/University of Cordoba, and CIBER Fisiopatologia Obesidad y Nutricion (CIBEROBN), Instituto de Salud Carlos III, Córdoba, Spain

Patricia Faria Di Pietro Department of Nutrition, Health Sciences Center, Federal University of Santa Catarina, Campus Trindade, Florianópolis/SC, Brazil

Dwight A. Dishmon Division of Cardiovascular Diseases, University of Tennessee Health Science Center, Memphis, TN, USA

Ligia J. Dominguez Geriatric Unit, Department of Internal Medicine DIBIMIS, University of Palermo, Italy

Victor Farah Division of Cardiovascular Diseases, University of Tennessee Health Science Center, Memphis, TN, USA

Antonio Garcia-Rios Lipids and Atherosclerosis Unit, IMIBIC/Reina Sofia University Hospital/University of Cordoba, and CIBER Fisiopatologia Obesidad y Nutricion (CIBEROBN), Instituto de Salud Carlos III, Córdoba, Spain

M. Garrido Department of Physiology (Neuroimmunophysiology and Chrononutrition Research Group), Faculty of Science, University of Extremadura, Badajoz, Spain

Jeffrey S. Greiwe Ausio Pharmaceuticals, LLC, Cincinnati, Ohio, USA

Erika Hosoi Research Team for Promoting the Independence of the Elderly, Tokyo Metropolitan Institute of Gerontology, Tokyo, Japan

Chao A. Hsiung Institute of Population Health Sciences, National Health Research Institutes, Miaoli County, Taiwan

Chih-Cheng Hsu Institute of Population Health Sciences, National Health Research Institutes, Miaoli County, Taiwan

Nikolay K. Isaev Lomonosov Moscow State University, A.N. Belozersky Institute of Physico-Chemical Biology, Moscow, Russia

Akihito Ishigami Molecular Regulation of Aging, Tokyo Metropolitan Institute of Gerontology, Tokyo, Japan

Hiroyasu Iso Public Health, Department of Social and Environmental Medicine, Graduate School of Medicine, Osaka University, Suita, Osaka, Japan

Richard L. Jackson Ausio Pharmaceuticals, LLC, Cincinnati, Ohio, USA

N.N. Kang Department of Nutritional Sciences, University of Toronto, Toronto, Canada

Nadezhda A. Kapay Department of Brain Research, Research Center of Neurology, Russian Academy of Medical Sciences, Pereulok Obukha 5, Moscow, Russia

Jozef Kedziora Department of Biochemistry, Collegium Medicum UMK in Bydgoszcz, Poland

Kornelia Kedziora-Kornatowska Department and Clinic of Geriatrics, Collegium Medicum UMK in Bydgoszcz, Poland

Hirofumi Koyama Department of Advanced Aging Medicine, Chiba University Graduate School of Medicine, Inohana, Chuo-ku, Chiba, Japan

Xi-Zhang Lin Department of Internal Medicine, College of Medicine, National Cheng Kung University, Tainan, Taiwan

Xiaoyan Liu University of Texas Health Science Center at San Antonio, Department of Cellular and Structural Biology, San Antonio, TX, USA, and The Preclinical Medicine Institute of Beijing, University of Chinese Medicine, Chao Yang District, Beijing, China

Jose Lopez-Miranda Lipids and Atherosclerosis Unit, IMIBIC/Reina Sofia University Hospital/University of Cordoba, and CIBER Fisiopatologia Obesidad y Nutricion (CIBEROBN), Instituto de Salud Carlos III, Córdoba, Spain

Konstantin G. Lyamzaev Lomonosov Moscow State University, A.N. Belozersky Institute of Physico-Chemical Biology, Moscow, Russia

Lucien C. Manchester University of Texas Health Science Center at San Antonio, Department of Cellular and Structural Biology, San Antonio, TX, USA

Koutatsu Maruyama Department of Basic Medical Research and Education, Ehime University Graduate School of Medicine, Shitsukawa, Toon, Ehime, Japan

M.S. Mekha Sri Sathya Sai Institute of Higher Learning, Anantapur, A.P., India

Maria Grazia Modena University of Modena and Reggio Emilia, Italy

Suhaila Mohamed Institute of BioScience, Universiti Putra Malaysia, Serdang, Selangor, Malaysia

Daichi Morikawa Department of Advanced Aging Medicine, Chiba University Graduate School of Medicine, Chuo-ku, Chiba, Japan, and Department of Orthopaedics, Juntendo University Graduate School of Medicine, Bunkyo-ku, Tokyo, Japan

Hidetoshi Nojiri Department of Orthopaedics, Juntendo University Graduate School of Medicine, Bunkyo-ku, Tokyo, Japan

Vinood B. Patel Department of Biomedical Science, Faculty of Science & Technology, University of Westminster, London, UK

Francisco Perez-Jimenez Lipids and Atherosclerosis Unit, IMIBIC/Reina Sofia University Hospital/University of Cordoba, and CIBER Fisiopatologia Obesidad y Nutricion (CIBEROBN), Instituto de Salud Carlos III, Córdoba, Spain

Pablo Pérez-Martinez Lipids and Atherosclerosis Unit, IMIBIC/Reina Sofia University Hospital/University of Cordoba, and CIBER Fisiopatologia Obesidad y Nutricion (CIBEROBN), Instituto de Salud Carlos III, Córdoba, Spain

Olga V. Popova Department of Brain Research, Research Center of Neurology, Russian Academy of Medical Sciences, Pereulok Obukha 5, Moscow, Russia

Ananda S. Prasad Department of Oncology, Wayne State University School of Medicine and Barbara Ann Karmanos Cancer Institute, Detroit, MI, USA

Victor R. Preedy Diabetes and Nutritional Sciences Division, School of Medicine, King's College London, Franklin-Wilkins Building, London, UK

C.U. Rajeshwari Sri Sathya Sai Institute of Higher Learning, Anantapur, A.P., India

A.V. Rao Department of Nutritional Sciences, University of Toronto, Toronto, Canada

L.G. Rao Department of Medicine, St Michael's Hospital and University of Toronto, Toronto, Canada

Russel J. Reiter University of Texas Health Science Center at San Antonio, Department of Cellular and Structural Biology, San Antonio, TX, USA

A.B. Rodríguez Department of Physiology (Neuroimmunophysiology and Chrononutrition Research Group), Faculty of Science, University of Extremadura, Badajoz, Spain

Sergio A. Rosales-Corral Centro de Investigacion Biomedica de Occidente, Instituto Mexicano Del Seguro Social, Guadalajara, Jalisco, Mexico, and University of Texas Health Science Center at San Antonio, Department of Cellular and Structural Biology, San Antonio, TX, USA

Joanna Rybka Department of Biochemistry, Collegium Medicum UMK in Bydgoszcz, Poland, and Life4Science Foundation, Bydgoszcz, Poland

Kyoko Saito Research Team for Promoting the Independence of the Elderly, Tokyo Metropolitan Institute of Gerontology, Tokyo, Japan

Dipayan Sarkar Department of Plant Sciences, Loftsgard Hall, NDSU, Fargo, ND, USA

Rahul Saxena Department of Biochemistry, School of Medical Sciences & Research, Sharda University, Greater Noida (UP), India

Irina N. Scharonova Department of Brain Research, Research Center of Neurology, Russian Academy of Medical Sciences, Pereulok Obukha 5, Moscow, Russia

Richard J. Schwen Ausio Pharmaceuticals, LLC, Cincinnati, Ohio, USA

Kalidas Shetty Department of Plant Sciences, Loftsgard Hall, NDSU, Fargo, ND, USA

Shuichi Shibuya Department of Advanced Aging Medicine, Chiba University Graduate School of Medicine, Chuo-ku, Chiba, Japan

Takahiko Shimizu Department of Advanced Aging Medicine, Chiba University Graduate School of Medicine, Inohana, Chuo-ku, Chiba, Japan

R.I. Shobha Sri Sathya Sai Institute of Higher Learning, Anantapur, A.P., India

David Simar Inflammation and Infection Research, School of Medical Sciences, Faculty of Medicine, University of New South Wales, Sydney NSW, Australia

P.M. Siu Department of Health Technology & Informatics, The Hong Kong Polytechnic University, Hung Hom, Kowloon, Hong Kong

Vladimir G. Skrebitsky Department of Brain Research, Research Center of Neurology, Russian Academy of Medical Sciences, Pereulok Obukha 5, Moscow, Russia

Vladimir P. Skulachev Lomonosov Moscow State University, A.N. Belozersky Institute of Physico-Chemical Biology, Moscow, Russia

John M. Starr Centre for Cognitive Ageing and Cognitive Epidemiology, Edinburgh, United Kingdom

Robert J. Starr School of Medicine and Dentistry, Polwarth Building, Foresterhill, Aberdeen, United Kingdom

Elena V. Stelmashook Department of Brain Research, Research Center of Neurology, Russian Academy of Medical Sciences, Pereulok Obukha 5, Moscow, Russia

Dun-Xian Tan University of Texas Health Science Center at San Antonio, Department of Cellular and Structural Biology, San Antonio, TX, USA

M.P. Terrón Department of Physiology (Neuroimmunophysiology and Chrononutrition Research Group), Faculty of Science, University of Extremadura, Badajoz, Spain

Carlos A. Torres-Ramos Department of Physiology, University of Puerto Rico Medical Sciences Campus, San Juan, Puerto Rico

Floor van Heesch Division of Pharmacology, Utrecht Institute for Pharmaceutical Sciences (UIPS), Faculty of Science, Utrecht University, Utrecht, The Netherlands

S. Wachtel-Galor Department of Health Technology & Informatics, The Hong Kong Polytechnic University, Hung Hom, Kowloon, Hong Kong

Karl T. Weber Division of Cardiovascular Diseases, University of Tennessee Health Science Center, Memphis, TN, USA

I-Chien Wu Institute of Population Health Sciences, National Health Research Institutes, Miaoli County, Taiwan

Tetsuji Yokoyama Department of Human Resources Development, National Institute of Public Health, Saitama, Japan

Elena M. Yubero-Serrano Lipids and Atherosclerosis Unit, IMIBIC/Reina Sofia University Hospital/University of Cordoba, and CIBER Fisiopatologia Obesidad y Nutricion (CIBEROBN), Instituto de Salud Carlos III, Córdoba, Spain

Dmitry B. Zorov Lomonosov Moscow State University, A.N. Belozersky Institute of Physico-Chemical Biology, Moscow, Russia

Preface

In the past few decades there have been major advances in our understanding of the etiology of disease and its causative mechanisms. Increasingly it is becoming evident that free radicals are contributory agents: either to initiate or propagate the pathology or add to an overall imbalance. Furthermore, reduced dietary antioxidants can also lead to specific diseases and preclinical organ dysfunction. On the other hand, there is abundant evidence that dietary and other naturally occurring antioxidants can be used to prevent, ameliorate or impede such diseases. The science of oxidative stress and free radical biology is rapidly advancing and new approaches include the examination of polymorphism and molecular biology. The more traditional sciences associated with organ functionality continue to be explored but their practical or translational applications are now more sophisticated.

However, most textbooks on dietary antioxidants do not have material on the fundamental biology of free radicals, especially their molecular and cellular effects on pathology. They may also fail to include material on the nutrients and foods which contain antioxidative activity. In contrast, most books on free radicals and organs disease have little or no text on the usage of natural antioxidants.

The series **Oxidative Stress and Dietary Antioxidants** aims to address the aforementioned deficiencies in the knowledge base by combining in a single volume the science of oxidative stress and the putative therapeutic usage of natural antioxidants in the diet, its food matrix or plants. This is done in relation to a single organ, disease or pathology. These include cancer, addictions, immunology, HIV, aging, cognition, endocrinology, pregnancy and fetal growth, obesity, exercise, liver, kidney, lungs, reproductive organs, gastrointestinal tract, oral health, muscle, bone, heart, kidney and the CNS.

In the present volume, **Aging: Oxidative Stress and Dietary Antioxidants,** holistic information is imparted within the structured format of two main sections:

1. **Oxidative Stress and Aging**
2. **Antioxidants and Aging**

The first section on **Oxidative Stress and Aging** covers the basic biology of oxidative stress, from molecular biology to physiological pathology. Topics include markers of frailty, skin aging, cardiovascular disease, the liver, arthritis and diabetes. The second section, **Antioxidants and Aging**, covers cellular and molecular processes of vegetarian diets, enteral nutrition, natural antioxidants in foods and the diet, herbs and spices, coenzyme Q10, vitamins C and D, S-equol, zinc, magnesium, tryptophan, melatonin-enriched foods and lycopene. There is also material on the aging processes, age-related pathologies and organ systems, including menopause, physical performance, skin, bone and osteoporosis, the brain and neurodegeneration, the cardiovascular system, diabetes, muscle, arthritis, inflammation, mitochondria and leukocytes. The aforementioned provide a detailed framework for understanding the relationships between aging, oxidative stress and dietary components. However, more scientifically vigorous trials and investigations are needed to determine the comprehensive properties of many of these antioxidants, food items or extracts, as well as any adverse properties they may have.

The series is designed for dietitians and nutritionists, and food scientists, as well as health care workers and research scientists. Contributions are from leading national and international experts including those from world-renowned institutions.

Professor Victor R. Preedy,
King's College London

SECTION 1

OXIDATIVE STRESS AND AGING

CHAPTER

1

Oxidative Stress and Frailty: A Closer Look at the Origin of a Human Aging Phenotype

I-Chien Wu, Chao A. Hsiung, Chih-Cheng Hsu

Institute of Population Health Sciences, National Health Research Institutes, Miaoli County, Taiwan

Xi-Zhang Lin

Department of Internal Medicine, College of Medicine, National Cheng Kung University, Tainan, Taiwan

List of Abbreviations

ATM ataxia-telangiectasia mutated
ATR ataxia telangiectasia and Rad3-related
BER base excision repair
BubR1 mitotic checkpoint serine/threonine-protein kinase BUB1 beta
CDC25 cell-division cycle 25
CHK1 checkpoint kinase 1
CHK2 checkpoint kinase 2
CREBH cyclic AMP response element binding protein hepatocyte
CuZnSOD copper/zinc superoxide dismutase (SOD1)
DDR DNA-damage response
DHEA dehydroepiandrosterone
DHEAS dehydroepiandrosterone sulfate
eNOS endothelial nitric oxide synthase
ER endoplasmic reticulum
IκB inhibitor of kappa B
IKK1 inhibitor of nuclear factor kappa-B kinase subunit alpha
IKK2 inhibitor of nuclear factor kappa-B kinase subunit beta
IL-2 interleukin-2
IL-6 interleukin-6
IL-8 interleukin-8
IRS-1 insulin receptor substrate-1
JNK kinases c-jun N-terminal kinases
MDA malondialdehyde
MnSOD Mn-superoxide dismutase (SOD2)
MPT mitochondrial permeability transition
mtDNA mitochondrialDNA
MTH1 mutT human homolog 1
NADPH reduced form of nicotinamide adenine dinucleotide phosphate
NER nucleotide excision repair
NF-κB nuclear factor-κB
nNOS neuronal nitric oxide synthase
NUDT5 Nudix (nucleoside diphosphate linked moiety X)-type motif 5
8-OHdG 8-hydroxy-2′-deoxyguanosine
ROS reactive oxygen species
TNF-α tumor necrosis factor-α
WRN Werner protein

INTRODUCTION

Oxidative stress, defined as a disturbance in the prooxidant-antioxidant balance leading to oxidative damage,[1] has a key role in aging. More importantly, an increasing amount of evidence suggests that oxidative stress acts causally in the pathogenesis of numerous age-dependent and age-related chronic diseases. Over the past few decades, frailty has been increasingly recognized as a major health problem for older adults. As a distinct pathologic state, frailty contributes to numerous poor health outcomes independently of diseases and disability, and it is characterized by clinical presentations which are well defined and easily identifiable. Because of years of research, we have a better understanding of the system-level pathogenesis of frailty. It is becoming clear that frailty may have its origin in the fundamental aging process. Oxidative stress could play a crucial role in the cellular-level pathogenesis of frailty. In this chapter, the relationship between oxidative stress and frailty is delineated. To address this issue comprehensively, we attempt to integrate the results from human studies and model organism experiments. After a brief overview of oxidative stress in aging, the better known system-level abnormalities associated with the frailty syndrome are introduced. We then discuss whether and how oxidative stress at cellular levels causes frailty. Finally, we present a model of frailty pathogenesis incorporating the current understanding of frailty at the levels of molecules, cells, organs, and systems.

Aging
http://dx.doi.org/10.1016/B978-0-12-405933-7.00001-9

OXIDATIVE STRESS AND AGING

Aging represents 'progressive deterioration during the adult period of life that underlies an increasing vulnerability to challenges and a decreasing ability of an organism to survive'.[2] The deterioration is due to progressive accumulation of unrepaired damage and has the following core features: intrinsicality, universality, progressiveness and irreversibility, and it is genetically programmed.[2] The literature suggests that oxidative stress is the major cause of somatic damage.[2] Denham Harman proposed the free-radical theory of aging in 1956, which states that aging results from random deleterious damage to tissue by free radicals. His theory is among the most acknowledged theories of aging.[3] Since then, an increasing amount of evidence has indicated that oxidative stress increases with age and contributes to numerous age-related pathologic processes.[4]

The laboratory model organism experiments provide direct evidence that supports the importance of oxidative stress in aging. Numerous mutations that extend the lifespan of yeast, worms, flies, and mice have elevated antioxidant defenses and reduced oxidative stress. In yeast, major mutations that extend replicative and/or chronologic lifespans involve Ras-AC-PKA or Tor-Sch9 signaling.[5] Lifespan extension associated with altered activities in these pathways has been shown to require the antioxidant enzyme superoxide dismutase (Mn-SOD), which scavenges superoxide free radicals.[5] In *Caenorhabditis elegans*, lifespan extension can be achieved by reducing the activities of insulin/IGF-like signaling pathways (e.g. *age-1* and *daf-2* mutants), thereby activating the Forkhead FoxO transcription factor *daf-16*.[6] Active DAF-16 promotes the transcription of major antioxidant genes, including genes encoding catalases, MnSOD, and CuZnSOD. These antioxidants are necessary for lifespan extension in these mutant worms.[6] As in yeast and worms, insulin/IGF1 signaling pathways affect longevity in mice. Acting downstream of IGF receptors, $p66^{Shc}$ enhances production of mitochondrial reactive oxygen species (ROS) by catalyzing redox reactions, which yield hydrogen peroxide.[7] Deleting $p66^{Shc}$ in mice results in decreased oxidative stress, which correlates with an increased lifespan.[7]

Results of human studies are congruent with the findings of model organism experiments. An age-related increase in oxidative damage to macromolecules has been observed in humans.[8] DNA variants in the genes that modulate oxidative stress were linked to longevity.[9,10]

FRAILTY

Definition

As an extreme phenotype of human aging, frailty is a state of increased vulnerability with a decreased ability to maintain homeostasis.[11] Although it can be compounded by disease or disability, this vulnerability is primarily age related and is caused by a reduced reserve capacity of interconnected physiologic systems that adapt to stressors, leading to a breakdown of homeostasis.[11] Despite the lack of a clear consensus, there are several operational definitions of frailty in the literature; these definitions are based on different theories on the underlying causes of frailty. Comprehensive reviews of the definition of frailty are beyond the scope of this article and can be found elsewhere.[11] Two commonly used definitions are discussed.

According to the operational definition of *frailty phenotype* proposed by Fried et al, a person is considered frail if three or more of the following five criteria are present: unintentional weight loss, muscle weakness, slow walking speed, low physical activity, and exhaustion (Table 1.1).[12] Older adults with one or two of the criteria are considered prefrail, whereas those without any criteria are considered robust.[12] Being the commonly cited operational definition in frailty research, the frailty phenotype is based on the assumption that frailty arises

TABLE 1.1 Frailty Phenotype According to Fried et al[12] [a]

Criteria	Frailty Characteristic	Measure
1	Weight loss (unintentional)	>10 lbs lost unintentionally in prior year (reported)
	Shrinking	
	Sarcopenia	
2	Muscle weakness	Grip strength below cutoff value,[12] adjusted for gender and body mass index
3	Exhaustion Poor endurance	Answering 'moderate or most of the time' to 'I feel that everything I do is an effort' or 'I cannot get going'.
4	Slow walking speed	Walking speed below cutoff value,[12] based on time to walk 15 feet, adjusting for gender and standing height.
5	Low physical activity	Kilocalories expended per week below cutoff value (383 kcal/wk in men; 270 kcal/wk in women)[12]

[a]*An individual is considered frail if three or more of the five criteria are present. People with one or two of the criteria are considered prefrail, whereas those without any criteria are considered robust.*

from unique pathologic processes that are independent of diseases and disability. Previous research has shown that the frailty phenotype is able to predict adverse health outcomes independently of disease and disability, and frail older adults are at greater risk compared with prefrail adults.[12] Moreover, there are clues that specific pathophysiologic processes lead to the development of frailty in the absence of disease.[13]

Unlike the Fried definition, Rockwood et al hypothesized that frailty arises from the accumulation of potentially unrelated diseases, subclinical dysfunctions, and disability, and represents an intermediary mechanism linking these conditions to poor health outcomes.[14] The concept of frailty being a distinct pathologic state, separate from diseases and disability, is less emphasized. Frailty is defined by a frailty index, which is created by counting the number of health deficits in an older adult. The health deficits can be any clinical symptom, sign, disease, disability, laboratory, imaging, or other examination abnormality.[14] Using this definition, frailty is associated with poor health outcomes in different populations.[15]

Clinical Significance

The prevalence of frailty is high. It is estimated that a minimum of 10–25% of people aged 65 years and older (and 30–45% of those aged 85 years and older) are frail.[12] Frailty is the core issue in healthy aging and geriatric medicine. In contrast to the younger population, the older population is characterized by a greater variation in health status, outcomes, or response to therapy, which cannot be explained by age and disease alone.[16] As a measure of biologic age, frailty permits superior risk prediction in older adults compared with chronologic age and diseases. Regardless of the operational definitions used, it has been repeatedly demonstrated that, compared with age and chronic diseases, frailty stratification is more strongly associated with an older adult's risk of poor outcomes, including infections, disabilities, institutionalization, and death.[12,15]

Organ and System Abnormalities Associated with Frailty

As described, frailty is caused by abnormal interconnected physiologic systems, which are essential for maintaining homeostasis. The key physiologic systems currently known to be involved in frailty pathogenesis include the musculoskeletal system (skeletal muscle), metabolism (adiposity, insulin activity), immune system (inflammation), endocrine system (insulin-like growth factor-1, dehydroepiandrosterone sulfate, and testosterone), and autonomic nervous system (Fig. 1.1).[11]

Sarcopenia

Body composition changes with age. An age-related loss of skeletal muscle mass is termed sarcopenia.[17]

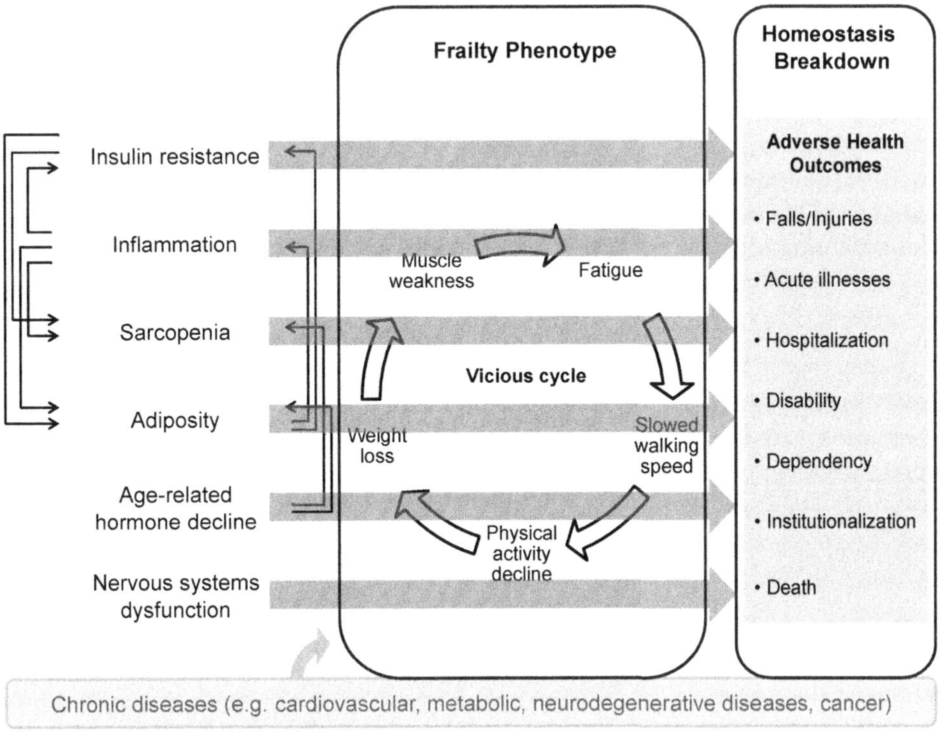

FIGURE 1.1 Organ and system abnormalities associated with frailty. As a human aging phenotype, frailty is characterized by an increased likelihood of homeostasis breakdown. Frailty, either alone or in the presence of diseases, predicts future adverse health outcomes in older adults. Previous research has suggested that several organ/system abnormalities are responsible for homeostasis breakdown at an advanced age, and frailty may represent the clinical manifestations of the pathogenic processes. The key pathogenic processes are (i) insulin resistance; (ii) inflammation; (iii) sarcopenia; (iv) adiposity; (v) age-related hormone decline; and (vi) nervous system dysfunction.[11] These processes are interrelated, and a vicious cycle typically develops. Subclinical diseases may have a role in frailty development.[16]

Sarcopenia is common among older people, with prevalence ranging from 9% to 18% over the age of 65 years.[17] Skeletal muscle contractions provide the power necessary for human mobility. In addition, skeletal muscles account for a large portion of human body mass and are essential for metabolism.[17] Thus, the loss of skeletal muscle mass with age could have significant effects on physical functions and health in old age. Sarcopenia has been demonstrated to lead to frailty.[18] Frailty can worsen the severity of sarcopenia through certain mechanisms (e.g. adiposity with lipid infiltration of muscle tissue), and a vicious cycle typically develops.[19]

Adiposity

Unlike skeletal muscle mass, fat increases with age.[20] In addition, aging is accompanied by fat redistribution, with accumulation inside and around skeletal muscles.[19] Older adults with a high body mass index (BMI) are more likely to be frail compared to those with a normal BMI.[21] Abdominal obesity with an elevated waist circumference also increases the risk of frailty in humans.[21] Adipose tissue, particularly visceral adipose tissue, is not only an organ specializing in the storage and mobilization of lipids, but is also a remarkable endocrine organ that regulates the entire body's energy metabolism by secreting numerous molecules.[22] Increasing abdominal fat is known to lead to insulin resistance.[22] Insulin resistance and related metabolic disorders have been shown to be major risk factors of frailty, possibly by altering the muscle metabolism.[13]

Inflammation

Inflammation is crucial in the pathogenesis of frailty. Inflammatory cytokines, including interleukin-6 (IL-6), increase with age, and related pathways are strongly implicated in aging.[23] In prospective cohort studies, older adults with higher baseline levels of inflammatory cytokines were more vulnerable to the future occurrences of frailty.[13] Activation of inflammation can directly contribute to muscle-mass loss and muscle dysfunction,[24] resulting in a frailty phenotype with a slow gait speed and low muscle strength.[25] Specifically, TNF-α is recognized to be involved in inflammation-related and age-related impairments in muscle mass and function.[26] In addition, inflammation can cause frailty by triggering global metabolic derangements, including increased adiposity, insulin resistance, and endothelial dysfunction.[13]

Age-Related Hormone Decline

Endocrine system activity is known to change markedly with age. Because of decreased releases of gonadotropin and decreased secretions at the gonadal level, a gradual decline in serum testosterone levels occurs during aging.[27] Testosterone plays a key role in maintaining body composition.[28] Low levels of serum testosterone have been shown to have an independent relationship with frailty.[29] Insulin-like growth factor-1 (IGF-1) is primarily produced by the liver; production is regulated by growth hormone secreted from the pituitary gland. The amount of growth hormone secreted from the pituitary gland declines gradually with age. In parallel, levels of circulating IGF-1 decrease with aging.[30] The primary function of IGF-1 is to promote growth and development, including muscle protein synthesis. As a crucial regulator of muscle mass, IGF-1 levels are associated with muscle strength and mobility.[31] Dehydroepiandrosterone (DHEA) and its sulfate form (DHEAS) are secreted from the adrenal gland. Adrenocortical cells that produce hormones decrease in activity in aging.[32] DHEAS is converted to androgenic and estrogenic steroid in peripheral tissues, and represents another hormone that has significant trophic effects on skeletal muscles.[32] Low circulating DHEAS levels have been shown to be independently linked to frailty.[33]

Autonomic Nervous System Dysfunction

The autonomic nervous system controls vital organ functions and has a crucial role in maintaining homeostasis. However, the functions of the autonomic nervous system change with age.[34] Although the sympathetic nervous system activities of the heart, skeletal muscles, and gut increase with age, epinephrine secretion from the adrenal medulla and cardiac vagal tone are markedly reduced with age.[34] Low heart-rate variability, a manifestation of autonomic nervous system dysfunction, is associated with frailty.[35]

OXIDATIVE STRESS AND FRAILTY

Frailty is a state of homeostasis breakdown in advanced age caused by sarcopenia, adiposity, insulin resistance, inflammation, age-related hormone decline, and nervous system dysfunction. What remains unclear is how these abnormalities in multiple systems occur. It is apparent that age is the greatest risk factor for frailty. Because oxidative stress increases with age, oxidative stress may be involved in frailty. A growing number of studies have suggested that cellular oxidative stress may cause multiple organ/system pathologies that lead to frailty and may trigger the vicious cycle leading toward failure of the homeostasis mechanism.

Oxidative Stress Is Associated with Frailty and Frailty Components

The association between oxidative stress and frailty has been consistently observed in studies of older adults; each study has used a different strategy to measure oxidative stress (Table 1.2). Oxidative stress represents a disturbance in the prooxidant-antioxidant balance, which leads to oxidative damage. Oxidative stress can be assessed by measuring antioxidants, ROS, or oxidative damage. Numerous studies have shown that low antioxidant levels predict the development of frailty. (A detailed discussion of how antioxidant levels are related to frailty can be found elsewhere in the book, and this topic is not addressed in this chapter.)

Oxidative Damage and Frailty

One strategy for measuring oxidative stress is to assess the damage that excessive prooxidants cause (the chemical fingerprint that prooxidants leave). Using oxidative damage as a biomarker of oxidative stress has an advantage over the measurement of antioxidants because the damage detected not only acts as an indicator of excessive ROS but can also potentially indicate the possible biologic consequences of oxidative stress.

Proteins can be damaged by ROS to yield oxidized amino acid residues, thereby forming proteins with carbonyl groups.[36] Howard et al demonstrated that high protein carbonyl levels were cross-sectionally associated with poor handgrip strength, which is a component of frailty.[36] Semba et al further showed that serum protein carbonyl levels independently predicted gait speed decline, which is another component of frailty.[37]

Lipids can also be oxidized, and certain oxidized lipids are cytotoxic. Isoprostanes are produced by lipid peroxidation (phospholipid peroxidation), and are among the best available biomarkers of lipid peroxidation.[38] Isoprostanes are biologically active.[38] Cesari et al noted that high levels of urinary 8-iso-prostaglandin $F_{2\alpha}$, an isoprostane, independently predicted death in older adults.[39] High levels of plasma oxidized low-density lipoprotein were shown to predict onset of mobility limitations.[40] Serviddio et al found that frailty was independently correlated with high plasma malondialdehyde (MDA), another marker of lipid peroxidation.[41]

Similarly, nucleic acid is subject to oxidative damage. Because of its central role in maintaining cell function, DNA damage could trigger numerous cellular abnormalities, and thus it has biologic significance. Oxidized DNA is known to facilitate disease pathogenesis.[42] ROS can cause various types of DNA damage; these include strand breakage, abasic sites, deoxyribose damage, and modifications of the DNA bases (purines and pyrimidines). The modified bases, particularly 8-hydroxy-2′-deoxyguanosine (8-OHdG), are the most commonly used markers of oxidative DNA damage.[43] Sensitive assays for detecting 8-OHdG are available. Furthermore, the presence of 8-OHdG could result in G-to-T transversion mutations and has deleterious biologic consequences.[42] The base 8-OHdG is extremely deleterious and is under strict regulation by multiple repair mechanisms.[42,43] First, oxidized

TABLE 1.2 Correlative Studies Supporting the Roles of Oxidative Stress in Frailty[a]

Studies	Oxidative Stress Biomarkers	Design	Population	Frailty Definition or Components	Key Findings
OXIDATIVE DAMAGE AND FRAILTY					
Howard et al[36]	Serum protein carbonyl levels	Cross-sectional study	672 women; aged ≥65	Muscle weakness (low grip strength)	Protein carbonyl levels were negatively correlated with grip strength after adjustment
Semba et al[37]	Serum protein carbonyl levels	Longitudinal study (3 years)	545 women; aged ≥65	Slow walking speed (<0.4 m/s)	Protein carbonyl levels were independently associated with decline in walking speed and incident slow walking speed
Serviddio et al[41]	Plasma malondialdehyde	Cross-sectional study	62 men and women; aged ≥65	Frailty phenotype according to Fried et al[12]	High plasma malondialdehyde levels were independently associated with frailty
Wu et al[43]	Serum 8-hydroxy-2′-deoxyguanosine	Cross-sectional study	90 men and women; aged ≥65	Frailty phenotype according to Fried et al[12]	High serum 8-OHdG levels were independently associated with frailty
REACTIVE OXYGEN SPECIES AND FRAILTY					
Baptista et al[45]	Whole blood cells superoxide anion production by NADPH oxidase	Cross-sectional study	280 men and women; aged >60	Slow walking speed (<0.8 m/s)	Persons within the highest tertile of superoxide anion production had higher adjusted odds ratio of being frail compared with those in the lower 2 tertiles

[a]*See the text for a detailed description.*

guanine nucleotide is removed from the cellular nucleotide pool and prevented from being incorporated into DNA through the actions of two key enzymes: mutT human homolog 1 (MTH1) and Nudix (nucleoside diphosphate linked moiety X)-type motif 5 (NUDT5). Once incorporated into DNA, the oxidized DNA damage is eliminated by either base excision repair (BER) or nucleotide excision repair (NER). Finally, if 8-OHdG escapes repair, and mispairing with adenine occurs, the mismatch repair system removes the adenine. Therefore, the observed 8-OHdG levels rely on the balance between levels of ROS and activities of DNA repair mechanisms. In a study of community-dwelling older adults, frailty was found to be associated with high serum 8-OHdG levels.[43] The observed link between oxidative DNA damage and frailty suggests the potential roles of oxidative DNA damage and DNA damage responses (e.g. DNA repair mechanisms, cell senescence, and apoptosis) in frailty, which is a human aging phenotype. Cell senescence and apoptosis are further described below.

Reactive Oxygen Species and Frailty

ROS are direct markers of oxidative stress.[44] However, because of their evanescent nature, they are difficult to measure directly in humans.[44] To examine the relationships between frailty and ROS, Baptista et al recently measured and compared the superoxide anion production by NADPH oxidase in whole blood cells of frail and healthy older adults in an *ex vivo* experiment.[45] Superoxide anion overproduction was noted in isolated white blood cells of frail older adults, which further confirmed the association between oxidative stress and frailty. More research is necessary.

Mechanisms Linking Oxidative Stress to Frailty

Existing evidence supporting the roles of oxidative stress in frailty is primarily from human observational studies. As described, oxidative stress is correlated with frailty. Based on these findings, we are unable to make causal links between oxidative stress and frailty. An experimental approach is necessary to test the causal relationship. However, a direct link between oxidative stress and frailty is biologically plausible. We review the possible mechanisms underlying this relationship, which should be addressed in future experimental research. One strategy to test the causality is to examine the effects of altering oxidative stress on the frailty pathology in animals by using genetic manipulation of the genes involved in antioxidant defense or damage repair. The supportive results of experiments using model organisms are shown in Table 1.3.

Oxidative stress could trigger the pathologic pathways underlying frailty through at least three interrelated mechanisms, such as causing oxidative damage, promoting cellular apoptosis and senescence, and generating inappropriate cellular signaling.

Oxidative Stress, Oxidative Damage and Frailty

Mitochondria are the primary source of ROS produced by cells. Approximately 0.4–4% of respired oxygen has been estimated to convert to superoxide radicals during the course of normal oxidative metabolism occurring in the mitochondria.[2] Hence, mitochondrial proteins are susceptible to oxidative damage.[8] In a mouse model lacking the mitochondrial superoxide dismutase (Mn-superoxide dismutase) (MnSOD), the activities of two essential enzymes involved in mitochondrial metabolic pathways, aconitase and succinate dehydrogenase, are severely decreased in multiple tissues.[46] Mammalian mitochondrial DNA (mtDNA) encodes polypeptides involved in the respiratory chain. Because it is attached to the inner mitochondrial membrane at a location in proximity to the site of ROS production, mtDNA is constantly at high risk of oxidative damage. Increased oxidative damage was noted in mtDNA isolated from mice that lacked MnSOD.[47]

Such oxidative damage can cause mitochondrial dysfunction, which has deleterious effects on both the mitochondria and other parts of the cell (Fig. 1.2). Because the mitochondrion is the powerhouse of the cell, mitochondrial dysfunction adversely affects energy-dependent cellular activity, resulting in a myriad of pathologies in metabolically active cells/tissues (e.g. neurons, muscle cells, pancreatic beta cells, and hematopoietic cells). For instance, skeletal muscle is the key insulin-responsive organ responsible for maintaining glucose homeostasis. The function of skeletal muscle mitochondria declines because of oxidative stress. Mitochondrial dysfunction raises intracellular levels of metabolites. Through activating protein kinase C, these metabolites can increase serine phosphorylation of IRS-1 and block insulin signaling.[48] This can best be shown in mtDNA-mutator mice. The integrity of mtDNA is normally maintained by the proofreading function of the inherent 3′–5′ exonuclease activity of mtDNA polymerase (PolgA). In these mtDNA-mutator mice, the proofreading function was defective, which resulted in the accumulation of mtDNA mutations and deletions during DNA replication.[49] The increase in mtDNA load is accompanied by abnormal mitochondrial functioning. Moreover, the mtDNA-mutator mice exhibit phenotypes overlapping with frailty (e.g. weight loss and decreased survival) and pathologies of frailty, including body composition changes, glucose intolerance, gonad degeneration, anemia, and cardiovascular abnormalities (Table 1.3).[49]

TABLE 1.3 Effects of Genetic Manipulations Related to Oxidative Stress and Stress Responses on Frailty Phenotypes in Rodent Models

Models	Description	Affected Cellular Process	Frailty-Like Phenotypes	References
Sod2$^{-/-}$ (Sod2^{tm1Cje}/Sod2^{tm1Cje}) (Sod2^{tm1Leb}/Sod2^{tm1Leb})	Homozygous targeted knock-out mice with mutation at the gene encoding intra-mitochondrial Mn-superoxide dismutase (SOD2)	Antioxidant defense maintaining the integrity of mitochondrial enzymes and function is defective, resulting in increased oxidative damage	Reduced lifespan; weight loss; muscle weakness; fatigue rapidly after exertion; sarcopenia with lipid accumulation in the skeletal muscle; anemia; neurodegeneration	46,47,54
Sod1$^{-/-}$ (Sod1^{tm1Cje}/Sod1^{tm1Cje}) (Sod1^{tm1Leb}/Sod1^{tm1Leb})	Homozygous targeted knock-out mice with mutation at the gene encoding CuZn superoxide dismutase (SOD1)	Antioxidant defense provided by the major superoxide scavenger in the cytoplasm, nucleus, lysosomes, and mitochondria intermembrane space is defective, resulting in increased oxidative damage	Reduced lifespan; weight loss; muscle weakness; abnormal gait; sarcopenia; gonadal degeneration	55
PolgA$^{mut/mut}$ (Polg$^{tm1.1Lrsn}$/Polg$^{tm1.1Lrsn}$)	Homozygous targeted knock-in mice with mutation at the gene encoding PolgA, the catalytic subunit of mtDNA polymerase	Defective proofreading of newly synthesized mtDNA leads to randomly accumulated somatic mtDNA mutations, mimicking age-related oxidative DNA damage accumulation	Reduced lifespan; weight loss; skeletal muscle abnormalities; gonadal degeneration; anemia; glucose intolerance	49
p53$^{+/m}$ (Trp53^{tm1Brd}/Trp53^{+})	Heterozygous targeted knock-out mice with mutation at the gene encoding p53	Truncated form of p53 augments the activity of wild type p53, resulting in increased senescence	Reduced lifespan; decreased stress tolerance; weight loss; abnormal gait; sarcopenia; anemia; gonadal degeneration	60
WRN$^{-/-}$ (Wrn^{tm1Lgu}/Wrn^{tm1Lgu})	Homozygous targeted mutation of the gene encoding WRN	DNA damage accumulates, resulting in accelerated senescence	Reduced lifespan; weight loss; insulin resistance; sarcopenia; gonadal degeneration	62,63
BubR1$^{H/H}$ (Bub1b^{tm1Jvd}/Bub1b^{tm1Jvd})	Homozygous targeted mutation of the gene encoding BubR1	Chromosome number instability due to BubR1defficiency leads to accelerated senescence	Reduced lifespan; weight loss; sarcopenia; gonadal degeneration	64,65

Mitochondrial dysfunction can cause a further increase in the mitochondria production of ROS, leading to a vicious cycle (Fig. 1.2).[2] The escalating ROS production raises the risk of widespread oxidative damage to other cellular components, including nuclear DNA, proteins, and lipids. Oxidative protein damage can impair the functioning of receptors, antibodies, signal transduction, and transport proteins. For example, the plasma membrane sodium pump, which uses energy released by ATP hydrolysis to export Na^{+} and import K^{+}, contains catalytically important SH groups and is susceptible to inactivation from oxidative damage.[50] The ER Ca^{2+}-uptake system is also vulnerable to oxidative damage. Proper functioning of these channels is critical to maintain cellular homeostasis, including keeping intracellular free calcium levels low and under tight regulation. By causing damage to these channels, oxidative stress can raise cytoplasmic Ca^{2+} levels, consequently activating calpains, which are Ca^{2+}-dependent cysteine proteases. Calpains are able to promote degradation of skeletal muscle actomyosin complexes by activating caspase-3 (through caspase-12), thus leading to muscle atrophy in frailty.[51] A rise in free Ca^{2+} can also stimulate endothelial nitric oxide synthase (eNOS) and neuronal nitric oxide synthase (nNOS), generating additional ROS and forming a vicious cycle.[52] Moreover, excess Ca^{2+} can trigger mitochondrial permeability transition (MPT), which opens pores in the inner membrane and reduces the membrane potential, thereby leading to cell death (apoptosis or necrosis). Oxidative damage to ER proteins or accumulation of misfolded proteins triggers the ER stress response.[53] The ER stress response inhibits insulin signaling and leads to insulin resistance through activating the Jun N-terminal kinase (JNK). In addition, during the stress response, transcription of acute-phase response genes is activated by cyclic AMP response element binding protein hepatocyte (CREBH); proteins involved in inflammation are secreted.

In MnSOD-knockout or MnSOD-heterozygous mice, where mitochondrial antioxidant defenses are deficient, severe mitochondrial damage is accompanied by damage to cellular macromolecules and increased apoptosis.[46,47,54] These animals exhibit various phenotypes and pathologies resembling frailty, including short survival time, muscle weakness, fatigue with poor endurance, sarcopenia, fat accumulation in the skeletal muscles, anemia, cardiovascular abnormalities, and neurodegeneration (Table 1.3).[54]

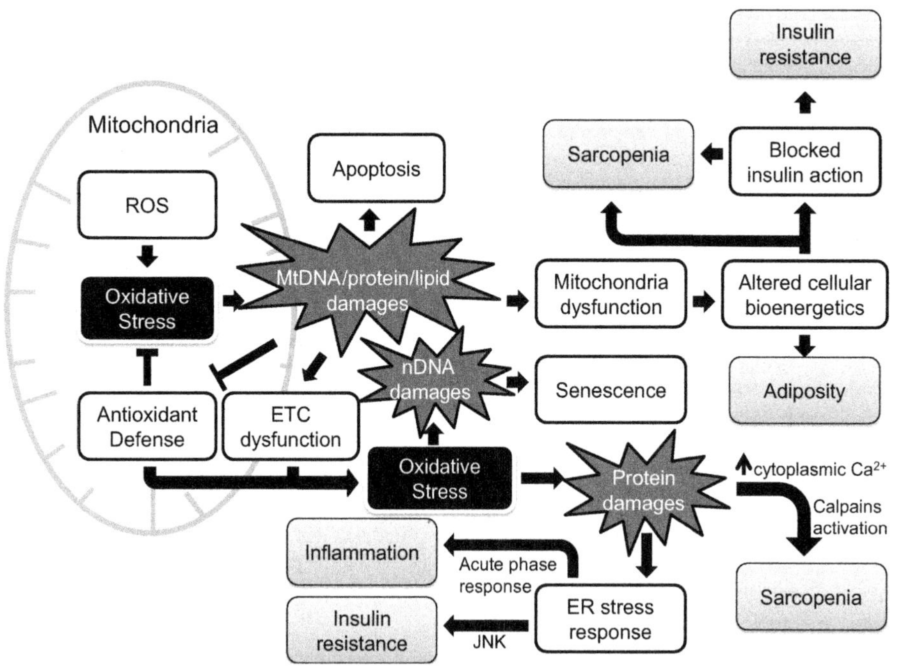

FIGURE 1.2 Oxidative stress, oxidative damage and frailty. Oxidative stress may cause frailty pathologies by generating oxidative damage. Mitochondria-generated ROS cause age-related accumulation of mitochondrial oxidative damage, leading to mitochondrial dysfunction and escalating ROS production, the latter causing further oxidative damage in other cellular components. Mitochondrial dysfunction can result in altered cellular bioenergetics, leading to insulin resistance, sarcopenia, and adiposity. Oxidative protein damage can cause defective cytoplasmic Ca^{2+} homeostasis, leading to sarcopenia. Oxidative protein damage may further trigger ER stress responses, resulting in acute-phase responses (inflammation) and inhibition of insulin signaling (through JNK activation). See the text for a detailed description. Arrows indicate activation; bars indicate inhibition.

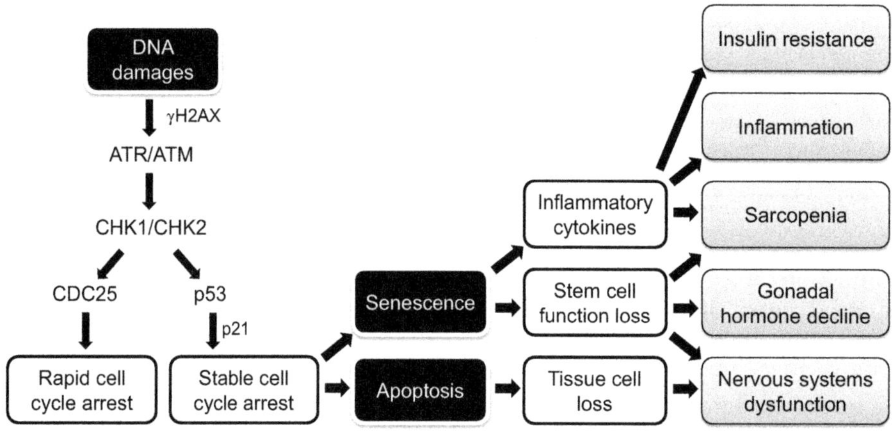

FIGURE 1.3 Oxidative stress, cellular senescence/apoptosis, and frailty. Cellular senescence and apoptosis can be initiated by severe oxidative DNA damage and contribute to frailty pathologies. DNA damage triggers cell cycle arrest and damage repair by activating ATR/ATM, CHK1/CHK2, CDC25, and p53. In the face of severe damage, cells undergo apoptosis or senescence. Apoptosis leads to neuronal cell loss and neurodegeneration. Stem cell senescence results in impaired regeneration in neural tissue, gonadal tissue, and skeletal muscle, thus contributing to the development of neurodegeneration, gonadal degeneration, and sarcopenia. Moreover, senescence cells secrete numerous pro-inflammatory cytokines and cause inflammation and insulin resistance. See the text for a detailed description. Arrows indicate activation; bars indicate inhibition.

Similar phenotypes and pathologies can be seen in mice lacking copper/zinc superoxide dismutase (SOD1) (CuZnSOD), another superoxide dismutase that provides both mitochondrial and cytoplasmic antioxidant defense (Table 1.3).[55]

Oxidative Stress, Cellular Senescence/Apoptosis, and Frailty

Oxidative DNA damage can induce stress responses that determine cell fates (Fig. 1.3).[56] Cells respond to DNA damage involving single- or double-stranded breaks by activating a signaling cascade known as DNA damage response (DDR) in an attempt to promptly repair the damage and maintain cell cycle arrest. The process involves two major protein kinases, ataxia telangiectasia and Rad3-related (ATR) and ataxia-telangiectasia mutated (ATM) proteins, which are recruited by sensors of the damage-specialized complex (ATR is recruited to damage sites by replication protein A and the heterotrimeric 9–1–1 complex composed of RAD9, RAD1, and HUS1, and ATM is primarily recruited by γH2AX, a process facilitated by DNA-damage checkpoint 1 and p53-binding protein 1). After recruitment, ATR and ATM activate downstream cell cycle checkpoint kinases, CHK1 and CHK2, which propagate the DNA damage signal to p53 and cell-division cycle 25 (CDC25). CDC25 inactivation leads to rapid cell cycle arrest, and the p53 activation results in stable cell cycle arrest through inducing the expression of p21. DDR is stopped, and cell proliferation is resumed when the DNA lesion is repaired. On the other hand, cells undergo apoptosis or cellular senescence if DNA damage is severe and remains unrepaired.

Apoptosis, also called programmed cell death, results in abnormal cell loss in tissues with limited regenerative capacity (e.g. neuron loss in neurodegenerative diseases) (Fig. 1.3).[4] Senescence is defined as a cell state of irreversible cell cycle arrest accompanied by characteristic morphologic and functional alterations.[56] Because the senescence state is characterized by a permanent inability to proliferate, senescent stem cells cease to proliferate, and the tissue regeneration capacity is suppressed. Senescence causes age-related diseases and dysfunction by diminishing the replicative capacities of neural cells, hematopoietic stem cells, skeletal muscle stem cells and pancreatic β-cells (Fig. 1.3).[57,58] Although senescence is confined to mitotic cells, recent studies have suggested that senescent cells are alive and actively secrete molecules that alter the structure and function of local tissue microenvironments and distant organs (Fig. 1.3).[56] IL-6 is among the major molecules secreted by senescent cells (e.g. senescent human monocytes and fibroblasts).[56] Inflammation with elevated circulating IL-6 contributes to frailty. Minamino et al[59] demonstrated that adipose tissue in obese mice had high levels of oxidative stress, DNA damage, and senescent cells. Senescent adipose tissue cells were noted to express and secrete proinflammatory cytokines, resulting in insulin resistance. Inhibition of senescence ameliorated proinflammatory cytokine expression and glucose intolerance associated with obesity.

Mutant mice with constitutively active p53 display features that resemble frailty phenotypes (shortened survival time and vulnerability to stress and weight loss) and pathologies of frailty (sarcopenia) (Table 1.3).[60] As a progeroid syndrome, Werner syndrome is a human autosomal recessive genetic disorder caused by loss-of-function mutations in a gene that codes for a member of the RecQ helicase family (Werner protein, WRN), which leads to ineffective DNA damage repair and telomere maintenance, thereby causing premature cell senescence.[61] Patients afflicted with Werner syndrome display premature aging phenotypes/pathologies similar to frailty, including shortened lifespan, accelerated development of atherosclerosis, arteriosclerosis, type 2 diabetes, hypogonadism, and body composition changes. These premature aging pathologies can be mostly recapitulated in mutant mice with deficient WRN (Table 1.3).[62,63]

Direct evidence showing the causal roles of cellular senescence in frailty was found in the study by Baker et al.[64] BubR1 protein is a mitotic checkpoint protein. In $BubR1^{H/H}$ mice, BubR1 insufficiency causes premature aging phenotypes because of accelerated senescence (Table 1.3).[65] In particular, $p16^{Ink4a}$-positive senescent cells accumulate in skeletal muscles, which leads to sarcopenia with loss of skeletal muscle mass and function in $BubR1^{H/H}$ mice.[64] To test the hypothesis that senescence causes the skeletal muscle to age prematurely, Baker et al used transgenic techniques to selectively eliminate $p16^{Ink4a}$-positive senescent cells in $BubR1^{H/H}$ mice. The clearance of senescent cells was found to block premature skeletal muscle aging and resulted in preserved muscle mass and function and a decreased expression of IL-6.[64]

Oxidative Stress, Cellular Signaling, and Frailty

A growing body of research suggests that ROS can act as a signaling molecule and actively takes part in intracellular signaling, intercellular signaling, and regulation at the level of the organism or whole body in response to stress.[66] Proper signaling functionality in which ROS participates requires highly regulated levels of oxidants. Oxidative stress with excessive amounts of oxidants can trigger aberrant signaling, which leads to pathologies.

Redox signaling generally involves reversible oxidation of –SH groups by ROS (primarily hydrogen peroxide), which alters the activity of target proteins (e.g. protein phosphatase, ion channels, p53, and redox-sensitive transcriptional factors). Tyr and Ser/Thr phosphatases (e.g. phosphatase and tensin homolog, calcineurin) represent major classes of signaling molecules that are oxidized and inactivated by ROS. Tyr and Ser/Thr phosphatase inactivation leads to increased phosphorylation of its target kinases (e.g. mitogen-activated protein kinases) and then activation of the signaling cascade. Therefore, oxidative stress with persistently high levels of oxidants can have large effects on the growth factor signaling networks. For instance, activated c-jun N-terminal kinases (JNK kinases) caused by oxidative stress-induced inactivation of phosphatase can lead to serine phosphorylation of IRS-1, thereby interfering with insulin signaling and causing insulin resistance.[67] Another example involves nuclear factor (NF)-κB, which is a transcriptional factor that remains inactive when bound to IκB. By inactivating phosphatase, oxidative stress can increase IκB phosphorylation on two serine residues (by the kinases IKK1 and IKK2) and cause IκB to dissociate from NF-κB. Free NF-κB enters the nucleus and increases the expression of cytokines (IL-2, IL-6, IL-8, and TNF-α) and acute-phase proteins.[68] After being secreted and bound to its receptor, TNF-α can further increase oxidative stress and activate NF-κB, thereby forming a positive feedback loop that amplifies inflammation and oxidative stress.[69]

UNANSWERED QUESTIONS

As described, existing evidence supporting the roles of oxidative stress in frailty is primarily from human observational studies. The causal relationship between oxidative stress and frailty in humans has not been firmly established. Experiments testing whether direct application of ROS could reproduce the frailty pathologies and phenotypes should be done. Studies should also be conducted to examine whether interventions that specifically reduce oxidative stress (or its downstream pathologies) have effects on frailty phenotype. It should be noted that ROS can act as signaling molecules under physiologic conditions, as described above. In this case, reducing ROS may not be entirely beneficial.

CONCLUSION AND FUTURE PERSPECTIVES

Increased oxidative stress during aging is a candidate etiology of frailty. Associations between frailty and high levels of oxidative stress biomarkers were repeatedly observed in older adults. Transgenic mice with high oxidative stress display pathologies and phenotypes resembling those of frailty. Increased oxidative stress can cause mitochondrial dysfunction, damage to macromolecules (DNA, proteins and lipids), ER stress, cellular apoptosis, cellular senescence, and aberrant cellular signaling. These cellular pathogenic states may in turn result in pathologies at organ/system levels, including insulin resistance, inflammation, sarcopenia, adiposity, age-related hormone decline, and nervous system dysfunction. These multisystem pathologies then lead to a vulnerable state at old age characterized by frailty phenotypes (Fig. 1.4).

Frailty is an emerging critical health issue worldwide. Frail older adults have shortened survival and are at high risk for numerous adverse health outcomes. They constitute older adults in need of health care, long-term care, and community and informal support services. Elucidating the underlying oxidative stress and its downstream pathogenic pathways (e.g. mitochondrial dysfunction and senescence) may pave the way for the development of the long-awaited frailty prevention and intervention strategies in aging populations. Meanwhile, more in-depth research of human and model organisms is required to define the specific roles of oxidative stress and related pathways in frailty pathogenesis, and to identify candidate therapeutic targets.

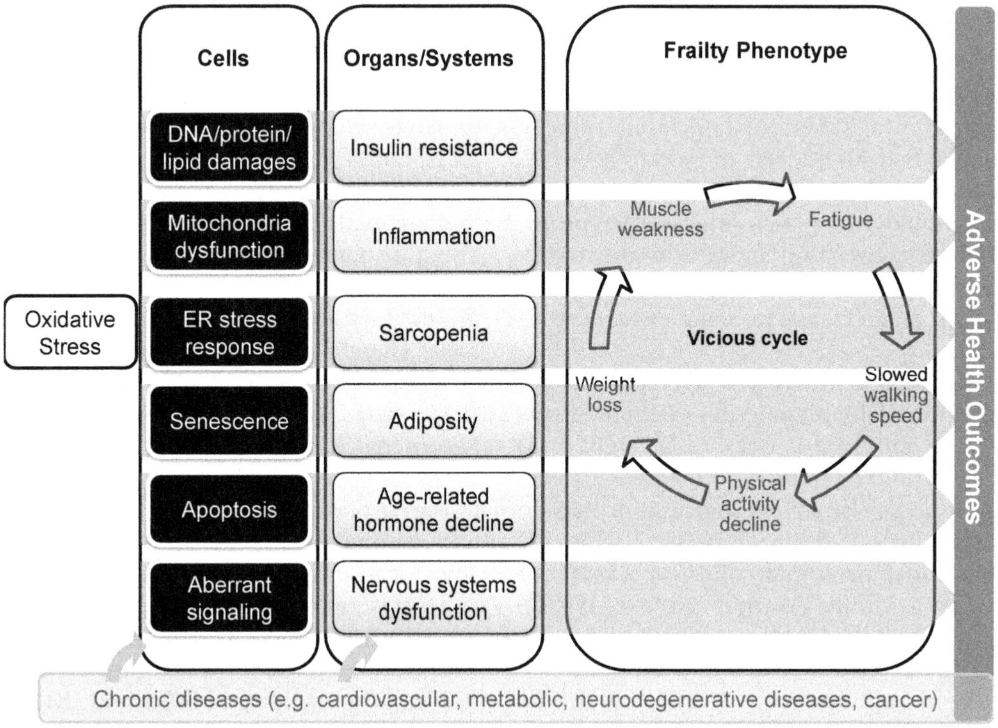

FIGURE 1.4 An oxidative stress model of frailty pathogenesis. An integrated model of frailty pathogenesis is presented. Excessive ROS cause cellular pathologic states, including mitochondrial dysfunction, damage to macromolecules (DNA, proteins and lipids), ER stress, cellular apoptosis, cellular senescence, and aberrant cellular signaling. These result in pathologies at organ/system levels, including insulin resistance, inflammation, sarcopenia, adiposity, age-related hormone decline, and nervous system dysfunction. These multisystem pathologies then lead to a vulnerable state at old age characterized by the frailty phenotype and a high risk of adverse health outcomes. Diseases may contribute to the pathogenic process at the cellular level (by increasing oxidative stress, or cellular damage, or by activating stress responses) and at the level of organs/systems.

SUMMARY POINTS

- Frailty is a state of age-related increased vulnerability with a decreased ability to maintain homeostasis and is characterized by insulin resistance, inflammation, sarcopenia, adiposity, age-related hormone decline, and nervous system dysfunction.
- Frailty independently predicts numerous adverse health outcomes in older adults and is viewed as a clinically relevant measure of biologic age.
- High levels of oxidative stress biomarkers (antioxidants, reactive oxygen species, or oxidative damage) predict the development of frailty in older adults.
- Transgenic mice with high oxidative stress (because of defective antioxidant defense or damage repair) and increased stress responses display pathologies and phenotypes resembling those of frailty.
- Based on this evidence, increased oxidative stress during aging is a candidate etiology of frailty. The potential cellular mechanisms through which oxidative stress cause frailty include mitochondrial dysfunction, damage to macromolecules (DNA, proteins and lipids) critical for maintaining homeostasis and muscle function, endoplasmic reticulum stress, cellular apoptosis, cellular senescence, and aberrant cellular signaling.
- Future research is required to firmly establish the role of oxidative stress in frailty pathogenesis.

References

1. Sies H. Oxidative stress: oxidants and antioxidants. *Exp Physiol* 1997;**82**(2):291–5.
2. Masoro EJ, Austad SN. *Handbook of the biology of aging*. 6th ed. New York: Academic Press; 2006.
3. Harman D. Aging: a theory based on free radical and radiation chemistry. *J Gerontol* 1956;**11**(3):298–300.
4. Lin MT, Beal MF. Mitochondrial dysfunction and oxidative stress in neurodegenerative diseases. *Nature* 2006;**443**(7113):787–95.
5. Fabrizio P, Liou LL, Moy VN, et al. SOD2 functions downstream of Sch9 to extend longevity in yeast. *Genetics* 2003;**163**(1):35–46.
6. Kenyon C. The plasticity of aging: insights from long-lived mutants. *Cell* 2005;**120**(4):449–60.
7. Migliaccio E, Giorgio M, Mele S, et al. The p66shc adaptor protein controls oxidative stress response and life span in mammals. *Nature* 1999;**402**(6759):309–13.
8. Stadtman ER. Protein oxidation and aging. *Science* 1992;**257**(5074): 1220–4.
9. Soerensen M, Christensen K, Stevnsner T, Christiansen L. The Mn-superoxide dismutase single nucleotide polymorphism rs4880 and the glutathione peroxidase 1 single nucleotide polymorphism rs1050450 are associated with aging and longevity in the oldest old. *Mech Ageing Dev* 2009;**130**(5):308–14.
10. Khabour OF, Abdelhalim ES, Abu-Wardeh A. Association between SOD2 T-9C and MTHFR C677T polymorphisms and longevity: a study in a Jordanian population. *BMC Geriatr* 2009;**9**:57.
11. Walston J, Hadley EC, Ferrucci L, et al. Research agenda for frailty in older adults: toward a better understanding of physiology and etiology: summary from the American Geriatrics Society/National Institute on Aging Research Conference on Frailty in Older Adults. *J Am Geriatr Soc* 2006;**54**(6):991–1001.
12. Fried LP, Tangen CM, Walston J, et al. Frailty in older adults: evidence for a phenotype. *J Gerontol A Biol Sci Med Sci* 2001;**56**(3):M146–156.
13. Barzilay JI, Blaum C, Moore T, et al. Insulin resistance and inflammation as precursors of frailty: the Cardiovascular Health Study. *Arch Intern Med* 2007;**167**(7):635–41.
14. Mitnitski AB, Mogilner AJ, Rockwood K. Accumulation of deficits as a proxy measure of aging. *Sci World J* 2001;**1**:323–36.
15. Mitnitski A, Song X, Skoog I, et al. Relative fitness and frailty of elderly men and women in developed countries and their relationship with mortality. *J Am Geriatr Soc* 2005;**53**(12):2184–9.
16. Fried LP, Kronmal RA, Newman AB, et al. Risk factors for 5-year mortality in older adults: the Cardiovascular Health Study. *JAMA* 1998;**279**(8):585–92.
17. Morley JE. Sarcopenia: diagnosis and treatment. *J Nutr Health Aging* 2008;**12**(7):452–6.
18. Delmonico MJ, Harris TB, Lee JS, et al. Alternative definitions of sarcopenia, lower extremity performance, and functional impairment with aging in older men and women. *J Am Geriatr Soc* 2007;**55**(5): 769–74.
19. Goodpaster BH, Carlson CL, Visser M, et al. Attenuation of skeletal muscle and strength in the elderly: the Health ABC Study. *J Appl Physiol* 2001;**90**(6):2157–65.
20. Hughes VA, Frontera WR, Roubenoff R, et al. Longitudinal changes in body composition in older men and women: role of body weight change and physical activity. *Am J Clin Nutr* 2002;**76**(2):473–81.
21. Hubbard RE, Lang IA, Llewellyn DJ, Rockwood K. Frailty, body mass index, and abdominal obesity in older people. *J Gerontol A Biol Sci Med Sci* 2010;**65**(4):377–81.
22. Ouchi N, Parker JL, Lugus JJ, Walsh K. Adipokines in inflammation and metabolic disease. *Nat Rev Immunol* 2011;**11**(2):85–97.
23. Ferrucci L, Corsi A, Lauretani F, et al. The origins of age-related proinflammatory state. *Blood* 2005;**105**(6):2294–9.
24. Mourkioti F, Kratsios P, Luedde T, et al. Targeted ablation of IKK2 improves skeletal muscle strength, maintains mass, and promotes regeneration. *J Clin Invest* 2006;**116**(11):2945–54.
25. Ferrucci L, Penninx BW, Volpato S, et al. Change in muscle strength explains accelerated decline of physical function in older women with high interleukin-6 serum levels. *J Am Geriatr Soc* 2002;**50**(12):1947–54.
26. Greiwe JS, Cheng B, Rubin DC, et al. Resistance exercise decreases skeletal muscle tumor necrosis factor alpha in frail elderly humans. *FASEB J* 2001;**15**(2):475–82.
27. Feldman HA, Longcope C, Derby CA, et al. Age trends in the level of serum testosterone and other hormones in middle-aged men: longitudinal results from the Massachusetts Male Aging Study. *J Clin Endocrinol Metab* 2002;**87**(2):589–98.
28. Bhasin S. Testicular disorders. In: Kronenberg HM, Melmed S, Polonsky KS, Larsen PR, editors. *Williams textbook of endocrinology*. 11th ed. Philadelphia: Saudners; 2008. pp. 645–99.
29. Wu IC, Lin XZ, Liu PF, et al. Low serum testosterone and frailty in older men and women. *Maturitas* 2010;**67**(4):348–52.
30. Corpas E, Harman SM, Blackman MR. Human growth hormone and human aging. *Endocr Rev* 1993;**14**(1):20–39.
31. Cappola AR, Bandeen-Roche K, Wand GS, et al. Association of IGF-I levels with muscle strength and mobility in older women. *J Clin Endocrinol Metab* 2001;**86**(9):4139–46.
32. Herbert J. The age of dehydroepiandrosterone. *Lancet* 1995;**345** (8959):1193–4.
33. Voznesensky M, Walsh S, Dauser D, et al. The association between dehydroepiandosterone and frailty in older men and women. *Age Ageing* 2009;**38**(4):401–6.
34. Seals DR, Esler MD. Human ageing and the sympathoadrenal system. *J Physiol* 2000;**528**(Pt 3):407–17.

35. Varadhan R, Chaves PH, Lipsitz LA, et al. Frailty and impaired cardiac autonomic control: new insights from principal components aggregation of traditional heart rate variability indices. *J Gerontol A Biol Sci Med Sci* 2009;**64**(6):682–7.
36. Howard C, Ferrucci L, Sun K, et al. Oxidative protein damage is associated with poor grip strength among older women living in the community. *J Appl Physiol* 2007;**103**(1):17–20.
37. Semba RD, Ferrucci L, Sun K, et al. Oxidative stress and severe walking disability among older women. *Am J Med* 2007;**120**(12):1084–9.
38. Fam SS, Morrow JD. The isoprostanes: unique products of arachidonic acid oxidation–a review. *Curr Med Chem* 2003;**10**(17):1723–40.
39. Cesari M, Kritchevsky SB, Nicklas B, et al. Oxidative damage, platelet activation, and inflammation to predict mobility disability and mortality in older persons: results from the Health, Aging, and Body Composition Study. *J Gerontol A Biol Sci Med Sci* 2012;**67**(6):671–6.
40. Cesari M, Kritchevsky SB, Nicklas BJ, et al. Lipoprotein peroxidation and mobility limitation: results from the Health, Aging, and Body Composition Study. *Arch Intern Med* 2005;**165**(18):2148–54.
41. Serviddio G, Romano AD, Greco A, et al. Frailty syndrome is associated with altered circulating redox balance and increased markers of oxidative stress. *Int J Immunopathol Pharmacol* 2009;**22**(3):819–27.
42. Cooke MS, Evans MD, Dizdaroglu M, Lunec J. Oxidative DNA damage: mechanisms, mutation, and disease. *FASEB J* 2003;**17**(10):1195–214.
43. Wu IC, Shiesh SC, Kuo PH, Lin XZ. High oxidative stress is correlated with frailty in elderly Chinese. *J Am American Geriatr Soc* 2009;**57**(9):1666–71.
44. Villamena FA, Zweier JL. Detection of reactive oxygen and nitrogen species by EPR spin trapping. *Antioxid Redox Signal* 2004;**6**(3):619–29.
45. Baptista G, Dupuy AM, Jaussent A, et al. Low-grade chronic inflammation and superoxide anion production by NADPH oxidase are the main determinants of physical frailty in older adults. *Free Radic Res* 2012;**46**(9):1108–14.
46. Li Y, Huang TT, Carlson EJ, et al. Dilated cardiomyopathy and neonatal lethality in mutant mice lacking manganese superoxide dismutase. *Nat Genet* 1995;**11**(4):376–81.
47. Williams MD, Van Remmen H, Conrad CC, et al. Increased oxidative damage is correlated to altered mitochondrial function in heterozygous manganese superoxide dismutase knockout mice. *J Biol Chem* 1998;**273**(43):28510–5.
48. Petersen KF, Dufour S, Befroy D, et al. Impaired mitochondrial activity in the insulin-resistant offspring of patients with type 2 diabetes. *N Engl J Med* 2004;**350**(7):664–71.
49. Trifunovic A, Wredenberg A, Falkenberg M, et al. Premature ageing in mice expressing defective mitochondrial DNA polymerase. *Nature* 2004;**429**(6990):417–23.
50. Matalon S, Hardiman KM, Jain L, et al. Regulation of ion channel structure and function by reactive oxygen-nitrogen species. *Am J Physiol Lung Cell Mol Physiol* 2003;**285**(6):L1184–1189.
51. Du J, Wang X, Miereles C, et al. Activation of caspase-3 is an initial step triggering accelerated muscle proteolysis in catabolic conditions. *J Clin Invest* 2004;**113**(1):115–23.
52. Alderton WK, Cooper CE, Knowles RG. Nitric oxide synthases: structure, function and inhibition. *Biochem J* 2001;**357**(Pt 3):593–615.
53. Martindale JL, Holbrook NJ. Cellular response to oxidative stress: signaling for suicide and survival. *J Cell Physiol* 2002;**192**(1):1–15.
54. Lebovitz RM, Zhang H, Vogel H, et al. Neurodegeneration, myocardial injury, and perinatal death in mitochondrial superoxide dismutase-deficient mice. *Proc Natl Acad Sci USA* 1996;**93**(18):9782–7.
55. Muller FL, Song W, Liu Y, et al. Absence of CuZn superoxide dismutase leads to elevated oxidative stress and acceleration of age-dependent skeletal muscle atrophy. *Free Radic Biol Med* 2006;**40**(11):1993–2004.
56. Campisi J, d'Adda di Fagagna F. Cellular senescence: when bad things happen to good cells. *Nat Rev Mol Cell Biol* 2007;**8**(9):729–40.
57. Krishnamurthy J, Ramsey MR, Ligon KL, et al. p16INK4a induces an age-dependent decline in islet regenerative potential. *Nature* 2006;**443**(7110):453–7.
58. Molofsky AV, Slutsky SG, Joseph NM, et al. Increasing p16INK4a expression decreases forebrain progenitors and neurogenesis during ageing. *Nature* 2006;**443**(7110):448–52.
59. Minamino T, Orimo M, Shimizu I, et al. A crucial role for adipose tissue p53 in the regulation of insulin resistance. *Nat Med* 2009;**15**(9):1082–7.
60. Tyner SD, Venkatachalam S, Choi J, et al. p53 mutant mice that display early ageing-associated phenotypes. *Nature* 2002;**415**(6867):45–53.
61. Yu CE, Oshima J, Fu YH, et al. Positional cloning of the Werner's syndrome gene. *Science* 1996;**272**(5259):258–62.
62. Chang S, Multani AS, Cabrera NG, et al. Essential role of limiting telomeres in the pathogenesis of Werner syndrome. *Nat Genet* 2004;**36**(8):877–82.
63. Lombard DB, Beard C, Johnson B, et al. Mutations in the WRN gene in mice accelerate mortality in a p53-null background. *Mol Cell Biol* 2000;**20**(9):3286–91.
64. Baker DJ, Wijshake T, Tchkonia T, et al. Clearance of p16Ink4a-positive senescent cells delays ageing-associated disorders. *Nature* 2011;**479**(7372):232–6.
65. Baker DJ, Jeganathan KB, Cameron JD, et al. BubR1 insufficiency causes early onset of aging-associated phenotypes and infertility in mice. *Nat Genet* 2004;**36**(7):744–9.
66. D'Autreaux B, Toledano MB. ROS as signalling molecules: mechanisms that generate specificity in ROS homeostasis. *Nat Rev Mol Cell Biol* 2007;**8**(10):813–24.
67. Kaneto H, Nakatani Y, Miyatsuka T, et al. Possible novel therapy for diabetes with cell-permeable JNK-inhibitory peptide. *Nat Med* 2004;**10**(10):1128–32.
68. Finkel T. Redox-dependent signal transduction. *FEBS Lett* 2000;**476**(1-2):52–4.
69. Haddad JJ. Pharmaco-redox regulation of cytokine-related pathways: from receptor signaling to pharmacogenomics. *Free Radic Biol Med* 2002;**33**(7):907–26.

CHAPTER

2

Skin Aging and Oxidative Stress

John M. Starr

Centre for Cognitive Aging and Cognitive Epidemiology, Edinburgh, United Kingdom

Robert J. Starr

School of Medicine and Dentistry, Polwarth Building, Foresterhill, Aberdeen, United Kingdom

List of Abbreviations

BCC basal cell carcinoma
DNA deoxyribonucleic acid
8-OHdG 8-hydroxy-2′-deoxyguanosine
UV ultraviolet light

INTRODUCTION

The skin is one of the largest organs of the human body. Like some other organs, it is directly exposed to a number of environmental toxins, but is almost unique in being exposed to the effects of sunlight. Skin aging can therefore be considered in terms of (1) intrinsic aging processes common to cells throughout the body, (2) the impact of chronic exposure to various environmental toxins, and (3) what is termed 'photoaging', the effects of exposure to UV light. Fortunately, compared to all other organs, the skin is relatively easy to gain access to for the study of cellular processes.

THE STRUCTURE OF HUMAN SKIN

The human skin is organized into two major layers, the deeper dermis and the superficial epidermis. The epidermis can be divided into five further layers, from most superficial to deepest:

- stratum corneum
- stratum licidum
- stratum granulosum
- stratum spinosum
- stratum basale

The stratum basale consists of columnar cells, keratinocytes, which divide and move gradually through subsequent layers to reach the stratum corneum. As they move towards the surface, the cells become gradually more and more flat. The stratum corneum consists of dead cells that are shed every 2 weeks. In addition to keratinocytes, there are three other types of specialized epidermal cells that make up only around 5% of the epidermal cell population. These are:

- melanocytes that produce the pigment melanin
- the Langerhans' cells that have primarily immune functions
- Merkel's cells found in touch-sensitive areas of the skin and associated with cutaneous nerves

Epidermal thickness varies up to 30-fold, from 0.05 mm on the eyelids to 1.5 mm on the palms and soles.

The thickness of the dermis also varies from around 0.3 mm on the eyelids to 3 mm over the back. The dermis has a far more complex structure than the epidermis. In general it has two layers, the more superficial papillary layer and the deeper reticular layer. But in addition it contains collagen, elastic tissue and reticular fibers that occur in both layers. The dermis is also penetrated by blood vessels and nerve cells, the former reaching superficially to supply the stratum basale of the epidermis. There are specialized nerve structures called Meissner's and Vater-Pacini corpuscles that are involved with light touch and pressure sensation. Hair follicles can be found throughout almost all of the human dermis, with the erector pili muscle attached to each follicle. Each follicle also may have sebaceous and apocrine glands associated with it. Finally, also widespread throughout the dermis are eccrine glands that produce sweat.

Aging
http://dx.doi.org/10.1016/B978-0-12-405933-7.00002-0

The dermis and the epidermis are bound together at a basement membrane. The stratum basale cells are attached to this membrane via anchoring filaments of hemidesmosomes. The papillary cells of the dermis attach to the membrane via anchoring fibrils made of collagen.

Below the dermis lies the so-called subcutaneous tissue that contains larger blood vessels and nerve cells along with adipocytes; loss of this tissue, rather than any intrinsic changes in either the dermis or epidermis, can change human appearance and result in changes associated with human aging. This is a very important consideration when choosing indices of skin aging because not all measures will necessarily relate entirely to changes occurring within the skin itself.

MEASURING SKIN AGING

The most common non-invasive method to measure skin aging is the counting of specific features from high-quality photographs, usually of the face. Guinot and colleagues validated such a scale in 361 white women aged 18 to 80 (mean age 43.5) years living in France.[1] They evaluated 62 facial skin characteristics on ordinal scales and found that only 33 of these were significantly associated with chronologic age. Factor analysis allowed them to exclude nine redundant items, leaving them with 24 items that they split into six groups by presumed etiology, but finally recommended using the total score. Allerhand and colleagues applied 16 of these scale items to photographs of 314 men and women of mean age 83.24 (range 82.04 to 84.57) years.[2] They identified 10 items that had good inter-rater reliability (Table 2.1) and performed confirmatory ordinal factor analysis with an oblique structure that proved to be consistent with the factor structure derived by Guinot and colleagues in their younger, female-only sample.

TABLE 2.1 Skin Aging Items and Their Relationship with Three Extracted Factors. ++ Represents a High Factor Loading, + Represents a Moderate Factor Loading

	Pigmented Spots	Wrinkles	Sagging
Pigmented spots cheek	++	+	
Pigmented spots forehead	++		
Fine lines forehead		+	
Wrinkles cheek		++	+
Wrinkles under eyes			+
Wrinkles upper lip		++	
Furrows between eyebrows			+
Nasolabial folds			+
Crows feet			++
Bags under eyes			

The three factors were best described as representing the number of pigmented spots, wrinkles and sagging. These last two factors were strongly correlated with each other ($r = 0.7$), but far less strongly with the factor representing pigmented spots. Interestingly, the 'bags under eyes' item did not have a loading >0.3 on any of the factors. In summary, the key finding from both studies using a facial skin aging scale is that skin aging is multidimensional. This multi-dimensionality may relate to the three major pathways to skin aging outlined above in the Introduction. It is unclear whether the same factors would summarize skin characteristics from parts of the body that are not exposed to sunlight on a regular basis.

As noted in the Introduction, skin is easily accessible for tissue sampling to examine the morphologic changes that occur with age in the dermis and epidermis. These changes can be measured from prepared slides, and some gross changes, such as skin thickness, can be measured by ultrasound. Under light microscopy there is progressive flattening of the epidermal junction, which is more undulating in youth, as humans age. This flattening results in a general thinning of the epidermis in older adults and also an increased tendency for epidermal–dermal shearing. Dermal thickness in non-light-exposed areas also decreases with increasing age and correlates with reduced collagen content.[3] This change reflects fibroblast function, but the dermis is a complex structure, and age-related changes in embedded blood vessels and cutaneous nerves also occur that are similar to changes in these structures seen elsewhere in the body. More specifically for skin, sebaceous glands produce less sebum from around 20 years of age so that the skin of older adults is more likely to be dry.

CELLULAR CORRELATES OF SKIN AGING

As noted above, the epidermis is a high-turnover tissue and this predisposes it not only to senescent effects, but importantly also to the generation of neoplastic cells. As the keratinocytes move out from the stratum basale they undergo proliferation, differentiation and finally apoptosis. In older people, replicative senescence, where cells remain viable and metabolically active but not capable of further replication, is more common. Langerhans' cells provide frontline immune responses to presenting antigens, migrating to local skin-draining lymph nodes where they facilitate the presentation of antigen to T-lymphocytes. Older adults have fewer Langerhans' cells[4] and those present have impaired migratory properties.[5] These observations have implications for increased susceptibility to skin infections and neoplasia in older adults.

The fibroblasts of the dermis are probably the most studied human cell type with regard to cellular senescence. Typical features are increased DNA double-stand breakage, telomere dysfunction and heterochromatinization of the nuclear genome which indicate that more than 15% of cells are in a senescent state in old age.[6] As expected, apoptotic cell death is increasingly likely with an increasing number of fibroblast divisions and is associated with impaired mitochondrial function.[7]

OXIDATIVE STRESS AND INTRINSIC SKIN AGING

There are limited *in vivo* data from humans suggesting that oxidative stress is associated with skin aging. An example is the finding that elevated serum 8-hydroxy-2′-deoxyguanosine (8-OHdG, a measure of oxidative DNA damage) occurs in older adults with more pigmented spots or facial sagging.[2] It is possible that collagen fragmentation, which is a feature of skin aging, increases serum 8-OHdG levels.[8] Another possibility is that there is some 'common cause' that drives both oxidative stress and skin aging independently. However, the redox status of a cell is a very basic influence on a whole range of processes and the only characteristics that might be considered more fundamental are genetic and epigenetic factors. But telomere length, the major indicator of genetic aging, is thought to be influenced by oxidative stress,[9] so invoking genetic factors as a 'common cause' would produce a paradoxical model. Given this, it is most parsimonious to consider a causal relationship between oxidative stress and skin aging with by far the biggest weight of evidence indicating that oxidative stress drives skin aging and not *vice versa*.

TABLE 2.2 Key Cellular Processes that Change with Increasing Levels of Oxidative Stress

Low oxidative state	Normal status for most organelles, although some, such as the endoplasmic reticulum, may prefer a higher oxidative resting state to facilitate protein folding and disulfide bridge formation.
Mildly increased oxidative state	Mild increases in intracellular Ca^{2+} levels and increased phosphorylation.
Moderately increased oxidative state (oxidative stress)	Further increase in intracellular Ca^{2+} levels and release of Fe^{2+}, Cu^{2+} and other transition metal ions. DNA damage may occur. Cell cycle halts to allow repair of DNA damage and antioxidant systems kick into action.
Highly oxidative state	Mitochondrial damage occurs. p53-related DNA damage induces apoptosis.
Very highly oxidative state (severe oxidative damage)	Shut down of caspases halts apoptosis leading to necrosis and release of transition metals etc. that damage surrounding tissue.

The skin and the lungs are the only organs substantially exposed to atmospheric oxygen. Aerobic metabolism draws on the availability of O_2 to increase energy generation. Aerobic metabolism occurs in the mitochondria and generates reactive oxygen species. It is an excess of these species leaking out of the mitochondria that results in oxidative stress. Table 2.2 summarizes some key cellular processes that vary according to redox status. Under normal conditions, atmospheric oxygen can supply the upper skin layers to a depth of 0.25–0.40 mm, which is the entire dermis and epidermis of the eyelids,[10] but only about one quarter of the depth of the palms and soles. Uptake in normal skin is unaffected by age,[10] but because the skin is thinner in older adults, more of the O_2 required is derived from the atmosphere rather than from the dermal blood supply. Those living at high altitude are more dependent on blood supply and are at increased risk of age-related skin changes because of higher UV light exposure (see below).

As outlined above, collagen is a key component of the human dermis, and the release of transition metals that occurs with increasing oxidative stress (Table 2.2) leads to metalloproteinase-1-induced collagen fragmentation.[8,11] Oxidative stress also increases the inversion of L- to D-form aspartyl residues in elastin typical of aging skin and related to its reduced elastic properties.[12] Hence, along with cellular changes resulting from oxidative stress, important changes to the skin tissue matrix also occur.

There are several cellular antioxidant defense systems that are present throughout the human body. Those enzymatic systems that are most important in the skin are superoxide dismutase,[13] which helps to combine the superoxide anion with two hydrogen ions to produce hydrogen peroxide and oxygen; catalase,[13] which catalyzes the breakdown of hydrogen peroxide to water and oxygen; and the glutathione system.[13] There are two main glutathione enzymes present in skin cells: glutathione peroxidase, with an enzyme activity in the epidermis 62% of that in the dermis, and glutathione reductase, with an enzyme activity level in the epidermis 32% of that in the dermis. Other, less active, antioxidant systems include thioredoxin reductase[14] and methionine sulfoxide reductase.[15] In addition to these enzymes, there are non-enzymatic compounds in the skin that have antioxidant properties (Table 2.3).

An increased number of pigmented spots is a skin aging characteristic associated with oxidative stress.[2] Nitric oxide radicals derived from keratinocytes promote pigmentation by inducing melaninogenic enzymes

TABLE 2.3 Non-Enzymatic Compounds in the Skin that Have Antioxidant Properties

- Ascorbic acid (vitamin C)
- Uric acid
- Glutathione
- Tocopherol (vitamin E)
- Ubiquinol
- Retinoids
- β-Carotene
- Protein thiols

tyrosinase and tyrosinase protein-1.[16,17] Ascorbic acid eliminates nitric oxide radicals and has an inhibitory effect on tyrosinase. There are various forms of tocopherol (vitamin E) with γ-tocopherol superior to α-tocopherol in its ability to inhibit melanogenesis by scavenging nitric oxide radicals.[18] Various skin cancers occur more commonly in older people. This is largely due to UV exposure (see below), but one study found that a single nucleotide polymorphism in *Nitric Oxide Synthase 1* oxidative stress gene increased the risk of malignant melanoma,[19] indicating that oxidative stress may play a part in predisposing skin cells to neoplasia. However, the exact role may differ between melanoma and non-melanoma skin cancer. In the latter it appears as if a reduction in antioxidant defense systems is the predisposing factor rather than increased levels of oxidative stress.[20]

ENVIRONMENTAL EXPOSURES ASSOCIATED WITH SKIN AGING

A fad for using arsenic-containing tonics in children in the 1930s saw a subsequent increase in people developing multiple basal cell carcinomas (BCC). In fact, even at relatively low levels, arsenic increases the risk of BCCs.[21] Arsenic predisposes to squamous cell cancer as well as to BCC, but not to malignant melanoma.[22] Arsenic can increase oxidative stress by acting as an electron carrier so that transition metals can cycle between oxidative states.[23] In addition, methylated arsenic can damage DNA directly and thus predispose to malignancy (see ref. 23 for a full review) and can induce apoptosis through the mitochondrial pathway. Notably, chronic arsenic exposure decreases antioxidant enzyme activity,[23] especially superoxide dismutase, catalase, glutathione peroxidase and glutathione reductase, all important in skin cells. This most likely explains the differences in arsenic risk between melanoma and non-melanoma skin cancer.

Cigarette smoking is the most widely recognized environmental source of carcinogens. It is associated with risk of squamous cell cancer of the skin, but not with risk of BCC.[24] Smoking does not increase the risk of melanoma, in fact, if anything, it may reduce susceptibility.[25] This limited impact of smoking on age-related neoplasia is in contrast with its major impact on skin aging.[26] The molecular basis of the premature skin aging effect of smoking is in the process of being elucidated. Smokers have higher concentrations than non-smokers of matrix metalloproteinase-1 mRNA in the dermis[27] and, as previously noted, this enzyme leads to increased collagen fragmentation and is, itself, influenced by oxidative stress.[8,11] There is clear evidence that smoking induces oxidative stress in other tissues, such as the heart.[28] How much of this effect is due to nicotine, which influences nitric oxide synthase levels, and how much to other moieties within cigarette smoke is unclear. Cigarette smoke contains free radicals, which may be able to diffuse into the skin from the atmosphere surrounding a smoker. This may explain, in part, why there is such a differential effect on facial skin aging.

If cigarette smoking can have a local effect on skin due to it containing free oxygen radical species, it is possible that other air pollutants may have this effect also. Ambient particulate matter in the nanosize range can cause oxidative stress because the large surface area to volume ratio of these particles makes them highly reactive towards biologic structures, and they may also carry a variety of transition metals that localize inside mitochondria, inducing oxidative stress. In addition, ambient particulate matter in the nanosize range may carry polycyclic aromatic hydrocarbons that are converted to quinones which, like arsenic, facilitate redox cycling and thus worsen oxidative stress. Skin aging was assessed in the SALIA study, which comprised 400 women residing either in industrialized cities of the Ruhr or in rural Borken, and traffic pollution correlated positively and significantly with the number of pigmented spots and with facial sagging characteristsics.[29] These relationships persisted after adjustment for potential confounding variables and would be consistent with skin aging changes related to oxidative stress.[2]

Less established, but biologically plausible, is a role for copper as an environmental toxin that may cause skin aging. Trace amounts of copper are, of course, essential for health, and failure to absorb copper results in the widespread abnormalities seen in Menke's disease, including brittle hair due to defective collagen and elastin polymerization. However, experimental data in dermal fibroblasts show that higher levels stimulate metalloproteinase-1 with all its consequences for collagen,[30] and it may be that it also affects mitochondrial function.

A major environmental determinant of both skin cancer and skin aging is UV light (listed in Table 2.4). It is of such importance that it requires a section of its own.

TABLE 2.4 Summary of the Effects of Known Environmental Exposures on Skin Aging and Skin Cancers

Environmental Exposure	Effects on Skin Aging	Association with Skin Cancer
Cigarette smoking	Direct effects of free radicals from cigarette smoke absorbed through the skin. Indirect effects via nitric oxide synthase and other possible pathways. Increased collagen fragmentation.	Increases basal cell carcinoma risk, but not squamous cell carcinoma or melanoma risk.
Ambient particulate matter	Nanosize particles highly reactive with cell membranes. May also carry transition metals and polycyclic aromatic hydrocarbons that are converted to quinones, facilitating redox cycling.	No associations identified.
Arsenic	Can act as an electron transporter for transition metals facilitating redox cycling. Decreases antioxidant enzyme activity.	Increases basal cell carcinoma and squamous cell carcinoma risk, but not melanoma risk.
Copper	Stimulates metalloproteinase-1 and thus collagen fragmentation. Probable direct effect on oxidative stress within mitochondria.	No associations identified.
UV light	Deactivates methionine sulfoxide reductase A and possibly other antioxidant enzymes. Major effect on pyrimidines.	All forms of skin neoplasia.

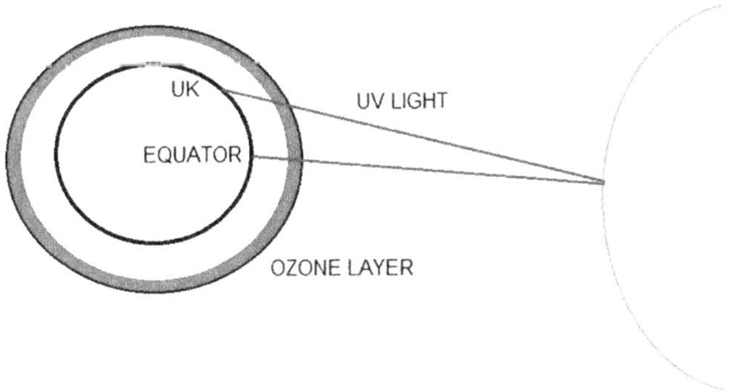

FIGURE 2.1 Illustration of how latitude affects the path of UV light from the sun through the ozone layer.

PHOTOAGING

The major contributing factor to skin aging is ultraviolet radiation (UV light). There are three types of UV light in the spectrum: UV A, B and C; however, only UV A and B have any bearing on skin aging, as UV C is absorbed in the atmosphere, leading to the formation of the ozone layer, and hence does not reach the Earth's surface. UV B has a wavelength of 280–320 nm and an energy of 3.94–4.43eV, and UV A has a wavelength of 320–400 nm and an energy of 3.10–3.94eV. UV A and UV B are both associated with direct DNA damage due to nucleic acid absorption of wavelengths within these ranges; UV B is absorbed more so in the epidermis and its basal layer, however UV A is able to penetrate further through the skin and hence also has the potential to disturb nucleic acids dermal cells.

In order to gauge the various factors influencing UV light damage to skin, it is important to look at several different layers which the radiation must penetrate prior to its random entry into a cell. First, before the UV radiation can even reach the skin it must first reach the Earth's atmosphere (Fig. 2.1); given the Earth's elliptical orbit, this is more likely at perihelion, in January, when the sun is 3.4% closer to the Earth than at aphelion, in July. Hence the Southern hemisphere receives more UV light during their summer than is received in the Northern hemisphere. Photoaging would thus be expected to be greater at a location in the Southern hemisphere which is equidistant from the equator in terms of latitude to a location in the Northern hemisphere. In essence, populations in the Southern hemisphere are at relatively greater risk of skin aging compared with those living in the Northern hemisphere.

Secondly, UV light which reaches the Earth's atmosphere is then subjected to two types of scattering which predominantly affect the amount of radiation reaching the Earth's surface. One type is Rayleigh scattering: this is isotropic scattering of photons by molecules in the Earth's atmosphere: N_2, O_2 and ozone (O_3, formed by absorption of UV C photons). The depth of the ozone layer is therefore, on average, greatest where there is the most UV C radiation. As O_3 has a molecular size similar to wavelengths in the UV B range, UV B radiation is twice as likely to be absorbed than UV A by the ozone layer. Hence, somewhere between ten and a hundred times more UV A reaches the Earth's surface than UV B depending on the depth of the ozone layer. This depth depends on the angle of incidence of the sun rays (Fig. 2.1), with areas closer to the equator having a shorter distance across the ozone layer for light to traverse. The average thickness of the ozone layer is around 50 km, so that, at the equinox, a location 45° latitude, such as Bordeaux in France, will have a traversal path of 70.2 km, 40% greater.

The other type of scattering is known as Mie's scattering; this is scattering brought about predominantly by water particles. Water particles are so large in comparison to the wavelengths of UV radiation that they scatter very large amounts of light. Large, dense formations of water droplets lead to very little light penetration, and this explains why cloud cover obscures sunlight to such a great extent. Nevertheless, a proportion of photons do get through, though the thicker the cloud cover, the darker it is during the day. This has major implications not only for skin aging, but also for vitamin D production. Another consideration is that the lower the altitude where someone lives, the greater is the chance of Rayleigh's scattering. Lower altitudes may well have greater cloud cover, and are thus subject to more Mie's scattering and hence, less exposure to UV radiation. Practically, this means that someone working in an observatory in the Atacama desert is at very high risk of photoaging.

Despite the considerable scattering, UV radiation is so abundant that skin is still exposed to a vast number of photons over years, most especially in areas that are not covered by clothing or thick hair. Once the light hits the surface of the skin it may (a) reflect off the skin, (b) be transmitted through into the skin matrix, or (c) both. To determine which course the photons will take, Fresnel's laws can be used; these laws determine the transmittance and reflection of radiation based upon (i) the refractive indices of the mediums involved (refractive index of air = 1.00 and the refractive index of the epidermis = ~1.36), (ii) the angle of incidence of the light on the epidermis, and (iii) the polarity of light (either s-polarized or p-polarized, dependent on electric field planes).

Given the non-uniformity of epidermal surfaces, there are a multitude of potential angles of incidence. Calculations are not simple to determine the amount of reflection, but it has been estimated that approximately 5% of photons are reflected from the surface of skin. However, further photons leave the skin due to scattering effects, similar to those within the atmosphere, within the skin matrix itself. UV radiation can be scattered or absorbed once inside the epidermis, and also the dermis in the case of UV A. The main particles which scatter light are lipids and proteins in the extracellular fluid, keratins and melanins in the epidermis, and collagens and elastins in the dermis. Mitochondria also cause scattering in both epidermis and dermis. This scattering is isotropic, hence the photons could move in any direction, even being scattered back out of the skin. Absorption can also occur, the main absorptive particles of UV light being melanin, keratin, nucleic acid and aromatic amino acids. Melanins absorb UV light greatest from a wavelength of ~300 nm, hence short-wavelength UV B light is scattered by melanin rather than absorbed. Keratin mostly absorbs short wavelengths of light, its peak absorption coming at around 280 nm and steadily falling from then on as the wavelength increases, dropping off just above the wavelengths of UV light. Absorption by these particles prevents absorption by nucleic acid molecules; hence, this is why sunlight exposure leads to either tanning or sunburn, depending on the amount of melanin. UV light-induced increase in melanins lies behind one of the characteristic features of skin aging, an increase in pigmented spots. However large amounts of melanin can also be harmful as melanins scatter short wavelengths of UV B light and this can lead to increased DNA damage. Table 2.5 summarizes the factors that influence the extent of UV radiation reaching skin cells.

Both UV A and UV B damage skin and predispose to neoplasia by direct DNA damage. In addition, UV damage to mitochondrial DNA represents a possible oxidative stress pathway to photoaging of the skin. However UV radiation can also affect oxidative stress directly. An example is the reduction of methionine sulfoxide reductase A activity and content in cultured human keratinocytes exposed to UV A radiation.[31] Methionine sulfoxide reductase levels were also reduced in sun-exposed skin compared to non-sun-exposed skin in 16 female subjects.[31] Methionine sulfoxide reductase activity reduction may lead to less oxygen free radical processing because these react with methionine in proteins to form methionine sulfoxide.

TABLE 2.5 Factors that Influence the Extent of UV Radiation Reaching Skin Cells

Factor Influencing UV Light Reaching Skin Cells	Pertinent Parameters
Distance of sun from earth	Perihelion occurs in January, so Southern hemisphere at greater relative risk.
Ozone layer	Absorbs both UV A and UV B.
Latitude	Determines the length of the path through the ozone layer – greater with higher latitudes.
Atmospheric scattering	Rayleigh scattering. Shorter wavelengths are scattered more than long ones. Hence high levels of UV scattering and also why the sky appears blue.
Water droplet scattering	Mie's scattering from larger particles in the atmosphere. Explains why thicker clouds appear darker.
Reflection of light from skin	Probably only 5% of light is reflected.
Scattering of light from skin surface	Probably larger effect than reflection.
Melanin content of skin	Absorbs UV A, but scatters UV B, so may actually worsen effects in epidermis (only UV A reaches the dermis).

Accumulation of oxidized proteins is a feature of the epidermis as it ages,[32] and in addition, proteins can be degraded by the cross-linked lipid peroxidation product, 4-hydroxy-2-nonenol.[33] UV light irradiation of the epidermis in a hairless rat model showed decreases in the number of Langerhans' cells, and this relates to changes in oxidative stress enzymes[34] — which may explain, in part, the reduced immune function of the skin that occurs with age.

Along with effects on enzymes that provide antioxidant defenses, UV light may well affect other enzyme systems that impact skin aging. One example is 11-beta-hydroxysteroid dehydrogenase, of which the 11-beta-HSD1 isoform that activates cortisol is expressed in human epidermal keratinocytes and dermal fibroblasts rather than the 11-beta-HSD2 isoform that inactivates cortisol. Many features of skin aging, such as dermal thinning, also occur in high glucocorticoid states, both in disease (e.g. Cushing's disease) and iatrogenically. 11-beta-HSD1 mRNA expression is higher in older adults and higher still in sun-exposed skin.[35] This demonstrates the uncertainty of the extent of any effect of photoaging being mediated via oxidative stress.

CONCLUSIONS

The skin is a large organ and has some unique features in terms of oxidative stress effects on its aging. First, it receives much of its oxygen directly from the atmosphere rather than from the blood supply; this means that free radical oxygen species can also be derived directly from the atmosphere, as in cigarette smoke, rather than being generated within cells. Second, it is exposed to other atmospheric pollutants, especially ambient particulate matter, to a far greater degree than other organs. Again, these represent an important source of generating free radical oxygen species. Third, the skin, unlike other organs, is exposed to UV light and this, too, may contribute to skin aging through oxidative stress pathways.

Skin aging is not monolithic. Increased numbers of pigmented spots probably relate most strongly to photoaging, although NO also promotes melaninogenic changes. Wrinkles and sagging reflect more the loss of collagen and elastin fibres from the dermal matrix, and these processes are influenced by oxidative stress. Wrinkling and sagging, however, both depend on changes to subcutaneous tissues that occur with age so that skin aging correlates with age-related changes in other organs. The degree to which oxidative stress contributes to an individual person's skin aging is difficult to ascertain given the multiple factors that influence skin aging, but skin is easily accessible and does afford an easy opportunity to study oxidative stress in a tissue other than blood.

SUMMARY POINTS

- Skin aging is multidimensional, relating to different pathways.
- Appearances of skin aging relate not only to changes within the skin, but also to changes, especially loss, of underlying subcutaneous tissues.
- Oxidative stress is determined by the balance between free radical oxygen species generation and processing of these species by enzymatic and non-enzymatic antioxidant systems in the skin.
- The most important antioxidant enzymes in the skin are superoxide dismutase, catalase, glutathione peroxidase and glutathione reductase.
- Important environmental exposures that affect skin aging through oxidative stress are cigarette smoking, ambient particulate matter, arsenic, copper and UV light.
- UV B penetrates the epidermis only, whereas UV A penetrates both epidermis and dermis.

References

1. Guinot C, Malvy D, Ambroisine J-M, Latreille L, et al. Relative contribution of intrinsic vs extrinsic factors to skin aging as determined by a validated skin age score. *Arch Dermatol* 2002;**138**:1454–60.
2. Allerhand M, Ooi ET, Starr RJ, et al. Skin aging and oxidative stress in a narrow-age cohort of older adults. *Eur Geriatric Med* 2011;**2**:140–4.
3. Shuster S, Black MM, McVitie E. The influence of age and sex on skin thickness, skin collagen and density. *Br J Dermatol* 1975;**93**:639–43.
4. Gilchrest BA, Murphy GF, Soter NA. Effect of chronologic aging and ultraviolet irradiation on Langerhans cells in human epidermis. *J Invest Dermatol* 1982;**79**:85–8.
5. Bhushan M, Cumberbatch M, Dearman RJ, et al. Tumour necrosis factor-alpha-induced migration of human Langerhans cells: the influence of aging. *Br J Dermatol* 2002;**146**:32–40.
6. Herbig U, Ferreira M, Condel L, et al. Cellular senescence in aging primates. *Science* 2006;**311**:1257.
7. Mammone T, Gan D, Foyouzi-Youssefi R. Apoptotic cell death increases with senescence in normal human dermal fibroblast cultures. *Cell Biol Int* 2006;**30**:903–9.
8. Fisher GJ, Quan T, Purohit T, et al. Collagen fragmentation promotes oxidative stress and elevates matrix metalloproteinase-1 in fibroblasts in aged human skin. *Am J Pathol* 2009;**174**:101–14.
9. Starr JM, Shiels PG, Harris SE, et al. Oxidative stress, telomere length and biomarkers of physical aging in a cohort aged 79 years from the 1932 Scottish Mental Survey. *Mech Aging Develop* 2008;**129**:745–51.
10. Stücker M, Struck A, Altmeyer P, et al. The cutaneous uptake of atmospheric oxygen contributes significantly to the oxygen supply of human dermis and epidermis. *J Physiol* 2002;**538**:985–94.
11. Schroeder P, Gremmel T, Berneburg M, Krutmann J. Partial deletion of mitochondrial DNA from human skin fibroblasts induces a gene expression profile reminiscent of photoaged skin. *J Inv Dermatol* 2008;**128**:2297–303.
12. Kuge K, Kitamura K, Nakaoiji K, et al. Oxidative stress induces the formation of D-aspartyl residues in elastin mimic peptides. *Chem Biodivers* 2010;**7**:1408–12.
13. Fang Y-Z, Yang S, Wu G. Free radicals, antioxidants, and nutrition. *Nutrition* 2002;**18**:872–9.

14. Schallreuter KU, Wood JM. Thioredoxin reductase – its role in epidermal redox status. *J Photochem Photobiol B Biol* 2001;**64**:179–84.
15. Ogawa F, Sander CS, Hansel A, et al. The repair enzyme peptide methionine-S-sulfoxide reductase is expressed in human epidermis and upregulated by UVA radiation. *J Invest Dermatol* 2006;**126**: 1128–34.
16. Roméro-Graillet C, Aberdam E, Clément M, et al. Nitric oxide produced by ultraviolet-irradiated keratinocytes stimulates melanogenesis. *J Clin Invest* 1997;**99**:635–42.
17. Sasaki M, Horikoshi T, Uchiwa H, Miyachi Y. Up-regulation of tyrosinase gene by nitric oxide in human melanocytes. *Pigment Cell Res* 2000;**13**:248–52.
18. Yoshida E, Watanabe T, Takata J, et al. Topical application of a novel, hydrophilic gamma-tocopherol derivative reduces photo-inflammation in mice skin. *J Invest Dermatol* 2006;**126**:1447–9.
19. Ibarrola-Villava M, Pena-Chilet M, Fernandez LP, et al. Genetic polymorphisms in DNA repair and oxidative stress pathways associated with malignant melanoma susceptibility. *Eur J Cancer* 2011;**47**:2618–25.
20. Sander CS, Hamm F, Elsner P, Thiele JJ. Oxidative stress in malignant melanoma and non-melanoma skin cancer. *Br J Dermatol* 2003;**148**:913–22.
21. Leonardi G, Vahter M, Clemens F, et al. Inorganic arsenic and basal cell carcinomas in areas of Hungary, Romania, and Slovakia: a case-control study. *Env Health Perspectives* 2012;**120**:721–6.
22. Guo HR, Yu HS, Hu H, Monson RR. Arsenic in drinking water and skin cancers: cell-type specificity. *Cancer Causes Control* 2001;**12**:909–16.
23. Flora SJS. Arsenic-induced oxidative stress and its reversibility. *Free Rad Biol Med* 2011;**51**:257–81.
24. Rollison DE, Iannacone MR, Messina JL, et al. Case-control study of smoking and non-melanoma skin cancer. *Cancer Causes Control* 2012;**23**:245–54.
25. DeLancey JO, Hannan LM, Gapstur SM, Thun MJ. Cigarette smoking and the risk of incident and fatal melanoma in a large prospective cohort study. *Cancer Causes Control* 2011;**22**:937–42.
26. Morita A. Tobacco smoke causes premature skin aging. *J Dermatol Sci* 2007;**48**:169–75.
27. Lahmann C, Bergemann J, Harrison G, Young AR. Matrix metalloprotease-1 and skin aging in smokers. *Lancet* 2001;**357**:935–6.
28. Varela Carver A, Parker H, Kleinert C, Rimoldi O. Adverse effects of cigarette smoke and induction of oxidative stress in cardiomyocytes and vascular endothelium. *Curr Pharmaceut Design* 2010;**16**:2551–8.
29. Vierkotter A, Schikowski T, Ranft U, et al. Airborne particle exposure and extrinsic skin aging. *J Invest Dermatol* 2010;**130**:2719–26.
30. Philips N, Hwang H, Chauhan S, et al. Stimulation of cell proliferation and expression of Matrixmetalloproteinase-1 and Interleukin-8 genes in dermal fibroblasts by copper. *Connective Tissue Res* 2010;**51**: 224–9.
31. Picot CR, Moreau M, Juan M, et al. Impairment of methionine sulfoxide reductase during UV irradiation and photoaging. *Exp Gerontol* 2007;**42**:859–63.
32. Bulteau AL, Petropoulos I, Friguet B. Age-related alterations of proteasome structure and function in aging epidermis. *Exp Gerontol* 2000;**35**:767–77.
33. Bulteau AL, Moreau M, Nizard C, Friguet B. Impairment of proteasome function upon UVA- and UVB-irradiation of human keratinocytes. *Free Radic Biol Med* 2002;**32**:1157–70.
34. Mulero M, Romeu M, Giralt M, et al. Oxidative stress-related markers and Langerhans cells in a hairless rat model exposed to UV radiation. *J Toxicol Env Health Part A* 2006;**69**:1371–85.
35. Tiganescu A, Walker EA, Hardy RS, et al. Localization, age- and site-dependent expression, and regulation of 11beta-hydroxysteroid dehydrogenase type 1 in skin. *J Invest Dermatol* 2011;**131**:30–6.

C H A P T E R

3

Cardiovascular Disease in Aging and the Role of Oxidative Stress

Fawaz Alzaid

Diabetes and Nutritional Sciences Division, School of Medicine, King's College London, Franklin-Wilkins Building, London, UK

Vinood B. Patel

Department of Biomedical Science, Faculty of Science & Technology, University of Westminster, London, UK

Victor R. Preedy

Diabetes and Nutritional Sciences Division, School of Medicine, King's College London, Franklin-Wilkins Building, London, UK

List of Abbreviations

3-NPA 3-nitropropionic acid
ADP adenosine diphosphate
AKT protein kinase B
AMP adenosine monophosphate
AMPK adenosine monophosphate-activated protein kinase
AP-1 activator protein 1
ATP adenosine triphosphate
CAT catalase
ERK extracellular signal regulated kinase
FADH flavin adenine dinucleotide
FOX forkhead box
FOXO O class forkhead box
GPx glutathione peroxidase
GST glutathione *S*-transferase
HIF-1 hypoxia-inducible factor 1
IGF1 insulin-like growth factor 1
LDLs low density lipoproteins
LKB1 liver kinase B1
mCLK1 mammalian CLOCK1
MiR1 microRNA 1
miRNA microribonucleic acid
Mn SOD manganese superoxide dismutase
mTOR mammalian target of rapamycin
mTORC1 mammalian target of rapamycin complex 1
mTORC2 mammalian target of rapamycin complex 2
NADH nicotinamide adenine dinucleotide
NADPH nicotinamide adenine dinucleotide phosphate
$p66^{Shc}$ 66 kDa splice variant of the Shc locus
PI3K phosphoinositide 3-kinase
PIT1 pituitary transcription factor 1
PKCB protein kinase C-β
PROP1 homeobox protein prophet of pituitary transcription factor 1
PTP permeability transition pore
RNA ribonucleic acid
ROS reactive oxygen species
SIR2 sirtuin
siRNA small interfering ribonucleic acid
SIRT sirtuin family member
TSC1 tuberous sclerosis complex 1
TSC2 tuberous sclerosis complex 2
UCP2 mitochondrial uncoupling protein 2
VEGF vascular endothelial growth factor

INTRODUCTION

Aging is an inevitable fate for any biologic system, and unfortunately the deterioration that occurs with aging predisposes individuals to numerous conditions. Cardiovascular diseases represent a large group of conditions with a dramatically increased occurrence with age. As well as the emotive impact on the family unit and communities, cardiovascular diseases claim a large number of lives and cause a significant detriment to the economy through healthcare costs and productivity losses. In Europe this amounts to 1.9 million deaths and a cost of €196 billion per year[1]. One-third of all deaths

Aging
http://dx.doi.org/10.1016/B978-0-12-405933-7.00003-2

in the USA are caused by cardiovascular disorders, with more than 83 million Americans currently living with one or more of these fatal conditions. Furthermore, cardiovascular disorders cost the American economy $444 billion a year, equating to one sixth of the country's spending on health.[2]

Biologic aging predisposes individuals to cardiovascular disease through age-related perturbation of systemic and/or cellular oxidative balance, i.e. discord between the rates of generation and clearance of reactive oxygen species (ROS). Due to this increased susceptibility, death related to cardiovascular events rapidly becomes more common with aging. For example, in the United Kingdom in 2009 there were approximately 180 000 deaths due to cardiovascular disease, 72% of which were in populations over the age of 75[3] (see Figs 3.1 and 3.2 for representative statistics related to cardiovascular diseases).

Oxidative stress has long been known to increase with the biologic aging process, and it independently causes a greater risk of cardiovascular disease. Biologic systems are equipped to neutralize endogenous oxidative stress and respond appropriately to oxidative challenges. Throughout biologic aging, these protective mechanisms decline and thus the oxidative theory of aging extends to pathophysiologic developments that predispose individuals to a much higher risk of conditions such as cardiovascular disease. The cardiovascular system is particularly sensitive to endogenous oxidative stress as the myocardium is particularly rich in mitochondria, and mitochondrial metabolism is the main source of cellular free radical generation.[4]

The genes that regulate cardiovascular physiologic function and responses to oxidative challenge in aging have been identified as *longevity genes*, a subset of which interacts extensively as the *longevity network*.[5] Biologic aging is also heavily attributed to the *telomere theory*. The telomere theory states that telomere shortening, which is accelerated by oxidative stress, is responsible for much of the age-related deterioration.[6] The implication of oxidative stress in age-related physiologic decline is well established and supported by data showing a marked increase in plasma potential for oxidative damage with age (see Fig. 3.3).[7] This increased oxidant potential is indicative of the increasing imbalance between the generation and clearance of ROS throughout the biologic aging process, contributing to the pathogenesis of age-related cardiovascular disease. Furthermore, as age increases, the antioxidant activity of several antioxidant enzymes in the cardiovascular system decreases (see Fig. 3.4a). The oxidative balance of the cardiovascular system is further disturbed by the increased oxidative stress from mitochondria and lipid peroxidation that occurs with aging (see Fig. 3.4b).

In this chapter we describe the physiologic changes that accompany aging and which predispose individuals to a decline in vascular function and the risk of cardiovascular diseases. We focus specifically on the effects of oxidative stress and the dual role it plays in aging and in the pathogenesis of cardiovascular disease.

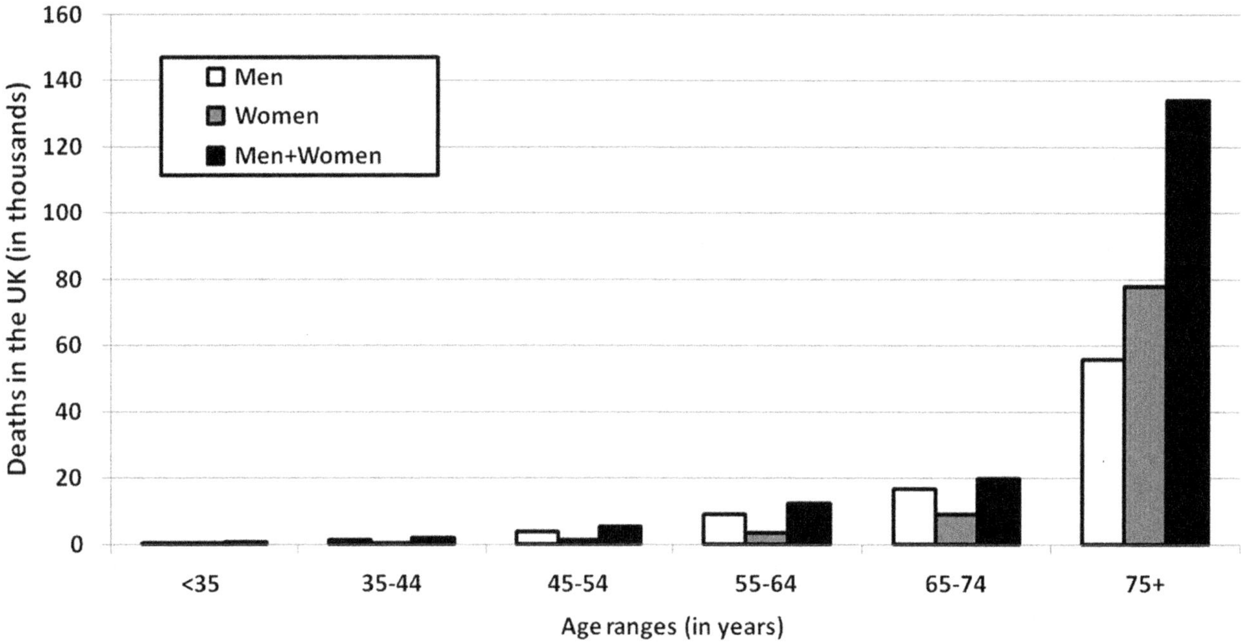

FIGURE 3.1 Rates of death from cardiovascular disease increase with age in the United Kingdom. Cardiovascular diseases represent the most common causes of death in the UK. This figure shows the deaths in the UK in 2009 that are due to cardiovascular disease. The risk of death from cardiovascular disease increases with age, with the highest risk group being those above 75 years of age; this trend is in both men and women. *Data adapted from Nichols et al 2012.*[1]

PHYSIOLOGY OF THE AGING CARDIOVASCULAR SYSTEM

The cardiovascular system is the most widely reaching system within the human body, delivering oxygenated blood and nutrients to all tissues. However, like any system, aging alone will cause a physiologic and functional decline that increases susceptibility to complications. In the cardiovascular system, the decline in functional capacity greatly increases the risk of hypertension and atherosclerosis, leading to life-threatening events such as myocardial infarction or stroke. Aging and the associated functional decline are greatly accelerated by systemic and local oxidative stress.

Figure 3.5 summarizes the physiologic changes that occur in an aging cardiovascular system. These effects

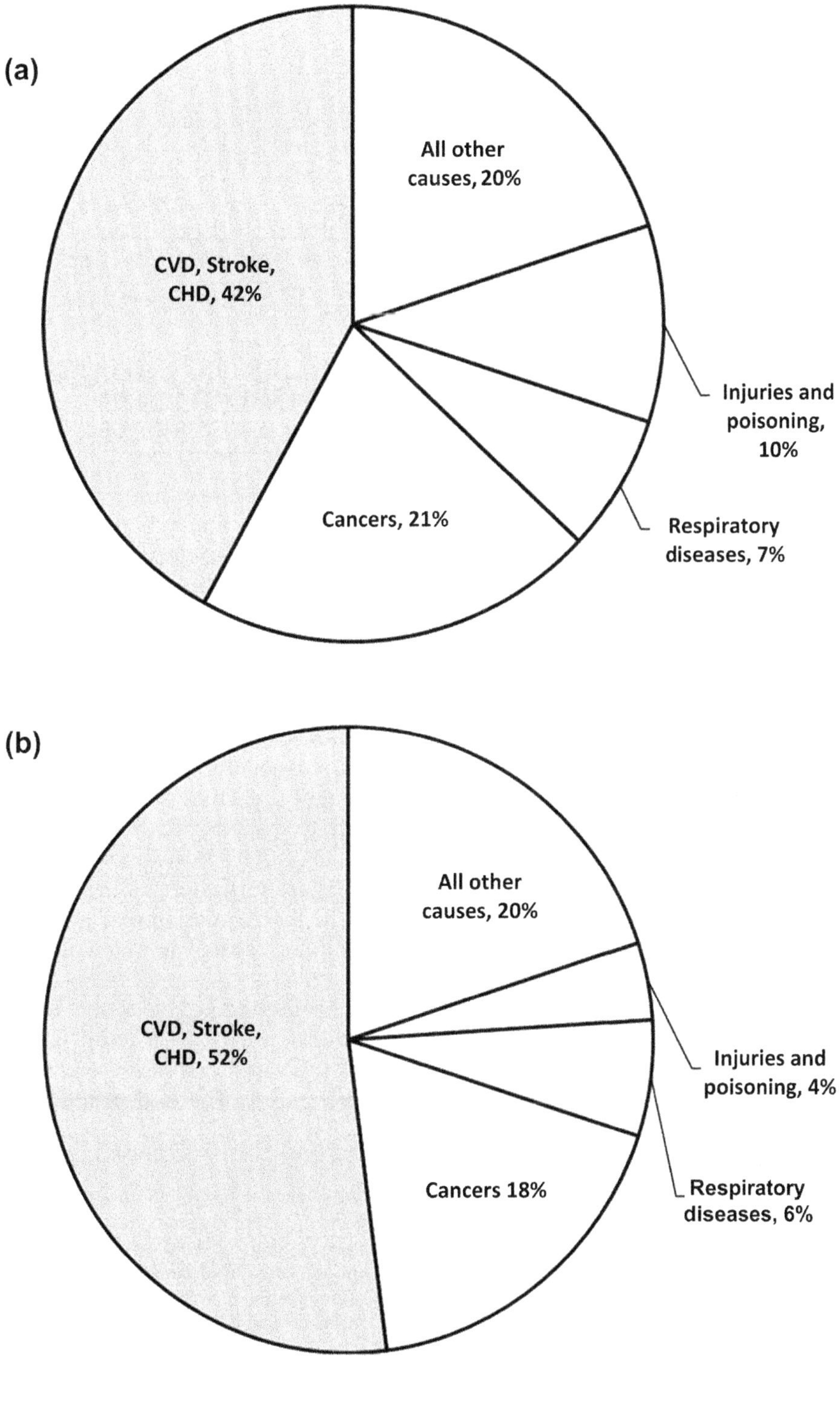

FIGURE 3.2 Impact of cardiovascular disease in Europe. (a) This shows the leading causes of death of men in Europe in 2012; cardiovascular disease, stroke and coronary heart disease are the leading causes of death. (b) This shows the leading causes of death of women in Europe in 2012; cardiovascular disease, stroke and coronary heart disease cause more than half of the deaths in Europe. *Data adapted from Scarborough et al 2011.*[3] CVD: cardiovascular disease; CHD: coronary heart disease.

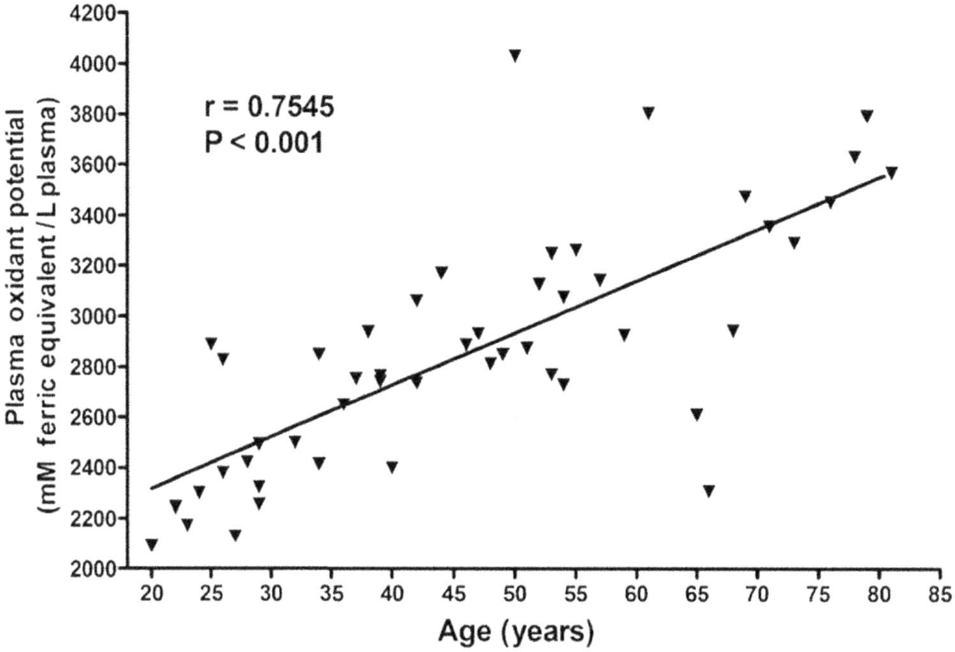

FIGURE 3.3 The potential for oxidative damage increases with age. Oxidative stress contributes to the biologic aging process and is a central part of the pathogenesis of cardiovascular disease. This figure shows the clear relationship between aging and the increased potential for oxidative stress (plasma oxidant potential). An increased potential for oxidative stress increases the risk of developing age-related pathologies, including cardiovascular disease. *Reprinted with permission from Mehdi & Rizvi (2013).*[7]

mainly cause a decline in endothelial function, left ventricle systolic reverse capacity and left ventricular diastolic function.[5] Interestingly, the effects of aging in either the vascular or cardiac system will induce a compensatory and potentially pathologic change in the other. For example, increased arterial stiffness causes accumulation of fibrotic tissues or hypertrophy in the myocardium.[5]

The main functional changes that occur with aging are: (1) a decline in heart rate, and (2) a decline in cardiac output. The decline in heart rate is due to cell loss in the sinoatrial node as well as impedance of electrical impulses from structural changes such as hypertrophy and fibrosis. These changes are due largely to an age-related decline in stress–response mechanisms.[5,8] The decline in cardiac output begins with hemodynamics adapting to meet changing requirements with age. The cardiovascular system's immediate response stimulates myocardial hypertrophy – which increases cardiac output accordingly. These adaptations subsequently lead to a decline in cardiac output and overall cardiovascular function. Myocardial hypertrophy, independently or as part of age-related compensation, is a risk factor for morbidity and mortality associated with cardiovascular disease.[5,8]

THE MOLECULAR BASIS OF OXIDATIVE STRESS AS APPLIED TO THE CARDIOVASCULAR SYSTEM

In any cell, a multitude of reactions require the transfer of electrons. Whenever electrons are exchanged, there is a change in oxidative state of the molecules involved in these reactions. Oxidation or reduction constitutes the gain or loss of electrons, respectively. These processes occur simultaneously and are termed redox reactions, where a reductant will be oxidized by donating an electron to an oxidant.[9]

Reactive oxygen species (ROS), generated as byproducts of the above redox reactions, contain unpaired electrons, making them highly reactive and potentially dangerous sources of oxidative stress. The main endogenous process that generates ROS is mitochondrial oxidative phosphorylation.[9] These mitochondrial reactions are particularly relevant to the cardiovascular system, as 45% of the heart's cellular volume is taken up by mitochondria.[4]

During oxidative phosphorylation, electrons from nicotinamide adenine dinucleotide (NADH) and flavin adenine dinucleotide (FADH) are processed via four mitochondrial enzyme complexes. The end product of

FIGURE 3.4 Decrease in antioxidant enzyme activity and increase in oxidative stress in the aging cardiovascular system; there is decreased activity of several antioxidant enzymes, namely, manganese superoxide dismutase (Mn SOD), catalase (CAT), glutathione peroxidase (GPx) and glutathione *S*-transferase (GST). (a) This shows the percentage decrease in activity of these enzymes between young and aged mouse hearts. As well as the decrease in these antioxidant enzymes, there is increased oxidative stress that occurs through increased lipid peroxidation and the generation of mitochondrial ROS. (b) This shows the increased production of ROS in hearts from aged versus young mice. Mn SOD: manganese superoxide dismutase; CAT: catalase; GPx: glutathione peroxidase; GST: glutathione *S*-transferase; MDA: malondialdehyde; RFU: relative fluorescence units; ROS: reactive oxygen species. *Data adapted from Sudheesh et al 2010.*[69]

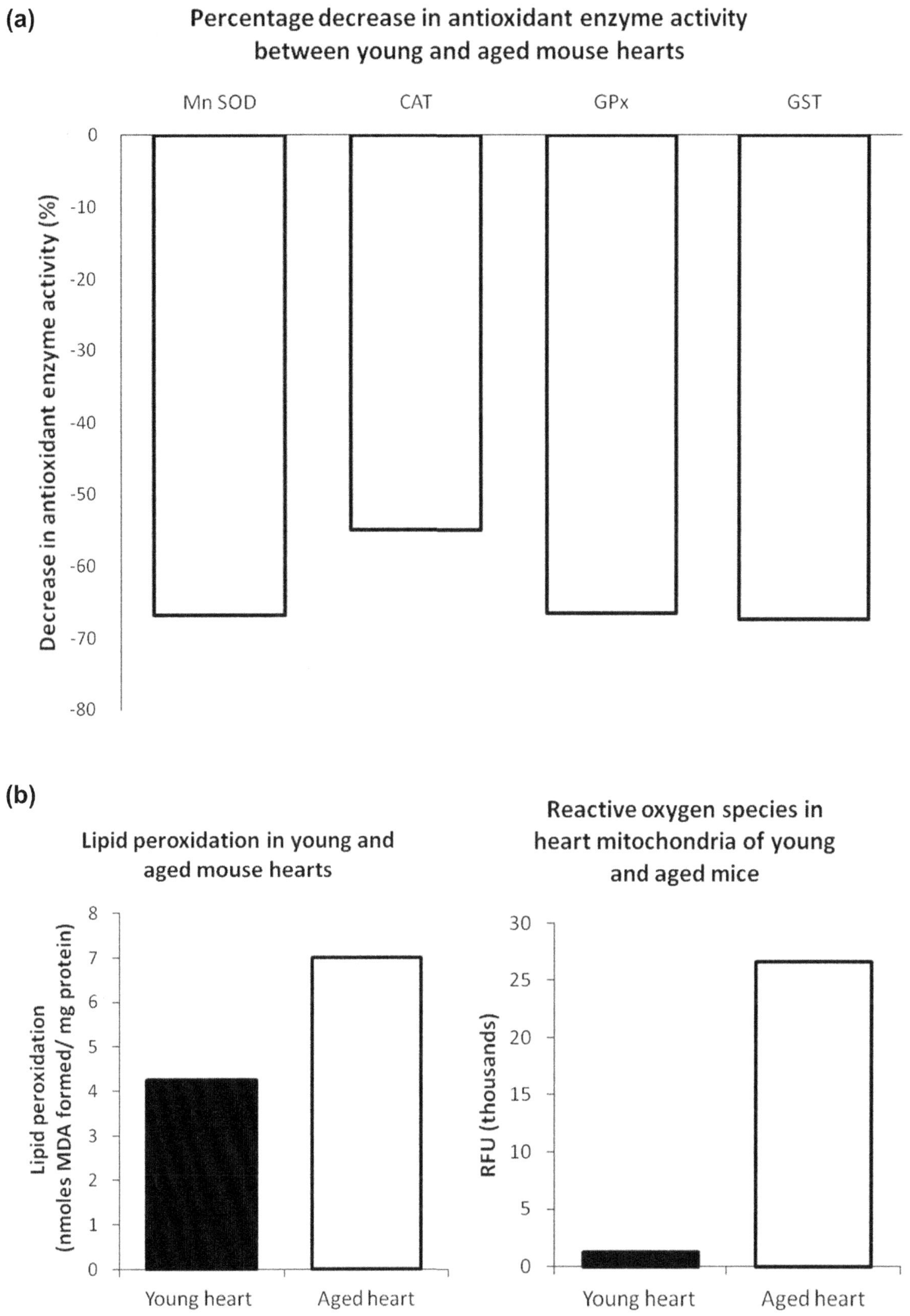
(a)
Percentage decrease in antioxidant enzyme activity between young and aged mouse hearts
Mn SOD
CAT
GPx
GST
Decrease in antioxidant enzyme activity (%)
0
-10
-20
-30
-40
-50
-60
-70
-80
(b)
Lipid peroxidation in young and aged mouse hearts
Lipid peroxidation (nmoles MDA formed/ mg protein)
0
1
2
3
4
5
6
7
8
Young heart
Aged heart
Reactive oxygen species in heart mitochondria of young and aged mice
RFU (thousands)
0
5
10
15
20
25
30
Young heart
Aged heart

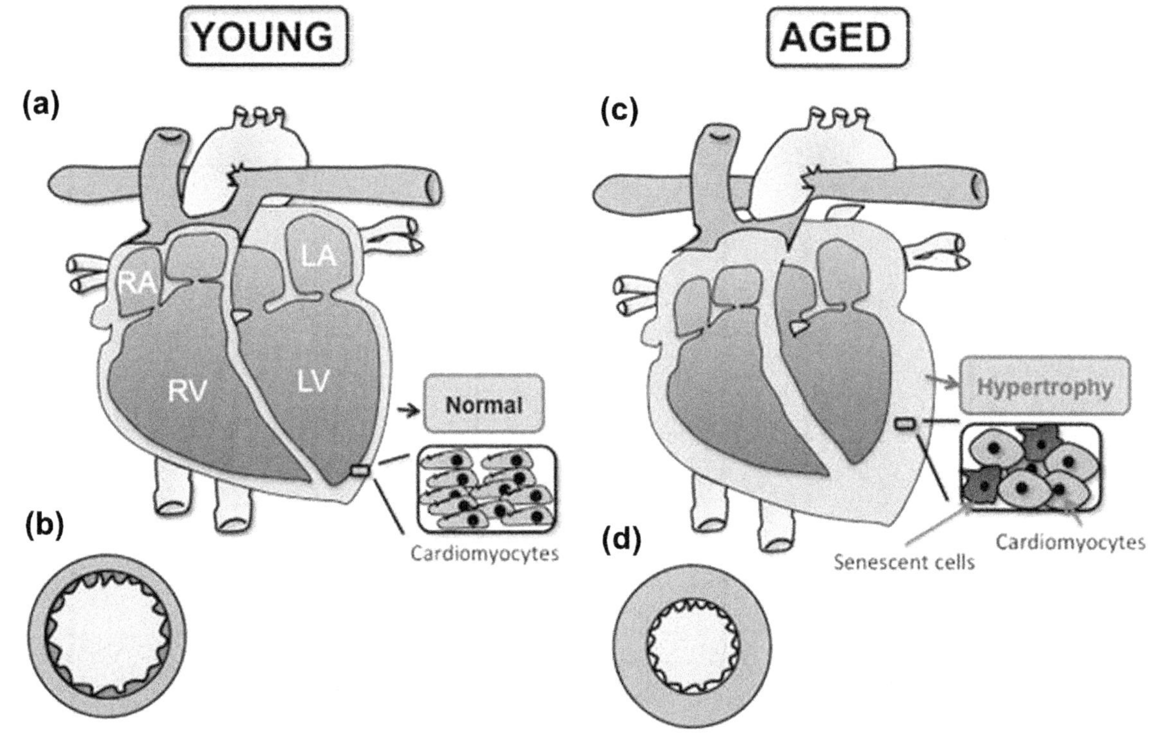

FIGURE 3.5 Age-related changes in the cardiovascular system. In the heart, several functional and structural changes occur with the aging process. Panel (a): Young heart with normal thickness of the myocardium and no compromised function. Panel (b): Young artery with normal sized lumen and vessel wall; adaptive contractility and relaxation is maintained. Panel (c): Aged heart with myocardial hypertrophy; cardiomyocytes are undergoing senescence and thus have compromised function. Hypertrophy and fibrotic tissues will lead to impeded signaling from the sinoatrial node. Panel (d): Aged artery with narrower lumen, increased wall thickness; adaptive contractility and relaxation is lost. RA: right atrium; LA: left atrium; RV: right ventricle; LV: left ventricle. *Reprinted with permission from Shaik et al 2013.*[68] Available from: http://www.intechopen.com/books/senescence-and-senescence-related-disorders/endothelium-aging-and-vascular-diseases.

this pathway is the production of adenosine triphosphate (ATP) from adenosine diphosphate (ADP). Harmful ROS are generated when electrons lost from mitochondrial complexes I and II form superoxide radicals. Furthermore, NADH and FADH are free to react with other redox compounds throughout the mitochondrial pathway.[9]

As vessels age, they produce more ROS, leading to functional impairment. Increased production of the superoxide anion with aging leads to increased reaction of this ROS with nitric oxide. This reaction will not only inhibit the biological activity of nitric oxide as a vasodilator, but also results in the formation of peroxynitrite. Peroxynitrite is a powerful membrane-permeable oxidant that inactivates several enzymes, including free radical scavengers, through substrate nitration. Age-dependent increases in nitric oxide synthase expression are associated with the peroxynitrite formation and are therefore implicated in age-related oxidative damage to vasculature.[9,10]

Age-related oxidative stress is also central to the pathologic changes leading to atherosclerosis, a starting point of more serious cardiovascular conditions. Endothelial cells and vascular smooth muscle cells produce reactive oxygen and nitrogen species which oxidize low-density lipoproteins (LDL). Oxidized LDL enter subendothelial spaces where they initiate atherosclerosis. The oxidation process will increase mitochondrial rupture and the release of proapoptotic molecules to the cytosol, increasing plaque cell apoptosis. Furthermore, the scavenging process for reactive nitrogen species increases endothelial dysfunction, smooth muscle cell proliferation, leukocyte adhesion and inflammatory responses. Thus, the generation and clearance of oxidative stress in vasculature is an extremely significant regulator of age-related cardiovascular decline on many levels.[5,9]

Alongside gross physiologic adaptation of the cardiovascular system, minute molecular and genetic mechanisms are known to mediate cardiovascular responses to oxidative stress and age-related decline. The three leading molecular mediators are the involvement of longevity genes and the longevity network (which are discussed later) and the telomere theory of aging that is also strongly associated with cardiovascular decline.[5,6]

Telomeres are structures of repeating nucleotides at the end of eukaryotic chromatids that are shortened during replication. This mechanism exists to protect valuable genetic data from being lost by incomplete replication. Telomere length, and thus protective capacity, are determined genetically, are decreased by oxidative stress and

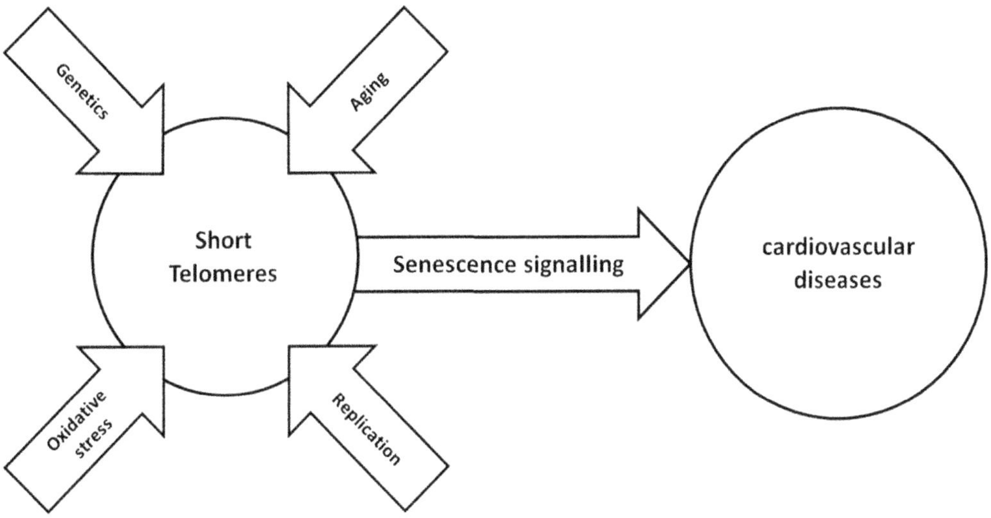

FIGURE 3.6 Factors that lead to telomere attrition and a higher risk of cardiovascular disease. The factors that affect telomere length help to explain interindividual variation in the risk of developing cardiovascular disease. Telomere length will vary genetically at birth and will shorten with age and exposure to oxidative stress (e.g. smoking or metabolic conditions such as obesity). Cell replication also shortens telomere length; when a critical length has been reached telomeres activate senescence signaling which, in turn, increases the risk of cardiovascular disease. *Adapted from Samani & van der Harst (2008).*[6]

vary with lifespan.[6] In conditions such as cardiovascular disease, with a significant multifactorial etiology involving both aging and metabolic oxidative stress, it is important to consider the role of telomere attrition. This is because the telomere theory is a candidate for distinguishing individual variability of risk and susceptibility, a factor that remains largely unknown, unlike population variability, which is attributed to known factors such as smoking or ethnicity.

When shortened to a critical length, telomeres will drive senescence signaling mechanisms. Dysregulation in the pathways of cell senescence are a strong underlying mechanism leading to cardiovascular decline. There have also been strong associations between systemic cell-type-specific telomere shortening and the development of cardiovascular disease. For example, the presence of shortened telomeres in circulating leukocytes is associated with the development of coronary artery disease. In observational studies, the presence of shortened telomeres precedes the development of clinically relevant disease, indicating a more causal role rather than a consequential effect of cardiovascular decline.[11,12]

Many interesting studies have investigated this hypothesis. For example, Cawthon et al[13] found that in subjects aged over 60, a shorter telomere length was strongly associated with a high cardiac mortality rate within a decade. In this study, the presence of other classic cardiovascular risk factors did not explain the higher mortality rate. Furthermore, as telomeres shorten with aging, the average loss of telomeres can be equated to the number of years. Those with coronary artery disease are, on average, 8–12 years older, in terms of telomere loss (i.e. similar average telomere length to healthy subjects actually 8–12 years older). Telomere attrition has been specifically associated with a number of cardiovascular pathologic events, including: congestive heart failure, peripheral vascular disease and carotid artery atherosclerosis.[6,11–13]

It is not surprising that there are continuously strengthening links between aging, cardiovascular disease and telomere shortening. Oxidative stress is certainly well known to be implicated in negatively affecting all three processes: accelerating telomere loss, accelerating aging and acting as a significant pathologic component of cardiovascular disease.[6] Whether a common causal pathway is responsible for the cardiovascular aging phenotype, for telomere attrition and for sensitivity to oxidative stress, is an extremely exciting area of current research. Figure 3.6 summarizes the factors that increase telomere attrition, senescence signaling and, subsequently, a higher risk of cardiovascular disease.

LONGEVITY GENES AND THE LONGEVITY NETWORK

Longevity genes regulate oxidative stress response pathways and biologic aging. These factors are major parts of cardiovascular disease pathology[5]. The list of genes includes the Sirtuins (SIR2), insulin-like growth factor-1 (IGF1), CLOCK1 (mCLK1), adenosine monophosphate-activated protein kinases (AMPK), $p66^{Shc}$, catalase and klotho; see Table 3.1 for a complete list of the genes discussed in the following section. A subset of the longevity genes interacts extensively and constitutes the longevity network, the functions and interactions of which are addressed first.

TABLE 3.1 Longevity Genes[a]

Official Gene Symbol	Name of Gene	Sequence ID (Known Human Transcript Variants)
SIRT1	Sirtuin 1	NM_001142498.1; NM_012238.4
SIRT3	Sirtuin 3	NM_00107524.2; NM_012239.5
SIRT6	Sirtuin 6	NM_001193285.1; NM_016539.2
SIRT7	Sirtuin 7	NM_016538.2
IGF1	Insulin-like growth factor 1	NM_000618.3; NM_001111283.1; NM_001111284.1; NM_00111285.1
FOXO1	Forkhead box O1	NM_002015.3
FOXO3	Forkhead box O3	NM_001455.3; NM_201559.2
PIT1	Pituitary-specific positive class 1 homeobox	NM_000306.2; NM_001122757.1
AMPK	Protein kinase, AMP-activated, alpha 1 catalytic subunit	NM_006251.1; NM_206907.3
mTOR	Mechanistic target of rapamycin	NM_004958.3
$p66^{Shc}$	Src homology 2 domain-containing transforming protein C1	NM_001113331.2; NM_011368.5
KL	Klotho	NM_004795
CAT	Catalase	NM_001752
CLK1	Circadian locomotor output cycles kaput	NM_001254772

[a]*These genes have been found to have a substantial link with age-related physiologic decline in cardiovascular function and with cellular and systemic responses to oxidative stress. A number of these genes (SIRT1, AMPK, IGF-1 and mTOR) form the longevity network, a network that regulates biologic aging and therefore the development of cardiovascular disease.*

Genes of the Longevity Network: SIR2, IGF-1, AMPK and mTOR

The Sirtuins

Sirtuins (SIR2) are a class of evolutionarily conserved enzymes. In mammals there are seven members of the SIR2 family with diverse cellular localizations. These proteins have a wide range of functions, namely, ribosyltransferase and deacetylase activity, as well as functions in aging, transcription, apoptosis, inflammation, DNA damage and repair, cell-cycle regulation, stress resistance and mitochondrial function.[14]

The Cardiovascular System and SIR2 Proteins

SIR2 family members 1, 3 and 7 (SIRT1, SIRT3 and SIRT7, respectively) have proven roles in the cardiovascular system. Cardiac-specific overexpression of SIRT1 in mice delays age-dependent cardiomyopathy and decreases stress-induced apoptosis. Knockout and constitutive overexpression is detrimental, increasing ischemia–reperfusion injury, cardiomyopathy, apoptosis and oxidative stress. SIRT1 also effects vasculature, inhibiting angiotensin-II-mediated vascular smooth muscle hypertrophy. It regulates angiogenesis and prevents arterial calcification and stiffness.[15–17] Knockout of mitochondrial SIRT3 and nuclear SIRT7 causes age-dependent, pressure-induced cardiac hypertrophy and inflammatory cardiomyopathy. Thus, SIRT3 and SIRT7 protect from oxidative damage as well as prevent cardiovascular remodeling and inflammation.[18,19]

Oxidative Stress and SIR2 Proteins

The roles of SIR2 proteins in oxidative stress are complex and are member-, site- and species-specific. SIRT1, SIRT6 and SIRT7 proteins have been strongly associated with mediating oxidative stress responses in mammals. Expression of SIRT1 ameliorates the aggravated oxidative stress seen in cardiovascular disease.[20] SIRT6 is expressed in a number of tissues, including the heart; its overexpression promotes resistance to oxidative stress and associated DNA damage. SIRT6 substrate specificity is very high, exerting deacetylase activity to maintain chromatin function, in turn promoting telomere stability. Therefore, SIRT6 prevents cell senescence through mediating resistance to oxidative stress, ameliorating DNA damage and preserving telomere length.[21] As for SIRT7, diminished expression in cardiomyocytes decreases oxidative stress resistance and increases apoptosis by up to 200%. This is mediated via hyperacetylation of the p53 protein and has been confirmed in haplotype mice with diminished SIRT7 expression.[19]

Insulin-Like Growth Factor 1

One of the initial genes to be identified as a longevity gene, insulin-like growth factor 1 (IGF1), encodes a peptide structurally similar to insulin with an important function in early growth and an anabolic effect in adults. IGF1 exerts its effects by binding to its receptor that is expressed on many cell types and tissues. From this binding site, IGF1 inhibits apoptosis and acts as a stimulator of cell growth and proliferation.[5,22] In mammals, the loss of this protein's function causes dwarfism, deficient growth or postnatal death from respiratory failure.[23]

The Cardiovascular System and IGF1

Several studies have implicated IGF1 signaling in the maintenance of cardiovascular health. In mice, cardiac-specific overexpression of IGF1 prevents cell death after myocardial infarction and reduces hypertrophy, ventricular dilation and diabetic cardiomyopathy. However, other overexpression studies have shown negative effects, including hypertrophy and diminished recovery after ischemia. Live knockout of mammalian IGF1 increases cardiomyocyte cell death under oxidative stress.[24-26] In simple organisms, with a primitive cardiovascular system, the induction of IGF1 delays age-related degeneration. Some studies report opposite effects in mammals. This difference is possibly explained by the complex autocrine and paracrine signaling pathways that are targeted by IGF1 and are not present in simpler organisms.[5,22]

Oxidative Stress and IGF1

Oxidative stress is associated with decreased plasma IGF1 in humans, leading to cardiac and mitochondrial dysfunction. When mice have cardiac-specific overexpression of IGF1 – or have IGF1 administered exogenously under conditions that induce cardiovascular disease – there is decreased apoptosis and mitochondrial dysfunction as well as lower systemic and myocardial oxidative stress. Furthermore, IGF1 promotes protective regulation of other mediators of aging and cardiovascular risk (e.g. FOXO and mTOR). Despite protective effects of IGF1 on cardiovascular function, remodeling and hypertrophy still occur due to preserved mitochondrial function and increased cell survival under pro-oxidant conditions.[25,26]

The interaction between IGF1 and its receptor regulates cell growth, transformation and survival under oxidative stress. When this interaction is perturbed (e.g. in haplotype mice and in myoblasts cultured from them) there is increased resistance to oxidative stress.[27] Studies in other types of cell show opposite effects; for example, increased IGF1 signaling confers protection from oxidative stress in microglial cells and in induced pluripotent stem cells.[28] Studies applying small interfering RNA (siRNA) have shown the complexity of these mechanisms. The effects of perturbing these mechanisms will vary upstream or downstream of the IGF1 signaling system[28]. Through the aforementioned studies, microRNA-1 (MiR1) targeting of the 3′-UTR region of the IGF1 gene has been shown to regulate the cytoprotective properties and prevent oxidative stress-induced apoptosis signaling.[28]

Forkhead Box Transcription Factors

Forkhead box (FOX) proteins are a large family of transcription factors regulating expression of genes that control cell growth, proliferation, differentiation and aging. Of particular interest to cardiovascular health and aging are the O class FOX proteins (FOXO) that also regulate metabolism and stress tolerance. They are post-translationally controlled by ubiquitination, phosphorylation and acetylation.[29]

The Cardiovascular System and FOXO Proteins

Maintaining finely tuned function of FOXO proteins is necessary for life. Deletion of the FOXO1 member is embryonically lethal and causes deficient cardiac and vascular growth. Interestingly, cardiac-specific overexpression of FOXO1 is also lethal and causes reduced heart size and myocardium thickness as well as heart failure and impaired cardiomyocyte proliferation.[29,30] A variant of FOXO3 is commonly found in centenarians, and is thought to delay aging and preserve cardiovascular health. Overexpression of the FOXO3a variant decreases cardiomyocyte size, whereas deficiency increases endothelial nitric oxide synthase expression and promotes postnatal angiogenesis and vessel formation.[31,32]

Oxidative Stress and FOXO Proteins

FOXO signaling maintains homeostasis and mediates responses to environmental pathologic changes, such as oxidative stress. FOXO3a is inhibited by the protein kinase B (AKT) pathway and upregulated by hypoxic conditions. FOXO3a is known to inhibit MYC-mediated adaptive mitochondrial metabolism. Under cellular oxidative stress, AKT is inhibited, which increases FOXO3a signaling. FOXO3a, via inhibition of MYC signaling, prevents increased mitochondrial metabolism under hypoxia. In this cascade, the knockdown of FOXO3a allows MYC signaling to increase mitochondrial oxidative stress generation.[32,33] Furthermore, under metabolic oxidative stress, FOXO3a signaling disrupts vascular function by facilitating degradation of vascular calcium/potassium channels. In mice, suppression of FOXO3a preserves cardiovascular function under metabolic oxidative stress.[34]

Adenosine Monophosphate-Activated Protein Kinase

Adenosine monophosphate (AMP)-activated protein kinase (AMPK) is involved in cellular energy homeostasis, glucose metabolism, lipid metabolism, cell growth, polarity, gene expression and autophagy. AMPK regulates cellular energy metabolism by mitophagy and mitogenesis: destruction of defective mitochondria and activation of mitochondrial biogenesis, respectively. AMPK is also a metabolic energy sensor for the AMP:ADP ratio, stimulating ketogenesis and hepatic fatty acid oxidation as well as inhibiting lipogensis, triglyceride and cholesterol synthesis.[35]

The Cardiovascular System and AMPK

In humans, hereditary syndromes due to AMPK mutations have a pathologic component in cardiovascular decline. AMPK mutations cause hypertrophic cardiomyopathy and ventricular pre-excitation. Diminished expression of AMPK exacerbates the effects of myocardial infarction, whereas overexpression is cardioprotective, ameliorating damage from ischemia–reperfusion injury and preventing pressure-overload hypertrophy.[36,37]

Oxidative Stress and AMPK

AMPK is a target of pharmaceutical regulation for its role in defense against mitochondrial oxidative stress. Endothelial mitochondrial ROS are implicated in age-related cardiovascular decline. Endothelial oxidative stress and AMPK expression increase during impaired vasodilation and hypertension from coronary artery disease and diabetes. Diabetes is strongly associated with systemic oxidative stress and is a leading co-morbidity with aging cardiovascular disease.[38]

Further observations of mitochondrial oxidative stress and AMPK regulation have been made in human saphenous vein endothelial cells. When these cells, under oxidative stress, are incubated with a mitochondrial-targeted antioxidant, AMPK expression decreases, providing evidence for the role of AMPK in mitochondrial oxidative stress defence.[39] It is necessary to maintain the delicate balance in production and clearance of such reactive species given their functions as signaling molecules in normal cardiovascular function and adaptability. Cardiovascular dysfunction occurs when their rate of production overtakes endogenous free radical scavenging processes. In this regard, AMPK's role as an energy sensor is clear as the major sources of oxidative stress are mitochondrial nicotinamide adenine dinucleotide phosphate (NADPH) oxidases. There is currently strong evidence to support AMPK's role as a suppressor of NADPH oxidase activity and therefore reactive species generation.[38,39]

AMPK also has a role in regulating adaptive angiogenesis under oxidative stress. This has been shown by rotenone treatment in coronary artery endothelial cells (rotenone increases oxidative stress and inhibits mitochondrial complex I). Vascular tube formation induced by vascular endothelial growth factor (VEGF) is inhibited by rotenone treatment. In the same process, the rotenone induction of oxidative stress increases expression of AMPK. Interestingly, knockdown of AMPK in rotenone treatment preserves VEGF-induced tube formation.[39]

Mammalian Target of Rapamycin (mTOR)

Rapamycin is an antifungal compound originally discovered in soil samples from Easter Island. It is clinically used as an immunosuppressant after organ transplantation.[40] Rapamycin inhibits the activity of the mammalian target of rapamycin (mTOR). The product of mTOR is a serine/threonine protein kinase that also integrates products of upstream pathways, including insulin and IGF1. These functions are mediated via two complexes (mTORC1 and mTORC2). The mTOR pathway is dysregulated in a number of human metabolic pathologies that induce oxidative stress, such as diabetes and obesity.[6]

The Cardiovascular System and mTOR

Cardiac inhibition of mTOR activity reverses pressure-overload hypertrophy, via inhibition of mTOR's control over cell size and protein translation.[41] The mTOR protein also interacts with AKT and phosphatidylinositide 3-kinase (PI3K) in the PI3K/AKT/mTOR pathway, which mediates hypoxia-induced angiogenesis. Rapamycin perturbation of the PI3K/AKT/mTOR pathway inhibits vessel growth, normally mediated by hypoxia-inducible factor 1 (HIF1) and VEGF.[42] The PI3K/AKT/mTOR pathway is central to many other signaling pathways, making these genes and their products vital to the understanding of oxidative damage, cardiovascular health and aging.

Oxidative Stress and mTOR

Oxidative stress influences targets upstream and downstream of the mTOR pathway. Sustained increased activity of the mTOR pathway is a strong causal candidate for cardiovascular decline with age and oxidative stress. Furthermore, mTOR's response to oxidative stress is associated with the inflammatory pathways that progress cardiovascular disease.[43] In human coronary artery endothelial cells, inducing oxidative stress by rotenone inhibits mTOR activity.[39] Furthermore, mTOR regulates arterial responses to oxidative stress alongside AMPK. Under oxidative stress, arterial contraction and compliance are compromised. When rapamycin is applied to mouse tissues, arterial contraction is preserved, indicating that mTOR mediates adaptive loss of contractile function under oxidative stress.[44] The effect of mTOR on cardiac remodeling is associated with, and possibly governed by, the body's renin–angiotensin–aldosterone axis, which is central to the oxidative stress response, maintaining blood pressure and maintaining arterial vasoconstriction. When renin is overexpressed in rats, there is a concomitant increase in mTOR activity, as well as metabolic, biochemical and physical manifestations of oxidative stress and of cardiovascular system decline.[45]

The Longevity Network: SIR2, IGF1, AMPK and mTOR

The longevity network comprises the interactions between SIRT1, IGF1, mTOR and AMPK (see Fig. 3.7).[5] SIRT1 is cardioprotective and increases stress resistance when mildly overexpressed. It regulates the hepatic AMPK pathway via the upstream liver kinase B1 (LKB1). IGF1 is regulated both directly by SIRT1 and via the mitochondrial uncoupling protein 2 (UCP2). SIRT1 also inhibits the mTOR pathway through the complexes formed by tuberous sclerosis 1 and 2 (TSC1, TSC2).[46]

IGF1 interacts with both mTOR and SIRT1 pathways. FOXO is a downstream effector of SIRT1's action on IGF1.[5] The mTOR pathway is stimulated by AKT signaling via IGF1 and it is inhibited by AMPK and phosphorylation of the TSC1 and TSC2 complex.[47] AMPK, in turn, activates SIRT1 and IGF1 signaling. SIRT1 activity increases with levels of nicotinamide

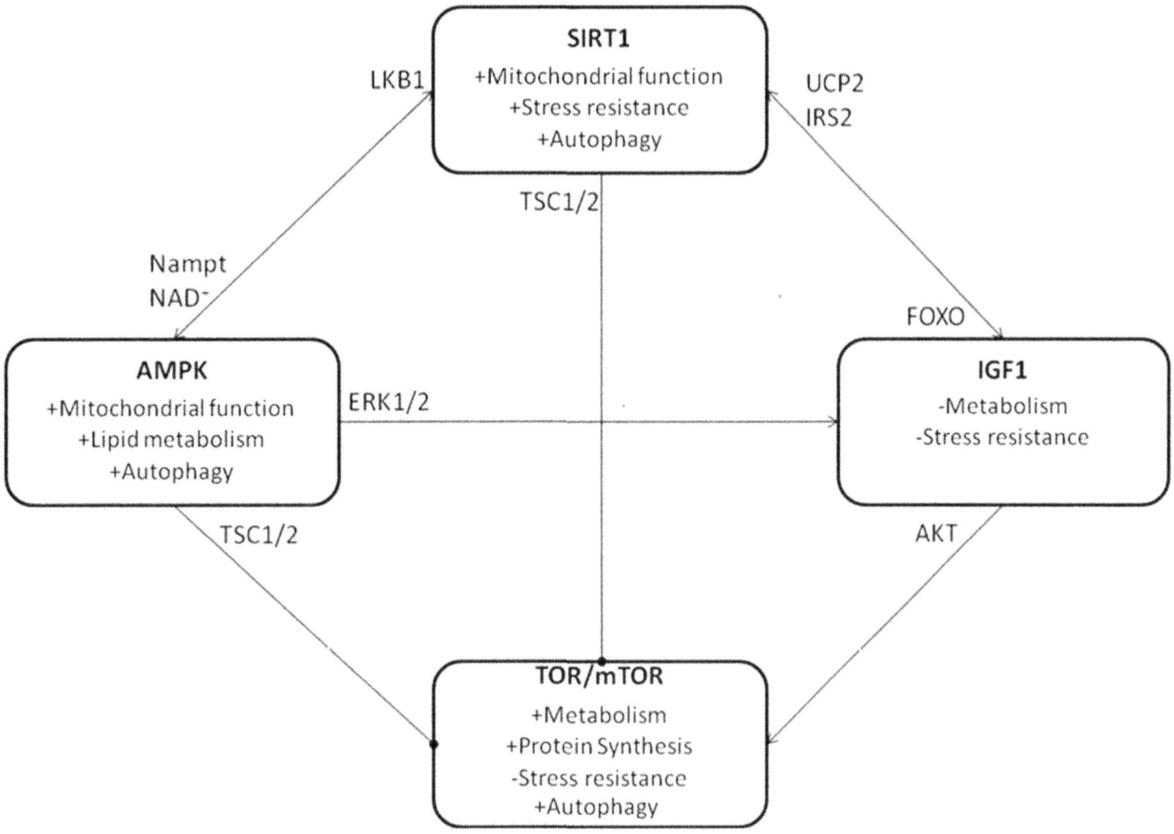

FIGURE 3.7 Interactions of the longevity network. The longevity network principally refers to positive and negative feedback mechanisms between four genes: SIRT1, IGF1, AMPK and mTOR. The signaling between these genes regulates biologic aging and numerous age-related diseases, including cardiovascular diseases. At the end of each line, an arrowhead indicates positive regulation/stimulation, and a round end indicates negative regulation or inhibition. SIRT1: Sirtuin1; TSC1/2: tuberous sclerosis 1/2; UCP2: mitochondrial uncoupling protein 2; IRS2: insulin receptor substrate 2; FOXO: O class forkhead box transcription factor; IGF1: insulin-like growth factor 1; AKT; protein kinase B; TOR/mTOR: target of rapamycin/mammalian target of rapamycin; ERK1/2: extracellular signal-regulated kinases 1/2; AMPK: adenosine monophosphate-activated protein kinase; Nampt: nicotinamide phosphoribosyltransferase; NADH: nicotinamide adenine dinucleotide; LKB: liver kinase B. *Adapted from North & Sinclair (2012).*[5]

phosphoribosyltransferase (Nampt) and nicotinamide adenine dinucleotide (NAD). In contrast, IGF1 is stimulated through the extracellular signal-regulated kinase cascade (ERK1/2).[5,47]

The longevity network has strong links to the oxidative stress response and aims to preserve both cellular and organ function throughout aging. In the longevity network mitochondrial function, metabolism and protein synthesis are modulated to preserve cellular function, whereas stress resistance and autophagy are stimulated to promote cellular senescence or proliferation to preserve organ function above cell fate.[5,46,47]

OTHER GENES AND PATHWAYS IN OXIDATIVE STRESS AND AGE-RELATED CARDIOVASCULAR DISEASES

CLOCK1

CLOCK1 (mCLK1) is a mitochondrial hydroxylase necessary for biosynthesis of ubiquinone, an endogenous antioxidant and cofactor in cellular redox pathways. Ubiquinone is present in all membranes and is responsible for the hydroxylation of 5-demethoxyubiquinone to 5-hydroxyubiquinone in the mitochondrial electron transport chain.[48]

The Cardiovascular System and mCLK1

Mutations causing partial or complete inactivation of mCLK1 are associated with delayed aging. Mice lacking mCLK1 are also protected from ischemia–reperfusion injuries, suggesting a role in vascular function and ischemic response. Studies show that partial loss of function impedes but does not stop certain developmental processes, including embryogenesis and the cell cycle (http://emboj.embopress.org/content/18/7/1783).[48]

Oxidative Stress and mCLK

In mCLK haplotype mice, aging is delayed, despite increased generation of mitochondrial ROS. As well as protection from this oxidative stress, haplotype mice also have enhanced immunity and suffer limited damage from other challenges. Biomarkers of oxidative stress in aging are also attenuated in the haplotype mouse. Seemingly, there is an initial increase in oxidative stress, with a subsequent higher protective capacity. A similar phenotype, without the longer lifespan, is observed in mice lacking antioxidant enzymes such as superoxide dismutase.[48,49] This marks mCLK-modulated mitochondrial oxidative stress as a 'safe' process to alter within the biologic aging framework. Given the enhanced immunity and oxidative stress resistance, mCLK is currently a valuable target for investigation.[48,49] In light of the above, the mCLK knockout mouse provides a genetic model to study mitochondrial oxidative stress in age-related cardiovascular decline. Work thus far is indicative of an increasingly important role for mitochondrial metabolism in cardiovascular aging.

Catalase

Catalase contains four porphyrin heme groups to facilitate degradation of hydrogen peroxide. Because of a high specific reactivity and easily detectable outcome, catalase is used analytically and industrially (e.g. in classifying bacteria).[50] Physiologically, catalase is a very common redox enzyme in nearly all aerobic organisms which protects the organism from oxidative stress. Despite this important function, catalase deficiency produces a largely normal phenotype. Animals and humans with acatalasia are only slightly more sensitive to oxidative stress.[51]

The Cardiovascular System and Catalase

When catalase is overexpressed in mitochondria, age-related cardiac damage is reduced and lifespan is increased by 20%.[52] Transgenic mice are resistant to hypertrophy, fibrosis, angiotensin II-mediated mitochondrial damage and G_q-α subunit-mediated heart failure.[52,53] In an *ex vivo* study of rat hearts, cardiac dysfunction occurs with ischemia and ischemia–reperfusion. When treated with catalase (and co-functioning superoxide dismutase), the ischemia–reperfusion damage is ameliorated. Other abnormalities from ischemia alone were also ameliorated (e.g. disruption of sodium/potassium ATPase, sodium–calcium exchange, and calcium uptake and release).[54]

Oxidative Stress and Catalase

Typically catalase is a large contributor to neutralizing metabolically by-produced hydrogen peroxide to water and oxygen, thus preventing accumulation of hydroxyl radicals. It is theorized that other redox proteins, such as superoxide dismutase, will compensate when catalase levels are low.[54] However, catalase also interacts with many important pathways related to oxidative balance (e.g. host and pathogen defense and alcohol metabolism). Catalase also forms part of myocardial local redox defense systems against systemic oxidative stress (e.g. in obesity or insulin resistance). Under such conditions, blocked angiotensin receptors increase antioxidant enzyme activity by 50–70%, with catalase having a marked increase.[55]

A significant role for catalase has also been demonstrated in oxidative stress-induced cardiac remodeling. Mice with cardiac-specific G_q-α subunit overexpression provide a model for structural changes in the cardiovascular system. In their system, oxidative stress also increases at a faster than normal rate. The increase in oxidative stress causes dilated cardiomyopathy, which progresses to heart failure. When G_q-α subunit overexpressing mice are cross-bred with cardiac-specific catalase overexpressing mice there is a reduction in age-related and oxidative stress-related structural changes. Myocyte hypertrophy, apoptosis and heart failure were all prevented in cross-bred mice overexpressing both the G_q-α subunit and catalase in the heart. This occurred without affecting the initial oxidative stress phenotype of G_q-α subunit overexpressing mice.[56]

Klotho

Klotho is a transmembrane protein, related to β-glucuronidases, that is highly expressed in specific kidney and brain regions. Klotho hydrolyzes steroid β-glucuronides and partially regulates systemic glucose metabolism and insulin sensitivity.[57] Overexpression of klotho delays aging, whereas deficiency results in a phenotype resembling human accelerated aging.[58,59]

The Cardiovascular System and Klotho

Klotho expression maintains cardiovascular health. In fact, low klotho expression is thought to be a leading cause of vascular degeneration in chronic renal failure patients. Deficiency is also associated with arteriosclerosis, impaired angiogenesis and impaired endothelium-dependent vasodilation.[5]

Soluble klotho is formed when the extracellular domain is shed into the circulation; its levels decline with age. Cell surface binding sites with which soluble klotho interacts remain unknown. It is, however, known that klotho binding results in perturbed intracellular insulin/IGF1 signalling. This is proposed as a mechanism for klotho's implication in healthy aging, as the aging phenotype of klotho-deficient mice is averted by perturbing the insulin/IGF1 signaling cascade. This explains the association of klotho single nucleotide polymorphisms (SNPs) with age-related cardiovascular decline.[5,60]

Oxidative Stress and Klotho

In vitro and *in vivo* research has provided evidence of klotho's potential to increase the endothelial layer's resistance to oxidative stress. This is achieved by maintaining signals for nitric oxide production. Defective klotho signaling also affects the heart, leading to fatal sinoatrial node dysfunction under stress.[6,61]

Pituitary Transcription Factor 1 and Prophet of Pituitary Transcription Factor 1

Pituitary transcription factor 1 (PIT1) is a pituitary-specific transcription factor responsible for normal pituitary development and for expression of hormones regulating mammalian growth.[62] Homeobox protein prophet of PIT1 (PROP1), another pituitary transcription factor, possesses transcriptional activation properties as well as DNA-binding properties. PROP1 expression leads to the development of specialized pituitary cells. The knockout of PIT1 in Snell/Ames dwarf mice inhibits the development of such anterior pituitary cells, in turn disturbing expression of other signaling peptides and contributing to the risk of diseases.[62]

Cardiovascular Disease and PIT1/PROP1

Mostly what is known about PIT1 and PROP1 is their effects on biologic aging and growth. Patients deficient in PIT1 and PROP1 typically have growth hormone deficiency. PROP1 deficiency also causes hypogonadism and deficiency of prolactin and thyroid-stimulating hormone.[62] Interestingly, Snell/Ames dwarf mice with PROP1 deficiency and delayed aging also exhibit lower cardiac collagen content and smaller cardiomyocytes. These properties may be beneficial with regard to preserving cardiovascular health in aging.[63]

Oxidative Stress and PIT1/PROP1

In Snell/Ames dwarf mice, where the PIT1/PROP1 signaling is perturbed, there is an altered response to oxidative stress. This altered response is thought to increase resistance to oxidative damage and contribute to healthy aging in this mouse.[64]

The dwarf mouse is known to be particularly resistant to mitochondrial oxidative stress. This was investigated by inducing mitochondrial oxidative stress through inhibition of mitochondrial complex II with 3-nitropropionic acid (3-NPA) treatment. The activator protein 1 (AP-1) transcription factor, which regulates transcriptional responses to numerous stimuli, contains c-Jun family proteins. Phosphorylation of Ser^{63} and Ser^{73} residues on the c-Jun protein is necessary to mediate apoptotic and/or proliferative responses to oxidative stress. In the dwarf mouse, after generation of mitochondrial oxidative stress by 3-NPA treatment, there is a lack of c-Jun Ser^{63} phosphorylation. This contrasts with the wild type rapid and robust phosphorylation of Ser^{63} and Ser^{73} residues. This mechanism allows cell survival in the face of mitochondrial oxidative stress in the dwarf mouse.[64]

$p66^{Shc}$

The Shc locus regulates several metabolic processes in mammals. Three splice variants are coded with molecular masses of 46, 52 and 66 kDa; they each carry a Src-homology 2 domain, a collagen-homology region and a phosphotyrosine-binding domain. The p66 splice variant has a unique N-terminal region with redox enzyme properties. It is actively involved in generating mitochondrial ROS. This N-terminal region is also part of the signaling cascade that translates oxidative signals to apoptosis.[5,65]

The Cardiovascular System and $p66^{Shc}$

Knockout of $p66^{Shc}$ delays aging, decreases generation of ROS, cardiac progenitor cell senescence, necrosis and DNA damage; vascular endothelial cell resistance to oxidative stress is increased.[66] Loss of $p66^{Shc}$ maintains left ventricular volume, reduces heart failure, and protects high-fat fed mice from atherogenesis.[66] In humans, $p66^{Shc}$ expression increases with age, under basal and disease-related conditions.[67] Generally speaking, $p66^{Shc}$ is a mediator of many cellular processes, namely apoptosis and stress response to maintain cardiovascular function.

Oxidative Stress and $p66^{Shc}$

A mechanistic model of $p66^{Shc}$ is that it oxidizes cytochrome *c*, preventing it from reducing oxygen radicals to water. Cytochrome *c* is at the final steps of mitochondrial oxidative phosphorylation. The presence of $p66^{Shc}$ at this point diverts electron flow to produce hydrogen peroxide rather than water. Hydrogen peroxide then opens the mitochondrial permeability transition pore (PTP) which increases mitochondrial membrane permeability to ions, solutes and water. This increased influx swells and ruptures mitochondria, releasing molecules such as cytochrome *c* into the cytosol. Cytochrome *c* acts as a proapoptotic factor in the cytosol. Therefore, $p66^{Shc}$ plays a vital role in mediating cellular generation of oxidants (e.g. hydrogen peroxide) as well as cellular apoptotic responses through downstream signaling.[9] See Figure 3.8 for the process through which $p66^{Shc}$ induces mitochondrial rupture and cellular apoptosis.

What remains unclear is the signaling between extracellular, intracellular, exogenous and endogenous oxidative stress and $p66^{Shc}$. It is hypothesized that $p66^{Shc}$ traffics between cytosolic and mitochondrial compartments following phosphorylation by protein kinase C-β (PKCB) in response to oxidative stress. Entry of $p66^{Shc}$ disrupts calcium signaling, thus rearranging and fragmenting the mitochondrial matrix, allowing the pathway of oxidative stress and apoptosis to continue. This provides a convenient explanation for $p66^{Shc}$ knockout animals having similar resistance to apoptotic signals responding to exogenous as well as endogenous

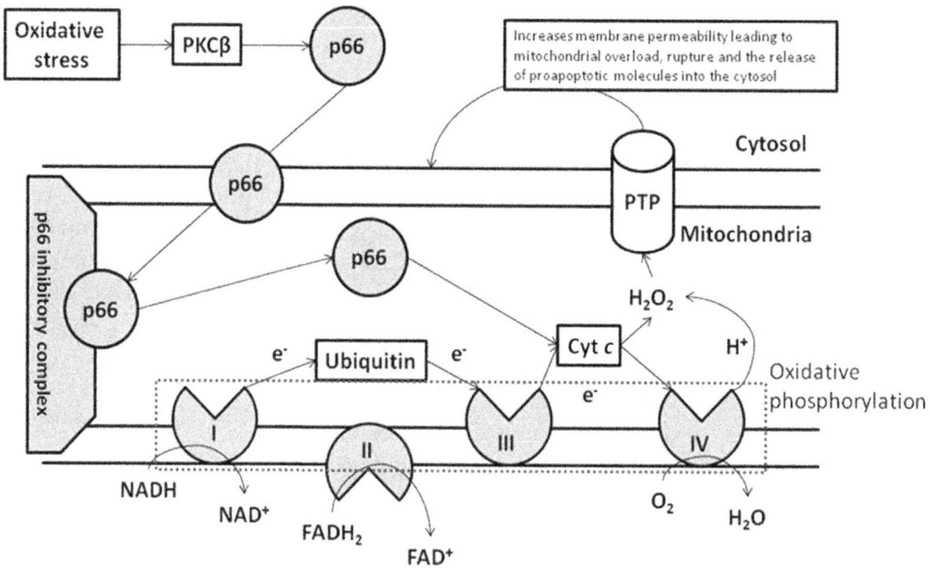

FIGURE 3.8 The role of $p66^{Shc}$ in apoptosis. Oxidative stress stimulates protein kinase C-β which phosphorylates $p66^{Shc}$. $p66^{Shc}$ can then traffic into the mitochondria where it is bound to an inhibitory complex. Pro-apoptotic stimuli destabilize $p66^{Shc}$, allowing it to oxidize cytochrome *c* between the third and fourth steps of oxidative phosphorylation. This catalyses the production of hydrogen peroxide from oxygen. Hydrogen peroxide induces the opening of the mitochondrial permeability transition pore, leading to an increase in mitochondrial membrane permeability to ions, solutes and water. Mitochondria then swell and rupture, releasing pro-apoptotic factors into the cytosol. *Adapted from Cosentino et al (2008).*[9]

oxidative stress. This creates a third role for $p66^{Shc}$ as an intermediary signaling molecule, as well as generating oxidative stress and regulating apoptosis.[9]

SUMMARY POINTS

- Cardiovascular disease is a large burden on the population, in terms of lives lost and economic loss.
- The aging portion of the population is increasing and is more likely to suffer the effects of cardiovascular decline leading to clinically relevant cardiovascular disease.
- Reduced cardiac antioxidant enzyme activity and increased generation of oxidative stress with aging contributes significantly to the risk of cardiovascular disease.
- The telomere theory states that telomere attrition, which is increased with oxidative stress and with age, significantly contributes to cardiovascular decline.
- Cardiomyocytes are particularly rich in mitochondria to meet the continuous and lifelong demand for energy and oxygen. Therefore the cardiovascular system is especially sensitive to mitochondrial generation of ROS.
- On the cellular level, mitochondrial oxidative stress is a significant mediator of age-related cardiovascular decline.
- SIR2, IGF1, mTOR and AMPK are key genes in the longevity network that regulate oxidative stress and associated changes to cardiovascular function.
- $p66^{Shc}$ is a powerful mediator of cellular oxidative balance. It functions as a signaling peptide for oxidative stress, as a contributor to mitochondrial generation of reactive oxygen species, and as an intermediary signaling molecule in apoptosis.
- PIT1 and PROP1 are pituitary-specific transcription factors that mediate important apoptosis signaling mechanisms in the cells that they target.
- Catalase is an important ubiquitously expressed antioxidant enzyme that controls apoptosis and hypertrophy responses in the myocardium.
- Klotho and mCLK regulate vascular health under oxidative stress and the generation of mitochondrial ROS, respectively.

References

1. Nichols M, Townsend N, Luengo-Fernandez R, et al. *European cardiovascular disease statistics 2012*. Sophia Antanapolis: European Heart Network, Brussels, European Society of Cardiology; 2012.
2. Centre for Disease Control. *Chronic diseaes prevention and health promotion*; 2010. Available at: http://www.cdc.gov/chronicdisease/resources/publications/AAG/dhdsp.htm. [accessed 30.08.13].
3. Scarborough P, Wickramasinghe K, Bhatnagar P, Rayner M. *Trends in coronary heart disease 1961–2011*. London: British Heart Foundation; 2011.
4. Kaasik A, Kuum M, Wilding J, et al. Mitochondria as a source of mechanical signals in cardiomyocytes. *Cardiovasc Res* 2010;**87**:83–91. http://dx.doi.org/10.1093/cvr/cvq039.
5. North BJ, Sinclair DA. The intersection between aging and cardiovascular disease. *Circ Res* 2012;**110**:1097–108. http://dx.doi.org/10.1161/CIRCRESAHA.111.246876.
6. Samani NJ, van der Harst P. Biological ageing and cardiovascular disease. *Heart* 2008;**94**:537–9. http://dx.doi.org/10.1136/hrt.2007.136010.
7. Mehdi MM, Rizvi SI. *N,N*-Dimethyl-*p*-phenylenediamine dihydrochloride-based method for the measurement of plasma oxidative capacity during human aging. *Anal Biochem* 2013;**436**:165–7. http://dx.doi.org/10.1016/j.ab.2013.01.032.
8. Antelmi I, de Paula RS, Shinzato AR, et al. Influence of age, gender, body mass index, and functional capacity on heart rate variability in a cohort of subjects without heart disease. *Am J Cardiol* 2004;**93**:381–5.
9. Cosentino F, Francia P, Camici GG, et al. Final common molecular pathways of aging and cardiovascular disease: role of the $p66^{Shc}$ protein. *Arterioscler Thromb Vasc Biol* 2008;**28**:622–8.

10. van der Loo B, Labugger R, Skepper JN, et al. Enhanced peroxynitrite formation is associated with vascular aging. *J Exp Med* 2000;**192**:1731–44.
11. Demissie S, Levy D, Benjamin EJ, et al. Insulin resistance, oxidative stress, hypertension, and leukocyte telomere length in men from the Framingham Heart Study. *Aging Cell* 2006;**5**:325–30.
12. Fitzpatrick AL, Kronmal RA, Gardner JP, et al. Leukocyte telomere length and cardiovascular disease in the cardiovascular health study. *Am J Epidemiol* 2007;**165**:14–21.
13. Cawthon RM, Smith KR, O'Brien E, et al. Association between telomere length in blood and mortality in people aged 60 years or older. *Lancet* 2003;**361**:393–5.
14. Costantini S, Sharma A, Raucci R, et al. Genealogy of an ancient protein family: the Sirtuins, a family of disordered members. *BMC Evol Biol* 2013;**13**:60. http://dx.doi.org/10.1186/1471-2148-13-60.
15. Hsu CP, Zhai P, Yamamoto T, et al. Silent information regulator 1 protects the heart from ischemia/reperfusion. *Circulation* 2010;**122**:2170–82. http://dx.doi.org/10.1161/CIRCULATIONAHA.110.958033.
16. Kawashima T, Inuzuka Y, Okuda J, Kato T, Niizuma S, Tamaki Y, Iwanaga Y, Kawamoto A, Narazaki M, Matsuda T, et al. Constitutive SIRT1 overexpression impairs mitochondria and reduces cardiac function in mice. *J Mol Cell Cardiol* 2011;**51**:1026–36. http://dx.doi.org/10.1016/j.yjmcc.2011.09.013.
17. Potente M, Ghaeni L, Baldessari D, Mostoslavsky R, Rossig L, Dequiedt F, Haendeler J, Mione M, Dejana E, et al. SIRT1 controls endothelial angiogenic functions during vascular growth. *Genes Dev* 2007;**21**:2644–58.
18. Hafner AV, Dai J, Gomes AP, et al. Regulation of the mPTP by SIRT3-mediated deacetylation of CypD at lysine 166 suppresses age-related cardiac hypertrophy. *Aging (Albany NY)* 2010;**2**:914–23.
19. Vakhrusheva O, Smolka C, Gajawada P, et al. Sirt7 increases stress resistance of cardiomyocytes and prevents apoptosis and inflammatory cardiomyopathy in mice. *Circ Res* 2008;**102**:703–10. http://dx.doi.org/10.1161/CIRCRESAHA.107.164558.
20. Wang F, Chen HZ, Lv X, Liu DP. SIRT1 as a novel potential treatment target for vascular aging and age-related vascular diseases. *Curr Mol Med* 2013;**13**(1):155–64.
21. Beauharnois NM, Bolivar BE, Welch JT. Sirtuin 6: a review of biological effects and potential therapeutic properties. *Mol Biosyst* 2013;**9**:1789–806. http://dx.doi.org/10.1039/c3mb00001j.
22. Arnaldez FI, Helman LJ. Targeting the insulin growth factor receptor 1. *Hematol Oncol Clin North Am* 2012;**26**:527–42. http://dx.doi.org/10.1016/j.hoc.2012.01.004.
23. Randhawa RS. The insulin-like growth factor system and fetal growth restriction. *Pediatr Endocrinol Rev* 2008;**6**:235–40.
24. Li Q, Li B, Wang X, et al. Overexpression of insulin-like growth factor-1 in mice protects from myocyte death after infarction, attenuating ventricular dilation, wall stress, and cardiac hypertrophy. *J Clin Invest* 1997;**100**:1991–9.
25. Delaughter MC, Taffet GE, Fiorotto ML, et al. Local insulin-like growth factor I expression induces physiologic, then pathologic, cardiac hypertrophy in transgenic mice. *FASEB J* 1999;**13**:1923–9.
26. Zhang Ym Yuan M, Bradley Km, et al. Insulin-like growth factor 1 alleviates high-fat diet-induced myocardial contractile dysfunction: role of insulin signaling and mitochondrial function. *Hypertension* 2012;**59**:680–93. http://dx.doi.org/10.1161/HYPERTENSIONAHA.111.181867.
27. Thakur S, Garg N, Adamo ML. Deficiency of insulin-like growth factor-1 receptor confers resistance to oxidative stress in C2C12 myoblasts. *PLoS One* 2013;**8**:e63838. http://dx.doi.org/10.1371/journal.pone.0063838.
28. Li Y, Shelat H, Geng YJ. IGF-1 prevents oxidative stress-induced apoptosis in induced pluripotent stem cells which is mediated by microRNA-1. *Biochem Biophys Res Commun* 2012;**426**:615–9. http://dx.doi.org/10.1016/j.bbrc.2012.08.139.
29. Hosaka T, Biggs 3rd WH, Tieu D, et al. Disruption of forkhead transcription factor (FOXO) family members in mice reveals their functional diversification. *Proc Natl Acad Sci USA* 2004;**101**:2975–80.
30. Furuyama T, Kitayama K, Shimoda Y, et al. Abnormal angiogenesis in Foxo1 (Fkhr)-deficient mice. *J Biol Chem* 2004;**279**:34741–9.
31. Zeng Y, Cheng L, Chen H, et al. Effects of FOXO genotypes on longevity: a biodemographic analysis. *J Gerontol A Biol Sci Med Sci* 2010;**65**:1285–99. http://dx.doi.org/10.1093/gerona/glq156.
32. Skurk C, Izumiya Y, Maatz H, et al. The FOXO3a transcription factor regulates cardiac myocyte size downstream of AKT signaling. *J Biol Chem* 2005;**280**:20814–23.
33. Peck B, Ferber EC, Schulze A. Antagonism between FOXO and MYC regulates cellular powerhouse. *Front Oncol* 2013;**3**:96. http://dx.doi.org/10.3389/fonc.2013.00096.
34. Lu T, Chai Q, Yu L, et al. Reactive oxygen species signaling facilitates FOXO-3a/FBXO-dependent vascular BK channel β1 subunit degradation in diabetic mice. *Diabetes* 2012;**61**:1860–8. http://dx.doi.org/10.2337/db11-1658.
35. Hardie DG. AMP-activated protein kinase: an energy sensor that regulates all aspects of cell function. *Genes Dev* 2011;**25**:1895–908. http://dx.doi.org/10.1101/gad.17420111.
36. Blair E, Redwood C, Ashrafian H, et al. Mutations in the gamma(2) subunit of AMP-activated protein kinase cause familial hypertrophic cardiomyopathy: evidence for the central role of energy compromise in disease pathogenesis. *Hum Mol Genet* 2001;**10**:1215–20.
37. Shibata R, Sato K, Pimentel DR, et al. Adiponectin protects against myocardial ischemia-reperfusion injury through AMPK- and COX-2-dependent mechanisms. *Nat Med* 2005;**11**:1096–103.
38. Mackenzie RM, Salt IP, Miller WH, et al. Mitochondrial reactive oxygen species enhance AMP-activated protein kinase activation in the endothelium of patients with coronary artery disease and diabetes. *Clin Sci (Lond)* 2013;**124**:403–11. http://dx.doi.org/10.1042/CS20120239.
39. Pung YF, Sam WJ, Stevanov K, et al. Mitochondrial oxidative stress corrupts coronary collateral growth by activating adenosine monophosphate activated kinase-α signaling. *Arterioscler Thromb Vasc Biol* 2013;**33**:1911–9. http://dx.doi.org/10.1161/ATVBAHA.113.301591.
40. Abraham RT, Wiederrecht GJ. Immunopharmacology of rapamycin. *Annu Rev Immunol* 1996;**14**:483–510.
41. McMullen JR, Sherwood MC, Tarnavski O, et al. Inhibition of mTOR signaling with rapamycin regresses established cardiac hypertrophy induced by pressure overload. *Circulation* 2004;**109**:3050–5.
42. Karar J, Maity A. PI3K/AKT/mTOR Pathway in angiogenesis. *Front Mol Neurosci* 2011;**4**:51. http://dx.doi.org/10.3389/fnmol.2011.00051.
43. Yang Z, Ming XF. mTOR signaling: the molecular interface connecting metabolic stress, aging and cardiovascular diseases. *Obes Rev* 2012;**2**:58–68. http://dx.doi.org/10.1111/j.1467-789X.2012.01038.x.
44. Gao G, Li JJ, Li Y, et al. Rapamycin inhibits hydrogen peroxide-induced loss of vascular contractility. *Am J Physiol Heart Circ Physiol* 2011;**300**:H1584–94. http://dx.doi.org/10.1152/ajpheart.01084.2010.
45. Whaley-Connell A, Habibi J, Rehmer N, et al. Renin inhibition and AT(1)R blockade improve metabolic signaling, oxidant stress and myocardial tissue remodeling. *Metabolism* 2013;**62**:861–72. http://dx.doi.org/10.1016/j.metabol.2012.12.012.
46. Ghosh HS, McBurney M, Robbins PD. SIRT1 negatively regulates the mammalian target of rapamycin. *PLoS One* 2010;**5**:e9199. http://dx.doi.org/10.1371/journal.pone.0009199.
47. Canto C, Gerhart-Hines Z, Feige JN, et al. AMPK regulates energy expenditure by modulating NAD+ metabolism and SIRT1 activity. *Nature* 2009;**458**:1056–60. http://dx.doi.org/10.1038/nature07813.
48. Liu X, Jiang N, Hughes B, et al. Evolutionary conservation of the clk-1-dependent mechanism of longevity: loss of mclk1 increases cellular fitness and lifespan in mice. *Genes Dev* 2005;**19**:2424–34.

49. Hekimi S. Enhanced immunity in slowly aging mutant mice with high mitochondrial oxidative stress. *Oncoimmunology* 2013;**2**:e23793.
50. Facklam R, Elliott JA. Identification, classification, and clinical relevance of catalase-negative, gram-positive cocci, excluding the streptococci and enterococci. *Clin Microbiol Rev* 1995;**8**:479–95.
51. Ho YS, Xiong Y, Ma W, et al. Mice lacking catalase develop normally but show differential sensitivity to oxidant tissue injury. *J Biol Chem* 2004;**279**:32804–12.
52. Schriner SE, Linford NJ, Martin GM, et al. Extension of murine life span by overexpression of catalase targeted to mitochondria. *Science* 2005;**308**:1909–11.
53. Dai DF, Johnson SC, Villarin JJ, et al. Mitochondrial oxidative stress mediates angiotensin II-induced cardiac hypertrophy and Galphaq overexpression-induced heart failure. *Circ Res* 2011;**108**:837–46. http://dx.doi.org/10.1161/CIRCRESAHA.110.232306.
54. Dhalla NS, Elmoselhi AB, Hata T, Makino N. Status of myocardial antioxidants in ischemia-reperfusion injury. *Cardiovasc Res* 2000;**47**:446–56.
55. Vazquez-Medina JP, Popovich I, Thorwald MA, et al. Angiotensin receptor-mediated oxidative stress is associated with impaired cardiac redox signaling and mitochondrial function in insulin-resistant rats. *Am J Physiol Heart Circ Physiol* 2013;**305**:H599–607. http://dx.doi.org/10.1152/ajpheart.00101.2013.
56. Qin F, Lennon-Edwards S, Lancel S, et al. Cardiac-specific overexpression of catalase identifies hydrogen peroxide-dependent and -independent phases of myocardial remodeling and prevents the progression to overt heart failure in G(alpha)q-overexpressing transgenic mice. *Circ Heart Fail* 2010;**3**:306–13. http://dx.doi.org/10.1161/CIRCHEARTFAILURE.109.864785.
57. Razzaque MS. The role of Klotho in energy metabolism. *Nat Rev Endocrinol* 2012;**8**:579–87. http://dx.doi.org/10.1038/nrendo.2012.75.
58. Kuro-o M, Matsumura Y, Aizawa H, Kawaguchi H, et al. Mutation of the mouse klotho gene leads to a syndrome resembling ageing. *Nature* 1997;**390**:45–51.
59. Kurosu H, Yamamoto M, Clark JD, et al. Suppression of aging in mice by the hormone Klotho. *Science* 2005;**309**:1829–33.
60. Pedersen L, Pedersen SM, Brasen CL, Rasmussen LM. Soluble serum Klotho levels in healthy subjects. Comparison of two different immunoassays. *Clin Biochem* 2013;**46**:1079–83. http://dx.doi.org/10.1016/j.clinbiochem.2013.05.046.
61. Takeshita K, Fujimori T, Kurotaki Y, et al. Sinoatrial node dysfunction and early unexpected death of mice with a defect of klotho gene expression. *Circulation* 2004;**109**:1776–82.
62. Kerr J, Wood W, Ridgway EC. Basic science and clinical research advances in the pituitary transcription factors: Pit-1 and Prop-1. *Curr Opin Endocrinol Diabetes Obes* 2008;**15**:359–63. http://dx.doi.org/10.1097/MED.0b013e3283060a56.
63. Helms SA, Azhar G, Zuo C, et al. Smaller cardiac cell size and reduced extra-cellular collagen might be beneficial for hearts of Ames dwarf mice. *Int J Biol Sci* 2010;**6**:475–90.
64. Madsen MA, Hsieh CC, Boylston WH, et al. Altered oxidative stress response of the long-lived Snell dwarf mouse. *Biochem Biophys Res Commun* 2004;**8**:998–1005.
65. Tomilov AA, Ramsey JJ, Hagopian K, et al. The Shc locus regulates insulin signaling and adiposity in mammals. *Aging Cell* 2011;**10**:55–65. http://dx.doi.org/10.1111/j.1474-9726.2010.00641.x.
66. Napoli C, Martin-Padura I, de Nigris F, et al. Deletion of the $p66^{Shc}$ longevity gene reduces systemic and tissue oxidative stress, vascular cell apoptosis, and early atherogenesis in mice fed a high-fat diet. *Proc Natl Acad Sci USA* 2003;**100**:2112–6.
67. Pandolfi S, Bonafe M, Di Tella L, et al. p66(shc) is highly expressed in fibroblasts from centenarians. *Mech Ageing Dev* 2005;**126**:839–44.
68. Shaik S, Wang Z, Inuzuka H, Wei W. Endothelium aging and vascular diseases, senescence and senescence-related disorders. In: Zhiwei Wang, editor. *InTech*. 2013. http://dx.doi.org/10.5772/53065. Available from: http://www.intechopen.com/books/senescence-and-senescence-related-disorders/endothelium-aging-and-vascular-diseases.
69. Sudheesh NP, Ajith TA, Ramnath V, Janardhanan. Therapeutic potential of *Ganoderma lucidum* (Fr.) P. Karst. against the declined antioxidant status in the mitochondria of post-mitotic tissues of aged mice. *Clin Nutr* 2010;**29**:406–12. http://dx.doi.org/10.1016/j.clnu.2009.12.003.

CHAPTER

4

Oxidative Stress, Aging and Mitochondrial Dysfunction in Liver Pathology

Sylvette Ayala-Peña

Department of Pharmacology and Toxicology, University of Puerto Rico Medical Sciences Campus, San Juan, Puerto Rico

Carlos A. Torres-Ramos

Department of Physiology, University of Puerto Rico Medical Sciences Campus, San Juan, Puerto Rico

List of Abbreviations

ANT adenine nucleotide translocase
AP site apurinic/apyrimidinic site
ATP adenosine triphosphate
BER base excision repair
GCL glutamate cysteine ligase
Gpx-1 glutathione peroxidase-1
4-HNE 4-hydroxynonenal
H_2O_2 hydrogen peroxide
MnSOD manganese superoxide dismutase
mtDNA mitochondrial DNA
NAFLD nonalcoholic fatty liver disease
NASH nonalcoholic steatohepatitis
NER nucleotide excision repair
8-oxo-dG 8-oxo-7,8-dihydroguanine
PCR polymerase chain reaction
QPCR quantitative PCR
ROS reactive oxygen species
T2D type 2 diabetes
TFAM mitochondrial transcription factor A

INTRODUCTION

Aging is the main risk factor for chronic diseases, as age increases the incidence and progression of pathology.[1] The liver plays a pivotal role in the process of aging by combining energy metabolism pathways with the detoxification of drugs and xenobiotics.[2] The function of the liver declines with age, and liver-related deaths are also increased, suggesting that aging of the liver may enhance disease susceptibility and predispose the elderly to exacerbated pathology. Aging not only leads to various morphologic and structural changes in the liver but it also triggers decreases in metabolic function.[3]

Aging increases the levels of oxidative stress in various organs, including the liver, and mitochondrial dysfunction seems to play a critical role in this process. Mitochondria not only play a central role in cellular energy production as generators of ATP, but they are also the main producers of endogenous reactive oxygen species (ROS). It has been suggested that the decline in liver function with advancing age may result from the increased generation of mitochondrial ROS. Many liver-associated diseases, such as alcoholic liver disease, nonalcoholic fatty liver disease, steatohepatitis and hepatocellular carcinoma, may result from the increased generation of ROS by dysfunctional mitochondria.[4] The mitochondrial theory of aging postulates that ROS, generated by the mitochondria, induce the accumulation of somatic mutations in the mitochondrial DNA (mtDNA) that results in impairment of the process of oxidative phosphorylation and ATP production.

Not only is the prevalence of chronic liver disease increasing in the elderly population,[5] but also the associated morbidity and risk of death are significantly greater in the aged population compared to the young. Therefore, understanding the molecular basis of liver disease is important for early diagnosis and treatment. Despite significant progress in the understanding of the processes underlying age-related liver dysfunction and disease, the molecular mechanisms by which aging may damage the liver remain uncertain. Substantial evidence suggests that oxidative stress and mitochondrial dysfunction may be critical contributors of age-associated liver pathology. In this chapter we focus on discussing the role of oxidative damage and mitochondrial dysfunction in the pathophysiology of the aging liver.

Aging
http://dx.doi.org/10.1016/B978-0-12-405933-7.00004-4

EVIDENCE FOR AGE-ASSOCIATED MORPHOLOGIC, STRUCTURAL AND FUNCTIONAL CHANGES IN THE LIVER

Aging of the liver manifests as both structural and functional alterations and is evident in humans and rodents. Early ultrasound studies suggest that liver from healthy individuals undergoes a significant 30% reduction in volume with increasing age (reviewed in ref. 6). Studies in aged mice and rats have also provided evidence of significant declines in hepatic weight similar to those observed in humans.[7] Consistent with impaired age-related changes in liver structure, hepatic blood circulation is significantly reduced 30–50% in rodents and human subjects and this correlates with increasing age (reviewed in ref. 6). In addition, sinusoidal blood flow in livers from aged mice undergoes a significant 35% reduction, which correlates with increased liver sinusoidal endothelial cell dysfunction.[8] There is substantial evidence in humans, baboons and rodents that the liver sinusoidal endothelium and the space of Disse undergo ultrastructural changes associated with the process of aging. A significant 40% increase in sinusoid endothelial thickness or pseudocapillarization is observed in rats and baboons, whereas a significant 60% increase is observed in surgical and postmortem human livers from subjects >60 years (reviewed in ref. 6). Increased deposition of cellular matrix and basal lamina in the space of Disse has also been documented.[6] Furthermore, electron microscopy analyses show a significant 40–80% age-related reduction in the number of sinusoidal fenestrations in rat, baboon and human livers (reviewed in ref. 6) and in mice.[9]

Interestingly, liver mitochondria also undergo morphologic changes during aging. Early studies show that aging is associated with increases in mitochondrial size in intact hepatocytes[10] and in mouse mitochondrial fractions,[11] and morphometric studies show that the numerical density of mouse liver mitochondria decreases with age.[12] The number of mitochondria per cell also significantly decreases in 24-month-old mouse mononucleate hepatocytes as determined by electron microscopic radioautography (reviewed in ref. 13).

Taken together these observations support the idea that age-associated changes in liver structure and function may contribute to impaired hepatic clearance of drugs, xenobiotics and endogenous substrates in the aging population, leading to disease susceptibility and age-related diseases (reviewed in ref. 6). Thus, aging results in morphologic and functional alterations in the liver (Fig. 4.1).

EVIDENCE FOR AGE-ASSOCIATED LOSS OF MITOCHONDRIAL BIOENERGETICS IN LIVER

Mitochondrial bioenergetics are essential for maintaining liver function, as these organelles are responsible for producing most of the energy required for appropriate cellular function through the generation of ATP. ATP is produced by the oxidation of carbohydrates and fats during oxidative phosphorylation when electrons donated to complexes I (NADH dehydrogenase) and II (succinate dehydrogenase) are sequentially transferred to coenzyme Q and then to complex III which in turn transfers them to cytochrome *c*. From cytochrome *c* the electrons are passed to complex IV (cytochrome oxidase) and finally to molecular oxygen to produce water. While driving the flow of electrons to molecular oxygen, complexes I, II, and IV extrude protons from the matrix to the inter-membrane space, thus creating a proton current/membrane potential at the inner mitochondrial membrane. The re-entry of protons back into the matrix through complex V (ATP synthase) drives the synthesis of ATP from ADP and inorganic phosphate. This flow of electrons through the mitochondrial respiratory chain represents the oxygen consumption rate. Mitochondria that are tightly coupled will generate the maximum levels of ATP when the electron transfer chain is highly efficient at pumping protons out of the inner mitochondrial membrane for use by the ATP synthase for ATP production. In the absence of oxygen, or when mitochondrial respiration is inefficient, cells switch to glycolysis for ATP synthesis and pyruvate is converted to lactic acid.

Mitochondrial dysfunction has been implicated as a major causative factor underlying the decline in physiologic functions associated with the process of aging. Substantial evidence in humans and rodents shows that aging induces the loss of mitochondrial function in liver. Livers from normal individuals over 50 years of age show defective mitochondrial respiration mostly associated with a deficiency in complex IV.[14] Deficient mitochondrial respiration and decreased ATP production are also evident in liver from aged rats and mice. Isolated liver mitochondria from old rats show a significant 25% and 33% reduction in complexes I and II, respectively, and a concomitant 41% reduction in ATPase activity.[7] A significant 30% decrease in mitochondrial membrane potential is observed in intact hepatocytes from old rats, as compared to young rats, and is associated with increased mitochondrial size.[10] Furthermore, impairment of ATP production is associated with decreased complex V activity in liver from old rats,[15] whereas mitochondrial membrane potential and the activity of complexes I, II and IV are reduced in liver from aged rats as compared to young animals.[16] Liver mitochondria from old rats show a reduction in mitochondrial membrane potential and ATP synthesis as well as increased proton leaks.[17]

Reductions in the activity of key mitochondrial metabolic enzymes emphasize the role of mitochondrial dysfunction in liver aging. Recent studies show that age-associated metabolomic perturbations in the mitochondrial electron transport chain and glucose and phospholipid metabolism are observed in mouse liver.[18]

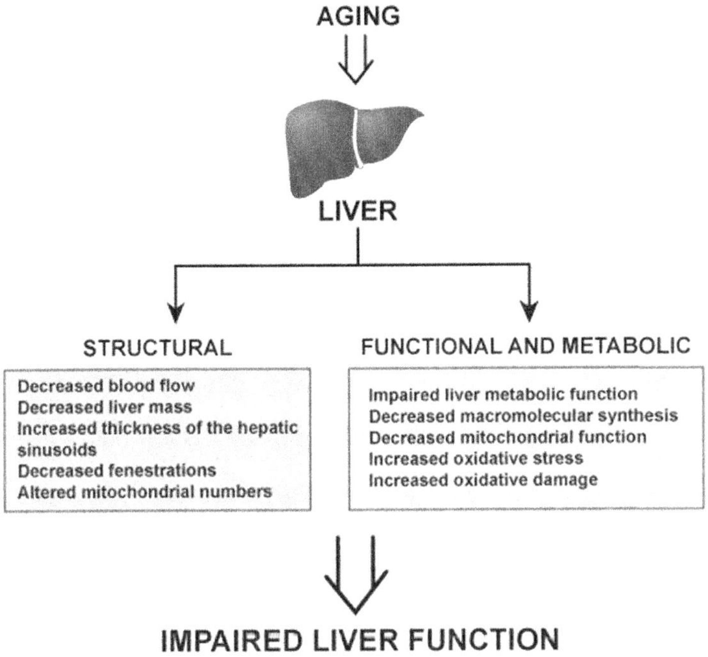

FIGURE 4.1 Age-associated structural and functional alterations in the liver. During aging a number of structural, functional and metabolic alterations have been described. These alterations may result in impaired liver function.

Moreover, the authors show that the expression of genes involved in oxidative phosphorylation, such as complex I subunit Ndufa2, complex IV, complex V subunit Atp5g1 and cytochrome *c*, was significantly reduced in liver from aged mice. Interestingly, proteomic analysis shows that protein expression levels of the mitochondrial tricarboxylic acid cycle isocitrate dehydrogenase and complex I protein subunits NDFS2, NDFS3 and NDUFV2 increase with age in rat liver whereas the levels of complex V beta subunit decrease with age.[19] Furthermore, the mitochondrial complex I NDUFA12 subunit exhibits a significant 1.8-fold increase in abundance with age whereas complex IV subunits II, Va and VIb show a 1.45-fold increase in isolated mouse liver mitochondria.[20] Interestingly, the authors show that the catalytic F_1 complex beta subunit of complex V exhibits an age-dependent decrease in protein abundance whereas the F_o subunit b of complex V shows a 1.25 increase in abundance with aging. In addition, aging also alters enzymes involved in fatty acid metabolism,[19] glycolysis and glycogen metabolism.[18] Interestingly, the liver of mice with extended longevity shows significant increases in the activities of complexes I–IV, and this correlates with increased expression of both mitochondrial- and nuclear-encoded complex IV proteins as compared to wild-type mice.[21] Moreover, the authors show that liver expression of the adenine nucleotide translocase (ANT), the mitochondrial protein complex responsible for fast exchange of ADP/ATP between cytosol and mitochondria, is significantly increased in the long-lived mice. These results suggest that increased mitochondrial function may contribute to longevity in the Ames dwarf mice. The fact that liver from Ames dwarf mice show increased levels of antioxidant enzymes[22] and decreased generation of mitochondrial hydrogen peroxide (H_2O_2) compared to wild-type mice further supports a major role for mitochondria and oxidative stress in aging of the liver. Consistent with this hypothesis, a senescence-accelerated mouse model exhibits significant down regulation of mitochondrial complex I proteins, decreased ATP levels and liver pathology.[23]

Collectively, the studies discussed above in humans and in rodents show that mitochondrial markers of energy metabolism are altered in liver aging and thus support a role for mitochondrial dysfunction in disease susceptibility and liver pathology during aging (Fig. 4.1). Despite these significant observations, it remains unclear how aging leads to mitochondrial dysfunction in the liver. A possible mechanism by which aging may contribute to mitochondrial dysfunction and tissue injury is by the induction of oxidative stress.

OXIDATIVE STRESS AND ANTIOXIDANT RESPONSES IN LIVER AGING

Mitochondria are the principal sources of endogenous ROS, which are generated under physiologic conditions as metabolic products of aerobic cellular metabolism. During oxidative phosphorylation, electrons that are donated to oxygen generate the superoxide anion, a short-lived oxygen radical that can be converted to H_2O_2 by the action of the mitochondrial antioxidant enzyme manganese superoxide dismutase (MnSOD/SOD2), and catalase and glutathione peroxidase convert H_2O_2 to water. In the presence of iron and other transition metals, H_2O_2 can be further converted to the highly reactive hydroxyl radical by means of the Haber–Weiss/Fenton reaction. The generation rates of mitochondrial ROS increase with age and macromolecules such as nucleic acids, lipids and proteins are their main targets and may contribute to the characteristic physiologic decline observed in aging. Thus, mitochondrial dysfunction leads to the overproduction of ROS, which in the aged liver may cause failure of mitochondrial bioenergetics, tissue dysfunction and pathology.

Aging leads to increased generation of ROS in intact hepatocytes from aged rats as compared to young animals and associates with lowered mitochondrial membrane potential and increased mitochondrial size.[24] Isolated hepatic mitochondria from old rats exhibit increased levels of H_2O_2 compared to young rats.[15] In addition, increased levels of mitochondria-generated H_2O_2, with concomitant increases in lipid peroxidation, reduced total antioxidant capacity and increased mitochondrial dysfunction is observed in liver from aging rats compared to young animals.[17] Interestingly, primary hepatocytes from old rats exhibit greater reductions in ATP levels and mitochondrial membrane potential after H_2O_2 and tert-butyl hydroperoxide treatment than do hepatocytes from young animals.[25]

Aging leads to increased lipid peroxidation and levels of protein carbonylation in liver mitochondria of rats, mice and nonhuman primates. Significant increases in the levels of protein carbonylation occur in liver from old mice relative to young animals, with the antioxidant enzyme Cu/ZnSOD containing the higher levels of protein carbonyls in the old animals.[26] Moreover, increased levels of lipid peroxidation products and carbonylated proteins in liver from old nonhuman primates (*Macaca mulatta*) compared to young rhesus monkeys further support the hypothesis that an age-associated increase in oxidative stress may contribute to liver aging.[27] Interestingly, dietary restriction, an intervention that extends both average and maximum lifespan, retards many age-associated diseases and reduces oxidative stress, attenuates protein carbonylation in liver and correlates with longevity in various mammalian species.[28] Moreover, levels of oxidatively damaged proteins, specifically, 4-hydroxynonenal (4-HNE) conjugated lipid peroxidation damage, significantly increase with age in mouse liver mitochondria[18,29] and dietary restriction significantly attenuates protein damage.[29]

Further evidence suggesting that aging involves a response to oxidative stress and antioxidant status in liver is depicted by a significant decrease in SOD2.[18] In addition, old primary hepatocytes fail to induce glutathione peroxidase-1 (Gpx-1) gene expression compared to young cells, suggesting that this altered antioxidant response may contribute to the increased susceptibility of old primary hepatocytes to oxidative stress.[25]

The tripeptide g-glutamylcysteinylglycine or glutathione is the most abundant thiol-reducing agent in mammalian tissues. The liver plays a central role in glutathione homeostasis because it is the organ with the highest glutathione content and also the major source of secretion to the plasma. Several studies have demonstrated age-dependent changes in glutathione content. Hazelton and Lang[30] showed significant decreases (20–30%) in reduced and total glutathione in senescent mice (>25 months old). Because these changes were mostly observed in liver, they may render mice particularly susceptible to toxicologic agents. Decreased glutathione levels during aging could be due in part to decreased affinity of the hepatic glutamate cysteine ligase (GCL), the rate-limiting step in glutathione biosynthesis, for its substrates glutamate and cysteine.[31]

Evidence suggesting that glutathione plays an important role in mitochondrial physiology comes from studies in mice harboring a homozygous null mutation in the Gpx-1 gene.[32] The product of the Gpx-1 gene uses glutathione as cofactor in the enzymatic conversion of H_2O_2 to water. Gpx-1 mutant mice not only exhibit a growth-deficient phenotype, but their livers show increased levels of lipid peroxidation, and liver mitochondria exhibit increased H_2O_2 production. Decreases in the glutathione antioxidant defense system have been described in non-human primates where glutathione peroxidase (GSH-Px) activity decreases 65% in old animals (>22 years old) as compared to young animals (<8 years old) (Fig. 4.2, panel A). Moreover, GSH-Px activity negatively correlates with age.[27] A significant 65% decrease in total SOD activity is observed in old monkeys versus middle-aged animals (Fig. 4.2, panel B). Interestingly, old monkeys show a significant 39% increase in catalase activity compared to young animals (Fig. 4.2, panel C), suggesting that aging modulates the activity of hepatic antioxidant systems. Moreover, evidence suggests that mtDNA becomes more susceptible to oxidative damage when glutathione levels decrease. A positive correlation has been observed between 8-oxo-dG levels in mtDNA and a high GSSG/GSH ratio in livers from both mice and rats.[33]

Overall, studies in various aging models suggest that antioxidant defenses may play a central role in liver function during aging. The emerging picture is one in which the liver in young organisms is in a state of balance between the production of oxidants and the antioxidants defenses (Fig. 4.3). During aging, an imbalance occurs due to: (1) increased production of oxidants, such as ROS, while the level of antioxidant defenses remains relatively stable; and/or (2) a decrease in the levels of antioxidant defenses while the levels of oxidants remain the same. A combination of both scenarios is also plausible.

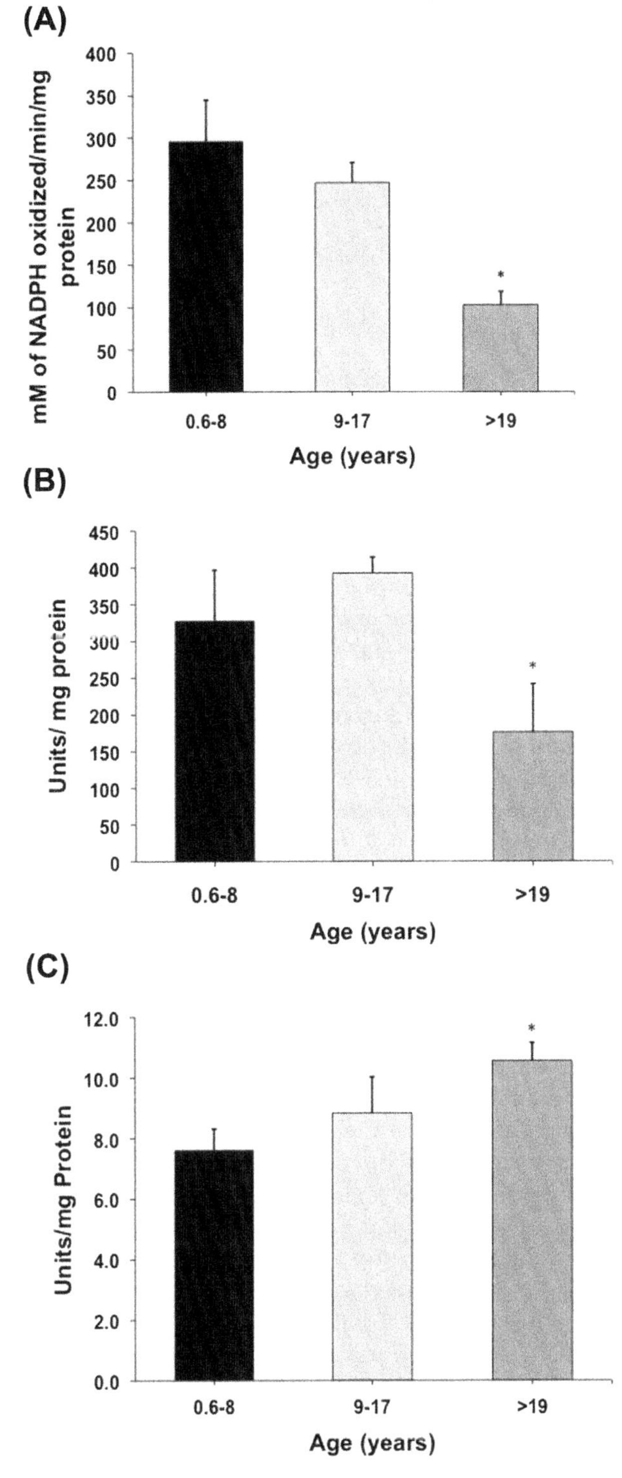

FIGURE 4.2 Age-associated activity of hepatic antioxidant enzymes. (A) Glutathione peroxidase (GHSPx) activity. One unit of GHSPx is equal to 1 mM of NADPH oxidize/min/mg of protein. n = 5 young, n = 4 middle age, and n = 3 old monkeys; *p <0.05 versus young. (B) Total SOD activity. Activity was expressed as the amount of enzyme that inhibits the oxidation of epinephrine by 50%, which is equal to 1 unit. n = 5 young, n = 2 middle age, n = 4 old monkeys. *p <0.05 versus middle age. (C) Catalase activity. One unit of activity is equal to the moles of H_2O_2 degraded/min/mg protein. n = 5 young, n = 3 middle age, and n = 3 old monkeys; *p <0.05 versus young. Data were expressed as mean ± SEM. Figure obtained from Castro, M del R et al 2012[27] with permission.

OXIDATIVE DAMAGE TO THE NUCLEAR AND MITOCHONDRIAL GENOMES IN LIVER AGING

DNA is constantly being damaged by exogenous (xenobiotics) and endogenous (ROS) agents that induce a variety of DNA lesions such as base modifications, abasic sites, DNA adducts, and single- and double-strand DNA breaks. If these lesions are not repaired they may lead to mutations, which can be deleterious for the organism. In addition, unrepaired DNA lesions may disrupt processes such as DNA replication and transcription, altering critical physiologic processes. A variety of DNA repair mechanisms have evolved to deal with DNA lesions. DNA bulky lesions that alter the DNA structure, such as those induced by UV light and some carcinogens, are mainly repaired by nucleotide excision repair (NER). Double-strand breaks are repaired by homologous recombination or by non-homologous end joining. Damage to DNA bases, such as alkylation and oxidative modifications, are repaired mostly by base excision repair (BER). It is important to point out that although these DNA repair pathways have been assigned to specific DNA lesions, *in vitro* and *in vivo* evidence indicates that cross talk among pathways may occur.[34]

Increases in DNA lesions have been reported in liver during aging. For example, levels of 8-oxodG, which is a common oxidative DNA lesion repaired mainly by BER, have been shown to increase in liver from aged rats and mice.[35,36] Interestingly, a significant 4.7-fold increase in mtDNA lesions is detected in liver of old rhesus monkeys (>19 years of age) compared to young (0.6–2 years of age) rhesus monkeys (Fig. 4.4, panel A), and the increased frequency of lesions significantly correlates with age.[27] Moreover, levels of apurinic/apyrimidinic (AP) sites, another type of DNA lesion repaired by BER, increase about 1.7-fold in liver from old rats (24 months old) as compared to young rats (4 months old).[37] The reason for such an increase is not fully understood, but two different (but not mutually exclusive) models could be invoked: increased oxidative stress and decreased DNA repair. Decreases in overall BER enzymatic activity (more than 50% decrease) have been observed in liver from aged mice.[38] However, studies designed to determine whether aging affects specific steps within BER show a complicated picture. For example, an age-dependent decrease in DNA polymerase β levels (one of the DNA polymerases involved in the repair synthesis step in BER) has been reported.[38] However, no age-related changes in nuclear 8-oxo-dG and UDG glycosylase activities have been detected.[39] On the other hand, an increase in AP-endonuclease activity (an enzyme that acts on AP sites as part of BER) in nuclear extracts from liver has been reported, suggesting a response to increased levels of oxidative stress during aging.[40] Thus, the repair response

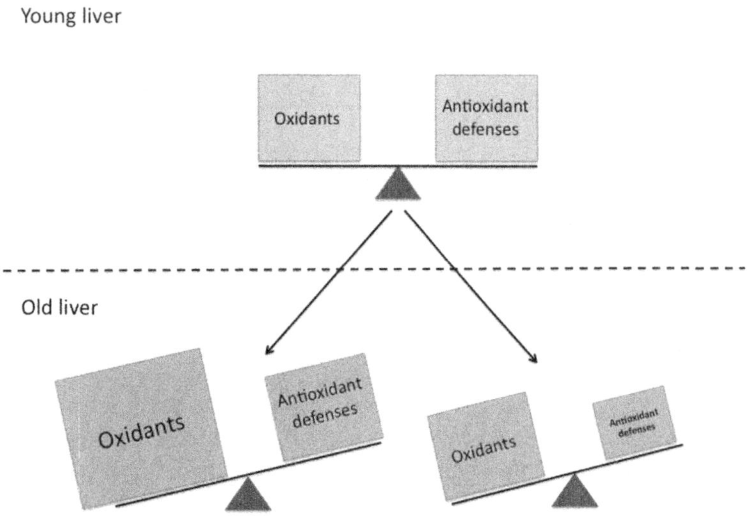

FIGURE 4.3 Oxidative stress and liver aging. In young animals the liver exerts a balance between oxidant production and antioxidant defenses. During aging at least two scenarios may occur: (1) the oxidant levels may increase and/or (2) the antioxidant defenses may decrease. In both instances the net result is an imbalance known as oxidative stress.

to DNA damage during aging is not only tissue-specific, but also damage- and subcellular compartment-specific (i.e. nuclear versus mitochondrial).

The nucleotide excision repair (NER) protein ERCC1 participates in the incision step 5′ to the DNA lesion. Recently, a mouse model with reduced ERCC1 expression (Ercc1$^{-/\Delta}$) was used to study liver aging.[9] This mouse model shows pseudocapillarization and premature loss of liver function, and electron microscopy studies show a significant reduction in the number of sinusoidal fenestrations in both the old and the progeroid mice. The authors also show increased levels of lipofuscin and lipid peroxidation in the progeroid liver similar to those found in old wild-type mice. Although most of the DNA lesions accumulating during liver aging may be oxidative in nature, NER may be an important backup repair pathway that complements BER, similarly to the studies performed in simple model organisms.[34]

Mitochondria are unique organelles in the sense that they contain their own genome. The mtDNA in higher eukaryotes consists of a 16569 bp circular molecule, containing 27 genes, encoding 13 polypeptides of the electron transport chain, 22 tRNAs and 2rRNAs. Strong evidence shows that mtDNA harbors more damage than nuclear DNA during aging, particularly in tissues with high metabolic rates such as the liver and brain.[35,42] Moreover, the mtDNA of cells cultured *in vitro* are also more susceptible than nuclear DNA to DNA-damaging agents such as H_2O_2 and alkylating agents.[41] Mitochondria exhibit robust BER, with both forms of BER (short patch and long patch) being detected by *in vitro* studies.[43] Studies to determine the activities of BER enzymes in liver during aging have shown an increase in 8-oxo-dG and AP endonuclease activities.[40,44] This could be a compensatory response to increased levels of DNA damage. On the other hand, mitochondrial import of DNA repair proteins is defective during aging, causing accumulation of DNA repair proteins in the exterior of the mitochondria, which overestimates the real enzymatic activity in the mitochondrial matrix.[45]

The most dramatic example of the role of mtDNA mutations in aging is observed in mice expressing a proofreading-deficient DNA polymerase γ.[46] Pol γ is a nuclear encoded protein that carries on both replicative and repair mtDNA functions. Mice carrying the proofreading-deficient mutant allele show high rates of mtDNA point mutations and deletions in the liver, brain and heart. These animals present signs of premature aging such as reduced lifespan, weight loss, alopecia, kyphosys, anemia and liver extramedullary hematopoiesis.

Besides qualitative changes in mtDNA, such as point mutations, deletions and rearrangements, mtDNA damage may lead to quantitative changes represented by changes in mtDNA copy number or abundance. Significant decreases (~50%) in mtDNA copy number have been reported in rat liver,[47] and in non-human primates a significant 20% decrease in liver mtDNA abundance is observed in middle-aged and old monkeys compared to young animals (Fig. 4.4, panel B).

Overall, oxidative damage to both the nuclear and mitochondrial genomes has been detected in liver of rodents and non-human primates. Oxidative lesions and mtDNA depletion may ultimately interfere with mitochondrial bioenergetics, resulting in reduced mitochondrial membrane potential and ATP levels, overproduction of ROS, and aggravated oxidative damage to the mitochondrial and nuclear genomes, thereby establishing a vicious cycle of ROS generation, mitochondrial dysfunction and macromolecular damage. Altered antioxidant defenses and deficient mtDNA repair may exacerbate this scenario (Fig. 4.5).

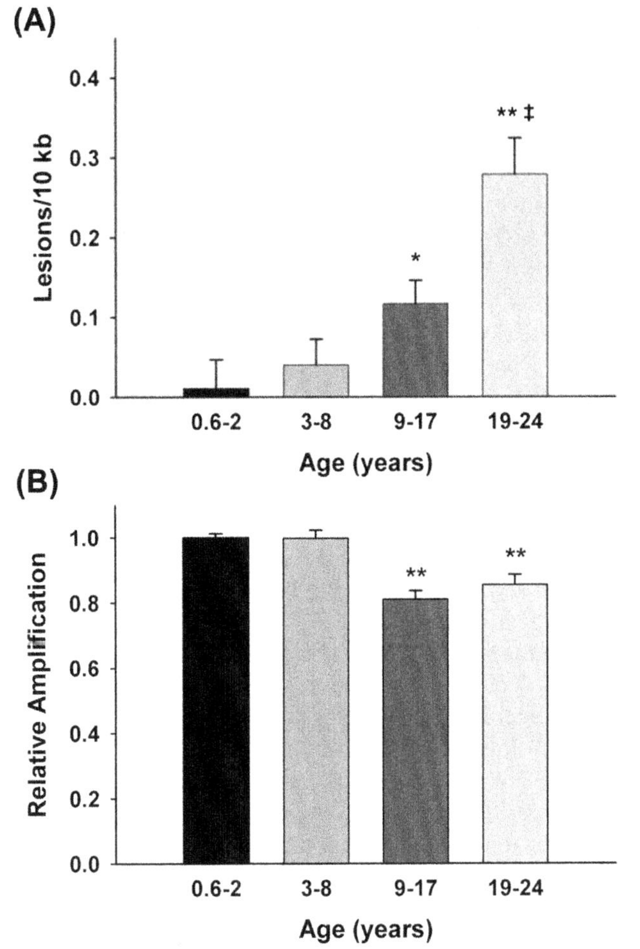

FIGURE 4.4 Mitochondrial DNA damage increases while its abundance decreases with age in liver from rhesus monkeys. Total DNA was isolated from liver obtained from 0.6- to 2.0-year-old (infant), 3- to 8-year-old (young), 9- to 17-year-old (middle-aged), and 19- to 24-year-old (aged) monkeys and analyzed by quantitative PCR (QPCR). (A) Frequency of mtDNA lesions per 10 kb per strand. *p <0.0001 and **p <0.0001 versus young, and ‡p <0.05 versus middle age. n = 5 infant, n = 7 young, n = 6 middle-aged, and n = 3 old; n = 4 QPCR analyses. (B) Relative mtDNA abundance during aging. **p <0.0001 versus young. Data were expressed as mean ± SEM. Figure obtained from Castro, M del R et al 2012[27] with permission.

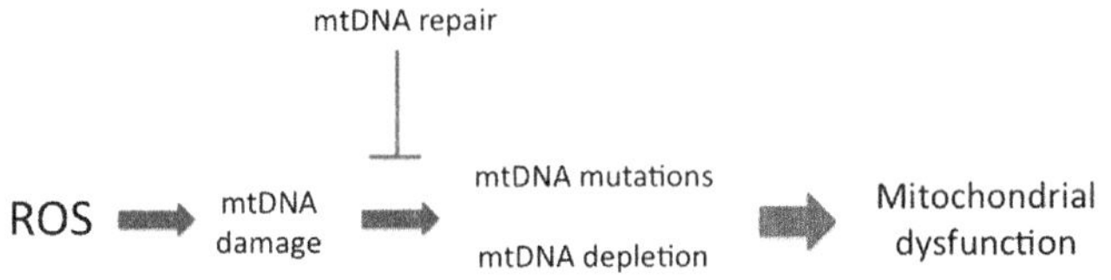

FIGURE 4.5 Proposed paradigm for the role of mtDNA damage and repair in liver mitochondrial dysfunction during aging. ROS generated from exogenous and endogenous sources can induce damage to liver mtDNA, which may result in mtDNA mutations and mtDNA depletion. This damage in turn may lead to decreased transcription of mitochondrial genes, ultimately resulting in mitochondrial dysfunction. DNA repair mechanisms may modulate this process by repairing mtDNA lesions.

MITOCHONDRIAL DYSFUNCTION AND LIVER-ASSOCIATED DISEASE

It is clear from the studies discussed above that age-associated impairment of mitochondrial bioenergetics associates with increased oxidative damage in the liver. While the liver integrates carbohydrate, fatty acid and protein metabolisms, age-associated impairment of liver mitochondrial function is related to insulin resistance, diabetes, and nonalcoholic fatty liver disease (NAFLD), a common liver disease characterized by the accumulation of lipids in hepatic cells that may precede liver cirrhosis and cancer.[4] Insulin resistance is a common feature in NAFLD and associates with oxidative stress and mitochondrial dysfunction. Insulin-resistant aged individuals exhibit significantly increased muscle and liver triglyceride content and a significant 40% reduction in liver mitochondrial function compared with young controls, suggesting that age-associated mitochondrial dysfunction may contribute to insulin resistance.[48] Moreover, patients suffering from nonalcoholic steatohepatitis (NASH), a stage in NAFLD, who are over 55 years of age exhibit more severe liver steatosis than do young patients, suggesting that age may influence the severity of steatohepatitis.[49] Whole-body nuclear magnetic resonance studies show that liver ATP levels are lower in patients with NASH after receiving a challenge with fructose to acutely deplete ATP levels, compared with age-matched healthy controls.[50] These data are consistent with the idea that patients with NASH exhibit impaired recovery of liver ATP stores and thus, impaired mitochondrial function. Interestingly, patients with NASH show a significant decrease in the activity of liver mitochondrial complexes I–V compared with control patients.[51] Paradoxically, other studies have shown either no change or increased activity of mitochondrial function in NASH (reviewed in ref. 4). Interestingly, patients with extrahepatic cholestasis exhibit decreased ATP levels in the liver.[52]

Human subjects with high intrahepatic triglyceride levels exhibit significantly increased hepatic gluconeogenesis compared to individuals with low triglyceride levels, suggesting that impaired suppression of glucose production by insulin is associated with increased lipid levels.[53] Other studies in patients with lipodystrophy show that increased hepatic triglyceride content is associated with severe hepatic steatosis and hepatic insulin resistance.[54] Moreover, the authors show a significant 2-fold induction in oxidative flux via the tricarboxylic acid cycle that correlated with hepatic triglyceride content. Genes involved in oxidative phosphorylation and key gluconeogenesis enzymes are upregulated in the livers of obese type 2 diabetes (T2D) patients and significantly correlated with the induction of genes involved in ROS production and antioxidant enzymes in the liver, suggesting that obesity may affect T2D pathophysiology by increasing the expression of genes involved in mitochondrial oxidative phosphorylation, liver gluconeogenesis and oxidative stress in the liver.[55]

Abnormal mitochondrial morphology is observed in patients with NASH, as demonstrated by the presence of megamitochondria and intramitocondrial crystalline inclusions.[56] Moreover, increased oxidative DNA damage demonstrated by immunohistochemical analysis shows that the modification of proteins by 4-HNE and 8-hydroxydeoxyguanosine increases with age in liver from patients with NAFLD.[57] Furthermore, levels of mtDNA damage measured as 8-hydroxydeoxyguanosine-positive expression were significantly increased in liver mitochondria from patients with steatosis and NASH.[56] Along these lines, non-human primates presenting increased age-associated liver steatosis also show an age-dependent increase in liver mtDNA damage, mtDNA depletion and increased oxidative stress compared to young monkeys.[27] Furthermore, patients with extrahepatic cholestasis present a significant increase in levels of 8-oxodG in the mtDNA and reduced mtDNA copy numbers and mitochondrial transcript levels.[52] Interestingly, the authors show that overexpression of mitochondrial transcription factor A (TFAM) attenuated mtDNA damage in a normal human cell line, further suggesting that damaged mtDNA may play a role in liver injury. Overall, these observations suggest that mitochondrial dysfunction and increased mitochondrial oxidative damage may play a significant role in the pathogenesis of liver disorders.

Activation of sirtuins (Sirts), a family of nicotinamide adenine dinucleotide (NAD)-dependent protein deacetylases, has been linked to the regulation of lipid and glucose metabolism during aging, lifespan extension and, more recently, to age-associated mitochondrial dysfunction and oxidative stress. Activation of Sirt1 extends mean and maximum lifespan, reduces liver steatosis, and increases mitochondrial biogenesis and insulin sensitivity in mice fed a high-fat diet, suggesting that regulation of energy homeostasis may improve liver health.[58,59] Moreover, feeding mice a high-fat diet reduces the activity of Sirt3, the main mitochondrial sirtuin that regulates mitochondrial function, leading to hyperacetylation of gluconeogenic, oxidative phosphorylation and antioxidant proteins, increased mitochondrial protein oxidation levels and fatty liver.[60] These results suggest that reduced SIRT3 activity, mitochondrial dysfunction and the accumulation of ROS may mediate high-fat-induced liver injury. In addition, $SIRT3^{-/-}$ mice develop a fatty liver and exhibit reduced ATP levels and mitochondrial respiration.

In summary, impairment of mitochondrial bioenergetics, increased ROS generation and oxidative damage to macromolecules are likely to contribute to age-associated liver pathology. In particular, mtDNA damage depicted as mtDNA oxidative lesions and mtDNA depletion may trigger liver dysfunction. We propose a model (Fig. 4.6) in which age-induced impairment of mitochondrial function due to the overproduction of mitochondrial ROS and the accumulation of oxidative damage, particularly to the mtDNA, may trigger liver dysfunction. Because the liver acts as a core for key metabolic functions, damaged mitochondria may alter liver response to glycolysis, gluconeogenesis, fat oxidation, oxidative

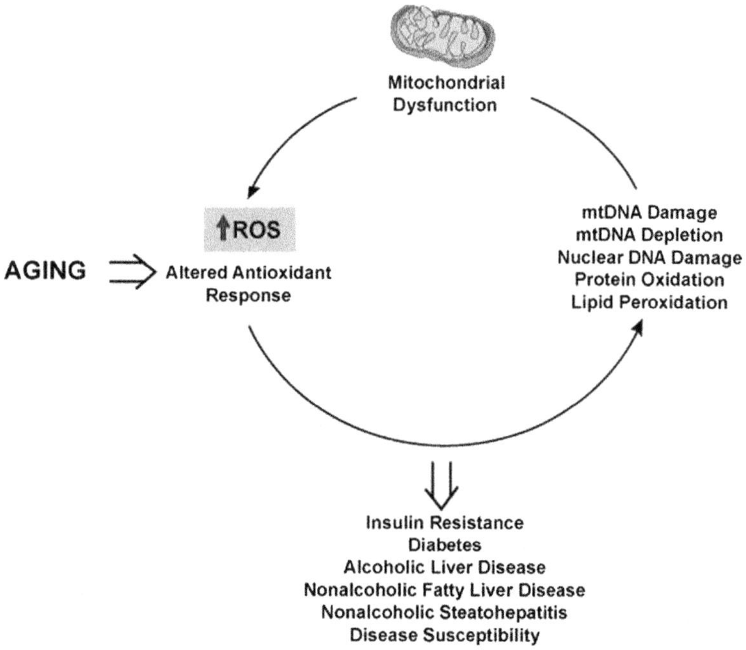

FIGURE 4.6 Proposed model for the role of mitochondria in liver aging. During aging an imbalance between ROS levels and antioxidant defenses leads to oxidative stress; this induces damage to macromolecules such as DNA, lipids and proteins. These lesions cause mitochondrial dysfunction, which leads to the generation of more ROS and a vicious cycle of oxidative damage. The loss of physiologic processes will render the aged liver more prone to develop disease.

phosphorylation and ATP synthesis, thus leading to insulin resistance, diabetes, and NAFLD among other liver-associated disorders. In addition, dysfunctional mitochondria may affect liver detoxification processes by altering the clearance of endogenous substrates, drugs and xenobiotics.

CONCLUSION

Mitochondria are fundamental to the metabolic activity of the liver. Aging causes liver structural and functional alterations and triggers events leading to increased oxidative stress and a decline in mitochondrial bioenergetics. Impairment of mitochondrial function in the aged liver in turn further elevates oxidative damage to mitochondrial and nuclear macromolecules. Increased impairment of mitochondrial bioenergetics during liver-associated disease could aggravate mitochondrial ROS production. Thus, aging induces in humans and rodents oxidative modifications in hepatic mitochondria that cause impaired mitochondrial bioenergetics that may contribute to liver dysfunction and pathology. Aging of the liver also affects the antioxidant response to oxidative stress. Furthermore, aging-induced modulation of DNA repair mechanisms, coupled with a decreased antioxidant response, may render liver tissue more sensitive to oxidative-induced increases in mtDNA damage. It is clear that more research is needed to explore this line of investigation. Thus, targeting mitochondrial oxidative damage due to impaired mitochondrial bioenergetics may be a valuable strategy for modifying age-associated liver disease.

SUMMARY POINTS

- During aging there are morphologic, structural and functional changes that result in impaired liver function.
- Aging leads to increased oxidative stress and impaired mitochondrial bioenergetics in the liver.
- An imbalance between the production of oxidants, particularly reactive oxygen species, and the antioxidant defense response, for example glutathione levels, occurs during aging, leading to a state of oxidative stress in aged organisms.
- Increases in oxidative damage in both the nuclear and mitochondrial genomes are observed in the liver during aging.
- Decreased mitochondrial DNA repair mechanisms, specifically base excision repair, have been documented; this can lead to mutagenesis, changes in mitochondrial DNA abundance and mitochondrial dysfunction.

References

1. Finkel T. Radical medicine: treating ageing to cure disease. *Nat Rev Mol Cell Biol* 2005;**12**:971–6.
2. Fabbrini E, Sullivan S, Kliein S. Obesity and nonalcoholic fatty liver disease: biochemical, metabolic, and clinical applications. *Hepatology* 2010;**51**(2):679–89.
3. Schmucker D. Age-related changes in liver structure and function: Implications for disease? *Exp Gerontol* 2005;**40**:650–9.
4. Begriche K, Massart J, Robin M-A, Bonnet F, Fromenty B. Mitochondrial adaptations and dysfunctions in nonalcoholic fatty liver disease. *Hepatology* 2013 (in press).
5. Frith J, Jones D, Newton J. Chronic liver disease in an ageing population. *Age Aging* 2009;**38**:11–8.
6. Le Couteur D, Cogger V, McCuskey R, et al. Age-related changes in the liver sinusoidal endothelium. A mechanism for dyslipidemia. *Ann NY Acad Sci* 2007;**1114**:79–87.
7. Modi H, Katyare S, Patel M. Ageing-induced alterations in lipid/phospholipid profiles of rat brain and liver mitochondria: implications for energy-linked functions. *J Membr Biol* 2008;**221**(1):51–60.
8. Ito Y, Sørensen K, Bethea N, Svistounov D, et al. Age-related changes in the hepatic microcirculation in mice. *Exp Gerontol* 2007;**42**:789–97.
9. Gregg S, Gutiérrez V, Robinson A, et al. A model of accelerated liver aging caused by a defect in DNA repair. *Hepatology* 2012;**55**(2):609–21.
10. Sastre J, Pallardo F, Pla R, et al. Aging of the liver: age-associated mitochondrial damage in intact hepatocytes. *Hapatology* 1996;**24**(5):1199–205.
11. Wilson P, Franks L. The effect of age on mitochondrial ultrastructure and enzymes. *Adv Exp Med Biol* 1975;**53**:171–83.
12. Herbener G. A morphometric study of age-dependent changes in mitochondrial population of mouse liver and heart. *J Gerontol A Biol Sci Med Sci* 1976;**31**(1):8–12.
13. Nagata T. Macromolecular synthesis in the livers of aging mice as revealed by electron microscopic radioautography. *Prog Histochem Cytochem* 2010;**45**:1–79.
14. Muller-Hocker J, Aust D, Rohrbach H, et al. Defects of the respiratory chain in the normal human liver and in cirrhosis during aging. *Hepatology* 1997;**26**(3):709–19.
15. Bellanti F, Romano A, Giudetti A, et al. Many faces of mitochondrial uncoupling during age: damage or defense? *J Gerontol A Biol Sci Med Sci* 2013 (in press).
16. Wu J-L, Wu Q-P, Peng Y-P, Zhang J- M. Effects of L-malate on mitochondrial oxidoreductases in liver of aged rats. *Physiol Res* 2011;**60**:329–36.
17. Garcia-Fernandez M, Sierra I, Puche J, et al. Liver mitochondrial dysfunction is reverted by insulin-like growth factor II (IGF-II) in aging rats. *J Translation Med* 2011;**9**:123–31.
18. Houtkooper R, Argmann C, Houten S, et al. The metabolic footprint of aging in mice. *Sci Rep* 2011;**1**:134.
19. Mussico C, Capelli V, Pesce V, et al. Rat liver mitocondrial proteome: changes associated with aging and acetyl-L-carnitine treatment. *J Proteom* 2011;**74**:2536–47.
20. Dani D, Shimokawa I, Komatsu T, et al. Modulation of oxidative phosphorylation machinery signifies a prime mode of anti-ageing mechanism of calorie restriction in male rat liver mitochondria. *Biogerontology* 2009;**11**:321–34.
21. Brown-Borg H, Jonhnson W, Rakoczy S. Expression of oxidative phosphorylation components in mitochondria of long-living Ames dwarf mice. *Age* 2012;**34**:43–57.
22. Brown-Borg H, Rakoczy S, Romanick M, Kennedy M. Effects of growth hormone and insulin-like growth factor-1 on hepatocyte antioxidant enzymes. *Exp Biol Med* 2002;**227**:94–104.
23. Liu Y, He J, Ji S, et al. Comparative studies of early liver dysfunction in senescence-accelerated mouse using mitochondrial proteomics approaches. *Mol Cell Proteomics* 2008;**7**(9):1737–47.

24. Hagen T, Yowe D, Bartholomew J, et al. Mitochondrial decay in hepatocytes from old rats: membrane potential declines, heterogeneity and oxidants increase. *Proc Natl Acad Sci USA* 1997;**94**:3064–9.
25. Sabaretnam T, Kritharides L, O'Reilly J, Le Couteur D. The effect of aging on the response of isolated hepatocytes to hydrogen peroxide and tert-butyl hydroperoxide. *Toxicology in Vitro* 2010;**24**:123–8.
26. Chaudhuri A, de Waal E, Pierce A, et al. Detection of protein carbonyls in aging liver tissue: a fluorescence-based proteomic approach. *Mech Ageing Develop* 2006;**127**:849–61.
27. Castro MdR, Suarez E, Kraiselburd E, et al. Aging increases mitochondrial DNA damage and oxidative stress in liver of rhesus monkeys. *Exp Gerontol* 2012;**47**:29–37.
28. Bhattacharya A, Leonard S, Tardif S, et al. Attenuation of liver insoluble protein carbonyls: indicator of a longevity determinant? *Aging Cell* 2011;**10**:720–3.
29. Li X-D, Rebrin I, Forster M, Sohal R. Effects of age and caloric restriction on mitochondrial protein oxidative damage in mice. *Mech Ageing Develop* 2012;**133**:30–6.
30. Hazelton G, Lang C. Glutathione contents of tissues in the aging mouse. *Biochem J* 1980;**188**(1):25–30.
31. Toroser D, Sohal R. Age-associated perturbations in glutathione synthesis in mouse liver. *Biochem J* 2007;**405**(3):583–9.
32. Esposito L, Kokoszka J, Waymire K, et al. Mitochondrial oxidative stress in mice lacking the glutathione peroxidase-1 gene. *Free Radic Biol Med* 2000;**28**(5):754–66.
33. de la Asuncion J, Millan A, Pla R, et al. Mitochondrial glutathione oxidation correlates with age-associated oxidative damage to mitochondrial DNA. *FASEB J* 1996;**10**(2):333–8.
34. Torres-Ramos CA, Johnson RE, Prakash L, Prakash S. Evidence for the involvement of nucleotide excision repair in the removal of abasic sites in yeast. *Mol Cell Biol* 2000;**20**:3522–8.
35. Hamilton M, Van Remmen H, Drake J, et al. Does oxidative damage to DNA increases with age? *Proc Natl Acad Sci USA* 2001;**98**(18):10469–74.
36. Mikkelsen L, Bialkowski K, Risom L, et al. Aging and defense against generation of 8-oxo-7,8-dihydro-2'-deoxyguanosine in DNA. *Free Radic Biol Med* 2009;**47**(5):608–15.
37. Atamna H, Cheung I, Ames BN. A method for detecting abasic sites in living cells: age-dependent changes in base excision repair. *Proc Natl Acad Sci USA* 2000;**97**:686–91.
38. Intano GW, Cho EJ, McMahan CA, Walter CA. Age-related base excision repair activity in mouse brain and liver nuclear extracts. *J Gerontol A Biol Sci Med Sci* 2003;**58**(3):B205–11.
39. de Souza-Pinto NC, Eide L, Hogue BA, et al. Repair of 8-oxodeoxyguanosine lesions in mitochondrial DNA depends on the oxoguanine dna glycosylase (OGG1) gene and 8-oxoguanine accumulates in the mitochondrial DNA of OGG1-defective mice. *Cancer Res* 2001;**61**(14):5378–81.
40. Szczesny B, Mitra S. Effect of aging on intracellular distribution of abasic (AP) endonuclease 1 in the mouse liver. *Mech Ageing Dev* 2005;**126**(10):1071–8.
41. Yakes FM, Van Houten B. Mitochondrial DNA damage is more extensive and persists longer than nuclear DNA damage in human cells following oxidative stress. *Proc Natl Acad Sci U S A* 1997;**94**(2):514–9.
42. Acevedo-Torres K, Berrios L, Rosario N, et al. Mitochondrial DNA damage is a hallmark of chemically induced and the R6/2 transgenic model of Huntington's disease. *DNA Repair (Amst)* 2009;**8**(1):126–36.
43. Xu G, Herzig M, Rotrekl V, Walter C. Base excision repair, aging and health span. *Mech Ageing Dev* 2008;**129**(7–8):366–82.
44. de Souza-Pinto NC, Hogue BA, Bohr VA. DNA repair and aging in mouse liver: 8-oxodG glycosylase activity increase in mitochondrial but not in nuclear extracts. *Free Radic Biol Med* 2001;**30**(8):916–23.
45. Szczesny B, Hazra TK, Papaconstantinou J, et al. Age-dependent deficiency in import of mitochondrial DNA glycosylases required for repair of oxidatively damaged bases. *PNAS* 2003;**100**(19):10670–5.
46. Trifunovic A, Wredenberg A, Falkenberg M, et al. Premature ageing in mice expressing defective mitochondrial DNA polymerase. *Nature* 2004;**429**(6990):417–23.
47. Barazzoni Rocco, Short KR, Sreekumaran Nair K. Effects of aging on mitochondrial DNA copy number and cytochrome *c* oxidase gene expression in rat skeletal muscle, liver, and heart. *J Biol Chem* 2000;**275**:3343–7.
48. Petersen K, Befroy D, Dufour S, et al. Mitochondrial dysfunction in the elderly: possible role in insulin resistance. *Science* 2003;**300**:1140–2.
49. Daryani N, Daryani N, Alavian S, et al. Non-alcoholic steatohepatitis and influence of age and gender on histopathologic findings. *World J Gastroenterol* 2010;**7**(16):4169–75.
50. Cortez-Pinto H, Chatam J, Chacko V, et al. Alterations in liver ATP homeostasis in human nonalcoholic steatohepatitis. A pilot study. *JAMA* 1999;**282**(17):1659–64.
51. Pérez-Carreras M, Del Hoyo P, Martín M, Rubio J, et al. Defective hepatic mitochondrial respiratory chain in patients with nonalcoholic steatohepatitis. *Hepatology* 2003;**38**(4):999–1007.
52. Xu S-C, Chen Y-B, Lin H, et al. Damage to mtDNA in liver injury of patients with extrahepatic cholestasis: the protective effects of mitochondrial transcription factor A. *Free Radic Biol and Med* 2012;**52**:1543–51.
53. Sunny N, Parks E, Browning J, Burguess S. Excessive hepatic mitochondrial TCA cycle and gluconeogenesis in humans with nonalcoholic fatty liver disease. *Cell Metabolism* 2011;**14**:804–10.
54. Petersen K, Oral E, Dufour S, et al. Leptin reverses insulin resistance and hepatic steatosis in patients with severe lipodystrophy. *J Clin Invest* 2002;**109**:1345–50.
55. Takamura T, Misu H, Matsuzawa-Nagata N, et al. Obesity upregulates genes involved in oxidative phosphorylation in livers of diabetic patients. *Obesity* 2008;**16**:2601–9.
56. Nomoto K, Tsuneyama K, Takahashi H, et al. Cytoplasmic fine granular expression of 8-hydroxydeoxyguanosine reflects early mitochondrial oxidative DNA damage in nonalcoholic fatty liver disease. *Appl Immunohistochem Mol Morphol* 2008;**16**:71–5.
57. Seki S, Kitada T, Yamada T, et al. In situ detection of lipid peroxidation and oxidative DNA damage in non-alcoholic fatty liver disease. *J Hepatol* 2002;**37**:56–62.
58. Baur J, Pearson K, Price N, et al. Resveratrol improves health and survival of mice on a high-calorie diet. *Nature* 2006;**444**:337–42.
59. Minor R, Baur J, Gomes A, et al. SRT1720 improves survival and healthspan of obese mice. *Sci Rep* 2011;**1**(70):1–10.
60. Kendrick A, Choudhury M, Rahman S, et al. Fatty liver is associated with reduced SIRT3 activity and mitochondrial protein acetylation. *Biochem J* 2011;**433**(3):505–14.

CHAPTER

5

Arthritis as a Disease of Aging and Changes in Antioxidant Status

Rahul Saxena

Department of Biochemistry, School of Medical Sciences & Research, Sharda University, Greater Noida (UP), India

List of Abbreviations

CAT catalase
COX-2 cyclooxygenase-2
CVD cardiovascular disease
GSHPx glutathione peroxidase
H_2O_2 hydrogen peroxide
HA hyaluronic acid
Hb hemoglobin
Hb^{3+} methemoglobin
HDL high-density lipoprotein
$HO_2\cdot$ hydroperoxyl radical
HOCl hypochlorous acid
IFN-γ interferon-γ
IL-1β interleukin-1β
LDL low-density lipoprotein
MDA malondialdehyde
NO nitric oxide
$O_2\cdot^-$ superoxide anion radical
OA osteoarthritis
OH· hydroxyl radical
8-OHdG 8-hydroxydeoxyguanosine
$ONOO\cdot^-$ peroxynitrite
PGE_2 prostaglandin E-2
PUFA polyunsaturated fatty acid
RA rheumatoid arthritis
RNS reactive nitrogen species
ROO· lipid peroxyl radical
ROS reactive oxygen species
SOD superoxide dismutase
TNF-α tumor necrosis factor-α

INTRODUCTION

Aging

'Death lays his icy hands on kings' – James Shirley quoted these words centuries ago.[1] Unfortunately, despite massive efforts, scientists have failed to reveal the secrets of immortality, and old age, followed by death, is an accepted fact of life. This has led to the enhanced interest of researchers in keeping people physically fit and useful to society. Accordingly, now the dream of extending life has shifted from fabled fountains of youth and biblical tales of long-lived patriarchs to the laboratory – where gerentologists explore the genes and organs involved in aging. As a result, life expectancy has dramatically increased worldwide, and it is estimated that the proportion of older people will be one third of the total population of the world by 2050. However, with increasing age, older people become more susceptible to various diseases.

The aging process is a normal, universal and inevitable biologic phenomenon resulting from a cascade of destructive events that lead to progressive morphologic and physiologic deterioration of the organs, often accompanied by the development of various chronic morbidities.[2] These morbidities include cardiovascular disease, diabetes, neurodegenerative diseases, musculoskeletal disorders, immunosenescence and endocrine dysfunction etc., and they have become common health problems among the geriatric population.

Arthritis

Among the different types of musculoskeletal disorders associated with aging, arthritis is a major public health problem worldwide and, indeed, the main cause of morbidity among older people. The spectrum of abnormalities ranges from pain and inflammation in joints to physical disability, identified as disease processes such as rheumatoid arthritis, osteoarthritis, psoriatic arthritis and systemic lupus erythematosus etc. The pathogenesis of various types of arthritis is multifactorial and includes several overlapping events. The two-hit hypothesis postulates that the aging of biomolecules in a living system, caused by various biologic and environmental factors,

Aging
http://dx.doi.org/10.1016/B978-0-12-405933-7.00005-6

is the first hit and that these alterations in the system make the body more susceptible to secondary insults, including pain and inflammation of joints, which later switch into the disease process. Thus, all of these hits are widely believed to be major contributors to the development of arthritis.

In normal mature cartilage, chondrocytes synthesize sufficient amounts of macromolecules to maintain the integrity of the matrix. It is conceivable that the development of arthritis involves various processes, including aging of biomolecules and the senescence of articular chondrocytes as well. Thus, the strong association between age and the increasing incidence of arthritis marks arthritis as an age-related disease. However, the exact mechanisms behind the genesis of arthritis with senescence are unclear. Denham Harman's free radical theory of aging suggests that the chronic production of endogenous free radicals and the subsequent cellular damage by these radicals could mediate many changes that are associated with cellular aging.[3] In addition to this theory, the seminal work of McCord on synovial fluid in rheumatoid arthritis (RA) patients[4] has gradually triggered intense research into the role of free radicals in aging and in age-mediated diseases such as arthritis as well.

THE CONCEPT OF OXYGEN TOXICITY AND FREE RADICALS

Today's concept of 'oxygen toxicity' is not restricted to only some diseases: it is now receiving much attention and aims at solving the unanswered questions relating to the pathogenesis of arthritis and other musculoskeletal disorders caused by oxygen toxicity and their possible prevention by antioxidant therapy.

The term oxygen toxicity primarily focuses on the stress or destructive changes caused by the reactive oxygen species (ROS) known as free radicals; these are generated during cellular metabolism as an integral part of our daily life. Free radicals are highly reactive molecules that possess an unpaired electron in their outer shell – making them very unstable and prone to react with biologic membranes and chemical species; they include e.g. superoxide anion ($O_2\cdot^-$) radical, hydroperoxyl radical ($HO_2\cdot$), hydroxyl radical ($OH\cdot$), hydrogen peroxide (H_2O_2), lipid peroxyl radical ($ROO\cdot$), urate radical, α-tocopherol radical and ascorbate radical, etc. In addition, nitric oxide (NO) and peroxynitrite ($ONOO\cdot^{--}$) radicals, generally documented as reactive nitrogen species (RNS), have also received a considerable amount of interest in the context of rheumatic diseases (Table 5.1). Excess production of ROS and RNS leads to oxidative stress, which has important implications for aging, inflammation, arthritis and a variety of other diseases.

TABLE 5.1 Different Types of Reactive Oxygen Species (ROS) in a Biologic System

Name of ROS	Symbol
Superoxide anion radical	$O_2\cdot^-$
Hydrogen peroxide	H_2O_2
Hydroxyl radical	$OH\cdot$
Hydroperoxyl radical	$HOO\cdot$
Alkoxyl radical	$RO\cdot$
Peroxyl radical	$ROO\cdot$
Nitric oxide	NO
Peroxynitrite anion	$ONOO^-$
Nitrogen dioxide radical	$NO_2\cdot$
Nitronium ion	NO_2^+
Singlet oxygen	$^1O_2$

Molecular oxygen can normally accept four electrons to produce two molecules of water. Incomplete reduction of oxygen gives rise to toxic and reactive intermediates. The univalent reduction of molecular oxygen produces $O_2\cdot^-$ which can be further reduced to hydrogen peroxide, hydroxyl radical ($OH\cdot$) and finally to a water molecule. The sequential univalent reduction steps of molecular oxygen are illustrated in Figure 5.1(A). In addition, nitric oxide (NO), a byproduct of the oxidative reaction catalysed by nitric oxide synthase from L-arginine, acts as a free radical and has an important role in the pathogenesis of inflammatory arthropathies. Superoxide anion, produced in excess during oxidative stress, reacts with nitric oxide to form the toxic product peroxynitrite anion ($ONOO^-$) and thereby amplifies the production of RNS, which consequently leads to progression of age-related disorders (Fig. 5.1B).

Chondrocytes are also damaged by H_2O_2, and it has been suggested that low concentrations of H_2O_2, $O_2^{\bullet-}$, or both, accelerate bone resorption by osteoclasts.[5] Formation of superoxides and H_2O_2 can be regulated by either enzymatic or non-enzymatic mechanisms, whereas no enzymes are required for the formation of hydroxyl radical. The hydroxyl radical may be formed either through transition-metal-ion-catalyzed Fenton's reaction or the Haber–Weiss reaction (Fig. 5.1C).

Hydroxyl radical is highly reactive and is capable of initiating deleterious reactions such as lipid peroxidation and DNA damage etc. It also produces hypochlorous acid (HOCl) by the action of the enzyme myeloperoxidase in neutrophils and macrophages. Hydroxyl radicals

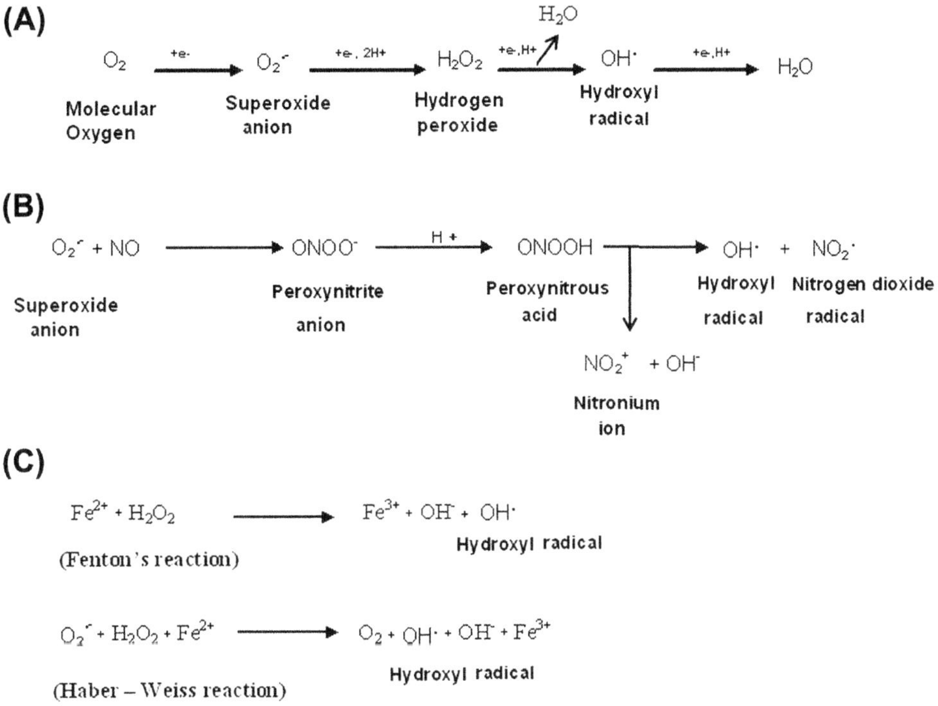

FIGURE 5.1 Sequential formation of reactive oxygen species (ROS) in a biological system. (A) Incomplete reduction of molecular oxygen and subsequent formation of superoxide anion radical, hydrogen peroxide and hydroxyl radical. (B) Reaction of superoxide anion with nitric oxide (NO) to form peroxynitrite anion followed by nitrogen dioxide radical, hydroxyl radical and nitronium ion. (C) Formation of potentially toxic hydroxyl radical via the Fenton reaction and the Haber–Weiss reaction.

degrade proteoglycans, and HOCl causes fragmentation of collagen.[6]

Sources of Free Radicals in Arthritis

A number of sources of free radicals have been identified in arthritic patients. These are well illustrated in Figure 5.2.

Leukocytes

Neutrophils, monocytes and macrophages contribute significantly to the production of ROS and RNS. Neutrophils comprise about 70% of blood leukocytes and provide the first line of defense. The activation of neutrophils by a soluble or phagocytic stimulus initiates a series of metabolic reactions collectively termed the 'respiratory burst', an important event in the pathogenesis of various sorts of arthritis. During this period a sudden and large increase in oxygen consumption occurs; this mostly accounts for the formation of superoxide anion through the action of the enzyme NADPH oxidase. The activation of NADPH oxidase may be induced by lipopolysaccharides, lipoproteins, and cytokines such as interferon (IFN)-γ, interleukin (IL)-1β and tumor necrosis factor (TNF)-α.[7] Several additional highly reactive oxygen-derived metabolites have been identified or predicted to exist as a result of the activation of phagocytic cells. These include hydroxyl radical, singlet oxygen and hypochlorous acid. These metabolites are capable of degrading cartilage proteoglycans and collagen gelatin, and they injure the structural matrix of synovial tissues in the development of arthritis.

Mitochondria

The mitochondrion is an important cellular organelle and is the main center for cellular oxidation reactions. Mitochondria damage the cells by producing reactive oxygen species from the respiratory chain. The production of reactive oxygen species by leukocyte mitochondria has been documented as one of the important sources of free radical production in the blood of rheumatoid arthritis patients and has been implicated in disease pathology.[8]

Purine Metabolism

Synovial tissue is rich in the cytosolic enzyme xanthine dehydrogenase, which does not produce superoxide anion. During hypoxic-reperfusion injury, as occurs in inflammatory synovitis, xanthine dehydrogenase is converted into xanthine oxidase – which catalyses the conversion of hypoxanthine to xanthine and then to uric acid.[9] Molecular oxygen acts as an electron acceptor in this reaction, leading to the production of superoxide anion.

Catecholamines

Catecholamines, produced by synovial tissues, play an important role in the appearance of pain or other clinical signs in rheumatoid arthritis.[10] These catecholamines are broken down by the enzyme monoamine oxidase, which involves oxidation and the subsequent production of excess electrons. Molecular oxygen can act as an electron acceptor and thus $O_2 \cdot^-$ and H_2O_2 are produced, thus increasing the formation of other free radicals.

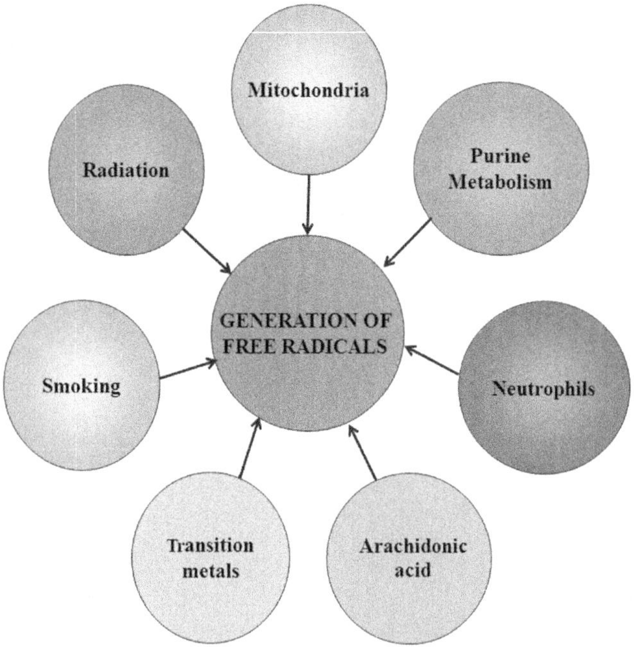

FIGURE 5.2 Generation of free radicals in arthritis. Formation of free radicals in arthritis involves various sources which include the electron transport chain in mitochondria, purine metabolism, neutrophils, arachidonic acid, transition metals, smoking and radiation etc.

Arachidonic Acid Metabolism

During inflammation, arachidonic acid is released in large amounts from chondrocytes; it can be metabolized in the lipoxygenase or cyclo-oxygenase pathway, leading to the synthesis of leukotrienes, prostaglandins or thromboxanes. These pathways are accompanied by the generation of reactive oxygen species. High concentrations of prostaglandin species have also been detected in the synovial fluid of rheumatoid arthritis (RA) and other inflammatory joint diseases.[11]

Transition Metals

Conditions such as hypoxic-reperfusion injury, ischemia and hemolysis etc. lead to the release of transition metal ions, e.g. iron and copper ions, which may further amplify free radical toxicity.[12]

Other Possible Sources

Some other possible sources of free radicals that enhance oxidative stress include smoking, alcoholism, radiation and the oxidation of ferrous myoglobin; it is suggested that these contribute to the development of arthritis with senescence.[13]

OXIDATIVE STRESS IN ARTHRITIS

Although free radicals are produced in large amounts in tissues under physiologic conditions, these are precisely controlled by several antioxidant protective mechanisms. Biologic tissues maintain a critical balance between antioxidant reserve and ROS or pro-oxidants, but in pathophysiologic conditions this balance is upset because of an increase in ROS production characterized by biomolecular deterioration and an alteration in the antioxidant defense system. This condition is known as oxidative stress (Fig. 5.3); it plays a crucial role in the development of degenerative diseases, including rheumatoid arthritis and osteoarthritis etc.

Free-Radical-Mediated Biomolecular Deterioration in Arthritis

Arthritis is an age-related disease, and increased production of free radicals followed by biomolecular deterioration represents a mechanism by which the aging process could directly contribute to chondrocyte damage and synovitis, and thereby enhance the progression of disease. The important cellular components affected by free radicals include membrane lipids, proteins, DNA, glucosaminoglycans and other biomolecules (Fig. 5.4); the activity of the free radicals then leads to the etiopathogenesis of arthritis, as indicated below.

Lipid Peroxidation

Lipid peroxidation is an autoxidation process initiated by the attack of free radicals (e.g. $OH\cdot$, $O_2^{\cdot-}$ and H_2O_2) on phospholipids or PUFA of the membranes of cellular or subcellular components, resulting in the formation of various sorts of aldehydes, ketones, alkanes, carboxylic acids and polymerization products. These products are highly reactive with other cellular components and the extracellular matrix; they serve as biomarkers of lipid peroxidation (Fig. 5.5). Among reactive aldehydes, malondialdehyde (MDA) is a toxic aldehydic end product of lipid peroxidation which mediates the oxidation of cartilage collagen, resulting in fragmentation, modification, aggregation and changes in protein conformation, and eventually leading to alterations in tissue functioning. In addition, chondrocyte-derived lipid peroxidation mediates collagen degradation.[14] Moreover, excess binding of these reactive aldehydes to matrix and cellular proteins may alter cellular function, membrane permeability and electrolyte balance, which further leads to fibrogenesis, matrix protein degradation and progressive deterioration of the biologic system associated with the progression of oxidative-stress-mediated arthritis. Various studies have documented the elevated level of lipid peroxidation markers in synovial fluid and their etiopathologic role in the development of arthritis with senescence.[5,6,15]

Protein Oxidation

Free-radical-mediated oxidatively modified forms of protein accumulate with senescence and have received considerable attention in exploring the molecular mechanism of arthritis progression. These ROS and RNS bring about the oxidation of side chains of amino acid residues,

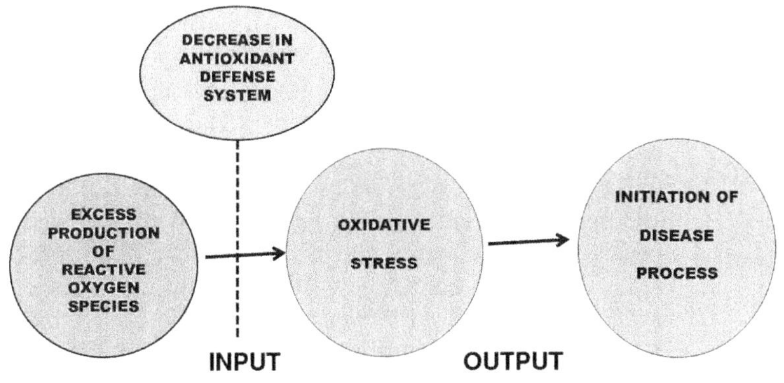

FIGURE 5.3 Overview of oxidative stress in arthritis: increase in free radical production, with subsequent decrease in the antioxidant defense system, leads to oxidative stress which, in turn, initiates the disease process.

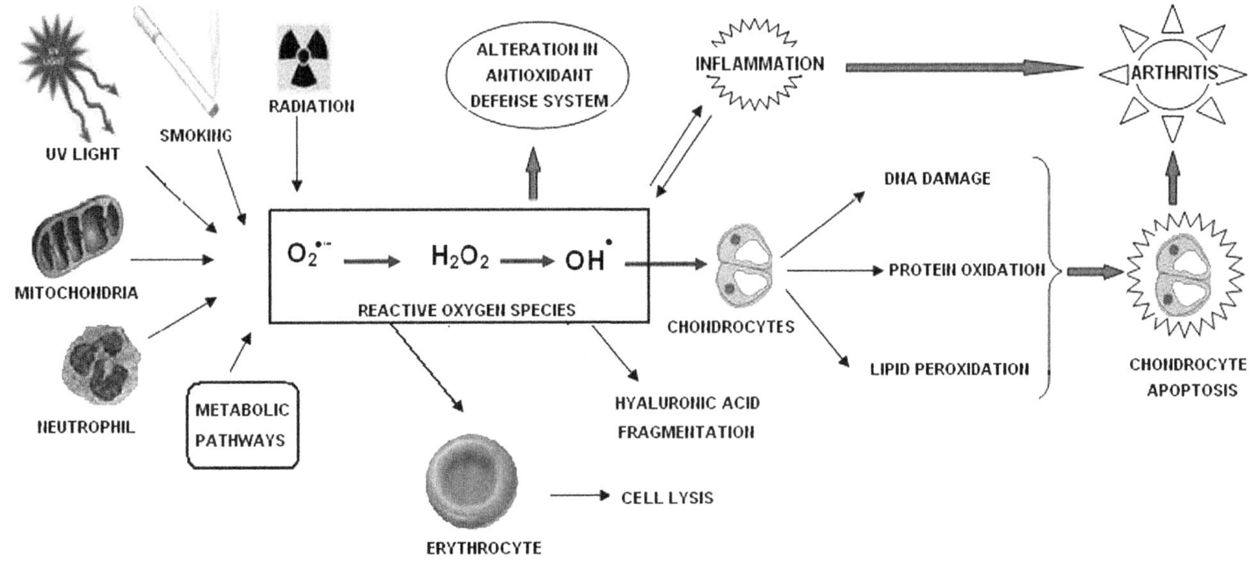

FIGURE 5.4 Culprit effect of reactive oxygen species in arthritis. Free radicals, generated by various sources, attack chondrocytes and mediate biomolecular destruction, such as lipid peroxidation, protein oxidation, DNA damage and fragmentation of hyaluronic acid – leading to chondrocyte apoptosis. Free radicals also cause alteration in the antioxidant defense system, erythrocyte lysis and inflammation. All of these culprits facilitate development of arthritis.

the formation of protein–protein cross linkages, and the oxidation of the protein backbone – resulting in denaturation, loss of function, aggregation and fragmentation of protein.[16] Radical protein reactions can impair the function of important cellular and extracellular proteins such as enzymes and connective tissue proteins, which further promotes age-related disease development.[17] In addition, *in vitro* studies on α_1-antitrypsin (inhibitor of neutrophil elastase) also showed that reactive oxygen species generated by polymorphonuclear leucocytes cause oxidation of methionine residues of α_1-antitrypsin and thereby make them biologically inactive.[18] Similarly, immunochemistry studies also showed the presence of 3-nitrotyrosine residues in human articular cartilage, suggesting that oxidative damage to proteins from reactive nitrite radicals may contribute to the development of age-related arthritis.[19] NO, or its derivatives, also interacts with the thiol group of proteins and glutathione to form nitrosothiols, blocking their catalytic activity. Several investigators also observed that the accumulation of oxidized proteins occurs in a number of tissues during aging and are thereby associated with reduced life expectancy via disease development.[20]

DNA Damage

DNA is another main target for free radical attack. Overproduced radicals can interact with DNA, break its strands, and delete the bases. Hydroxyl radical reacts with the DNA base deoxyguanosine, resulting in the formation of 8-hydroxydeoxyguanosine (8-OHdG) which increases misincorporation of DNA bases, leading to somatic mutation.[21] In addition, a role for free-radical-mediated DNA damage has been implicated in genome instability, cell proliferation, cell differentiation and chondrocyte apoptosis in arthritis patients.[22] Much evidence indicates that urinary 8-OHdG is not only a biomarker of generalized, cellular oxidative stress but also serves as a risk factor for age-related complications.[23] Similarly, elevated levels of DNA

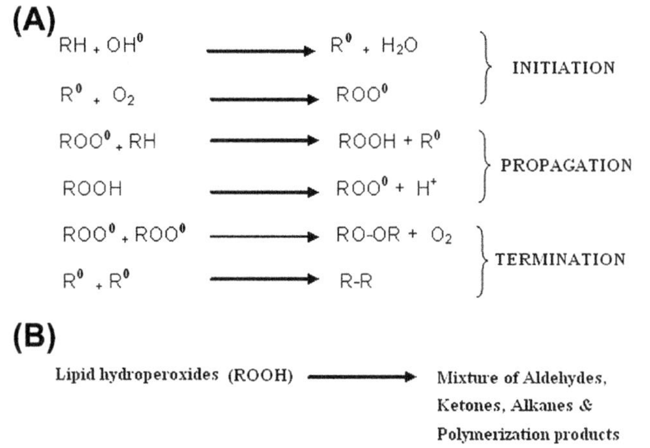

FIGURE 5.5 Lipid peroxidation in the chondrocyte: (A) Chain reaction of lipid peroxidation is initiated by the attack of free radicals (e.g. $OH^{\cdot}$, $O_2{\cdot}^-$ and H_2O_2) on phospholipid or PUFA of cell membrane to form carbon centered radical ($R^{\cdot}$). It reacts with $O_2{\cdot}^-$ to form lipid peroxyl radical ($ROO^{\cdot}$) which attacks another lipid molecule to generate lipid hydroperoxide (ROOH) and thereby propagate the chain reaction. This chain reaction terminates when a lipid peroxyl radical ($ROO^{\cdot}$) or carbon centered radical ($R^{\cdot}$) reacts with a similar radical to form non-radical products (adduct). (B) Lipid hydroperoxide (ROOH), being unstable, spontaneously decomposes into a complex mixture of aldehydes, ketones, alkanes and polymerization products, and enhances the lipid- peroxidation-mediated toxicity.

damage in peripheral lymphocytes, measured by the alkaline comet assay method, have been assessed and found to be associated with increased oxidative stress and decreased antioxidant capacity in RA patients.[24]

Effect on Hemoglobin

Chronic inflammatory rheumatic diseases are complicated by hematologic abnormalities, including anemia. In patients with RA, the prevalence of anemia ranges from 30% to 60%. It has been proposed that superoxide radical, produced on neutrophil stimulation, enters the intracellular compartment through anionic membrane channels and directly oxidizes oxyhemoglobin ($Hb^{2+}O_2$) to methemoglobin (Hb^{3+}), as shown in Eq. (1), leading to erythrocyte lysis. Similarly, H_2O_2 reacts with oxyhemoglobin and converts it into methemoglobin, as shown in Eq. (2). In addition, H_2O_2 also diffuses through the cell membrane and reacts with Hb^{3+}, forming the H_2O_2-methemoglobin complex (Hb^{3+}-H_2O_2), as shown in Eq. (3), thereby facilitating erythrocyte lysis.[25] This proposed mechanism of cell injury by phagocyte-derived oxygen metabolites offers a novel mechanism by which free radicals lead to the development of hematologic abnormalities in arthritic patients.

$$2\,H^+ + O_2^{-} + Hb^{2+}O_2 \rightarrow Hb^{3+} + O_2 + H_2O_2 \quad (1)$$

$$H_2O_2 + 2\,Hb^{2+}O_2 \rightarrow 2\,Hb^{3+} + 2\,O_2 + 2\,OH^- \quad (2)$$

$$H_2O_2 + Hb^{3+} \rightarrow Hb^{3+} - H_2O_2 \text{ complex} \quad (3)$$

Effect on Hyaluronic Acid

Hyaluronic acid (HA), a carbohydrate polymer, is a critical constituent of normal synovial fluid and is responsible for the viscoelastic nature of synovial fluid; it is thus an important contributor to joint homeostasis. Hyaluronic acid concentration also varies with age. In addition to lipid peroxidation and protein oxidation, reactive oxygen species also react with glycosaminoglycan hyaluronic acid, inducing fragmentation and reducing viscoelasticity of the synovial fluid.[26] *In vitro* studies have shown that $O_2{\cdot}^-$ produced by neutrophils causes depolymerization of hyaluronate and thereby increases its susceptibility to degradation by β-*N*-acetyl glucosaminidase A.[27] Furthermore, it has been observed that chronic oxidative stress originating from ROS produced intracellulary in synovial T lymphocytes has a significant effect on HA – and thus plays a key role in the perpetuation of arthritis.[28]

Alteration in the Antioxidant Defense System in Arthritis

Several protective mechanisms have evolved in the biologic system that serve to protect the biomolecules, membranes and cells by preventing the generation, or by scavenging, the free radicals. These include antioxidant enzymes, e.g. superoxide dismutase, catalase, glutathione peroxidase, ceruloplasmin and paraoxonase, and putative non-enzymic antioxidants, e.g. vitamin C, vitamin E, β-carotene, uric acid, glutathione, ubiquinone, bilirubin and glucose etc. Oxidative stress ensues when ROS evade or overwhelm the antioxidant protective mechanisms of synovial tissues. Alterations in the levels of these antioxidants, with subsequent biomolecular deterioration via increased ROS production, can cause degradation of cartilage collagen, loss of homeostasis in chondrocytes leading to impaired chondrocyte function, destructive changes in extracelluar matrix, synovitis and cartilage aging, and thereby perpetuate the development of arthritis with senescence.[29]

Antioxidant Enzymes

Superoxide dismutase (E.C.1.15.1.1), a non-heme-containing intracellular enzyme, occurs in both cytosol (Cu-Zn-SOD) and mitochondria (Mn-SOD); it catalyzes the dismutation of $O_2^{\bullet -}$ to H_2O_2 and O_2, as shown in Eq. (4), and therefore protects chondrocytes from potentially injurious superoxide anion. In our previous studies, the diminished activity of SOD in arthritic patients and elderly subjects could be explained on the basis of either (i) progressive enzyme inactivation by the product of the dismutation reaction, i.e. H_2O_2, or (ii) an increase in the glycosylation of SOD with aging.[6,30] Similarly, decreased erythrocyte SOD activity in RA patients due to utilization of these enzymes by free radicals during the detoxification process has been reported[31]

$$2\,O_2^{\cdot -} + 2H^+ \xrightarrow{\text{SUPEROXIDE DISMUTASE}} H_2O_2 + O_2 \quad (4)$$

Superoxide anion is scavenged by another blue-colored copper-containing enzyme, ceruloplasmin (E.C. 1.16.3.1), which mimics the action of SOD. It functions as a serum antioxidant enzyme by virtue of its ferroxidase activity (i.e. conversion of Fe^{2+} to Fe^{3+}), and thus it can halt the 'Haber reaction'. In our previous study, the serum ceruloplasmin level was found to be increased in arthritic patients, suggesting that ceruloplasmin protects the tissues against the deleterious effects of oxygen free radical and compensates for the loss of SOD activity, which occurs due to augmented oxidative stress.[6] Although age-related changes in ceruloplasmin level have been the subject of intensive investigation, its ambiguous property as an antioxidant enzyme as well as an acute phase protein is well reported.[32]

In addition, catalase (CAT) (E.C. 1.11.1.6), a heme-containing enzyme, and glutathione peroxidase (GPx) (E.C. 1.11.1.9), a selenium-containing enzyme, play a crucial role in the final detoxification of H_2O_2 in blood and tissues, as shown in Eqs 5 and 6. Both enzymes spontaneously react with, and scavenge, many forms of ROS, prevent oxidation of lipids and phospholipids, maintain the intracellular redox milieu, replenish a number of crucial antioxidants (vitamins E and C), and thereby prevent free-radical-mediated biomolecular destruction in arthritic patients.[6] Depleted levels of GSHPx and CAT in RA and OA patients have been observed in previous studies, clarifying their role in protecting chondrocytes from augmented oxidative stress.[6,15] Conversely, unchanged and elevated levels of these antioxidant enzymes in RA patients have been documented.[33,34]

$$2\,H_2O_2 \xrightarrow{\text{CATALASE}} O_2 + 2\,H_2O \quad (5)$$

$$\underset{\text{(Reduced glutathione)}}{2\,GSH} + H_2O_2 \xrightarrow{\text{GLUTATHIONE PEROXIDASE}} \underset{\text{(Oxidized glutathione)}}{GSSG} + 2\,H_2O \quad (6)$$

Recently, much attention has been given to paraoxonase (PON) (E.C. 3.1.8.1) in arthritic patients. This is a calcium-dependent arylesterase synthesized primarily in the liver and secreted into the serum as an HDL-associated enzyme; this enzyme prevents oxidation of low-density lipoprotein (LDL), is responsible for the anti-atherogenic property of high-density lipoprotein (HDL), and contributes significantly as an antioxidant enzyme in the antioxidant defense system of the body.[9] Alteration in PON activity, due to increased production of reactive aldehydes, has a significant effect in the development of cardiovascular disease (CVD). In previous studies, plasma PON activity was found to be decreased significantly in patients with rheumatoid arthritis and osteoarthritis. This can be explained on the basis of inactivation of the enzyme due to interaction of oxidized lipids with the PON free sulphydryl group. It reflects the increasing susceptibility of arthritic patients to develop CVD with senescence.[22,35]

Non-enzymic Antioxidants

The oxidant scavenging role of antioxidant enzymes is well supported by cooperative action of non-enzymatic antioxidants such as vitamin C, E, β-carotene and uric acid etc. These non-enzymic antioxidants have a significant role in reducing oxidative-stress-mediated cascade associated with age-related biomolecular deterioration. Vitamin C, an effective, water-soluble exogenous antioxidant, acts as first line of defense against free radicals in the blood. Vitamin C stimulates type II collagen and aggrecan synthesis (cartilage protein and proteoglycan) in articular cartilage. Moreover, it also acts as a synergist to regenerate α-tocopherol by reducing the α-tocopheroxyl radical, produced in the membranes via the reaction of α-tocopherol with free radicals. In our previous studies, plasma vitamin C levels were found to be significantly low in arthritic patients. This is basically due to its protective and radical scavenging action.[2,35] Similarly, a marked reduction in vitamin C levels in arthritis patients was observed by many researchers and suggests that vitamin C supplementation may prevent arthritis progression.[31,36] An additional possible mechanism through which ascorbate may reduce the risk of age-related complications is its protective effect on Na^+- K^+ ATPase by the protection of biomembranes from peroxidative damage; it therefore plays a significant role in maintaining electrolyte balance. Free radicals also serve as mediators of vascular injury, and under physiologic conditions these radicals inactivate and destroy endothelial NO (a well known vasodilator) and thereby lead to increased blood pressure. Plasma ascorbate scavenges oxygen free radicals and increases the availability of endothelial NO, thereby reducing the risk of stroke and CVD in arthritic and elderly subjects.[35,37]

Vitamin E, a universal lipophilic, chain-breaking antioxidant and a stabilizer of biologic membranes, prevents accumulation of free radicals and decreases oxidative stress. Vitamin E prevents the release of lysosomal enzymes (aryl sulfatase A and acid phosphatase) by inhibiting lipid peroxidation-mediated membrane-bound phospholipid degradation, decreases their activities, and thereby reduces the destruction of human articular chondrocytes (i.e. the key step in OA progression). In addition, the role of vitamin E in the inhibition of chondrocyte-derived lipid peroxidation-mediated collagen degradation has been well documented.[35] Vitamin E also replenishes the antioxidant enzyme activity in arthritic patients not only by quenching $O_2^{\bullet -}$ and preventing the loss of SOD activity but also by virtue of its role in modulating the enzyme system that generates free radicals and in regulating the protein expression of the enzyme at the transcriptional or post-transcriptional level. Decreased levels of vitamin E in arthritic patients, as observed in our previous studies, could not only be due to its free radical scavenging action but also to its action in maintaining the body's antioxidant reserve, which prevents cartilage degradation and the associated risk of developing arthritis. Vitamin E reduces the risk of CVD in arthritic patients not only by inhibiting LDL oxidation but also by inhibiting smooth muscle proliferation, platelet adhesion and aggregation.[6,35] Vitamin E also exerts anti-inflammatory and analgesic properties by inhibiting the release of arachidonic acid from membrane phospholipids which otherwise might be utilized for prostaglandin synthesis by cyclooxygenase and contribute to the inflammatory process in arthritic elderly patients. In addition, the glutathione-sparing action of vitamin E – by scavenging NO, by inhibiting lipid peroxidation, and by preventing the utilization of catalase against augmented oxidative stress – has also been well documented in OA elderly patients.[6,38] However, non-significant effects of vitamin E supplementation on SOD activity have been reported.[39]

Similarly, vitamin A, another fat-soluble antioxidant, is believed to play a protective role against oxidative damage as it is specifically carried on LDL particles, quenches singlet oxygen, and competitively spares selenium in metabolic reactions.[40] Low levels of plasma vitamin A in arthritic patients, as observed in previous studies, could be explained as it contributes to the prevention of LDL oxidation in free radical scavenging action in cooperation with other antioxidants and in the bone remodeling process. A deficiency of vitamin A in the elderly makes them susceptible to infection and increases the risk of night blindness, xerophthalmia, CVD and other age-related complications which can be prevented by improving the blood and LDL-β carotene status. Despite data linking the antioxidant role of these vitamins, the pro-oxidant properties of these vitamins also play a controversial role.[41]

Uric acid is one of the major endogenous, preventive and chain-breaking antioxidants in human plasma; it contributes about 65% of the free radical scavenging action, stabilizes ascorbate, protects erythrocytes from peroxidative damage, and inhibits free radical damage to DNA and oxidative degradation of hyaluronic acid.[35] In our recent study, plasma uric acid levels were found to be significantly high in arthritis patients; this reflects the idea that the body is trying to protect itself from the deleterious effects of free radicals with a continuous increase in chondrocyte destruction, by increasing uric acid production. Similarly, the positive correlation of hyperuricemia with obesity and hypertension (risk factors of CVD) in OA patients is well documented.[42] Despite data linking the antioxidant role of uric acid in arthritis, its controversial role in promoting LDL oxidation and as a danger signal of increasing risk for arthritis through inflammasome activation are well documented; this authenticates the fact that uric acid has a dual action in arthritis and needs further research in this direction.[43,44]

Oxidative Stress and Chondrocyte Cell Death

An evolving concept that is gaining acceptance is that human chondrocytes actively produce free radicals which are toxic to the extracellular matrix of synovial fluid and induce apopotic cell death in articular chondrocytes. Articular chondrocyte cell death is common in both the age-related diseases, i.e. osteoarthritis and rheumatoid arthritis. During the pathogenesis of OA, chondrocyte cell death induced by oxidative stress represents a mechanism by which the aging process could directly contribute to the progression of arthritis.[45] In addition to biomolecular deterioration and depletion in antioxidant reserves, free radicals participate in disease pathology by inducing several alterations and thereby lead to chondrocyte cell apoptosis. In this context, NO, which is produced in large amounts by chondrocytes upon proinflammatory cytokine stimulation, has long been considered as the primary inducer of chondrocyte apoptosis mediated by caspase-3 and tyrosine kinase activation. NO-induced chondrocyte death signaling in human osteoarthritis also includes DNA fragmentation via COX-2 induction-mediated PGE_2 production.[46]

An additional mechanism by which free radicals perpetuate arthritis involves the decreased responsiveness of chondrocytes to the anabolic and anticatabolic role of insulin-like growth factor 1 (IGF-1). IGF-1, a small 70-amino-acid polypeptide synthesized in the liver in response to binding of pituitary growth hormone, is also synthesized locally by chondrocytes; it acts in both an autocrine and a paracrine manner, and stimulates matrix production by cartilage cells. It has been shown that inflammatory cytokines produced due to augmented oxidative stress, particularly interleukin-1 (IL-1), inhibit cartilage matrix production and stimulate catabolic activity in cartilage and appear to play a key role in the development of OA. IGF-1 exerts a beneficial effect against cartilage degradation by inhibiting the activity of IL-1 to degrade proteoglycan and thereby limits OA progression. NO inhibits tautophosphorylation of the IGF-1 receptor (IGF-IRβ) in chondrocytes and thereby interferes in the binding of IGF-1 to IGF-IRβ – which plays a pivotal role in decreasing the chondrocyte survival promoting capacity of IGF-1. Therefore, the reduced responsiveness of chondrocytes to IGF-1 and increased apoptotic cell death in chondrocytes could be a direct consequence of increased oxidative stress in arthritic cartilage.[47]

Oxidative Stress and Inflammation in Arthritis

Free-radical-mediated oxidative stress has been described as an important mechanism underlying destructive proliferative synovitis in arthritic patients. ROS, RNS and their intermediates serve as mediators of inflammation in inflammatory and arthritic disorders by enhancing various culprit events such as inhibition of glycolytic enzymes, reduction of antioxidant reserves in synovial fluid and activation of proteolytic enzymes to degrade cartilage.[48] In addition to promoting cytotoxicity, free radicals can also function as secondary messengers to activate transcription factors, such as NFκB, and induce gene expression. NFκB is an important transcription factor controlling the transcription of a number of inflammatory cytokine genes. The production of inflammatory cytokines, such as interleukin 1 (IL-1b) and tumor necrosis factor alpha (TNF-α), driven by NFkB, enhances inflammation and the destruction of cartilage and bone in both OA and RA, not only by amplifying the action of matrix metalloproteinases, such as collagenases, but also by enhancing the production of ROS and RNS.[7,49,50]

In this context, an excess amount of NO, produced in arthritic cartilage upon inducible nitric oxide synthase (iNOS) stimulation by cytokines, has also been reported in the synovial fluid and serum of arthritic patients.[51] NO was also found to be involved in the development of inflammatory arthritic lesions not only by inhibiting the synthesis of cartilage matrix macromolecules, and by inducing chondrocyte death, but also by reducing the synthesis of IL-1 receptor antagonist in chondrocytes, which facilitates the enhanced IL-1β effect on chondrocytes. Moreover, diffusion of NO from the superficial layer of cartilage to the deeper zone plays a significant role in increasing the level of matrix metalloproteinase (MMP-13) and cyclooxygenases.[52] Inhibition of iNOS reduces synovial inflammation and thereby regulates the progression of destructive proliferative synovitis with senescence. Recently, a marked elevated level of inflammatory markers (IL-6 and CRP), along with altered levels of MDA and antioxidant enzymes in elderly patients with knee OA, has been documented; this indicates that oxidative stress persists along with inflammation and may have a significant role in destructive and inflammatory arthropathies.[6]

Therapeutic Interventions

Treatment of arthritis is becoming a major medical issue with the aging of the world's population. This disease is responsible for a significant portion of the financial costs, and its management involves multimodal therapeutic intervention because no complete cure has been found. In addition, arthritis management requires drugs that could slow down, stop or even avoid synovitis and joint degradation. Many of the recommended interventions present only symptom-modifying effects, and a few structure-modifying effects. In this context, a wide range of antioxidants, both natural and synthetic, have been proposed for use in the treatment of arthritis. Considerable attention has been devoted to the potential use of α-tocopherol and trace elements in reducing oxidative stress and thereby preventing arthritis and age-related complications.[53] In clinical studies, vitamin E has been found to be effective in the symptomatic treatment of arthritis not only by inhibition or reduction of free-radical-mediated biomolecular deterioration but also by amelioration of altered antioxidant reserves and levels of inflammatory markers.[6] However, there is also a controversial report on the inefficacy of vitamin E in the management of arthritis.[54] Furthermore, despite the massive efforts of researchers and various therapeutic triumphs, we know little about the genesis of arthritis with senescence, or its prevention, and there is a need for further research and the discovery of novel markers for early intervention in the geriatric population.

SUMMARY POINTS

- Aging is a universal and inevitable biologic phenomenon which occurs as a result of cumulative damage incurred by various deterministic factors, including environmental stresses, lifestyle and genetic programming etc.
- Free radicals, such as superoxide anion, hydroxyl radical and nitric oxide, are highly reactive, often participate in chain reactions that multiply their impact, and are continually generated as a byproduct of numerous redox reactions taking place inside the biologic system.
- Excessive production of free radicals followed by biomolecular deterioration and alteration in the antioxidant defense system significantly contribute to the development of oxidative stress.
- Oxidative stress acts as a cornerstone of lipid peroxidation, DNA damage, protein cross-linking, synovial fluid toxicity, induction of inflammation, apoptotic articular chondrocyte death and cartilage destruction which facilitate the initiation and progression of age-mediated arthritis.
- A daily diet rich in antioxidants should be increased with advancing age in order to sustain or postpone the free-radical-mediated biomolecular deterioration of the chondrocytes or musculoskeletal tissues and to replenish the antioxidant reserve as well; this may be a crucial step in the prevention of arthritis and other age-related complications.

References

1. http://www.bartleby.com/100/pages/page209.html.
2. Saxena R, Lal AM. Culprit effect of altered total antioxidant status and lipid peroxidation mediated electrolyte imbalance on aging. *J Indian Acad Geriat* 2007;**3**(4):137–44.
3. Chatterjee M, Shinde R. Biochemistry of aging. In: Chatterjee M, Shinde R, editors. *Text book of medical biochemistry*. 7th ed. New Delhi: Jaypee Brothers Medical Publishers; 2008. pp. 793–8.
4. McCord JM. Free radicals and inflammation. *Science* 1974;**185**:529–31.
5. Halliwell B. Oxygen radicals, nitric oxide and human inflammatory joint disease. *Ann Rheum Dis* 1995;**54**:505–10.
6. Bhattacharya I, Saxena R, Gupta V. Efficacy of vitamin E in knee osteoarthritis management of North Indian geriatric population. *Ther Adv Musculoskel Dis* 2012;**4**(1):11–9.
7. Filippin LI, Vercelino R, Marroni NP, Xavier RM. Redox signaling and the inflammatory response in rheumatoid arthritis. *Clin Exp Immuno* 2008;**152**:415–22.
8. Miesel R, Murphy MP, Kroger H. Enhanced mitochondrial radical production in patients with rheumatoid arthritis correlates with elevated levels of tumor necrosis factor alpha in plasma. *Free Radic Res* 1996;**25**(2):161–9.
9. Woodruff T, Blake DR, Freeman J, et al. Is chronic synovitis an example of reperfusion injury? *Ann Rheum Dis* 1986;**45**:608–11.
10. Capellino S, Cosentino M, Wolff C, et al. Catecholamine-producing cells in the synovial tissue during arthritis: modulation of sympathetic neurotransmitters as new therapeutic target. *Ann Rheum Dis* 2010;**69**(10):1853–60.
11. Brodie MJ, Hensby CN, Parke A, Gordon D. Is prostacyclin the major pro-inflammatory prostanoid in joint fluid? *Life Sci* 1980; **27**:603–8.
12. Halliwell B, Gutteridge MC. Oxygen toxicity, oxygen radicals and transition metals and disease. *Biochem J* 1984;**219**:1–14.
13. Mahajan A, Tandon VR. Antioxidants and rheumatoid arthritis. *J Ind Rheumatol Assoc* 2004;**12**:139–42.
14. Shah R, Raska KJ, Tiku ML. The presence of molecular markers of in vivo lipid peroxidation in osteoarthritic cartilage. *Arthritis Rheum* 2005;**52**:2799–807.
15. Aryaeian N, Djalai M, Shahram S, et al. Beta-carotene, vitamin E, MDA, glutathione reductase and arylesterase activity levels in patients with active rheumatoid arthritis. *Iranian J Publ Health* 2011;**40**(2):102–9.
16. Berlett BS, Stadtman ER. Protein oxidation in aging, disease, and oxidative stress. *J Biol Chem* 1997;**272**(33):20313–6.
17. Stadtman ER. The status of oxidatively modified proteins as a marker of aging. In: Esser K, Martin GM, editors. *Molecular aspects of aging*. Chichester: Wiley; 1995. pp. 129–44.
18. Carp H, Janoff A. In vitro suppression of serum elastase inhibitory capacity by reactive oxygen species generated by phagocytosing polymorphonuclear leucocytes. *J Clin Invest* 1979;**63**:793–7.
19. Loeser RF, Carlson SS, del Carlo M, Cole A. Detection of nitrotyrosine in aging and osteoarthritic cartilage: correlation of oxidative with the presence of interleukin-1β and with chondrocyte resistance to insulin like growth factor1. *Arthritis Rheum* 2002;**46**: 2349–57.
20. Davies KJA. Protein damage and degradation of oxygen radicals. *J Biol Chem* 1987;**262**:9895–902.
21. Kuchino Y, Mori F, Kasai H, et al. Misreading of DNA templates containing 8-hydroxy deoxyguanosine at the modified base and at adjacent residues. Nature 327, 77–79.
22. Altindag O, Kocyigit A, Celik N, et al. DNA damage and oxidative stress in patients with osteoarthritis: a pilot study. *Rheumatism* 2007;**22**:60–3.
23. Wu LL, Chiou CC, Chang PY, Wu JT. Urinary 8-OHdG: a marker of oxidative stress to DNA and a risk factor for cancer, atherosclerosis and diabetes. *Clin Chim Acta* 2004;**339**(1-2):1–9.
24. Altindag O, Karakoc M, Kocyigit A, et al. Increased DNA damage and oxidative stress in patients with rheumatoid arthritis. Clin Biochem 40:167–171.
25. Weiss SJ. The role of superoxide in the destruction of erythrocyte target by human neutrophils. *J Biol Chem* 1980;**255**:9912–7.
26. Greenwald RA, Moy Wai W. Effect of oxygen free radicals on hyaluronic acid. *Arthritis Rheum* 1980;**23**:455–63.
27. Parellada P, Planas JM. Synovial fluid degradation induced by free radicals: in vitro action of several free radical scavengers and anti-inflammatory drugs. *Biochem Pharm* 1978;**27**:535–7.
28. Remans PHG, van Oosterhout M, Smeets TJM, et al. Intracellular free radical production in synovial T lymphocytes of patients with rheumatoid arthritis. *Arthritis Rheum* 2005;**52**:2003–9.
29. Surapaneni KM, Venkataramana G. Status of lipid peroxidation, glutathione, ascorbic acid, vitamin E and antioxidant enzymes in patients with osteoarthritis. *Ind J Med Sci* 2007;**61**:9–14.
30. Saxena R, Lal AM. Effect of aging on antioxidant enzyme status and lipid peroxidation. *J Ind Acad Geriatrics* 2006;**2**:53–6.
31. Karatas F, Ozates I, Canatan H, et al. Antioxidant status and lipid peroxidation in patients with rheumatoid arthritis. *Indian J Med Res* 2003;**118**:178–81.
32. Klipstein GK, Grobee DE, Koster JF. Serum ceruloplasmin as a coronary risk factor in the elderly. The Rotterdam study. *Br J Nutr* 1999;**81**:139–44.
33. Mulherin DM, Thurnham DI, Situnayake RD. Glutathione reductase activity, riboflavin status, and disease activity in rheumatoid arthritis. *Ann Rheum Dis* 1996;**55**(11):837–40.
34. Biemond P, Swaak AJ, Koster JF. Protective factors against oxygen free radicals and hydrogen peroxide in rheumatoid arthritis synovial fluid. *Arthritis Rheum* 1984;**27**(7):760–5.
35. Gupta V, Saxena R, Bhattacharya I, Sunita. Assessment of coronary heart disease risk in knee osteoarthritic North Indian elderly. *J Ind Acad Geriartics* 2012;**8**:64–71.
36. Surapaneni KM, Venkataramana G. Status of lipid peroxidation, glutathione, ascorbic acid, vitamin E and antioxidant enzymes in patients with osteoarthritis. *Ind J Med Sci* 2007;**61**:9–14.
37. Tousoulis D, Davis G, Toutouzas P. Vitamin C increases nitric oxide availability in coronary atherosclerosis. *Ann Intern Med* 1999;**131**: 156–7.
38. George W. γ-tocopherol: an efficient protector of lipids against nitric oxide initiated peroxidative damage. *Nutrition Rev* 1997; **55**(10):376–8.
39. Li RK, Cowan DB, Mickle DAG, et al. Effect of vitamin E on human glutathione peroxidase expression in cardiomyocytes. *Free Radic Biol Med* 1996;**21**:419–26.
40. Olivieri O, Stanzial AM, Girelli D, et al. Selenium status, fatty acids, vitamin A and E, and aging: the Nove study. *Am J Clin Nutr* 1994;**60**:510–7.
41. Nishikimi M, Yagi K. Biochemistry and molecular biology of ascorbic acid biosynthesis. In: Harris JR, editor. *Subcellular biochemistry, vol XXV*. New York: Plenum Press; 1996. pp. 17–25.
42. Magliano M. Obesity and arthritis. *Menopause Int* 2008;**14**: 149–54.
43. Denoble AE, Huffman KM, Stabler TV, et al. Uric acid is a danger signal of increasing risk for osteoarthritis through inflammasome activation. *Proc Natl Acad Sci USA* 2011;**108**:2088–93.
44. Schlotte V, Sevanian A, Hochstein P, Weithmann KU. Effect of uric acid and chemical analogues on oxidation of human low density lipoprotein in-vitro. *Free Radic Bio Med* 1998;**25**:839–47.
45. Carlo MD, Loeser RF. Increased oxidative stress with aging reduces chondrocyte survival: correlation with intracellular glutathione levels. *Arthritis Rheum* 2003;**48**(12):3419–30.
46. Notoya K, Jovanovic DV, Reboul P, et al. The induction of cell death in human osteoarthritis chondrocytes by nitric oxide is related to the production of prostaglandin E2 via the induction of cyclooxygenase-2. *J Immunol* 2000;**165**(6):3402–10.

47. Studer RK, Levicoff E, Georgescu H, et al. Nitric oxide inhibits chondrocyte response to IGF-I: inhibition of IGF-IRβ tyrosine phosphorylation. *Am J Physiol Cell Physiol* 2000;**279**:961–9.
48. Greenwald RA. Oxygen radicals, inflammation, and arthritis: pathophysiological considerations and implications for treatment. *Semin Arthritis Rheum* 1991;**20**(4):219–40.
49. Khansari N, Shakiba Y, Mahmoudi M. Chronic inflammation and oxidative stress as a major cause of age related diseases and cancer. *Recent Patents on Inflammation & Allergy Drug Discovery* 2009;**3**:73–80.
50. Conner EM, Grisham MB. Inflammation, free radicals and antioxidants. *Nutrition* 1996;**12**(4):274–7.
51. Mc Innes IB, Leung BP, Field M, et al. Production of nitric oxide in the synovial membrane of rheumatoid and osteoarthritis patients. *J Exp Med* 1996;**184**:1519–24.
52. Zaragoza C, Balbin M, Lopez-Otin C, Lamas S. Nitric oxide regulates matrix metalloprotease-13 expression and activity in endothelium. *Kidney Int* 2002;**61**:804–8.
53. Kerimova AA, Atalay M, Yusifov EY, et al. Antioxidant enzymes: possible mechanism of gold compound treatment in rheumatoid arthritis. *Pathophysiology* 2000;**7**(3):209–13.
54. Brand C, Snaddon J, Bailey M, Cicuttini F. Vitamin E is ineffective for symptomatic relief of knee osteoarthritis: a six month double blind, randomised, placebo controlled study. *Ann Rheum Dis* 2001;**60**:946–9.

CHAPTER

6

Diabetes as a Disease of Aging, and the Role of Oxidative Stress

Dipayan Sarkar, Kalidas Shetty

Department of Plant Sciences, Loftsgard Hall, NDSU, Fargo, ND, USA

List of Abbreviations

HDL high-density lipoprotein
IDF International Diabetes Federation
LDL low-density lipoprotein
NO nitric oxide
ROS reactive oxygen species
WHO World Health Organization

INTRODUCTION

The prevalence of type 2 diabetes mellitus has increased rapidly in the last few decades. According to the latest (September 2012) estimate of the World Health Organization (WHO), 347 million people worldwide have diabetes, and almost 90% have type 2 diabetes.[1] It has been estimated that death from diabetes will increase by two thirds between 2008 and 2030. In the USA, 8.3% of the population, including children and adults, has diabetes, and the epidemic of diabetes is increasing alarmingly in both the developed and developing countries.[2] According to the International Diabetes Federation (IDF), life expectancy has increased significantly worldwide during the last century, but the risk and incidence of type 2 diabetes is also increasing at an alarming rate.[3] Men and women 75 years of age have a life expectancy of 10 and 13 years respectively.[4] Around 4.6 million people die due to diabetes each year, and half of those people are over 60 years of age.[5] By the year 2030, people of 60 and over are going to outnumber those aged 15 and below.[3] In developing countries, the number of people with adult diabetes is going to increase by 69% between 2010 and 2030, and it is going to double for those over 60 years of age.[4,5] Overall, the occurrence and incidence of type 2 diabetes are more prevalent in the aged population worldwide. Approximately 40% of people in the USA with known diabetes are older than 65, and many older people meet criteria for prediabetes according to the American Diabetes Association (ADA).[6] It has been estimated that 3 to 6 years of life are lost by diabetics over the age of 65.[4] The risk of early aging is even more prevalent in the population with undiagnosed diabetes.[3] Elderly people with undiagnosed diabetes may experience sudden blindness, heart failure, and the development of Alzheimer's disease or other dementia-related diseases. On the one hand, an increase in the incidence of diabetes is due to the aging of the population, particularly in Europe, but diabetes and the associated oxidative stress itself also induces aging in the younger population.

TYPE 2 DIABETES AND AGING

Diabetes mellitus is a chronic metabolic disorder caused by an improper balance of glucose homeostasis. It has been characterized by relative insulin deficiency due to impaired insulin production combined with peripheral insulin resistance.[2] The primary cause of fasting hyperglycemia is due to an elevated rate of basal hepatic glucose production in the presence of hyperinsulinemia; by contrast, postprandial (after a meal) hyperglycemia is due to the impaired suppression of hepatic glucose production by insulin and decreased insulin-mediated glucose uptake by muscle.[7] 'Metabolic syndrome' or 'Syndrome X' associated with diabetes and cardiovascular diseases includes hypertension, dyslipidemia, insulin resistance, hyperinsulinemia, glucose intolerance and obesity.[8] Type 2 diabetes patients experience different metabolic and physiologic disorders involving microvascular complications (retinopathy, nephropathy and neuropathy) and macrovascular complications (heart attacks, stroke and peripheral vascular disease).[9] These complications can be induced by the oxidative

Aging
http://dx.doi.org/10.1016/B978-0-12-405933-7.00006-8

stress associated with type 2 diabetes (Fig. 6.1). All of these vascular complications have a significant impact on the aging process, and they act as major determinants of the lifespan in different organisms. The pathophysiology of insulin impairment and the development of type 2 diabetes have a close association with the mechanism and process of aging. Imbalances in glucose homeostasis as a result of type 2 diabetes slowly affect and cause degeneration of different organs, which eventually leads to aging.

Aging involves natural senescence of multiple organ systems, including the kidney, the autonomic nervous system and the heart[10] (Fig. 6.2). Both aging and diabetes are closely associated with the impairment of the autonomic nervous system, including endothelial dysfunction and the impairment of autonomic neurons.[10,11] Damage to autonomic ganglia and the loss of parasympathetic peripheral nerves are the result of vascular endothelial dysfunction. The occurrence and incidence of cardiovascular diseases are more prevalent in diabetic patients compared to their non-diabetic counterparts.[9] Cardiovascular disorders increase the chances of mortality in diabetic patients and slowly induce aging-related complications. Both diabetes and aging involve a loss of function in the cardiovascular system.[10] Structure and function of vasodilators and vasoconstrictors significantly influence the pathophysiology of aging and diabetes. The inability of blood vessels to adequately dilate is a common mechanism in many alterations in aging and diabetes.[10] A reduction in the ability of vascular endothelial cells to produce nitrous oxide (NO), a potent vasodilator, can reduce resting and post-ischemic blood flow in subjects both with aging and diabetes.[12] Under stress conditions, almost 30% of patients with diabetes showed autonomic impairment.[13] Arterial wall stiffening and a decrease in myocardial compliance is associated with the normal aging process.[14] Diabetic patients also manifest arterial stiffening and diastolic dysfunction at a younger age.[14]

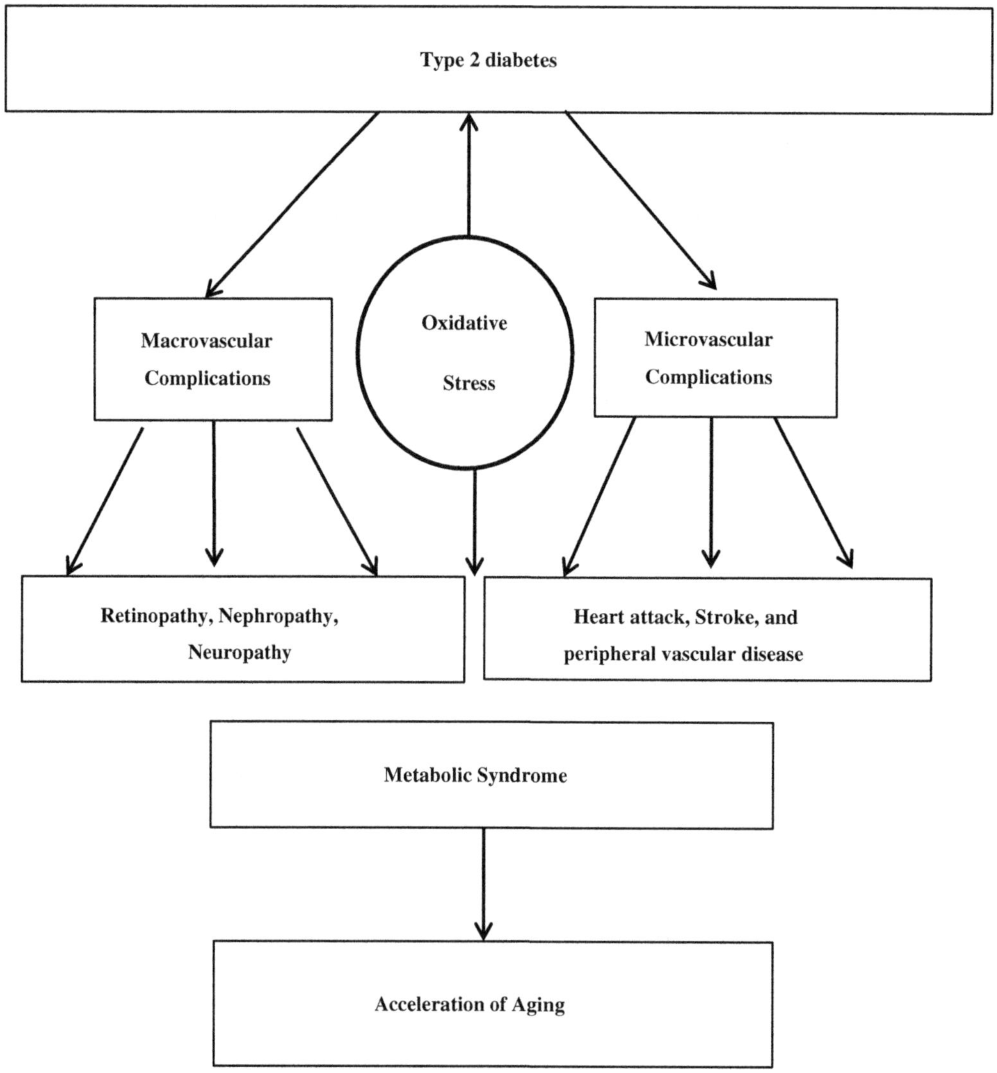

FIGURE 6.1 Acceleration of aging through microvascular and macrovascular complications in diabetic patients involving oxidative stress.[9]

Arterial stiffness results in increased systolic blood pressure, a decrease in diastolic blood pressure, and increased pulse pressure – which eventually lead to the development of isolated systolic hypertension.[14,15] The common occurrence of diastolic dysfunction in patients with diabetes is also due to the progressive reduction in the arterial and myocardial elasticity which accompanies the aging process.[15] The Maillard reaction or advanced glycation of proteins is predominant with normal aging, but occurs at an accelerated rate in diabetic individuals.[14] The loss of vascular and cardiac compliance in aging and diabetes involves collagen cross-linking.[14,16] Non-enzymatic glycosylation of myosin in cardiac and skeletal muscle increased with age in the diabetic rat.[14–18]

Not only cardiovascular diseases, but also cerebrovascular diseases, such as vascular dementia, are more common in older diabetic patients. Patients with chronic diabetes experience not only metabolic disorders – they also suffer with significant cognitive impairment[19] (Fig. 6.3). Type 2 diabetes increases the incidence of cognitive impairment in many patients, and the deficit is more prominent in patients older than 70 years.[20] Glucose can act as a neural fuel for the brain, and thus, breakdown of glucose homeostasis leads to cognitive impairment. Studies have shown that working memory, verbal declarative memory and executive function reduce in older participants, due to impaired glucose tolerance, while a deficit of verbal memory was also found in young participants.[21] In general, cognitive impairment in the diabetic patient increases with age. Diabetes-associated obesity, hypertension and hypocholesterolemia also lead to impaired cognition. As diabetes is closely associated with the incidence of clinical depression, improved glucose tolerance with diabetic treatment can reverse the process and improve cognitive function.[22] Impaired phospholipid metabolism and impaired fatty acid-related signal transduction processes may be the primary cause of depression and may connect with other diseases, such as diabetes and process of aging.[23] The major reason for cognitive dysfunction in diabetic patients is the vascular

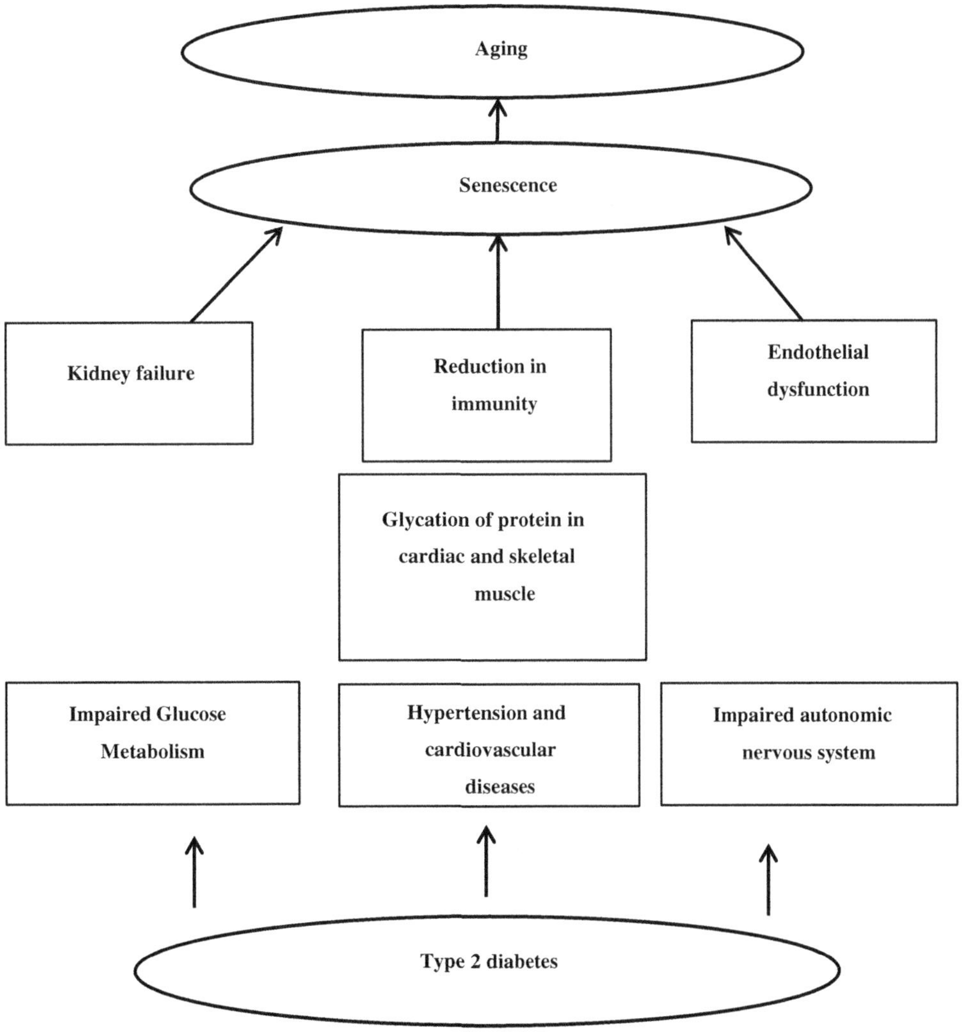

FIGURE 6.2 Interaction and impact of type 2 diabetes in senescence and aging.

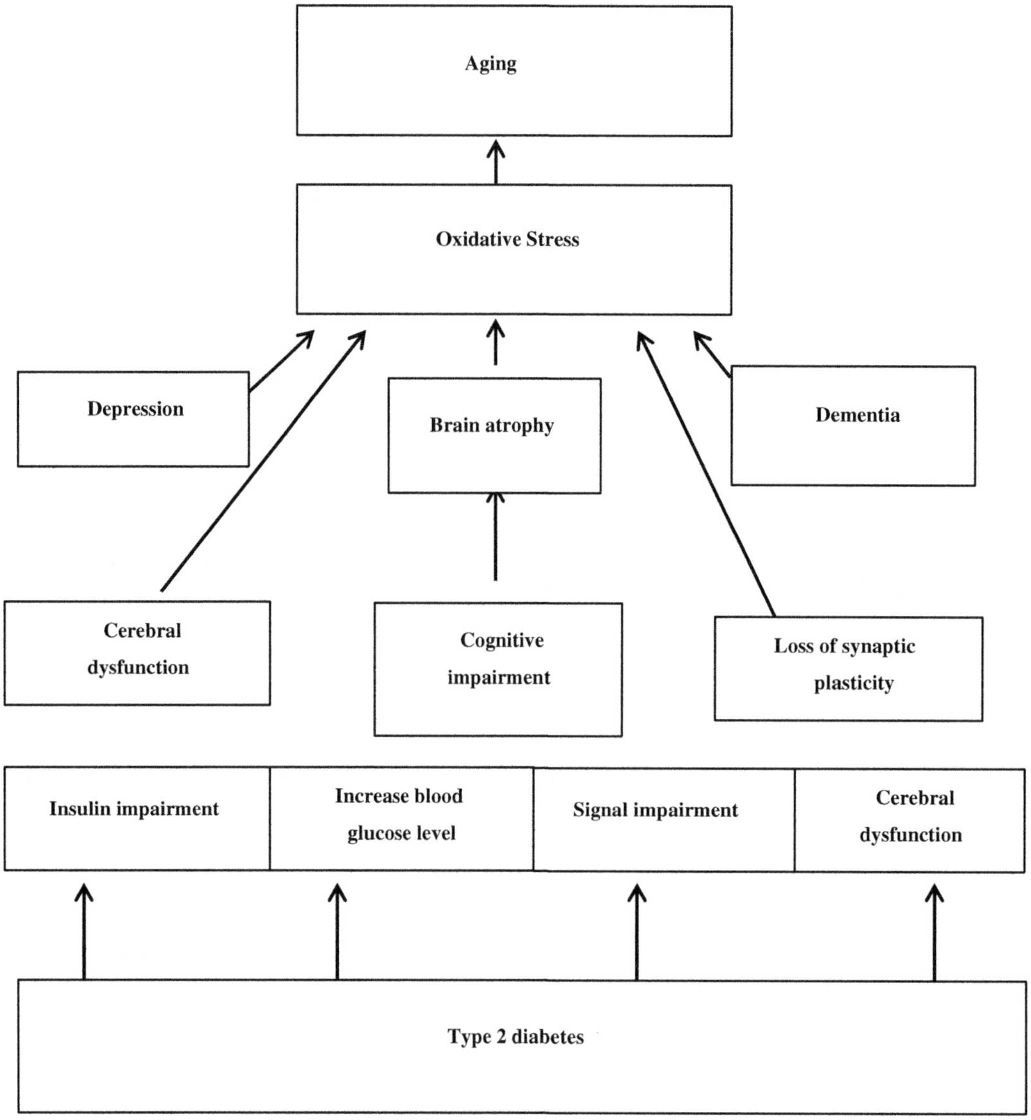

FIGURE 6.3 Cognitive impairment in diabetes and aging.

disease. Better glycemic control improves cognition, and episodes of hyperglycemia can be restricted through cognitive benefits in type 2 diabetes.[22] Diabetes is also linked with brain atrophy. One study reported 40% higher brain atrophy in diabetic patients compared to control participants.[24] Another study has shown that diabetes, in conjunction with hypertension, was associated with cortical brain atrophy. Brain atrophy may be an early indication of impaired glucose tolerance and diabetes.[20]

Diabetes and aging both affect learning and hippocampal synaptic plasticity.[25] Type 2 diabetes develops in different phases in most patients, and cognitive decline is also somewhat associated with these stages. Both glucose tolerance and cognitive function change with aging.[26] Diabetic patients who have experienced slight cognitive decline may become more susceptible to cognitive dysfunction due to impairment in fasting glucose. Poor glucose tolerance is associated with delayed recall and worsening of recognition as well as worsening of verbal fluency and dichotic listening.[26] Insulin resistance and a higher insulin level also influence cognitive deficits. Poorer cognitive performances were observed in cases of insulin resistance and hyperinsulinemia. A decrease in insulin transport in the brain, due to hyperglycemia and obesity-induced diabetes, has been observed in mice.[27] Insulin resistance within the brain may cause neurons to be functionally insulin deficient. Insulin-sensitive glucose transporters in various brain regions also play a significant role in cognitive function. There is a lack of evidence to determine the precise role of insulin sensitivity in cognitive impairment. Diabetes can also affect the central nervous system.[20] Acute cerebral dysfunction was observed in hyper- and hypoglycemic episodes. Neuroimaging has revealed the process of accelerated aging in diabetic patients.[28] The effects of diabetes and aging interact with each other and affect cognition, synaptic plasticity and glutamatergic neurotransmission in rats.[25,28]

Like humans, diabetic rats also show end-organ damage which affects the eyes, kidneys, heart and blood vessels and the peripheral and central nervous system.[27] Insulin-like growth factor has neuroprotective functions in the brain and it also regulates other growth hormones.[29] Impairment in insulin production and function can impede cellular growth in different organs and can lead to senescence related to aging. It has a close association with β-cell dysfunction and the development of insulin resistance.[29] Insulin signaling pathways and signal transduction are also involved in such mechanisms. Chronic diabetes can impair the function of other hormones, and it results in failure of the endocrine system. Endocrine disruption may lead to many aging-related types of pathogenesis. Insulin signaling adjusts the growth rate and regulates the lifespan under different environmental conditions in model organisms.[29] Diabetes also causes immunosenescence and produces progressive deterioration in the ability to respond to infections.[30] A higher prevalence of infectious diseases and the breakdown of the immune system are closely associated with aging and diabetes. Impaired β-cell development and a reduction in immune function can be the result of both insulin impairment and aging. Recent evidence suggests that a gene called SIRT1 is associated with both age-related diseases and the development of type 1 diabetes.[31] Non-enzymatic reaction of carbohydrates resulting in the formation of glycated end products can damage cellular protein, which eventually leads to the pathogenesis of type 2 diabetes and obesity. In normal aging, increasing amounts of advanced glycation end products can be detected, but they increase further in patients with diabetes.[17,18]

Generation of reactive catalysts such as copper (Cu) is common both in normal lens and idiopathic cataract associated with aging.[32] Tissue damage associated with age and diabetes involves such ocular inflammation through metal-catalysed enediol oxidation.[32] Diabetic retinopathy is a leading cause of blindness in the USA; it affects mostly adults of working age. Several mechanisms, including neovascularization – leading to vitreous hemorrhage or retinal detachment, macular edema and retinal capillary nonperfusion – are involved in loss of vision in diabetic patients[33] (Fig. 6.4). Patients suffering with long-standing chronic diabetes along with hypertension mostly experience retinopathy. According to one estimate, 4.1 million people over 40 years of age have diabetic retinopathy, and 1 in 132 persons has vision-threatening diabetic retinopathy in the USA, and it is increasing at an alarming rate across different races and ethnic groups.[33]

THE ROLE OF OXIDATIVE STRESS IN HUMAN DISEASE

The balance between ROS formation and antioxidant molecules and associated cellular enzymatic systems is critical to maintaining cellular homeostasis. As a part of aerobic life, animal cells are susceptible to oxidative stress. The pathophysiology of many human diseases involves the generation of ROS and oxidative damages to the cells and tissues.[34] A demand for excessive cellular energy (ATP) usually results in incomplete reduction of oxygen in the mitochondria and the generation of ROS. While an oxidizing environment favors cell death, a reducing environment favors cell proliferation. Environmental stress, pathogenic attack and oxidative stress-related diseases can easily break down redox homeostasis in the cell and can induce dysfunction of oxidative phosphorylation in the mitochondria. Perturbation of the pro-oxidant/antioxidant balance can lead to the alteration of cellular function and oxidative damage to macromolecules (lipids, proteins or DNA). Degeneration of cellular organelles due to oxidative stress leads to the severity and complexity of many noncommunicable chronic diseases, including diabetes and associated aging processes. The pathogenesis of vascular degeneration involves oxidative stress either by triggering or exacerbating the biochemical processes accompanying the metabolic syndrome.[35] Chronic antioxidant deficiency may favor the propagation of oxidative alterations from intra- to extracellular spaces and from confined to distant sites, and induce a systemic oxidative stress state. ROS not only damage cellular structures, but also contribute to cellular aging, mutagenesis, carcinogenesis, coronary heart disease and apoptosis.[36] Oxidative stress plays a major role in damage to cells and tissues associated with diabetes and aging.

OXIDATIVE STRESS IN TYPE 2 DIABETES AND AGING

Oxidative stress plays a key role in the development of diabetes and aging, particularly in the pathogenesis of vascular complications.[10] Excess ROS production significantly influences insulin signal transduction and insulin resistance.[34,37] Failure in glucose homeostasis can result in a multi-symptom disorder encompassing obesity, hyperglycemia, impaired glucose tolerance, hypertension and dyslipidemia, generally known as metabolic syndrome.[38] Type 2 diabetes is the leading cause of blindness and end-stage renal failure, and the risk of heart disease and stroke are two to four times more frequent in a person with diabetes.[37] There is a lack of evidence to establish the exact role of type 2 diabetes in the development of aging. However, extension of the

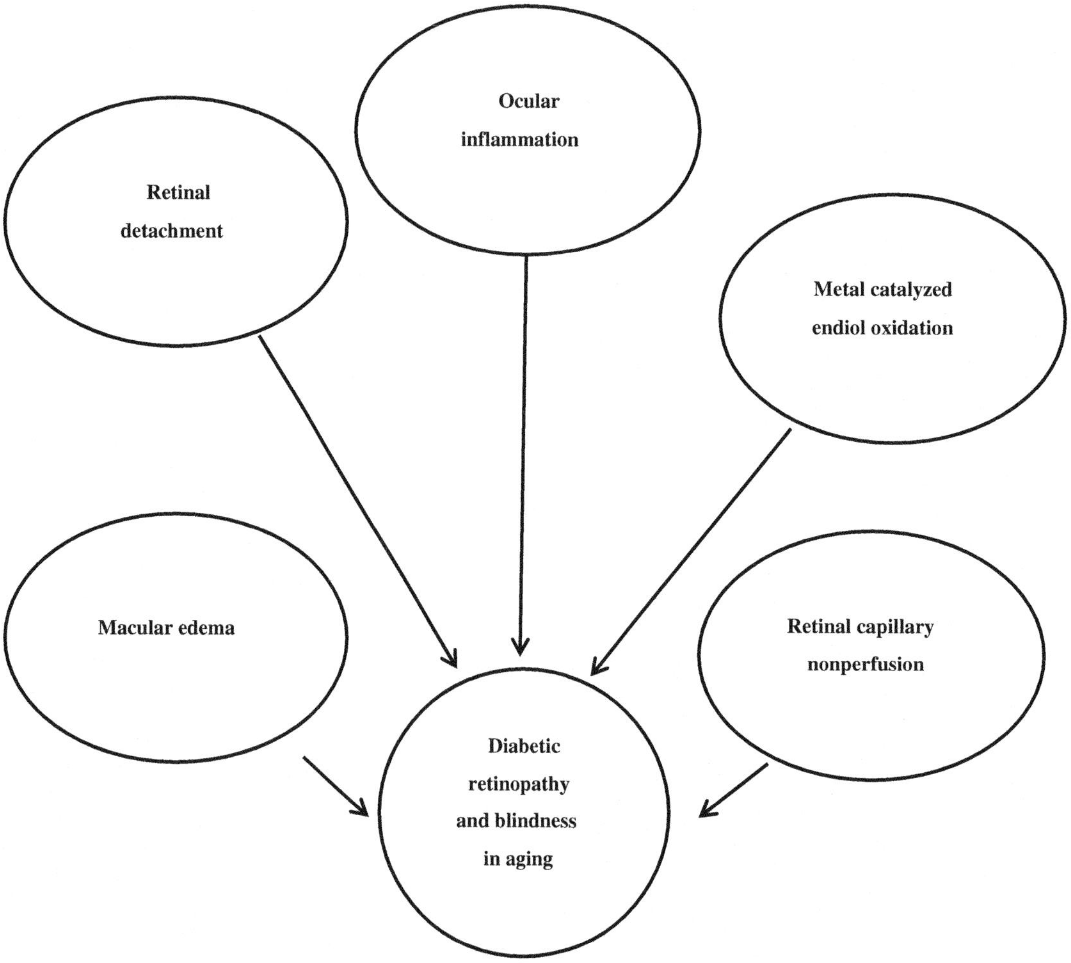

FIGURE 6.4 Mechanism of diabetic retinopathy and aging.[33]

lifespan has been observed with genetic manipulation of insulin signaling in many species.[29] Better regulation of glucose with dietary antioxidants – such as resveratrol, and potentially other dietary phenolics – protects cells against oxidative stress and helps to manage age-related diabetes.

The generation of excess free radicals is one reason for the prevalence of hypertension in diabetes.[39] Studies have shown that the simultaneous reduction of the antioxidant defense system is the important risk factor for cardiovascular complications and the increased rate of mortality in the older population.[40,41] Hyperglycemia triggers oxidation-linked stress in capillary endothelial cells and alters the transportation of glucose inside the cell. A positive correlation exists between mitochondrion-linked oxidative stress and insulin resistance.[42,43] Multiple pathways lead to insulin resistance, and the alteration of several of these pathways is required before insulin resistance becomes clinically evident. Mitochondrial oxidative stress could be a key element for the development of insulin resistance as it results from the failure of metabolic regulation in times of higher energy demand and leads to aging.[34] The high nutrient intake pattern, particularly from soluble carbohydrates, or overload with such macronutrients, can cause excessive mitochondrial oxidation and enhanced ROS formation. Inflammatory states, endoplasmic reticulum stress and endocrine dysregulation are multiple processes involved in obesity-associated oxidative stress that is also associated with such excessive macronutrient intake.[34] An increase in redox catalysts and substances prone to generation of peroxide and free radicals, such as monosaccharides, may cause inappropriate oxidation in both diabetes and aging. Insulin signaling mediates starvation-induced dauer formation and longevity in *Caenorhabditis elegans*.[29] Calorie restriction and periodic starvation have been associated with lifespan extension in many organisms.

In addition, it is suggested that malnutrition or nutrient deficiency in the early stage of life is also associated with the later, more rapid, development of glucose impairment and the impact on subsequent cardiovascular disease under a high intake of soluble carbohydrate.[37,38,44] Many studies have shown that early exposure to famine,

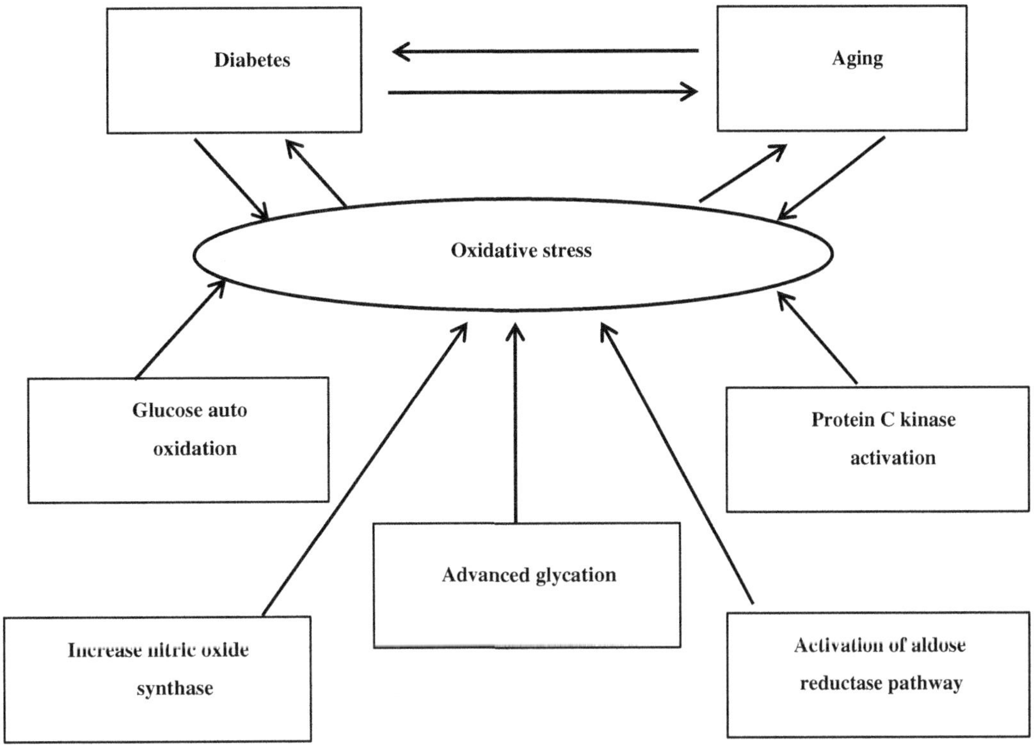

FIGURE 6.5 The role of oxidative stress in diabetes and aging.[38]

or a lower birthweight, can lead to a higher rate of impaired glucose tolerance and type 2 diabetes, which has a significant role in aging too.[37] The generation of ROS in a stress-mediated environment may result from the reduced activity of electron transport complexes in the mitochondria. Altered mitochondrial function in a limited energy environment can have a deleterious effect on cells, such as β-cells, that have a higher energy requirement. Damage to β-cells can have long-term implications and can lead to insulin resistance and type 2 diabetes, respectively.[37]

Dysfunction of β-cells – due either to pre-exposure to nutrient deficiency or over-nutrition with refined carbohydrate and fat intake in the later stages of life – is strongly associated with oxidative stress and ROS formation. In the diabetic state, such oxidative stress decreases insulin biosynthesis and secretion, leading to the aggravation of type 2 diabetes and acceleration of the process of aging.[38] Studies have shown that increased fat intake can shift the cellular redox environment to a more oxidized state in the skeletal muscles.[37] High consumption of fat and prolonged exposure to fatty acids can result in increased ROS production in β-cells and the development of hepatic insulin resistance.[37] NADPH oxidase is also involved in the generation of higher levels of ROS in response to saturated fatty acids in muscle cells.[38]

Several mechanisms – including glucose auto-oxidation, advanced glycation end-product formation, abnormal arachidonic acid metabolism and its coupling to cyclooxygenase catalysis, protein kinase C activation and the increase in activity of nitric oxide (NO) synthase and activation of the aldose reductase pathway – are involved in oxidative stress-associated hyperglycemia, and all of these facilitate aging in both younger and older populations[38] (Fig. 6.5). In hyperglycemia, the mitochondrial electron transport chain generates a larger amount of superoxides and NO, which ultimately induce the formation of strong oxidants such as peroxynitrite and subsequent DNA damage.[45] These processes result in an acute endothelial dysfunction in diabetic blood vessels and significantly contribute to the development of long-term cardiovascular disease – particularly in older adults.[46]

The oxidation of low-density lipoprotein (LDL) is another complication associated with hyperglycemia. A higher level of blood glucose causes glycation or glycoxidation of LDL, which increases serum oxidative stress, lowering the capacity of high-density lipoprotein (HDL) to protect against oxidation of LDL.[47] Cholesterol accumulation in diabetic patients may result from the arterial cell oxidative stress caused by hyperglycemia.[45] Insulin can act as a vasodilator and is thus characterized as 'anti-atheromatic' and anti-inflammatory; a balanced insulin supply and signaling can reduce vascular complications. Mitigating vascular diseases is key to restricting the aging process in diabetic patients. Hyperinsulinemia-induced atherosclerosis increases the production of ROS and reduces insulin sensitivity. The anti-inflammatory properties of insulin are also related to pro-inflammatory cytokines.[48] Different pathways – such as non-enzymatic glycosylation and the electron transport chain in oxidative phosphorylation – are involved in the generation of ROS. Continuous exposure to oxidative stress creates glucose toxicity and can potentially cause the failure of different organs through tissue damage in diabetic patients.[45] The development of type 2 diabetes is a multistage process and thus requires systematic preventative measures such as a balanced diet, exercise, and a healthy lifestyle to control this epidemic. The pathogenesis of diabetes is closely related to that of aging, but it also involves very complex mechanisms and needs extensive investigation to identify the biochemical rationale more accurately.

CONCLUSION

With the significant increase in health care costs, the severity of diabetes has a major impact on the elderly population worldwide. The burden is not just on health: it also disturbs the economic and social stability of the disease-affected population. Type 2 diabetes-induced aging is associated with oxidative stress, which can seriously affect healthy lifestyle. This can act as a major cause of health disparities in many communities. Overall it is clear that combinations of diet, lifestyle and the environment have major roles in the further pathogenesis of type 2 diabetes-related aging. This requires special attention for the development of dietary disease-intervention programs. Functional food with beneficial bioactives can help to mitigate many type 2 diabetes-induced physiologic and biochemical irregularities in diabetic patients. Food with a high level of dietary antioxidants can slow down the cellular degeneration and potentially restrict the process of aging. A higher intake of fruits and vegetables is inversely correlated with the development of risk for diabetes and aging. Dietary management, along with a healthy lifestyle, can prevent the early development of type 2 diabetes and may slow the process of aging by providing protection against oxidative stress and by maintaining cellular homeostasis.

SUMMARY POINTS

- Type 2 diabetes, along with cardiovascular diseases, is prevalent in aging populations worldwide and significantly increases the chances of mortality.
- Glucose impairment accelerates cellular degeneration and promotes the process of aging.
- Different macrovascular and microvascular complications are closely associated with both type 2 diabetes and aging.
- Type 2 diabetes not only impairs metabolic regulation, it also damages cognitive functions.
- The pathophysiology of both type 2 diabetes and aging involves the generation of ROS and the development of oxidative stress in cellular compartments.
- Patients with chronic diabetes experience impairment in different organs, including kidney, heart, bone and nervous system.
- Prevention and better management of type 2 diabetes, with better management of oxidative stress, is important for healthy living with increasing life expectancy worldwide.

References

1. WHO Fact Sheet. http://www.who.int/mediacentre/factsheets/fs312/en/; Sept, 2012.
2. Green A, Hirsch NC, Krøger PS. The changing world demography of type 2 diabetes. *Diabetes/Metab Res Rev* 2003;**19**:3–7.
3. International Diabetes Federation. *Diabetes and aging: hand in hand.* http://www.idf.org/diabetes-and-aging-hand-hand; April, 2012.
4. Fagot-Campagna A, Bourdel-Marchasson I, Simon D. Burden of diabetes in an aging population: prevalence, incidence, mortality, characteristics and quality of care. *Diabetes Metab* 2005;**31**: 5S35–5S52.
5. Shaw JE, Sicree RA, Zimmet PZ. Global estimates of the prevalence of diabetes for 2010 and 2030. *Diab Res Clin Pract* 2010;**87**:4–14.
6. Halter JB. Diabetes mellitus in an aging population: the challenge ahead. *J Gerontol* 2012;**67**:1297–9.
7. DeFronzo RA. Pharmacologic therapy for type 2 diabetes mellitus. *Ann Intern Med* 1999;**131**:281–303.
8. Gluckman PD, Hanson MA. The developmental origins of metabolic syndrome. *Trends Endocrinol Metab* 2004;**15**:183–7.
9. Klein R. Hyperglycemia and microvascular and macrovascular disease in diabetes. *Diab Care* 1995;**18**:258–68.
10. Petrofsky J, Lee S, Cuneo M. Effects of aging and type 2 diabetes on resting and post occlusive hyperemia of the forearm; the impact of Rosiglitazone. *BMC Endocr Disord* 2005;**5**:1–7.
11. Accurso V, Shamsuzzaman AS, Somers VK. Rhythms, rhymes and reasons – spectral oscillation in neural cardiovascular control. *Auton Neurosci* 2001;**20**:41–6.
12. Lacolley PJ, Lewis SJ, Brody MJ. Role of sympathetic nerve activity in the generation of vascular nitric oxide in urethane-anesthetized rats. *Hypertension* 1991;**17**:881–7.
13. Petrofsky JS, Besonis C, Rivera D, et al. Heat tolerance in patients with diabetes. *J Appl Res* 2003;**3**:28–34.
14. Aronson D. Cross-linking of glycated collagen in the pathogenesis of arterial and myocardial stiffening of aging and diabetes. *J Hypertens* 2003;**21**:3–12.
15. Avolio AP, Chen SG, Wang RP, et al. Effects of aging on changing arterial compliance and left ventricular load in northern Chinese urban community. *Circulation* 1983;**68**:50–8.
16. Paul RJ, Bailey AJ. Glycation of collagen: the basis of its central role in the late complications of aging and diabetes. *Int J Biochem Cell Biol* 1996;**28**:1297–310.
17. Brownlee M, Cerami A, Vlassara H. Advanced glycosylation end products in tissue and the biochemical basis of diabetic complications. *New Engl J Med* 1988;**318**:1315–21.
18. Nass N, Bartling B, Navarette Santos A, et al. Advanced glycation end products, diabetes and aging. *Z Gerontol Geriat* 2007;**40**: 349–56.

19. Awad N, Gagnon M, Messier C. The relationship between impaired glucose tolerance, type 2 diabetes, and cognitive function. *J Clin Exp Neuropsychol* 2004;**26**:1044–80.
20. Messier C. Impact of impaired glucose tolerance and type 2 diabetes on cognitive aging. *Neurobiol Aging* 2005;**26S**:S26–30.
21. Messier C, Tsiakas M, Gagnon M, et al. Effect of age and glucoregulation on cognitive performance. *Neurobiol Aging* 2003;**24**:985–1003.
22. Biessels GJ, Van der Heide LP, Kamal A, et al. Aging and diabetes: implications for brain functions. *Eur J Pharmacol* 2002;**441**:1–14.
23. Horrobin DF, Bennett CN. Depression and bipolar disorder: relationship to impaired fatty acid and phospholipid metabolism and to diabetes, cardiovascular disease, immunological abnormalities, cancer, aging and osteoporosis. Prostaglandins. *Leukot Essent Fatty Acids* 1999;**60**:217–34.
24. Araki Y, Nomura M, Tanaka H, et al. MRI of brain in diabetes mellitus. *Neurobiology* 1994;**36**:101–3.
25. Kamal A, Biessels GJ, Duis SEJ, Gispen WH. Learning and hippocampal synaptic plasticity in streptozotocin-diabetic rats: interaction of diabetes and aging. *Diabetologia* 2000;**43**:500–506.
26. Artola A. Diabetes, stress- and aging-related changes in synaptic plasticity in hippocampus and neocortex – the same metaplastic process? *Eur J Pharmacol* 2008;**585**:153–62.
27. Lamport DJ, Lawton CL, Mansfield MW, Dye L. Impairments in glucose tolerance can have negative impact on cognitive function: a systematic research review. *Neurosci Biobehav Rev* 2009;**33**(3): 394–413.
28. Segovia G, Porras A, Del Arco A, Mora F. Glutamatergic neurotransmission in aging: a critical perspective. *Mech Aging Develop* 2001;**122**:1–29.
29. Hafen E. Cancer, type 2 diabetes, and aging: news from flies and worms. *Swiss Med Weekly* 2004;**134**:711–9.
30. Aw D, Silva AB, Palmer DB. Immunosenescence: emerging challenges for aging population. *Immunology* 2007;**120**:435–46.
31. Chang HC, Guarente L. SIRT1 mediates central circadian control in the SCN by a mechanism that decays with aging. *Cell* 2013; **153**:1448–60.
32. Hunt JV, Dean RT, Wolff SP. Hydroxil radical production and autooxidative glycosylation. *Biochem J* 1988;**256**:205–12.
33. Eye Disease Prevalence Research Group. The prevalence of diabetic retinopathy among adults in the United States. *Arch Opthalmol* 2004;**122**:552–63.
34. Bisbal C, Lambert K, Avignon A. Antioxidants and glucose metabolism disorders. *Curr Opin Clin Nutr Metab Care* 2010;**13**:439–46.
35. Hanhineva K, TörröneN R, Bondia-Pons I, et al. Impact of dietary polyphenols on carbohydrate metabolism. *Int J Mol Sci* 2010;**11**:1365–402.
36. Kaneto H, Katakami N, Kawamori D, et al. Involvement of oxidative stress in pathogenesis of diabetes. *Antioxid Redox Signal* 2007;**9**:355–66.
37. Anderson EJ, Lustig ME, Boyle KE, et al. Mitochondrial H_2O_2 emission and cellular redox state link excess fat intake to insulin resistance in both rodents and humans. *J Clin Invest* 2009;**119**: 573–81.
38. Inoguchi T, Sonta T, Tsubouchi H, et al. Protein kinase C-dependent increase in reactive oxygen species (ROS) production in vascular tissues of diabetes: role of vascular NAD(P)H oxidase. *J Am Soc Nephrol* 2003;**14**. S277–S232.
39. Gey KF, Puska P, Jordan P, Moser UK. Inverse correlation between plasma vitamin E and mortality from ischemic heart disease in cross-cultural epidemiology. *Am J Clin Nutr* 1991;**53**:326S–34S.
40. Baynes JW. Role of oxidative stress in development of complications in diabetes. *Diabetes* 1991;**40**:405–12.
41. Ceriello A, Quatraro A, Giugliano D. New insight of non-enzymatic glycosylation may lead to therapeutic approaches for the prevention of diabetic complications. *Diabetic Med* 1992;**9**:297–9.
42. Ceriello A, Quatraro A, Giugliano D. Diabetes mellitus and hypertension: the possible role of hyperglycemia through oxidative stress. *Diabetologia* 1993;**36**:265–6.
43. Hoehn KL, Salmon AB, Hohnen-Behrens C, et al. Insulin resistance is a cellular antioxidant defense mechanism. *Science* 2009;**106**:17787–92.
44. Simmons RA. Developmental origins of diabetes: the role of oxidative stress. *Best Practice Res Clin Endocrinol Metab* 2012;**26**:701–8.
45. Ceriello A. New insights on oxidative stress and diabetic complications may lead to a 'causal' antioxidant therapy. *Diabetes Care* 2003;**26**:1589–96.
46. Ceriello A. Postprandial hyperglycemia and diabetes complications: is it time to treat. *Diabetes* 2005;**54**:1–7.
47. Kaplan M, Aviram M. Oxidized low density lipoprotein: atherogenic and proinflammmatory characteristics during macrophage foam cell formation. An inhibitory role for nutritional antioxidants and serum paraoxonase. *Clin Chem Lab Med* 2005;**37**:777–87.
48. Albacker T, Carvalho G, Schricker T, Lachapelle K. High-dose insulin therapy attenuates systematic inflammatory response in coronary artery bypass grafting patients. *Ann Thoracic Surgery* 2008;**86**:20–7.

SECTION 2

ANTIOXIDANTS AND AGING

CHAPTER

7

Oxidative Stress and Antioxidants in Elderly Women

Brunna Cristina Bremer Boaventura, Patricia Faria Di Pietro
Department of Nutrition, Health Sciences Center, Federal University of Santa Catarina, Campus Trindade, Florianópolis/SC, Brazil

List of Abbreviations

CAT catalase
FRAP ferric reducing ability of plasma
GPx glutathione peroxidase
GSH reduced glutathione
H_2O_2 hydrogen peroxide
IL-6 interleukin 6
MDA malondialdehyde
NADPH nicotinamide adenine dinucleotide phosphate
ORAC oxygen radical absorbance capacity
RDW red cell distribution width
SOD superoxide dismutase
TAS total antioxidant status
TBARS thiobarbituric acid-reactive substances
TEAC trolox equivalent antioxidant capacity

INTRODUCTION

The aging process is associated with biochemical alterations and an increase in free radical production caused by mitochondrial dysfunction.[1] The main theory postulated is that damage caused to DNA, protein and lipids by free radicals increases on aging and that this damage accumulates and contributes to senescence.[2] A decrease in the length of telomeres has also been proposed as a biomarker of biologic aging and is favored by mitochondrial dysfunction.[3,4] These age-related changes lead to an increase in oxidative stress, which is believed to be associated with many pathologic conditions and disease processes in elderly persons.[5]

The period after menopause may be considered the beginning of the aging process in women. Menopause results from a series of endocrinologic variations caused by declining estrogen levels.[6] Once women begin ovarian senescence, estrogen production becomes irregular, antioxidant protection is decreased, and oxidative stress is assumed to increase (Figs 7.1 and 7.2).[6] Lower exposure to endogenous estrogen may be associated with a higher risk of age-related disease in elderly women.[7]

Researchers have investigated whether oxidative stress, as implied by oxidative damage to proteins, in elderly women is associated with greater mortality.[8] The study was conducted with 746 moderately to severely disabled elderly women, aged 65 and older, from the Women's Health and Aging Study I.[8] It was reported that greater oxidative stress, as indicated by elevated serum protein carbonyl concentrations, was associated with a greater risk of death in the sample of elderly women studied.[8] Oxidative damage to proteins can lead to a loss of the structural integrity of cells, compromise cellular function, and increase susceptibility to proteolysis.[2]

Considering that diseases related to oxidative stress are a major cause of death in old age, it is important to develop effective prevention strategies. Because diet can modulate levels of oxidative stress,[9] the consumption of dietary antioxidants may contribute to healthy aging in women.

ROLE OF ESTROGEN IN OXIDATIVE STRESS AND ANTIOXIDANT DEFENSE IN ELDERLY WOMEN

Estrogens have been reported to lower vascular oxidative stress by modulating the expression and function of NADPH oxidases as well as antioxidant enzymes, such as SOD, GPx and CAT, providing protection against oxidative stress during the reproductive stage.[6] However, although estrogens are known to exert an antioxidant effect, they are poor free radical scavengers.[10] Considering that menopause is linked to the loss of estrogen production by the ovaries, and that estrogen has a regulatory role on many organs, the age-related deficit of this hormone

Aging
http://dx.doi.org/10.1016/B978-0-12-405933-7.00007-X

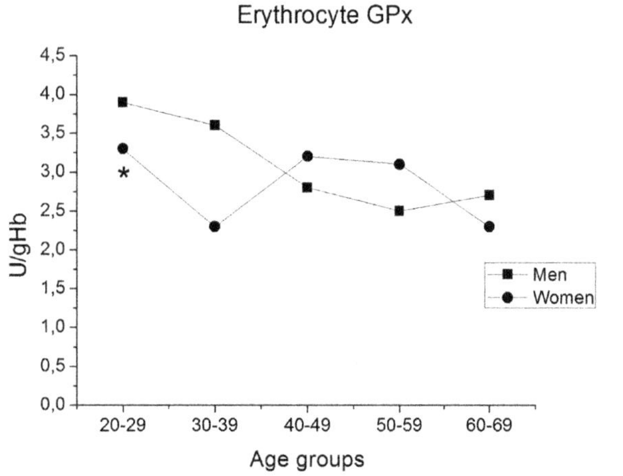

FIGURE 7.1 Variations in the mean values of the erythrocyte GPx activities as a function of age for men and women.[43] *P <0.05, significantly different from age group 60–69. Reprinted from Kasapoglu M and Ozben T, Experimental Gerontology 2001, 36(2), 209-220, Copyright 2013, with permission from Elsevier.

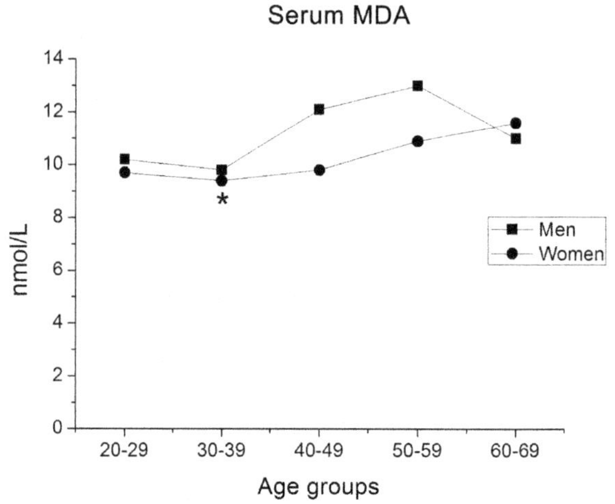

FIGURE 7.2 Variations in the mean values of the serum MDA levels, a marker of lipid peroxidation, as a function of age for men and women.[43] *P <0.05, significantly different from age group 50–59. Reprinted from Kasapoglu M and Ozben T, Experimental Gerontology 2001, 36(2), 209-220, Copyright 2013, with permission from Elsevier.

triggers many physiopathologic reactions.[11] These reactions include degenerative processes, such as acceleration of arteriosclerosis and skin aging, which may impair the health of older women.[11] Moreover, the postmenopausal decline in circulating estrogen is accompanied by a deterioration in the functioning of the immune system with the overproduction of proinflammatory cytokines.[12]

The mechanisms of the processes observed in postmenopausal women are linked not only to estrogen loss but also to high levels of oxidative stress, which play a key causal role in the menopause-related symptoms and pathologic processes and are often accompanied by glutathione deficiency and health loss.[11]

The influence of menopause as a risk factor for oxidative stress has been analyzed.[6] A study was conducted with 94 premenopausal and 93 postmenopausal women. Using logistic regression to control pro-oxidant variables, it was found that menopause was the main risk factor for oxidative stress. This study showed that the extent of lipid peroxidation is increased, and that the activity of the antioxidant enzyme GPx and the total antioxidant status (TAS) are decreased, in postmenopausal women compared with premenopausal women.[6] Table 7.1 shows the degree of oxidative stress related to menopause status, where a statistically significant increase in the percentage of severe oxidative stress in postmenopausal women was found in comparison with the premenopausal women.

TABLE 7.1 Degree of Oxidative Stress Related to Menopause Status[6]

Degree of Oxidative Stress	Premenopausal Women (n = 94)	Postmenopausal Women (n = 93)	OR (95% CI)	P
Without	27 (29%)	17 (18%)	1	
Low	43 (46%)	35 (38%)	1.29 (0.57–2.94)	0.503
Moderate	16 (17%)	25 (27%)	2.48 (0.95–6.55)	0.039
Severe	8 (9%)	16 (17%)	3.18 (1.00–10.35)	0.027

χ^2 for trend = 7.26, P = 0.007.
OR, odds ratio.
Reprinted from Sánchez-Rodríguez MA et al, Menopause (the Journal of the North American Menopause Society) 2012, 19(3), 361-367, Copyright 2013, with permission from Wolters Kluwer Health.

ROLE OF TELOMERE LENGTH IN OXIDATIVE STRESS AND ANTIOXIDANT DEFENSE IN ELDERLY WOMEN

Telomere maintenance might be a key factor in relation to the cumulative effects of genetic, environmental and lifestyle factors involved in aging and aging-related diseases.[13] Various nutrients influence telomere length via mechanisms that reflect their role in cellular functions, including DNA repair and chromosome maintenance, DNA methylation, inflammation, oxidative stress and the activity of the enzyme telomerase (Fig. 7.3). Damage to telomeric DNA due to either oxidative stress or nucleotide precursors results in shorter telomeres.[14] In this context, antioxidant nutrients can reduce the erosion of telomeres through the potential to influence the regulation of telomere length.[14]

Although gene expression of oxidative stress and telomere shortening may be epiphenomena of an underlying aging process, allelic variation in oxidative stress genotypes is fixed and may contribute to individual differences in both telomere length and biologic aging.[15] An

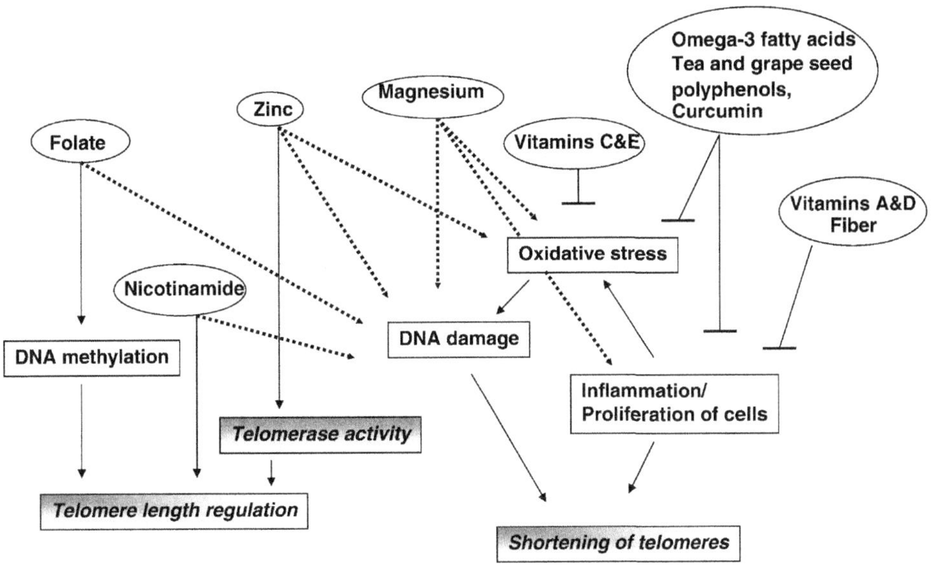

FIGURE 7.3 Potential mechanisms behind the influence of nutrients on telomere length.[14] Dietary nutrients and oxidative stress influence telomere length by various mechanisms that reflect their role in cellular functions. The dashed line indicates the effect of deficiency of a nutrient. Reprinted from Paul L, Journal of Nutritional Biochemistry 2011, 22(10), 895-901, Copyright 2013, with permission from Elsevier.

investigation of elderly persons demonstrated that oxidative genes that are associated with shorter telomeres are also associated with the impairment of physical biomarkers of aging.[15] The authors hypothesized that pathways involving altered Cu/Fe metabolism, glutathione transferase and mitochondrial function are likely to be centered on sirtuins, providing a possible functional link between redox biology and telomere biology.[15] Hence, associations between telomere length and physical biomarkers of aging may, in part, reflect cellular redox status as underlying common causes.[15]

The relationship between estimated endogenous estrogen exposure and telomere length was evaluated in a sample of postmenopausal women at risk of cognitive decline.[7] The authors found that greater endogenous estrogen exposure, as measured by a longer duration of reproductive years, was related to longer telomere length.[7]

The telomerase molecule, which synthesizes telomeres, contains an estrogen-responsive element in its promoter region, which determines that telomerase activity is higher in females than in males.[16] Longer telomeres in women have been ascribed to the ability of estrogen to upregulate telomerase and at the same time reduce oxidative stress.[7] In addition, it has been observed that telomerase is also regulated by glutathione.[16]

DIETARY ANTIOXIDANT THERAPIES IN ELDERLY WOMEN

Because diet can modulate levels of oxidative stress, and the consumption of dietary antioxidants, provided by supplements or foods, may contribute to healthy aging in women, intervention studies were conducted to evaluate the benefits of antioxidant intake in relation to oxidative stress biomarkers.

Considering that telomere length may be a marker of biologic aging, and also that multivitamin supplements, which contain several antioxidant micronutrients, may affect telomere length by modulating oxidative stress, a study was conducted to examine whether multivitamin use is associated with longer telomeres in the women involved in what has been called the Sister Study's Women.[4] Cross-sectional analysis performed with 586 women, between 35 and 74 years of age, indicated that the average telomere length was shorter in older women in comparison with younger women. However, in older women the average telomere length increased with a weekly multivitamin intake when compared to older women who did not take multivitamins. The researchers observed an association between multivitamin use and longer telomeres. Additionally, it was verified that higher intakes of vitamin C and E from foods were each associated with longer telomeres. These results provided the first epidemiologic evidence that multivitamin use is associated with longer telomere length in women.[4]

Considering that antioxidant nutrients contribute to the protection afforded by fruits, vegetables and red wine against diseases of aging, a study was conducted to investigate the response, in terms of serum total antioxidant capacity, following the consumption of strawberries (240 g), spinach (294 g), red wine (300 ml) or vitamin C (1250 mg) in elderly women.[17] It was shown that the total antioxidant capacity of serum determined as ORAC, TEAC and FRAP increased significantly by 7–25% during the 4-hour period following the consumption of red wine, strawberries, vitamin C or spinach.[17] It was suggested by the authors that the consumption of antioxidant foods such as strawberries, spinach or red wine, which are rich in phenolic compounds, can increase the serum antioxidant capacity in elderly women.[17]

The current consumption by elderly persons of a series of 26 common antioxidant-rich foods in relation to serum TAS has been evaluated.[18] This investigation was conducted with older men and women enrolled in the Quebec Longitudinal Study on Nutritional and Successful Aging, *NuAge*. Although women had a lower total energy intake, they ingested more antioxidant foods and antioxidant supplements than did the men. In addition, the number of different antioxidant foods consumed daily was positively correlated with serum TAS only in women.[18] The TAS biomarker is considered to be an overall indication of the body's capacity to eliminate reactive oxygen species and thus it is used to estimate the antioxidant/pro-oxidant balance and oxidative stress.[18] These findings highlight the relation between the daily consumption of easily accessible fruits and vegetables, whole grains and other common food items and an improvement in the antioxidant status of elderly women.[18]

In another study the relationship between food intake and antioxidant biomarkers in elderly women was investigated and a positive association between plasmatic thiol levels and carotenoid-rich vegetables was observed.[19] Thiols are known to be efficient antioxidants due to their ability to react with free radicals, protecting cells against oxidative damage.[20] An important component in the analysis of thiols is GSH, which is the most important and representative intracellular antioxidant. It has been shown that age is associated with the generation of reactive species and also with GSH homeostasis disorders.[21]

Antioxidant nutrients, including selenium and the carotenoids, are important constituents of the antioxidant defense system. Selenium is an important component of antioxidant defense as it is a component of selenoproteins such as GPx, and carotenoids are found primarily in fruits and vegetables and act as antioxidants.[22] In the Women's Health and Aging Studies I and II, the authors observed that low serum selenium and total carotenoid concentrations are associated with an increased risk of death in older women living in the community.[22] The underlying biologic mechanism by which low serum selenium and carotenoid levels could contribute to an increased risk of mortality may be related to increased oxidative stress in elderly women.[22] The relationship between the fruit and vegetable intake, physical activity levels and all-cause mortality of the same elderly population evaluated in the Women's Health and Aging Studies I and II has also been investigated.[23] It was found that low total serum carotenoids, associated or not with low physical activity, strongly predicted earlier mortality. In this context, an enhancement in the fruit and vegetable intake, i.e. foods which are rich sources of carotenoids, could improve the survival of elderly women.[23]

The consumption of dietary carotenoids has also been associated with a decrease in oxidative stress in postmenopausal women who underwent breast cancer treatment.[24] In fact, postdiagnosis increases in fruit and vegetable intake, which result in greater total plasma carotenoid concentrations, may offer protection against recurrent breast cancer disease, partially due to a reduction in oxidative stress.[25]

Considering the role of carotenoids in oxidative stress, the protective effect of carotenoid supplementation against DNA damage in postmenopausal women was investigated.[26] DNA damage is an important biomarker of oxidative stress and is positively associated with age. Before starting the carotenoid supplementation, the subjects of the study consumed a low fruit and vegetable diet for 2 weeks, which promoted a decrease in plasma carotenoid concentrations. The women studied were divided into five treatment groups: placebo, mixed carotenoids, lutein, β-carotene and lycopene. At the end of the study, all carotenoid-supplemented groups showed significantly lower endogenous DNA damage than at baseline, whereas the placebo group did not show any significant change. It was demonstrated that either combined or isolated carotenoids resulted in beneficial effects against DNA damage in older women.[26]

Moreover, low concentrations of serum carotenoids were strongly associated with the frailty status of older women.[27] Frailty is a geriatric syndrome characterized by a multisystem reduction in physiologic reserve and vulnerability to stressors.[27] This condition is associated with weight loss, weakness, low exercise tolerance, slow walking speed and low physical activity.[27] These factors may contribute to functional impairment because they increase susceptibility to fractures.

The oxidative stress that underlies physiologic aging is a pivotal pathogenic mechanism of age-related loss of bone tissue and strength.[28] Loss of estrogens also accelerates the effects of aging on bone tissue by decreasing its defense against oxidative stress.[28] Thus, osteoporosis is a major public health problem that is characterized by a decrease in bone mass associated with aging, especially in postmenopausal women.[29] Women are particularly affected by osteoporosis owing to the important role that estrogens play in bone turnover, as they stimulate bone formation and inhibit bone resorption.[30] Oxidative stress parameters, such as lipid and protein oxidation, are associated with bone loss in postmenopausal women.[31] Additionally, plasma levels of the dietary antioxidants vitamin A, C and E, and the activity of the enzymatic antioxidants GPx and SOD, are markedly decreased in aged osteoporotic women.[32]

A randomized controlled intervention study has been carried out to determine whether lycopene can act as an antioxidant and decrease oxidative stress parameters, resulting in decreased bone turnover markers, thus reducing the risk of osteoporosis in postmenopausal women.[33] After 1 month without lycopene consumption, 60 women, 50–60 years old, were divided equally into four groups receiving different treatments: (1) regular tomato juice, (2) lycopene-rich tomato juice, (3) tomato Lyc-O-Mato® lycopene capsules, or (4) placebo capsules, twice daily, to provide total lycopene intakes of 30, 70, 30, and 0 mg/day, respectively, for 4 months. It was reported that the women who consumed regular tomato juice, lycopene-rich tomato juice or lycopene capsules had higher serum lycopene levels compared to the placebo.[33] Decreased lipid peroxidation after lycopene supplementation was evident as shown by a considerable decrease (11.93 ± 2.16 %) in TBARS, which was significantly greater than the slight decrease (0.37 ± 5.25 %) observed in placebo-supplemented participants (Fig. 7.4). In addition, lycopene treatments resulted in increased antioxidant capacity, decreased protein oxidation levels and decreased levels of the bone resorption marker N-telopeptide when compared to the placebo group.[33] Therefore, it is suggested that the antioxidant lycopene is beneficial because it increases the antioxidant capacity and reduces oxidative stress parameters and bone resorption in postmenopausal women.

In another study, the effect of antioxidant supplements on bone mineral density in elderly women was evaluated.[34] The antioxidant supplementation consisted of a daily intake of 600 mg of vitamin E and 1000 mg of vitamin C. The authors found that antioxidant therapy exerted a beneficial effect on bone mineral density in postmenopausal women to the same extent as resistance training. No interactive effect was observed when combining resistance training and antioxidant supplements. The placebo group showed a bone loss corresponding to the typical average loss for postmenopausal women.

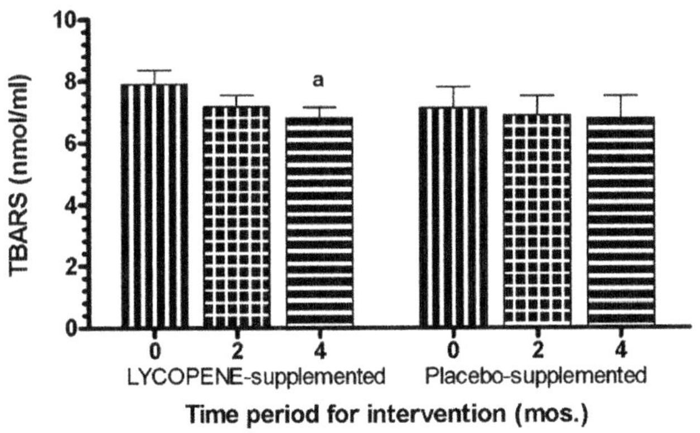

FIGURE 7.4 Variation in lipid peroxidation of postmenopausal female participants supplemented with lycopene (tomato lycopene Lyc-O-Mato® capsules or two types of tomato juice) or placebo after 4 months of supplementation.[33] [a]Indicates significant change (P <0.001). There was no significant change in participants consuming placebo. Reprinted from Mackinnon ES et al, Osteoporosis International 2011, 22(4), 1091-1101, Copyright 2013, with permission from Springer.

These findings suggest that antioxidant supplementation may reduce the damaging effects of oxidative stress on bone mass by decreasing the upregulated osteoclastic differentiation and increasing the downregulated osteoblastic differentiation.[34]

Oxidative stress is also one of the earliest events associated with a decline in cognitive function, which plays a key role in Alzheimer's disease pathogenesis. It has been demonstrated that elderly women with Alzheimer's disease have increased levels of oxidative stress indicators, such as TBARS and carbonyl concentrations.[35] In addition, this study revealed that the GSH concentration and the activity of the enzymatic antioxidant GPx are decreased in elderly women with Alzheimer's disease. The reduction in the GPx activity could indicate the failure of the antioxidant system to produce enough GSH to avoid oxidative stress produced by free radicals.[35] The authors explain that there is a defect in the antioxidant defense system in that it is incapable of responding to increased free radical production, which may lead to oxidative damage and the development of the pathologic alterations that characterize the neurodegenerative disorder of patients with Alzheimer's disease.

The relation between high-dose antioxidant supplements and cognitive function was investigated in community-dwelling elderly women who participated in the Nurses' Health Study.[36] To estimate overall cognitive performance, a global score was calculated by combining the results from different cognitive function assessments. A modest effect of high-dose supplementation with antioxidant vitamins E or C was demonstrated.[36] However, the subjects currently taking vitamins E and C supplements and those taking vitamin E alone had higher mean global scores than the elderly women who had never taken either vitamin E or C.[36] When the long duration of current supplement use was examined (more than 10 years), the global score was higher among the women who had taken vitamins E or C, and there was a significant trend of increasingly higher mean scores with increasing durations.[36] For the women taking vitamin E alone, there was weak evidence of increasing benefits with increasing duration of use.[36] It was reported that 6 weeks of vitamin E supplementation promoted an increase in the reactive scavenger activity of red blood cells, and a decrease in TBARS levels.[37] However, vitamin E supplementation did not influence the enzymatic antioxidant parameters of GPx and SOD.[37]

Menopause is associated with increased risk for ischemic heart disease and cerebrovascular disease, which collectively are the main causes of morbidity and mortality in women.[7] The total antioxidant capacity of a diet takes into account all antioxidants and the synergistic effects of their combination.[8] In the case of the population-based prospective Swedish Mammography Cohort, comprised of middle-aged and elderly women, the total antioxidant capacity originating from diet and the risk of myocardial infarction were evaluated.[8] The subjects completed a food-frequency questionnaire and the total dietary antioxidant capacity was calculated based on oxygen radical absorbance capacity values.[8] The authors observed that a diet with a high total antioxidant capacity, based on fruits, vegetables, coffee, and whole grains, was associated with a lower incidence of myocardial infarction. Oxidative stress is strongly associated with both aging and cardiovascular diseases. Age-associated changes in cardiovascular structure/function are implicated in the significantly increased risk of cardiovascular disease in older persons.[38,39] Oxidative stress triggers the production of IL-6, and antioxidant micronutrients play a critical role in decreasing this inflammatory response. Thus, a cohort study was carried out on 619 elderly women to evaluate the association of some dietary antioxidants with IL-6 and mortality.[40] A robust inverse cross-sectional and longitudinal relationship between several specific carotenoids and serum IL-6 was identified. Furthermore, a strong cross-sectional inverse relationship was observed between selenium and IL-6 and increased mortality among elderly women with lower selenium levels. These findings suggest that specific antioxidant nutrients, which attenuate oxidative stress damage, may play an important role in suppressing IL-6 in older disabled women.[40] Another investigation on 786 elderly women from the same cohort study demonstrated that the serum selenium level is an independent predictor of red cell distribution width (RDW).[41] RDW is theoretically affected by oxidative damage and inflammation; however, the study did not find an association with protein carbonyl levels and RDW.[41]

The inflammation and oxidative stress parameters were also evaluated after a dietary intervention with soymilk.[42] In this study, 31 subjects were randomly assigned to consume three servings of vanilla soymilk or reduced fat dairy milk daily for 4 weeks. Although isoflavones possess estrogenic and antioxidant properties, the authors reported no significant effect of soymilk on any of the antioxidant enzymes or inflammatory markers analyzed.[42]

SUMMARY POINTS

- The period after menopause may be considered the beginning of the aging process in women, which is caused by a series of endocrinologic variations caused by the decline of estrogen levels.
- Lower exposure to endogenous estrogen may be associated with a higher risk of age-related disease in elderly women.
- Once women begin ovarian senescence, estrogen production becomes irregular, antioxidant protection is decreased, and oxidative stress is assumed to increase.
- The mechanisms of the processes shown by elderly women are linked not only to estrogen loss but also to high levels of oxidative stress, which play a key causal role in the menopause-related symptoms and pathologic processes.
- The consumption of dietary antioxidants, provided by supplements or foods, may contribute to healthy aging in women.

References

1. Sanz A, Pamplona R, Barja G. Is the mitochondrial free radical theory of aging intact? *Antioxid Redox Signal* 2006;**8**(3–4):582–99.
2. Semba RD, Ferrucci L, Sun K, et al. Oxidative stress is associated with greater mortality in older women living in the community. *J Am Geriatr Soc* 2007;**55**(9):1421–5.
3. Keefe DL, Liu L, Marquard K. Telomeres and aging-related meiotic dysfunction in women. *Cell Mol Life Sci I* 2007;**64**(2):139–43.
4. Xu Q, Parks CG, DeRoo LA, et al. Multivitamin use and telomere length in women. *Am J Clin Nutr* 2009;**89**(6):1857–63.
5. Floyd RA, Towner RA, He T, et al. Translational research involving oxidative stress and diseases of aging. *Free Radic Biol Med* 2011;**51**(5):931–41.
6. Sánchez-Rodríguez MA, Zacarías-Flores M, Arronte-Rosales A, et al. Menopause as risk factor for oxidative stress. *Menopause* 2012;**19**(3):361–7.
7. Lin J, Kroenke CH, Epel E, et al. Greater endogenous estrogen exposure is associated with longer telomeres in postmenopausal women at risk for cognitive decline. *Brain Res* 2011;**1379**:224–31.
8. Rautiainen S, Levitan EB, Orsini N, et al. Total antioxidant capacity from diet and risk of myocardial infarction: a prospective cohort of women. *Am J Med* 2012;**125**(10):974–80.
9. Bokov A, Chaudhuri A, Richardson A. The role of oxidative damage and stress in aging. *Mech Ageing Develop* 2004;**125**(10–11):811–26.
10. Perez E, Wang X, Simpkins JW. Role of antioxidant activity of estrogens in their potent neuroprotection. In: Qureshi GA, Parvez SH, editors. *Oxidative stress and neurodegenerative disorders*. Amsterdam: Elsevier; 2007. pp. 503–24.
11. Miquel J, Ramirez-Bosca A, Ramirez-Bosca JV, Alperi JD. Menopause: a review on the role of oxygen stress and favorable effects of dietary antioxidants. *Arch Gerontol Geriatr* 2006;**42**(3):289–306.
12. Chakrabarti S, Lekontseva O, Davidge ST. Estrogen is a modulator of vascular inflammation. *Iubmb Life* 2008;**60**(6):376–82.
13. Lin J, Epel E, Blackburn E. Telomeres and lifestyle factors: roles in cellular aging. *Mutation Res* 2012;**730**(1–2):85–9.
14. Paul L. Diet, nutrition and telomere length. *J Nutr Biochem* 2011;**22**(10):895–901.
15. Starr JM, Shiels PG, Harris SE, et al. Oxidative stress, telomere length and biomarkers of physical aging in a cohort aged 79 years from the 1932 Scottish Mental Survey. *Mech Ageing Develop* 2008;**129**(12):745–51.
16. Viña J, Borrás C. Women live longer than men: understanding molecular mechanisms offers opportunities to intervene by using estrogenic compounds. *Antioxid Redox Signal* 2010;**13**(3):269–78.
17. Cao GH, Russell RM, Lischner N, Prior RL. Serum antioxidant capacity is increased by consumption of strawberries, spinach, red wine or vitamin C in elderly women. *J Nutr* 1998;**128**(12):2383–90.
18. Khalil A, Gaudreau P, Cherki M, et al. Antioxidant-rich food intakes and their association with blood total antioxidant status and vitamin C and E levels in community-dwelling seniors from the Quebec longitudinal study NuAge. *Exp Gerontol* 2011;**46**(6):475–81.

19. Boaventura BCB, Di Pietro PF, de Assis MAA, et al. Antioxidant biomarkers and food intake in elderly women. *J Nutr Health Aging* 2012;**16**(1):21–5.
20. Sen CK. Redox signaling and the emerging therapeutic potential of thiol antioxidants. *Biochem Pharmacol* 1998;**55**(11):1747–58.
21. Zhu YG, Carvey PM, Ling ZD. Age-related changes in glutathione and glutathione-related enzymes in rat brain. *Brain Res* 2006;**1090**(1):35–44.
22. Ray AL, Semba RD, Walston J, et al. Low serum selenium and total carotenoids predict mortality among older women living in the community: The Women's Health and Aging Studies. *J Nutr* 2006;**136**(1):172–6.
23. Nicklett EJ, Semba RD, Xue QL, et al. Fruit and vegetable intake, physical activity, and mortality in older community-dwelling women. *J Am Geriatr Soc* 2012;**60**(5):862–8.
24. Thomson CA, Stendell-Hollis NR, Rock CL, et al. Plasma and dietary carotenoids are associated with reduced oxidative stress in women previously treated for breast cancer. *Cancer Epidemiol Biomark Prevent* 2007;**16**(10):2008–15.
25. Rock CL, Flatt SW, Natarojan L, et al. Plasma carotenoids and recurrence-free survival in women with a history of breast cancer. *J Clin Oncol* 2005;**23**(27):6631–8.
26. Zhao XF, Aldini G, Johnson EJ, et al. Modification of lymphocyte DNA damage by carotenoid supplementation in postmenopausal women. *Am J Clin Nutr* 2006;**83**(1):163–9.
27. Michelon E, Blaum C, Semba RD, et al. Vitamin and carotenoid status in older women: associations with the frailty syndrome. *J Gerontol A Biol Sci Med Sci* 2006;**61**(6):600–7.
28. Manolagas SC. From estrogen-centric to aging and oxidative stress: a revised perspective of the pathogenesis of osteoporosis. *Endocrine Rev* 2010;**31**(3):266–300.
29. Cummings SR, Melton LJ. Epidemiology and outcomes of osteoporotic fractures. *Lancet* 2002;**359**(9319):1761–7.
30. Weitzmann MN, Pacifici R. Estrogen deficiency and bone loss: an inflammatory tale. *J Clin Invest* 2006;**116**(5):1186–94.
31. Cervellati C, Bonaccorsi G, Cremonini E, et al. Bone mass density selectively correlates with serum markers of oxidative damage in post-menopausal women. *Clin Chem Lab Med* 2012. http://dx.doi.org/10.1515/cclm-2012-0095. published online May 23.
32. Maggio D, Barabani M, Pierandrei M, et al. Marked decrease in plasma antioxidants in aged osteoporotic women: results of a cross-sectional study. *J Clin Endocrinol Metab* 2003;**88**(4):1523–7.
33. Mackinnon ES, Rao AV, Josse RG, Rao LG. Supplementation with the antioxidant lycopene significantly decreases oxidative stress parameters and the bone resorption marker N-telopeptide of type I collagen in postmenopausal women. *Osteoporosis Int* 2011;**22**(4):1091–101.
34. Chuin A, Labonte M, Tessier D, et al. Effect of antioxidants combined to resistance training on BMD in elderly women: a pilot study. *Osteoporosis Int* 2009;**20**(7):1253–8.
35. Puertas MC, Martinez-Martos JM, Cobo MP, et al. Plasma oxidative stress parameters in men and women with early stage Alzheimer type dementia. *Exp Gerontol* 2012;**47**(8):625–30.
36. Grodstein F, Chen J, Willett WC. High-dose antioxidant supplements and cognitive function in community-dwelling elderly women. *Am J Clin Nutr* 2003;**77**(4):975–84.
37. Park OJ, Kim HYP, Kim WK, et al. Effect of vitamin E supplementation on antioxidant defense systems and humoral immune responses in young, middle-aged and elderly Korean women. *J Nutr Sci Vitaminol* 2003;**49**(2):94–9.
38. Lakatta EG, Levy D. Arterial and cardiac aging: major shareholders in cardiovascular disease enterprises – Part II: the aging heart in health: links to heart disease. *Circulation* 2003;**107**(2):346–54.
39. Lakatta EG, Levy D. Arterial and cardiac aging: major shareholders in cardiovascular disease enterprises – Part I: aging arteries: a 'set up' for vascular disease. *Circulation* 2003;**107**(1):139–46.
40. Walston J, Xue Q, Semba RD, et al. Serum antioxidants, inflammation, and total mortality in older women. *Am J Epidemiol* 2006;**163**(1):18–26.
41. Semba RD, Patel KV, Ferrucci L, et al. Serum antioxidants and inflammation predict red cell distribution width in older women: The Women's Health and Aging Study I. *Clin Nutr* 2010;**29**(5): 600–604.
42. Beavers KM, Serra MC, Beavers DP, et al. Soymilk supplementation does not alter plasma markers of inflammation and oxidative stress in postmenopausal women. *Nutr Res* 2009;**29**(9):616–22.
43. Kasapoglu M, Ozben T. Alterations of antioxidant enzymes and oxidative stress markers in aging. *Exp Gerontol* 2001;**36**(2):209–20.

CHAPTER

8

Antioxidants, Vegetarian Diets and Aging

S. Wachtel-Galor, P.M. Siu, I.F.F. Benzie

Department of Health Technology & Informatics, The Hong Kong Polytechnic University, Hung Hom, Kowloon, Hong Kong

List of Abbreviations

CVD cardiovascular disease
DNA deoxyribonucleic acid
FRAP assay ferric reducing/antioxidant power assay
Gpx glutathione peroxidase
GSH glutathione
MLSP mean life-span potential
PUFA polyunsaturated fatty acids
ROS reactive oxygen species
SOD superoxide dismutase
UV radiation ultraviolet radiation

INTRODUCTION

Aging is an intrinsic process that causes a gradual but relentless impairment of normal biologic functions and the ability to resist physiologic stresses, resulting in functional decline and an increased risk of morbidity and mortality. There are many complex interacting mechanisms and changes associated with aging, including shortened and dysfunctional telomeres, accumulation of mutations, inflammation, disturbed hormonal pathways, mitochondrial dysfunction, cellular senescence, apoptosis and altered gene expression. The underlying cause or trigger of such age-related changes is not yet clear, but oxidative stress, which causes oxidation-induced damage to biomolecules, such as nucleotide bases, proteins and polyunsaturated fatty acids (PUFA), is believed to be a key factor. Oxidation-induced changes to organelles, cells and structures accompany chronologic aging, and their effects on the structure and function of cellular components and functional units, such as DNA, enzymes and membranes, have a deleterious impact on gene expression, feedback mechanisms, signalling pathways and, thereby, affect the integrity and function of cells, tissues, organs and homeostatic systems. The importance of oxidative stress as an etiological factor in aging is supported by the finding that agents or disorders that accelerate physiologic aging, such as cigarette smoking and diabetes, are associated with increased oxidative stress; those who habitually take a diet rich in antioxidants, such as vegetarians, have lower oxidative stress and, provided that the diet is not deficient in iron, vitamin B12 or sulfur, vegetarians generally live longer, healthier lives than do non-vegetarians.

Currently there is no known method of arresting or reversing human aging. However, there is some evidence that antioxidants can modulate or slow the deleterious effects of aging and lower the risk of age-related disease. In the following sections, aging and oxidative stress will be discussed along with associations between oxidative stress, dietary antioxidants, vegetarian diets and age-related diseases. Furthermore, the way in which dietary antioxidants may work to oppose oxidative stress and prolong the human healthspan will be addressed briefly.

HUMAN AGING: WHY AND HOW DOES IT OCCUR, AND WHAT ARE THE CONSEQUENCES OF AGING?

Aging is the result of post-maturity changes that lead to a slow but continuous decline in all physiologic systems[1–3] (Fig. 8.1). There are many theories as to why and how we age. Conceptually, they range from aging as a result of a gene-based 'programme' of self-destruction, to aging and decline as the result of a purely random sequence of damaging events within cells.[1–3] Damage is widespread and of diverse types, and although the origin of the aging process is not clear, there is a 'common soil' of inter-acting inflammatory and oxidative processes that appears to drive the deleterious age-related changes that lead to physiologic decline, dysfunction, disease and death.[1,3]

Aging is a virtually universal process in multi-celled, sexually reproducing organisms, and the commonality

Aging
http://dx.doi.org/10.1016/B978-0-12-405933-7.00008-1

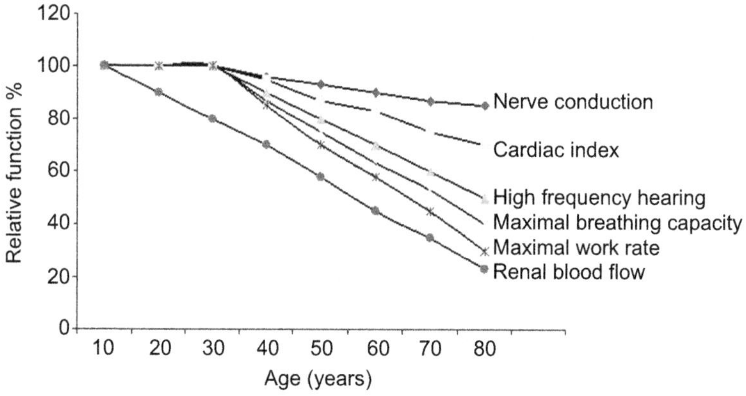

FIGURE 8.1 All physiologic systems show a decline with age. *Redrawn from Baker & Frozard in Biology of ageing, Arking R (ed) Sunderland MA: Sinauer Associates Inc; 1998, p 7 [reference 1].*

of such a biologic feature implies a reproductive benefit during very early evolutionary development.[3] Furthermore, because aging remains a feature of most life forms, clearly there has been no effective selection pressure against it, even though it brings functional decline, disability and death. Proposed biologic advantages (which in evolutionary terms act at the species level rather than at the individual level) of aging include limiting population numbers, which would act as a 'resource-allocation' measure favoring the young. However, this would be an advantage only if older members of the population lost their reproductive ability, which is the case in very few species, and even then is true only in females. Human females cease to reproduce around the age of 50 years, and yet most women continue to live for many years beyond menopause. Indeed, the life expectancy of women exceeds that of men by around 8 years.[4]

Another theory as to why aging occurs is related to internal resource allocation.[3] Maintenance and repair of tissues is metabolically expensive, and so is reproduction, but biologic resources are limited. Biologic success at species level relies on reproduction. Preferential allocation of resources to reproduction in early adult life promotes such success, but likely accompanies poor maintenance and repair, accumulation of damage, and functional decline – the characteristics of aging. But why has there been no effective selection pressure against this? The reason is likely to lie in the short post-maturity lifespan of most species in their native habitats – in which death is from external agents, such as starvation or trauma, rather than the end result of aging. This was true of our early ancestors also, but due to rapid medical and social developments in the past 100–200 years, most of us now live to become truly 'old', showing the outward signs and suffering the internal consequences of aging. This is a very recent development in the history of humankind, and is bringing unprecedented challenges to our society in terms of dealing with the huge and increasing burden of chronic disease and disability in our aging population.[4,5]

Humans, as with other mammals, age at broadly similar rates, although there are genetic disorders (progerias) that greatly accelerate certain aspects of aging.[1–3] Different species show similar age-related changes, but have vastly different mean life-span potentials (MLSPs), ranging from a few days or weeks to many decades. The human MLSP is not known, but it is at least 122 years, which is the confirmed age at death of the longest lived individual to date. The average life expectancy, which is the probable, as opposed to potential, lifespan, has increased dramatically in the past 100 years in most areas of the world as a result of improved antenatal and neonatal care, antibiotics, vaccinations, education, nutrition and other public health measures.[2–6] The life expectancy in Hong Kong is 86.9 years for women (currently the longest in the world) and 78.4 years for men, but is still far short of the MLSP.[4] The shortfall is due to age-related diseases, the most common of which are cancer, cardiovascular disease, diabetes and dementia.[4,5] These non-communicable degenerative diseases account for around one-third of annual deaths worldwide[5] (Fig. 8.2). Furthermore, 75% of people 65 years of age and over suffer from one or more age-related diseases such as cataract, hearing loss, osteoporosis, rheumatoid arthritis and sarcopenia.[5] These compromise quality of life and contribute to the huge economic and social burden of aging in our society. There is a strong link with oxidative stress in all of these diseases, and strategies to promote functional longevity that focus on lowering oxidative stress are of great interest for their potential value in opposing oxidative stress, slowing the onset of age-related decline and disease, and promoting functional longevity.

OXIDATIVE STRESS AND AGING

Oxidative stress can be defined as a pro-oxidant shift in redox balance that causes deleterious oxidation-induced changes in biologic components, pathways and structures.[2,6–8] These include lipids, DNA, proteins, signaling

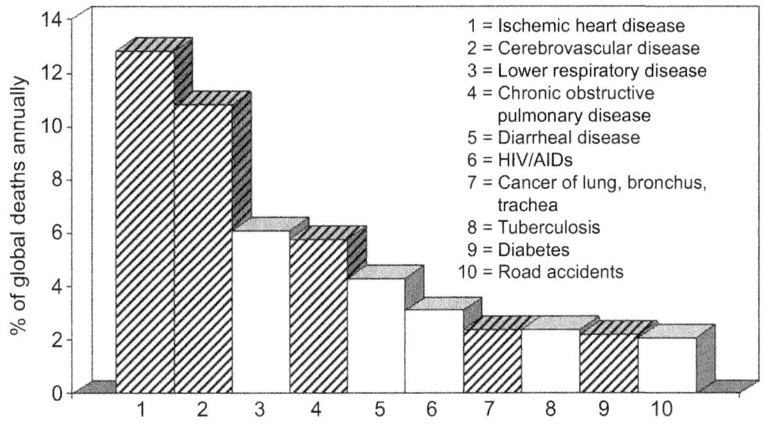

FIGURE 8.2 Ten leading causes of death globally. These account for 52% of total deaths, with 34% of total deaths annually due to age-related, non-communicable disease (hatched bars). *Data from World Health Organization* www.who.int/mediacentre/factsheets.

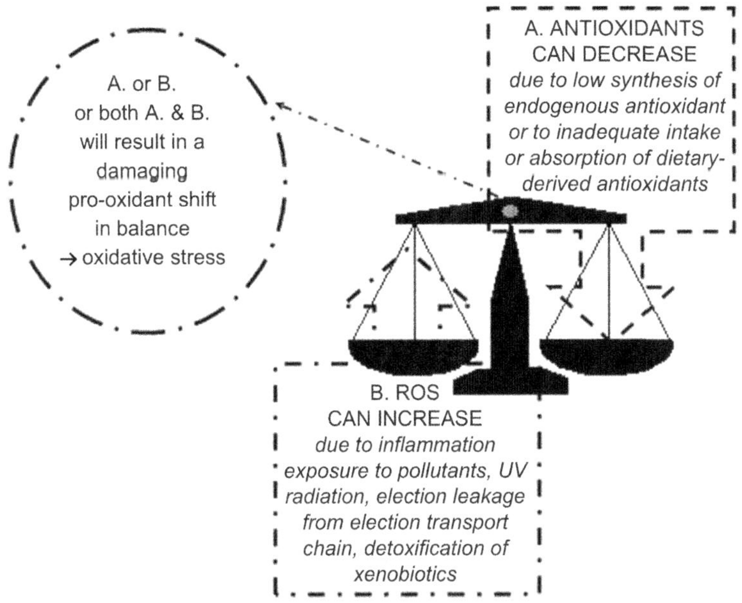

FIGURE 8.3 Oxidative stress is a consequence of a pro-oxidant imbalance in oxidants and antioxidants caused by a relative or absolute lack of antioxidant defense.

pathways, gene expression, membranes, receptors and mitochondria, changes in which lead to cellular dysfunction and increased risk of disease – the characteristics of aging. A damaging pro-oxidant shift can be the result of increased amounts of reactive oxygen species (ROS), less effective removal of ROS and poor defense against the action of ROS.[6–9] Therefore, the antioxidant side of the redox balance is also important (Fig. 8.3).

Human tissues are unavoidably, ubiquitously and most probably constantly exposed to ROS, which are of various forms, from different sources and of varied reactivity.[6–9] If the ROS contains an unpaired electron in an orbital, as is the case with superoxide and the highly reactive hydroxyl radical, it can be referred to as an 'oxygen free radical', but non-radical forms of ROS exist, including hydrogen peroxide and the much more reactive peroxynitrite, which is formed from the interaction of nitric oxide and superoxide.[6,7] Many ROS are of endogenous origin. Some are produced purposefully – for example, hydrogen peroxide for cellular signaling, and nitric acid for vasodilation. Others are produced accidentally but unavoidably – for example, superoxide is produced as a result of electron leakage from the mitochondrial electron transport chain and during detoxification of drugs and other xenobiotics.[6–9] ROS are also generated *in vivo* by pollutants such as tobacco smoke, and by ultraviolet (UV) and ionizing radiation. Large amounts of ROS are produced in acute and chronic inflammation, as a result of post-ischemic reperfusion, by hyperhomocysteinemia, and also by increased oxygen utilization, such as occurs during aerobic exercise.[6–9] Therefore, oxidative stress is a constant threat to, and a common reality within, cells. Cellular defense against, and response to, stress is important, and if defense is compromised, and damage not repaired, there will be an accumulation of damage and cellular dysfunction.[2,3,6–11] Older adults are reported to have impaired capacity to deal with oxidative insults, and to have increased susceptibility to oxidative stress.[12] Therefore, conceptually at least, if defenses and response to oxidative challenge are impaired, then oxidative stress will increase, and age-related decline and risk

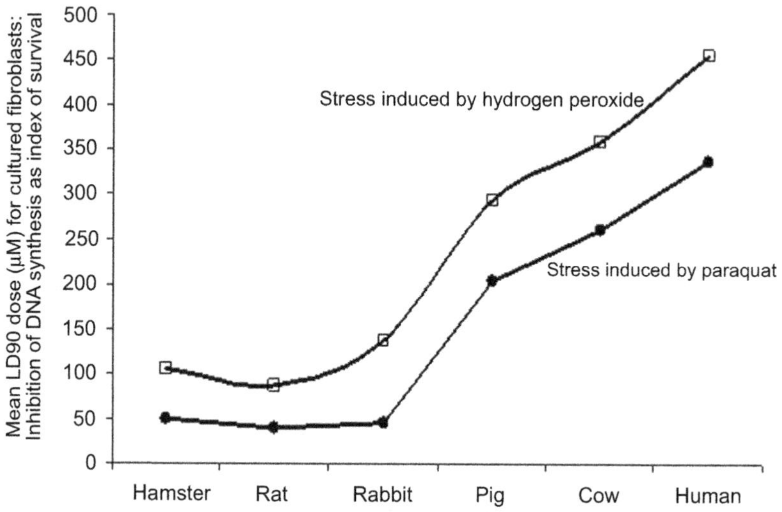

FIGURE 8.4 Positive correlation between maximal life-span potential (MLSP) of different species of mammal and the resistance of their cells to oxidative stress induced by two different agents. *Based on data from Kapahi P. et al., Free Radic Biol Med 1999;26:495–500 [reference 13].*

of disease will also increase. Conversely, lowering oxidative stress would slow age-related decline and promote healthy aging.

Does oxidative stress cause biologic aging *per se*, or is it rather the driver of age-related pathologic changes? The association between aging and disease is so strong that it is difficult to separate the biologic from the pathologic effects of aging. However, there is evidence from animal and cell culture studies showing that oxidation-induced damage (oxidative stress) increases with age and that cells of older people are less able to resist or adapt to oxidant challenge.[2,7–12] Furthermore, there is intriguing evidence showing that cells from mammalian species with different MLSP have different levels of resistance to oxidant challenge[13] (Fig. 8.4).

In the aging human, oxidation-induced changes to skin cause loss of collagen, wrinkling and aging of skin. Oxidation-induced damage to DNA causes mutation, genomic defects, and changed phenotypic expression of cells, leading to apoptosis or carcinogenesis. Oxidation of proteins changes their structure and function, resulting, for example, in clouding of the ocular lens (senile cataract). Oxidation of lipids and membranes can accelerate atherosclerosis, cause cellular dysfunction, or loss of receptor recognition/function resulting in, for example, atherosclerosis, type 2 diabetes, or immune dysfunction. Oxidation-induced apoptosis of post-mitotic tissue leads to loss of skeletal muscle (sarcopenia) and bone mass (osteoporosis) and to neurologic and cognitive decline (dementia).[2,7–16] Taking sarcopenia as an example, the way in which oxidative stress is proposed to be involved to induce age-related myofiber atrophy and loss of muscle fiber is outlined in Figure 8.5. Based on extensive studies of aging of human muscle, the pathogenesis of sarcopenia is suggested to be multifactorial.[15] The age-related decline of muscle mass and strength is attributed to the concurrent alteration of multiple pathways in relation to chronic inflammation, mitochondrial DNA damage, mitochondrial dysfunction, aberrant apoptosis, and catabolic shift of protein balance.[15,16] Many of the pathways that are implicated in the etiology of sarcopenia have been shown to be associated with the disruption of cellular redox balance, which is in support of the proposition that oxidative stress has an important role in the development of adverse phenotypic changes in muscle with aging.

There is little doubt that oxidative stress is closely associated with aging and age-related decline, although oxidative stress has not been confirmed as the root cause or trigger of aging. Current evidence places oxidative stress, and resistance or response to this, more clearly as a key determinant of healthspan, rather than causing aging *per se* or determining the MLSP.[7,9–14,17] This being the case, the potential role of antioxidants is more likely to be in relation to extending healthspan through lowering oxidative stress and addressing the processes that cause it, such as lack of antioxidant defense against ROS, elevated homocysteine, and inflammation.[2,3,6–17] Extending the MLSP is a much less important goal than extending healthspan and prolonging functional longevity. Strategies to extend healthspan would enable more of us to remain healthy in our old age rather than, as is all too commonly the case, succumbing to oxidative stress-related diseases when only 50–60% of the way through our potential lifespan.

LOWERING OXIDATIVE STRESS: THE POTENTIAL ROLE OF DIETARY ANTIOXIDANTS IN INCREASING HEALTHSPAN

Oxidative stress can be lowered by preventing situations or processes that generate ROS, such as chronic infections, inflammation, obesity, diabetes, post-ischemic reperfusion, hyperhomocysteinemia, exposure to tobacco smoke and other pollutants, ultraviolet A radiation, drugs and other xenobiotics.[6–9] Due to their ubiquitous nature

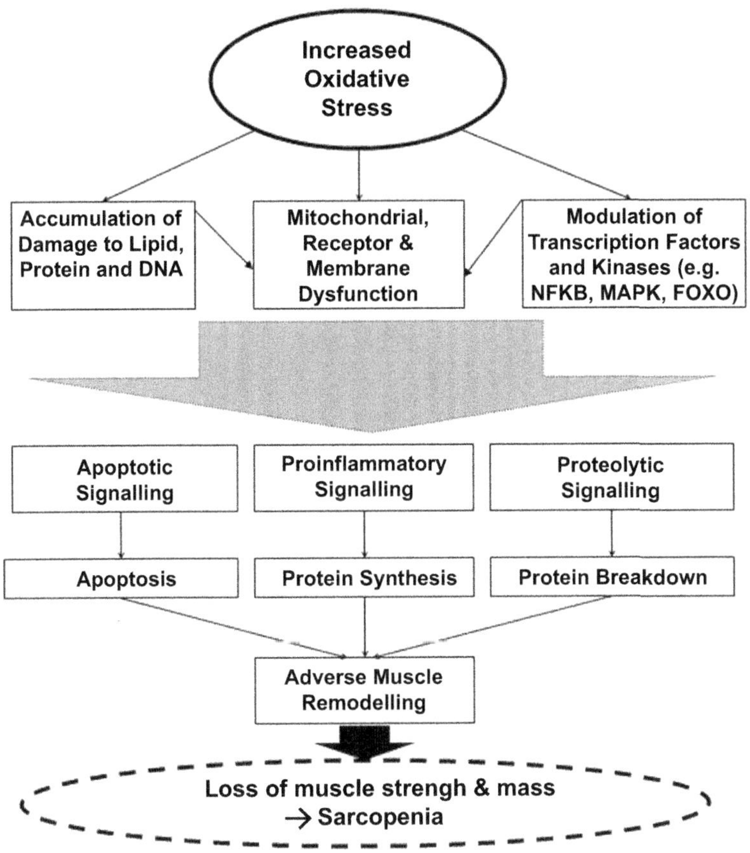

FIGURE 8.5 Proposed relationship between oxidative stress and sarcopenia. *Adapted from Meng SJ and Yu LJ. Int J Mol Sci 2010;11:1509–1526 [reference 16].*

and to the physiologically important role of some ROS, exposure cannot be completely prevented, nor indeed is it desirable.[6,7,9,18] Still, the real and constant threat of oxidative stress raises questions of how the burden of ROS can be lowered and their deleterious effects minimized. The focus here is on the role of antioxidants, particularly diet-derived antioxidants.

To deal with the constant and unavoidable threat of ROS, biologic antioxidant defense systems exist to oppose the development of oxidative stress. Biologic antioxidants of various types evolved to protect the early oxygen-producing organisms that were responsible for the increasingly oxygenic atmosphere that accompanied their development almost 2 billion years ago and from which current life forms evolved.[6,9] All aerobic organisms produce antioxidant defense factors, but chlorogenic plants are very rich in antioxidants. These are needed to protect plant cells against the high amounts of oxygen produced *in situ*.[6,7,9] The human endogenous antioxidant defense system comprises a range of diverse compounds that help to limit the generation and action of ROS. They act through enzymatic action, quenching activity, and by direct interaction with ROS.[6,7,9] The system is malleable to some extent, and higher levels of antioxidants are found in cells of those exposed to increased ROS on a regular basis, for example trained athletes.[7,10] This upregulation of antioxidant defenses is a cytoprotective adaptation to increased production of ROS, but the physiologic response is limited. Large adaptive increases in endogenous antioxidants such as superoxide dismutase (SOD), glutathione peroxidase (Gpx), bilirubin and glutathione (GSH) do not occur in healthy adults. The reason for the limited adaptive response in endogenous antioxidants is not clear. It may be related to metabolic cost, although it is noted that the true and real time changes in the intracellular redox balance are not known, and we lack adequate tools to clarify this, so the degree of cellular response and adaptability remain to be confirmed. Still, it is known that our cells and biologic fluids contain exogenous antioxidants that are active against ROS, and their action is believed to complement our effective but imperfect endogenous antioxidants, forming an integrated antioxidant defense system.[6,7,9,19] These additional components of our antioxidant defense system are diet-derived, and plasma concentrations can be modulated by changing dietary intake, with potential benefits in terms of lowered oxidative stress.

ANTIOXIDANTS IN FOOD

Diet-derived antioxidants include water-soluble ascorbic acid (vitamin C), the lipophilic tocopherols and tocotrienols ('vitamin E', which consists mainly of α-tocopherol in humans), the colorful carotenoids, and the huge family of polyphenols that include catechins, anthocyanins, quercetin, and resveratrol, among others.[6,7,9,19] Plant-based foods are rich in these antioxidants, though berries and other fruits, vegetables, teas, herbs and spices are diverse in their antioxidant profiles and content. For example, strawberries and kiwi fruit are very rich in ascorbic acid (vitamin C) but low in vitamin E, which is found in vegetable oils and seeds; teas and cocoa are low in vitamin C but have a very high catechin (flavan-3-ols) content; red wine contains relatively high amounts of resveratrol; carrots and pumpkin contain carotenoids; berries are rich in various types of anthocyanins.[19,20] Due to the enormous number and different types of individual antioxidants in foods, their measurement is difficult, and a commonly used approach is to measure the 'total' antioxidant content (or activity, capacity or power) of foods. There are several methods available for this, but a widely adopted method is the ferric reducing/antioxidant power (FRAP) assay, which assesses the combined reductive capacity of the redox active components in a sample.[21] There is now a very large database of the total antioxidant content of fresh and dried fruits, vegetables, herbs and spices, as well as teas, wines, juices and other beverages and foods.[21–24] These data can be used to aid in dietary planning in order to promote higher antioxidant intake. The typical antioxidant content (as the FRAP value) of some common fruits and vegetables is given in Table 8.1.

TABLE 8.1 Typical Total Antioxidant Content (as the FRAP Value[21]) of Some Common Fruits and Vegetablesa

	FRAP Value mmol/100 g Fresh Wet Weight
Apple (green)	0.63
Apple (red)	0.42
Banana	0.42
Kiwi	0.82
Lemon	1.04
Mango	0.51
Orange	0.94
Plum	0.93
Strawberry	1.59
Celery	0.16
Kale	1.04
Lettuce	0.09
Mange tout	0.49
Potato	1.44
Turnip	0.36

[a]*Data from reference 23: Szeto, Tomlinson & Benzie. Brit J Nutr 2002;87:55–59.*

It is noted that the bioavailability of all diet-derived antioxidants is limited, and in some cases bioavailability is extremely low.[9,25–27] There is also rapid post-absorption catabolism and excretion of some antioxidant phytochemicals, with the body handling these compounds as xenobiotics (as indeed they are) to be detoxified and removed.[19,25–27] Consequently, plasma concentrations are in the undetectable to nanomolar range for polyphenolic antioxidants such as anthocyanins, catechins and carotenoids. The plasma concentration of α-tocopherol is typically in the range of 20–40 μmol/L, and even the better absorbed ascorbic acid is generally below 60 μmol/L in fasting plasma, with post-ingestion levels rarely exceeding 100 μmol/L even after a dose of 1 g or more.[6,9,25–27] It is noted that there is extensive colonic transformation of some unabsorbed polyphenolic antioxidants in teas and berries, and these are found in plasma in small amounts 7 hours or more after ingestion, but the antioxidant activity and biologic significance of these is not clear.[27] Still, despite low bioavailability, regular and frequent intake of foods that are rich in antioxidants strongly associates with increased antioxidant status, lowered oxidative stress, and lowered risk of age-related disease. For example, a landmark study of 11,000 British subjects followed for 17 years showed that daily consumption of fresh fruit increased antioxidant levels, and was associated with ~50% lower mortality from heart disease, stroke, and all causes,[28] and a study of 10,499 diabetic subjects with a mean follow-up of 9 years showed that intake of total vegetables, legumes and fruit was inversely associated with cardiovascular mortality.[29] Overall, convincing inverse correlations are seen in different populations, and in different regions across the world, between intake of, and plasma levels of, antioxidants, especially vitamin C and carotenoids, vegetarian diets, and the risk of development of stroke, diabetes, cancer, macular degeneration and impaired brain function and mood.[20,28–35]

It is important to note that the relationship between dietary antioxidants and health is seen only when antioxidant intake is from whole foods, not purified antioxidant supplements.[36] Food is the key and, as noted, plant-based foods in general are a rich source of various types of antioxidant and these are effective in opposing oxidative stress and improving antioxidant balance. Intake of at least five portions of fresh fruits and vegetables daily is recommended for health.[37] Vegetarians, with their habitually plant-rich diet, are likely to reap the most benefit, and in the following section we examine some of the evidence that vegetarians have less oxidative stress, less age-related disease, and a longer healthspan.

VEGETARIAN DIET, ANTIOXIDANTS, OXIDATIVE STRESS AND HEALTHSPAN

Vegetarianism is a lifestyle choice that is made for various reasons, including religious requirements, concerns for animal welfare and, increasingly, a desire to benefit from the health-promoting effects of plant-based foods. Vegetarian diets range in the extent to which foods of animal origin are excluded. Some allow dairy products or fish, while the most extreme form, the vegan diet, permits no animal-based foods of any kind.[20] In general, high consumption of plant-based foods, a feature of the vegetarian diet, leads to high plasma levels of exogenous antioxidants (most notably ascorbic acid), lower oxidative stress, lower inflammation (a driver of oxidative stress) and improved lipids and glycemic control,[20,38–42] although a note of caution must be mentioned here in regard to possible drawbacks of a purely plant-based diet. Following a strict vegetarian diet can lead to deficiencies in vitamin B12, folic acid and sulfur-containing compounds, leading to elevated plasma homocysteine levels.[20,38,43] Hyperhomocysteinemia leads to increased production of superoxide and inflammatory cytokines, causing increased oxidative stress.[43] Indeed, elevated homocysteine and inflammation are reportedly the main determinants of oxidative stress in the elderly.[17] Therefore, deficiency of vitamin B12, folic acid and sulfur-containing compounds must be avoided if the benefits of a vegetarian diet are to be realized. It must also be noted that the health benefits of a vegetarian diet are the result of many interacting biologic effects and factors, such as better lipid profile, lower blood pressure, less obesity, and lower intake of preservatives, salt, calories and carcinogens.[20] Still, a higher intake of antioxidants in plant-based foods likely makes an important contribution to the modulation of oxidative stress in vegetarians and the promotion of functional longevity.

Antioxidant Status and Oxidative Stress in Vegetarians

Higher antioxidant status, lower oxidative stress and less inflammation are found in vegetarians and others who consume larger amounts of plant-based foods containing carotenoids, ascorbic acid, α-tocopherol and polyphenols. For example, in a study of 30 long-term vegetarians in Hong Kong, fasting plasma ascorbic acid levels were ~50% higher than those found in 30 age- and sex-matched omnivores (mean (SD) 91(21.0) vs. 62(17) μmol/L, $p < 0.01$).[42] In a Swedish study of over 700 men,[41] those in the highest quintile of ascorbic acid had less inflammation, as evidenced by significantly lower ($p < 0.05$) plasma C-reactive protein (CRP) and interleukin 6 (IL6), and had lower oxidative stress, as evidenced by significantly ($p < 0.05$) lower urinary F2 isoprostanes. An Indian study of 50 vegetarians and 50 non-vegetarians, all healthy, reported lower lipid peroxidation (as malondialdehyde) and higher antioxidants in the vegetarians.[39] In a Slovakian study of young (20–30 years) and old (60–70 years) vegetarian and non-vegetarian women, no differences in oxidative stress biomarkers were seen across the younger age dietary groups, but in the older age groups the vegetarians ($n = 33$) showed lower oxidative stress (evidenced by less DNA damage and less lipid peroxidation, $p < 0.05$) and a higher antioxidant status (higher plasma ascorbic acid and β carotene, $p < 0.01$) than the non-vegetarians ($n = 34$).[44] Protein carbonyls, another biomarker of oxidative stress, were also lower, though not significantly so, in the older group of vegetarians compared to the older non-vegetarians. The key point in this study is that the age-associated increase in oxidative stress seen in omnivores was not seen in the elderly vegetarians.[44] This indicates that age-associated increases in oxidative stress are indeed avoidable and, by extension, so is the functional decline and disease caused by oxidative stress.

Intake of Plant-Based Antioxidants, Risk of Age-Related Disease and Healthspan

As noted, strong, inverse correlations are consistently reported between intake and plasma levels of diet-derived antioxidants and the risk or incidence of various age-related diseases, including type 2 diabetes, cardiovascular disease (CVD) and cancer. Oxidative stress is believed to underlie the development of these disorders, and vegetarians, whose antioxidant intake is high, have a lower risk for these diseases. Provided that their diet is well balanced, vegetarians generally live longer, healthier lives.[20]

Type 2 diabetes is characterized by intermittent or continuous hyperglycemia, and is a state of increased oxidative stress, with depleted antioxidants, increased oxidation-induced DNA damage, inflammation, accelerated aging and increased risk of CVD and cancer.[45,46] Ascorbic acid is obtained exclusively from the diet in humans, and higher intake, such as in the vegetarian diet, promotes better ascorbic acid status.[6,9,20] A large prospective study in the UK of apparently healthy subjects showed that, in over 12 years of follow-up, those in the highest quintile of plasma ascorbic acid had a 60% lower incidence of diabetes than those in the lowest quintile ($p < 0.05$), and it was estimated that an increase of 20 μmol/L in plasma ascorbic acid was associated with a 20% decrease in the risk of type 2 diabetes developing.[31] As well as aiding in prevention of type 2 diabetes, a high intake of plant-based foods may help in the management of the disease and prevention of its long-term complications. This was examined in a 24-week study of 74 type 2 diabetes patients who were allocated randomly to a vegetarian diet or a control (non-vegetarian diabetic diet) of matched calorie content.[47] The vegetarian group showed better results in terms of reduced medication, lowered body fat (visceral and subcutaneous) and increased insulin sensitivity, and their plasma adiponectin increased while leptin decreased. The group on the vegetarian diet was reported also to have higher levels of ascorbic acid, superoxide dismutase and GSH, and lower oxidative stress.[47]

CVD is the cause of death in ~70% of diabetes patients; it is a leading cause of morbidity and mortality even in the absence of diabetes, and it is the major cause of death worldwide.[5] The main underlying cause of CVD is the vicious cycle of inflammatory and oxidative processes that promote atherosclerosis and endothelial dysfunction.[48] A meta-analysis and systematic review of CVD in vegetarians reveals ~30% less mortality from CVD in vegetarians ($p < 0.05$).[33] A longitudinal study of 44,561 men and women in England showed that those who had adhered to a vegetarian diet for 5 or more years had a 32% lower chance ($p < 0.001$) of hospitalization or death from CVD than non-vegetarians.[34] Population-based studies of fruit and vegetable intake and plasma ascorbic acid have also shown significant long-term reduction in stroke and cancer.[28–35] Oxidative stress is also involved in cognitive decline, macular degeneration, osteoporosis and sarcopenia, all of which are very common causes of age-related morbidity. A higher intake of dietary antioxidants has been related to improved bone health and muscle strength, and to better mood and brain function.[7,15–20,49] In brief, epidemiologic data showing that a high intake of antioxidant-rich plant-based foods promotes healthy aging are strong and convincing. Observational and experimental evidence points to oxidative stress as a causal factor in age-related decline and disease, and indicates that antioxidant phytochemicals are modulators of this and promoters of functional longevity. Still, it must be noted that, while the evidence supports food choices as a key determinant of healthspan, and the antioxidant-rich vegetarian diet as a means to promote functional longevity, the molecular links between dietary antioxidants and health have not yet been established with any certainty. Direct antioxidant action may be responsible for some of the benefits seen with vegetarian diets, but the idea that direct antioxidant action is responsible for the control of all the complex, interacting processes leading to aging and disease is overly simplistic. It is important to note that the only deficiencies associated with dietary antioxidants are for vitamin C and vitamin E. There is no recommended daily intake for the other diet-derived antioxidants – including catechins, anthocyanins, resveratrol or carotenoids – even though there is observational and experimental evidence that a higher intake of these in teas, berries, other fruits and vegetables is beneficial. Furthermore, the extremely low plasma levels of many diet-derived antioxidants, even when intake is high, combined with their rapid conjugation and removal from the body, raises questions as to the direct impact of antioxidants and the advisability of taking excessively high levels.[18,50] These issues are discussed briefly in the next section, and another suggested mechanism of phytochemical action is briefly discussed, with a focus on redox balance and redox-driven cytoprotective adaptations that may be induced by diet-derived antioxidants.

ANTIOXIDANTS AND HEALTH: MOLECULAR CONNECTIONS AND RESEARCH NEEDS

There are several ways in which antioxidants work within our integrated defense system.[6,7,9,51] In relation to the phytochemical antioxidants of interest, some, like the catechins, can bind transition metal ions, thereby preventing radical formation by Haber-Weiss chemistry, and the carotenoids can absorb energy and 'quench' singlet oxygen. However, most diet-derived antioxidants, including ascorbic acid, the vitamin E family and the polyphenols, are redox active.[6,7,9,19,21,22] They react directly with ROS, inactivating them by means of an electron-donating, reducing action. In the process, the 'antioxidant' becomes oxidized and, depending on the biochemical environment and its redox potential, the 'antioxidant' could now act as an oxidant.[7,18,52] It is known that, in the presence of oxygen, redox-active antioxidants can autoxidize, generating hydrogen peroxide.[7,18,53] This pro-oxidant action has been shown clearly *in vitro* for various phytochemicals, including ascorbic acid, α-tocopherol and epigallocatechin gallate (EGCG), and requires us to revisit how we view dietary antioxidants and their action.[50–54]

Hydrogen peroxide is of relatively low reactivity, and intracellular antioxidant enzymes can destroy it rapidly, but it is an important cell signaling agent, acting through a series of redox 'switches' to activate redox-sensitive transcription factors and triggering an array of cytoprotective adaptations. One of these redox-sensitive routes is via Keap-1/Nrf2/ARE.[53] The transcription factor Nrf2 is bound to cysteine-rich Keap-1 in the cytosol and presented for rapid ubiquitination. With a pro-oxidant shift in redox tone, cysteine resides in Keap-1 are oxidized, and Nrf2 is released and its half-life extended, enhancing its translocation into the nucleus and binding to the antioxidant response element (the ARE, also referred to as the electrophile response element). This activates the ARE and leads to upregulation of genes that produce antioxidant enzymes, phase II detoxification enzymes, GSH and haem oxygenase 1 (HO-1), and the base-excision repair enzyme hOGG1.[18,52,53]

The combination of improved endogenous antioxidant defenses, DNA repair and detoxification systems would be expected to lower oxidative stress and the risk of mutagenesis and apoptosis. In addition, HO-1 has been described as a 'therapeutic funnel' because the products of its catalytic action on heme breakdown have powerful cytoprotective effects.[18,55] Bilirubin is a powerful antioxidant, release of iron from heme increases ferritin levels and carbon monoxide has anti-inflammatory effects.[55] Therefore, given that oxidative stress and inflammation together form an interacting and vicious cycle that leads to cellular damage, aging and degenerative disease, the subtle changes of redox balance induced by phytochemicals could bring significant and widespread effects on health.

This 'redox hormesis' mechanism of phytochemical action has not been confirmed to occur *in vivo*, but, if true, it would help to answer important questions as to why antioxidant-rich diets are beneficial to long-term health, and why there is such limited bioavailability of phytochemical antioxidants. In this redox hormesis scenario, regular intake of antioxidants is good for health because, apart from providing some additional direct protection against ROS, small but regular induction of a pro-oxidant shift in redox tone exerts small but regular stresses on the cell that induce beneficial adaptation – leading to lower oxidative stress and inflammation, a kind of oxidative conditioning or 'polishing' of the cell.[18,53] This leads to the hypothesis that the cells of those whose diet lacks a sufficiency of these redox-active phytochemicals will have fewer antioxidant defenses (endogenous as well as diet-derived), more oxidative stress and inflammation, poorer DNA repair and will be less able to withstand other stresses, predisposing them to accelerated age-related changes. However, excessive intake of antioxidants could be harmful by tipping the balance too much and too suddenly to the pro-oxidant side, with the damaging outcome of oxidative stress.[50] This concept is represented in Figure 8.6.

This redox hormesis concept has yet to be confirmed, and redox mechanisms may not be the only means by which phytochemicals act at the molecular level. There are reports on non-redox action of phytochemicals in signalling pathways, in driving epigenetic changes and as controllers of anti- and pro-apoptotic factors that determine the function and fate of cells.[56–59] It is also reported that a transient accumulation of oxidized guanine in DNA, which is a sign of oxidative stress, is advantageous to the cell.[57] If confirmed, this supports the concept that small, pro-oxidant swings in balance, and subsequent cytoprotective cellular adaptations to these, create a dynamic molecular mechanism that underlies the health benefits of antioxidant phytochemicals. This is an important area of scientific research because of its potential for promotion of healthy aging, and offers also a rich area for research in the food and agriculture fields and in

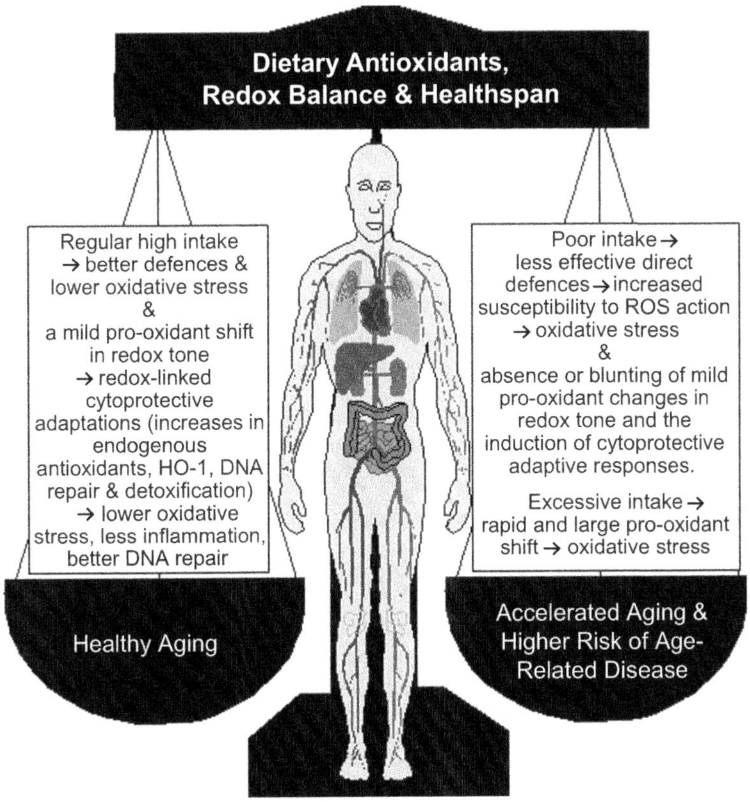

FIGURE 8.6 Concepts of redox hormesis; dietary antioxidants, redox balance and healthspan.

guiding production and responsible marketing of antioxidant-rich foods designed to promote health.[18,56,56–60]

In conclusion, in relation to dietary antioxidants and aging, further study of the hormetic interactions of phytochemicals with cell signalling pathways and genetic regulators of cell function and fate is needed before healthspan can be linked clearly to the molecular action of phytochemical antioxidants. Only then can dietary recommendations for healthy aging be based on our confirmed molecular requirements for antioxidants from plant-based foods, rather than, as is currently the case, on observational and biomarker associations between oxidative stress, antioxidants and age-related disease. Still, even though exact mechanisms of action are yet to be revealed, these associations are strong enough to conclude that an antioxidant-rich diet, and a well-balanced vegetarian diet in particular, extends healthspan.

SUMMARY POINTS

- Increased oxidative stress and inflammation form a 'common soil' that accelerates aging and increases the risk of age-related disease.
- Vegetarians in general have a lower risk of age-related disease and longer healthspan than non-vegetarians.
- High intake of plant-based, antioxidant-rich foods is associated with lower oxidative stress, increased antioxidant defense and healthy aging.
- Many diet-derived antioxidants can act as pro-oxidants *in vitro*, and so both sides of the cellular 'redox balance' could be affected by phytochemical intake, affecting the overall outcome.
- Small, pro-oxidant changes in redox balance caused by antioxidant phytochemicals could induce mild stresses that lead to protective adaptations in the cell which lower oxidative stress and inflammation and promote DNA repair.
- If the diet is poor in such phytochemicals, or if responses to mild changes in redox tone are blunted, then accelerated aging and increased risk of age-related disease could occur due to poor defense, cytoprotection and repair.
- In contrast, regular, mild, pro-oxidant swings induced by 'antioxidant' phytochemicals and the consequent cytoprotective adaptations could be responsible for at least some of the long-term health benefits of antioxidant-rich diets and the longer healthspan of vegetarians.
- Further research is needed to map phytochemical action to molecular changes responsible for the long-term health effects of dietary antioxidants in order to more clearly understand the role of dietary antioxidants and vegetarian diets in the promotion of healthy aging and to better inform the general public and the food and agriculture industry in relation to dietary antioxidants and healthspan.

References

1. Arking R. *Biology of ageing*. Sunderland MA: Sinauer Associates; 1998. p 7.
2. Beckman KB, Ames BN. The free radical theory of aging matures. *Physiol Rev* 1998;**78**:54–81.
3. Kirkwood TBL. The odd science of aging. *Cell* 2005;**120**:437–47.
4. WHO, World Health Organization. www.who.int/mediacentre/factsheets and http://www.whi.int/healthinfo/global_disease/en/index/html.
5. *CIA World Factbook*. 2012; www.theodora.com/wfbcurrent/hong_kong_people.html.
6. Benzie IFF. Evolution of antioxidant defence mechanisms. *Eur J Nutr* 2000;**39**:53–61.
7. Halliwell B, Gutteridge JM. *Free radicals in biology and medicine*. Oxford: Oxford University Press; 2007.
8. Salmon AB, Richardson A, Pérez VI. Update on the oxidative stress theory of aging: does oxidative stress play a role in aging or healthy aging? *Free Radic Biol Med* 2010;**48**:642–55.
9. Benzie IFF. Evolution of dietary antioxidants. *J Comp Biochem Physiol* 2003;**136**:113–26.
10. Calabrese V, Cornelius C, Dinkova-Kostova AT, et al. Cellular stress responses, hormetic phytochemicals and vitagenes in aging and longevity. *Biochim Biophys Acta* 2012;**1822**:753–83.
11. Wanagat J, Dai DF, Rabinovitch P. Mitchondrial oxidative stress and mammalian healthspan. *Mech Ageing Develop* 2010;**131**:527–35.
12. Davies SS, Traustadóttir T, Stock AA, et al. Ischemia/reperfusion unveils impaired capacity of older adults to restrain oxidative insult. *Free Radic Biol Med* 2009;**47**:1014–8.
13. Kapahi P, Boulton ME, Kirkwood TB. Postive correlation between mammalian lifespan and cellular resistance to stress. *Free Radic Biol Med* 1999;**26**:495–500.
14. Sinclair DA, Oberdoerffer P. The ageing epigenome: damaged beyond repair? *Ageing Res Rev* 2009;**8**:189–98.
15. Doria E, Buonocore D, Focarelli A, Marzatico F. Relationship between human aging muscle and oxidative system pathway. *Oxid Med Cell Longevity* 2012. http://dx.doi.org/10.1155/2012/830257.
16. Meng SJ, Yu LJ. Oxidative stress, molecular inflammation and sarcopenia. *Int J Mol Sci* 2010;**11**:1509–26.
17. Ventura E, Durant R, Jaussent A, et al. Homocysteine and inflammation as main determinants of oxidative stress in the elderly. *Free Radic Biol Med* 2009;**46**:737–44.
18. Benzie IFF, Wachtel-Galor S. Vegetarian diets and public health: biomarker and redox connections. *Antiox Redox Signal* 2010;**13**:1575–91.
19. Benzie IFF, Strain JJ. Diet and antioxidant defence. In: Caballero B, Allen L, Prentice A, editors. *The encyclopedia of human nutrition*. 2nd ed. London: Academic Press; 2005. pp. 131–7.
20. Benzie IFF, Wachtel-Galor S. Biomarkers of long term vegetarian diets. *Adv Clin Chem* 2009;**47**:169–220.
21. Benzie IFF, Strain JJ. The ferric reducing ability of plasma (FRAP) as a measure of 'antioxidant power': the FRAP assay. *Anal Biochem* 1996;**239**:70–76.
22. Carlsen MH, Halvorsen BL, Holte K, et al. The total antioxidant content of more that 3100 foods, beverages, spices, herbs and supplements used worldwide. *Nutr J* 2010;**9**(3). http//www.nutritionj.com/content/9/1/3.
23. Szeto YT, Tomlinson B, Benzie IFF. Total antioxidant power and ascorbic acid content of fresh fruits and vegetables: implications for dietary planning. *Br J Nutr* 2002;**87**:55–9.

24. Benzie IFF, Szeto YT. Total antioxidant capacity of teas by the ferric reducing/antioxidant power (FRAP) assay. *J Agric Food Chem* 1999;**47**:633–7.
25. Benzie IFF, Wachtel-Galor S. Bioavailability of antioxidant compounds from fruits. In: Skinner M, Hunter D, editors. *Bioactives in fruit: health benefits and functional foods*. Chichester, UK: John Wiley; 2013.
26. Fung ST, Ho C, Choi SW, et al. *Comparison of catechin profiles in human plasma and urine after single dosing and regular intake of green tea*; 2012. PMID 23110850.
27. Serafini M, Del Rio D, Yao DN, et al. Health benefits of tea. In: Benzie IFF, Wachtel-Galor S, editors. *Herbal medicine biomolecular and clinical aspects*. Boca Raton, FL: Taylor and Francis Group; 2011. pp. 239–62.
28. Key TJ, Appleby PN, Davey GK, et al. Mortality in British vegetarians: review and preliminary results from EPIC-Oxford. *Am J Clin Nutr* 2003;**78**(Suppl):533S–8S.
29. Nöthlings U, Schulze MB, Weikert C, et al. Intake of vegetables, legumes, and fruit, and risk for all-cause, cardiovascular, and cancer mortality in a European diabetic population. *J Nutr* 2008;**138**:775–81.
30. Myint PK, Luben RN, Welch AA, et al. Plasma vitamin C concentrations predict risk of incident stroke over 10y in 20 649 participants of the European Prospective Investigation into Cancer Norfolk prospective population study. *Am J Clin Nutr* 2008;**87**:64–9.
31. Harding AH, Wareham NJ, Bingham SA, et al. Plasma vitamin C level, fruit and vegetable consumption, and the risk of new-onset type 2 diabetes: the European prospective investigation of cancer – Norfolk prospective study. *Arch Intern Med* 2008;**168**:1493–9.
32. Tantamango-Bartley Y, Jaceldo-Siegle K, Fan J, Fraser G. Vegetarian diets and the incidence of cancer in a low-risk population. *Cancer Epid Biomarkers Prev* 2012. http://dx.doi.org/10.1158/1055-9965.EPI-12-1060.
33. Huang T, Yang B, Zheng J, et al. Cardiovascular disease mortality and cancer incidence in vegetarians: a meta-analysis and systematic review. *Ann Nutr Metab* 2012;**60**:233–40.
34. Crowe FL, Appleby PN, Travis RC, Key TJ. Risk of hospitalization or death from ischemic heart disease among British vegetarians and nonvegetarians: results from the EPIC-Oxford cohort study. *Am J Clin Nutr* 2013. http://dx.doi.org/10.3945/ajcn.112.044073.
35. Benzie IFF. Antioxidants: observational studies. In: Caballero B, et al., editor. *The encyclopedia of human nutrition*. 2nd ed. London, UK: Academic Press; 2005. pp. 117–30.
36. Bjelakovic G, Nikolova D, Gluud LL, et al. Mortality in randomized trials of antioxidant supplements for primary and secondary prevention systematic review and meta-analysis. *JAMA* 2007;**297**:842–57.
37. World Cancer Research Fund/American Institute for Cancer Research. *Food, nutrition, physical activity and the prevention of cancer: a global perspective*. Washington, DC: AICR; 2007.
38. Craig WJ, Mangeles AR, American Dietetic Association. Position of the American Dietetic Association: vegetarian diets. *J Am Diet Assoc* 2009;**109**:1266–82.
39. Somannavar MS, Kodliwadmath MV. Correlation between oxidative stress and antioxidant defence in South Indian urban vegetarians and non-vegetarians. *Eur Rev Med Pharmacol Sci* 2012;**16**:351–4.
40. Kim MK, Cho SW, Park YK. Long-term vegetarians have low oxidative stress, body fat, and cholesterol levels. *Nutr Res Pract* 2012;**6**:155–61.
41. Helmersson J, Ärnlöv J, Larsson A, Basu S. Low dietary intake of β-carotene, α-tocopherol and ascorbic acid is associated with increased inflammatory and oxidative stress in a Swedish cohort. *Br J Nutr* 2009;**101**:1775–82.
42. Szeto YT, Kwok TCY, Benzie IFF. Effects of a long-term vegetarian diet on biomarkers of antioxidant status and cardiovascular disease risk. *Nutr* 2004;**20**:863–6.
43. Ingenbleek Y, McCully KS. Vegetarianism produces subclinical malnutrition, hyperhomocysteinaemia and atherogenesis. *Nutr* 2012;**28**:148–53.
44. Krajčovičová-Kudláčková M, Valachovičová M, Paukoá V, Dušinská M. Effects of diet and age on oxidative damage products in healthy subjects. *Physiol Res* 2008;**57**:647–51.
45. Choi SW, Benzie IFF, Ma SW, et al. Acute hyperglycaemia and oxidative stress: direct cause and effect? *Free Radic Biol Med* 2008;**44**:1217–31.
46. Choi SW, Benzie IFF, Lam CSY, et al. Inter-relationships between DNA damage, ascorbic acid and glycaemic control in Type 2 diabetes mellitus. *Diab Med* 2005;**22**:1347–53.
47. Kahleova H, Matoulek M, Malinska H, et al. Vegetarian diet improves insulin resistance and oxidative stress markers more than conventional diet in subjects with Type 2 diabetes. *Diab Med* 2011:28549–59.
48. American Heart Association. Arteriosclerosis, thrombosis and vascular biology. http://atvb.ahajournals.org/cgi/content/full/27/5/1177.
49. Nieves JW. Skeletal effects of nutrients and nutraceuticals, beyond calcium and vitamin D. *Osteoporosis Int* 2012. PMID 23152094.
50. Benzie IF, Wachtel-Galor S. Increasing the antioxidant content of food: a personal view on whether this is possible or desirable. *Int J Food Sci Nutr* 2012;**63**(Suppl 1):62–70.
51. Serafini M, Villano D, Spera G, Pellegrini N. Redox molecules and cancer prevention: the importance of understanding the antioxidant network. *Nutr Cancer* 2006;**56**:232–40.
52. Carocho M, Ferreira CFR. A review on antioxidants, prooxidants and related controversy: natural and synthetic compounds, screening and analysis methodologies and future perspectives. *Food Chem Toxicol* 2013;**51**:15–25.
53. Surh YJ, Kundu JK, Na HK. Nrf2 as a master redox switch in turning on the cellular signaling involved in the induction of cytoprotective genes by some chemoprotective phytochemicals. *Planta Med* 2008;**74**:1526–39.
54. Packer L, Cadenas E. Oxidants and antioxidants revisited: new concepts of oxidative stress. *Free Radic Res* 2007;**41**:951–2.
55. Soares MP, Bach FH. Heme oxygenase-1: from biology to therapeutic potential. *Trends Mol Med* 2009. http://dx.doi.org/10.1016/j.molmed.2008.12.004.
56. Virgili F, Marino M. Regulation of cellular signals from nutritional molecules: a specific role for phytochemicals, beyond antioxidant activity. *Free Radic Biol Med* 2008;**45**:1205–16.
57. Radak Z, Boldogh I. 8-oxo-7,8-dihydroguanine: links to gene expression, aging, and defense against oxidative stress. *Free Radic Biol Med* 2010;**49**:587–96.
58. Ribaric S. Diet and aging. *Oxid Med Cell Longev PMID* 2012. 22928085.
59. Page MM, Robb EL, Salway KD, Stuart JA. Mitochondrial redox metabolism: aging, longevity and dietary effects. *Mech Ageing Dev* 2010;**131**:242–52.
60. Finley JW, Kong AN, Hintze KJ, et al. Antioxidants in foods: state of the science important to the food industry. *J Agric Food Chem* 2011;**59**:6837–46.

CHAPTER

9

Enteral Nutrition to Increase Antioxidant Defenses in Elderly Patients

José Eduardo de Aguilar-Nascimento

Department of Surgery, Julio Muller University Hospital, Federal University of Mato Grosso, Cuiaba, Mato Grosso, Brazil

List of Abbreviations

AGEs advanced glycation end-products
CI confidence interval
DNA deoxyribonucleic acid
EN enteral nutrition
HIV human immunodeficiency virus
ICU intensive care unit
IL-6 interleukin 6
mRNA messenger ribonucleic acid
NO nitric oxide
OS oxidative stress
pg picogram
ROS reactive oxygen species
RR risk ratio
SOD superoxide dismutase
WMD weighted mean difference
WP whey protein

INTRODUCTION

Aging is associated with a pro-oxidant state and a decline in endothelial function. Environmental and genetic factors, including elevated oxidant stress (OS), cumulative DNA damage, altered gene expression, telomere shortening, and energy utilization, are among the postulated mechanisms of senescence and aging in mammals.[1] Whether enteral feeding containing ingredients that may increase antioxidant defenses can reverse this decrement in oxidative stress is not well known. Moreover, aging can influence the formation of advanced glycation end-products (AGEs) which accelerate the aging process.[2] The Western diet is high in AGEs, which have been found to be proinflammatory. AGEs have been implicated in various chronic diseases characterized by sustained oxidant stress and low-level inflammatory injury.[3,4] AGEs are formed spontaneously in the organism via non-enzymatic glycol-oxidative reactions between reducing sugars, proteins, and lipids. However, some industrial processed food in Western culture is not only rich in fat but needs heating, irradiation, and ionization, to become safer, flavorful, and colorful. This combination has been found to contribute significantly to the production and accumulation of AGEs in the body.[5]

In this context it is easy to believe that patients exposed for years to a Western diet may accumulate large amounts of AGEs and, consequently, be prone to a pre-inflammatory state that can become worse after an acute or subchronic injury.[6] Therefore, severe oxidative stress conditions – such as trauma, acute respiratory distress syndrome, sepsis, etc. – may be aggravated by the existence of a pre-inflammatory state.

ENTERAL NUTRITION IN ELDERLY PATIENTS

Aging is associated with a decline in skeletal muscle mass that accelerates the process of sarcopenia and directly affects the health of elderly people. The daily ingestion of 25–30 g of high-quality protein per meal, supplemented with leucine, is necessary to prevent sarcopenia in the elderly.[7] In the elderly, an acute condition that necessitates admission to an intensive care unit (ICU) may aggravate the loss of lean body mass; therefore, both the appropriate amount and the type of protein in the formulation of the enteral diet may play a role in the outcome.[8]

Nutritional therapy should begin as early as possible in the ICU, provided that the patient is hemodynamically stable.[9]

Enteral nutrition has been preferred over total parenteral nutrition in the past decade for several reasons, including lower morbidity, lower costs, a shorter length

Aging
http://dx.doi.org/10.1016/B978-0-12-405933-7.00009-3

of hospital stay, and the preservation of both the mucosal barrier and immune competence. Enteral nutrition is the preferred route and is recommended by international guidelines;[10] it should be started in the first 24 hours following admission – and no later than 48 hours.

ENTERAL NUTRITION WITH ANTIOXIDANT PROPERTIES

Critical illness is characterized by hyperinflammation, cellular immune dysfunction, oxidative stress, and mitochondrial dysfunction. Critical conditions increase the formation of reactive oxygen species (ROS). ROS are normally produced during physiologic processes such as cellular respiration and inflammatory defense mechanisms but are significantly increased in critical conditions.[11] Massive increases in ROS and other reactive species can lead to oxidative stress, promoting cell injury and death.[12] The trend to increased oxidative stress during critical illness is associated with activation of neutrophils and monocytes, the production of nitic oxide (NO), and the release of iron and copper ions and metalloproteins.[11,12]

The endogenous antioxidant defense system in humans includes various extracellular and intracellular antioxidants which are able to protect tissues from ROS. Some trace micronutrients, such as zinc, copper, iron, manganese, and selenium, have a role in the antioxidant defenses. They are said to improve the activity of superoxide dismutase (SOD) and glutathione peroxidase.[13] Glutathione levels in critically ill patients are usually decreased.[14,15] Glutathione (L-γ-glutamyl-L-cysteineylglycine) is an endogenous tripeptide, and is one of the most important scavengers in humans, with many protective and metabolic functions in cellular metabolism, such as counteracting ROS (Fig. 9.1). Glutathione is derived from glutamate, cysteine, and glycine, and, because whey protein is rich in cysteine, whey protein may favor the synthesis of glutathione (Fig. 9.2).[11,13,14] Critically ill patients usually experience an oxidative stress status that is associated with a decrease in serum glutathione levels and an increased rate of protein catabolism.[16] Various randomized trials have been published and, as can be seen in Table 9.1, the results show a trend toward less mortality with antioxidant therapy.[17–18]

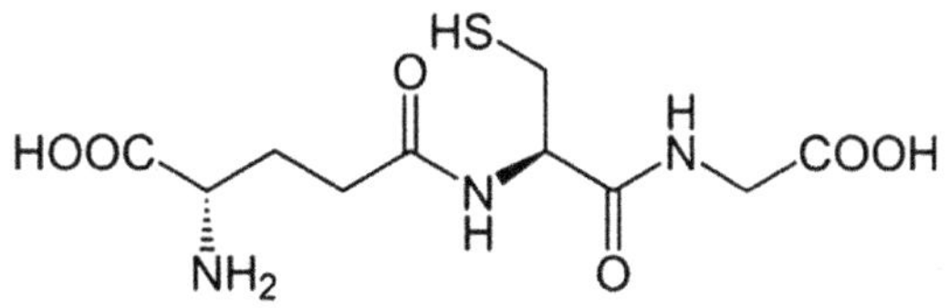

FIGURE 9.1 The glutathione molecule.

ENTERAL NUTRITION WITH WHEY PROTEIN

There is experimental and clinical evidence that whey protein, which is rich in leucine – as opposed to slowly digested protein such as casein – increases the synthesis of protein in the aged.[19] Whey protein, known as fast-protein, is formed by soluble protein fractions extracted from dairy milk (Fig. 9.3); it contains a high concentration of essential amino acids, and is the richest known source of branched-chain amino acids, particularly leucine (up to 14 g/100 g protein).[20] Leucine enhances the reversible phosphorylation of proteins that control mRNA function and, thus, stimulates muscle protein synthesis.[21] Whey protein may have antioxidant properties owing to the rich concentration of cysteine, which is essential for the production of endogenous glutathione, one of the most important organic scavengers (Figs 9.1 and 9.2).[22] Whey protein may also have anti-inflammatory properties. The composition of whey includes other important proteins such as alpha-lactalbumin, beta-lactoglobulin, and lactoferrin that may inhibit the formation of interleukin 6

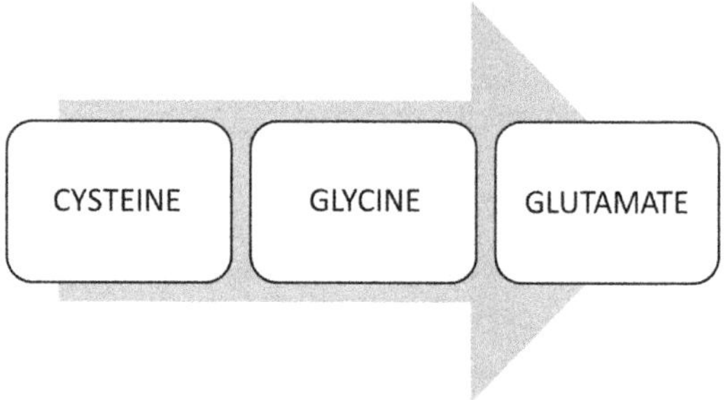

FIGURE 9.2 Glutathione is a tripeptide formed from three amino acids: cysteine, glycine, and glutamate.

TABLE 9.1 Mortality in Various Randomized Trials of Critically Ill Patients Receiving Either Antioxidants or Standard Therapy

		Mortality (%)	
Study	N	Antioxidant	Control
Young et al 1996[17]	68	12	26
Berger et al 1998[18]	20	10	0
Angstwurm et al 1999[19]	42	33	52
Crimi et al 2004[20]	224	44	68
Angstwurm et al 2007[21]	249	40	50
Berger et al 2007[22]	200	14	11
Valenta et al 2011[23]	150	25	32
Manzanares et al 2011[24]	31	33	44
Andrews et al 2011[25]	502	33	33

(IL-6) and various other inflammatory mediators.[23] Whey protein's rapid digestibility, and its contents, which are rich in both leucine and cysteine, may play a role in modulating the acute-phase protein response and increasing glutathione release in a short period of time.[18,19]

In the elderly, there is some evidence that the nitrogen balance is improved with whey protein supplementation.[24] The improvement of the oxidative stress may theoretically improve the balance of protein anabolism/catabolism, with comprehensive benefits to the critically ill patient. Supplementation with whey protein consistently increased plasma glutathione levels in patients with advanced HIV infection.[25]

A recent randomized trial investigated the effects of an early enteral formula containing whey protein, in comparison to a standard enteral formula containing casein as the protein source, on the levels of glutathione and inflammatory markers in aged patients with acute ischemic stroke. The authors randomized 31 elderly patients (12 males and 19 females; median age = 74 [range 65–90] years old) with ischemic stroke to receive early nasogastric feeding (35 kcal/kg/d and 1.2 g of protein/kg/d) with either a formula containing casein (casein group, n =16) or another isocaloric and isonitrogenous formula containing hydrolyzed whey protein (WP group, n = 15) for 5 days. The findings showed that while albumin levels dropped only in the casein group ($P < 0.01$), serum IL-6 decreased (62.7 ± 47.2 to 20.6 ± 10.3 pg/dL; $P = 0.02$) and glutathione increased (32.2 ± 2.1 to 39.9 ± 6.8 U/G Hb; $P = 0.03$) only in the WP group (Fig. 9.4). Serum IL-6 was lower ($P = 0.03$) and glutathione was higher ($P = 0.03$) in whey protein-fed patients than in the casein group (Fig. 9.5). The authors concluded that an enteral formula containing whey protein may decrease inflammation and increase antioxidant defenses in elderly patients with ischemic stroke, compared to a casein-containing formula.[26]

THE ROLE OF VITAMINS AND MICRONUTRIENTS WITH ANTIOXIDANT PROPERTIES

Oxidative stress affects a wide spectrum of critically ill patients. Trace elements, such as copper, manganese, zinc, iron, and selenium, are required for antioxidant defenses, especially for the activity of SOD, catalase, and glutathione peroxidase. Some vitamins – such as E, C, and β-carotene – are also implicated in activity against oxidative stress.[13] Thus, the addition of such nutrients in enteral formulas seems to be attractive, and various randomized trials have reported accordingly.[27,28]

Recently, Manzanares et al performed a meta-analysis of 21 randomized trials that evaluated important clinical endpoints with a supplementation of antioxidant micronutrients (vitamins and trace elements) versus placebo in critically ill patients.[29] Overall, a total of 2531 patients were included in the analysis. The inclusion of antioxidant micronutrients in the enteral formula was associated with a significant reduction in mortality (risk ratio (RR) = 0.82, 95% confidence intervals (CI) 0.72 to 0.93, $P = 0.002$). There was also a trend toward a reduc tion in infectious complications (RR = 0.88, 95% CI 0.76 to 1.02, $P = 0.08$). Furthermore, antioxidants significantly decreased the duration of mechanical ventilation (weighted mean difference (WMD) = −0.67, 95% CI −1.22 to −0.13, $P = 0.02$) in four trials that looked at this endpoint. However, there was no difference in length of stay.

Among all micronutrients, selenium is probably the most effective in the composition of enteral nutrition

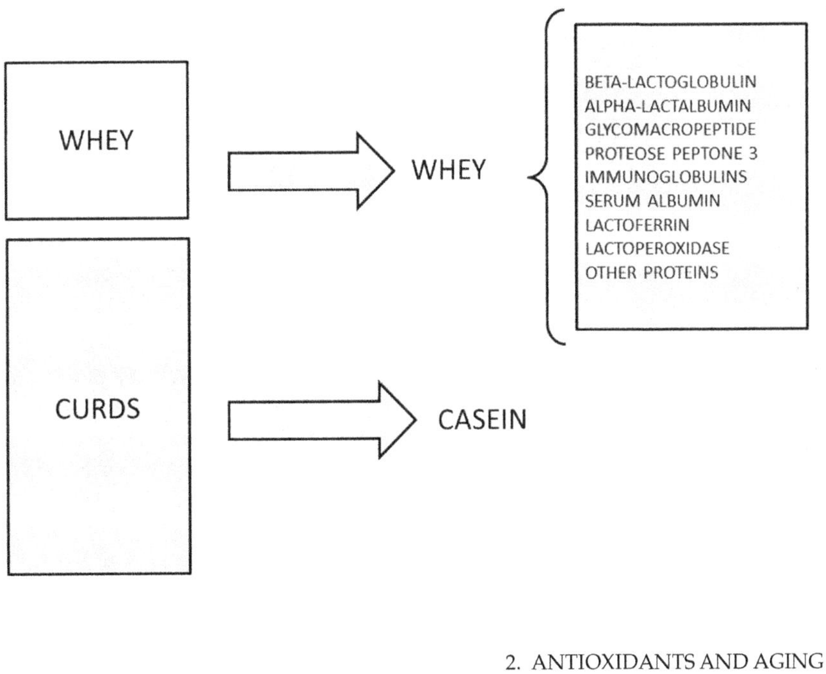

FIGURE 9.3 The composition of whey protein.

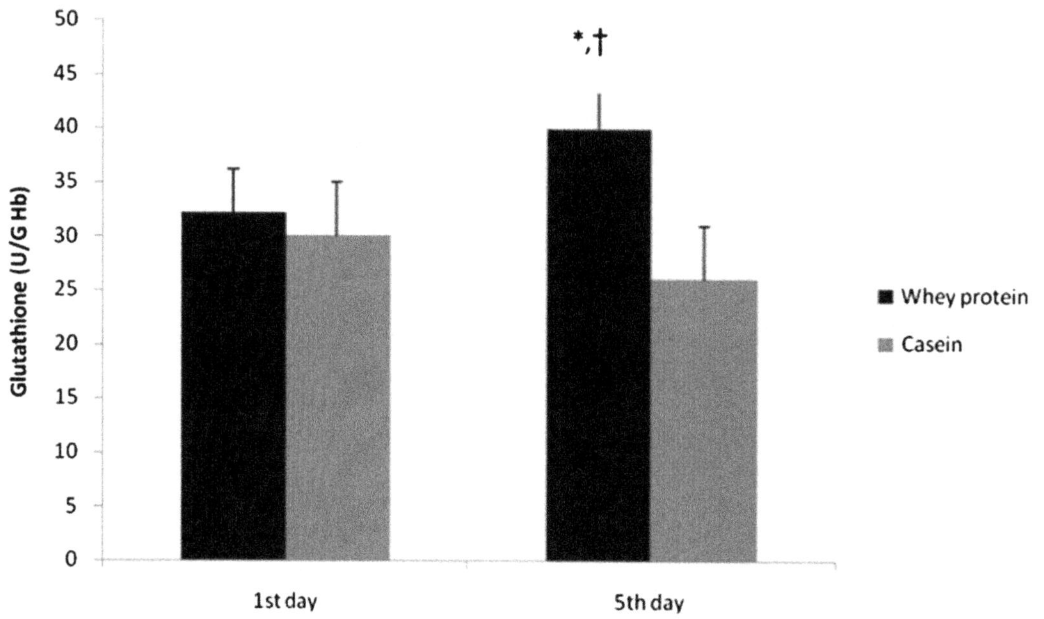

FIGURE 9.4 The increase in glutathione after 5 days of enteral nutrition containing whey protein in elderly critically ill patients.[33] Data are expressed as the mean and SEM. Whey protein group > casein group on the 5th day ($P < 0.05$). I thank Elsevier and the journal *Nutrition* for permission to use this figure.[33]

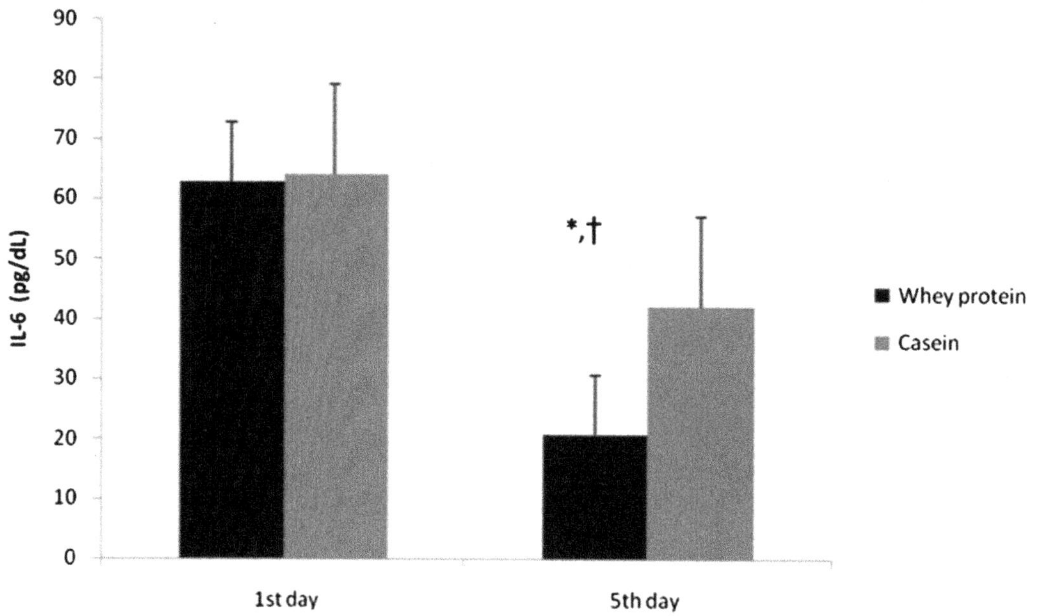

FIGURE 9.5 Serum IL-6 on the first and fifth days after enteral feeding with either casein or whey protein in elderly critically ill patients.[33] Data are expressed as the mean and SEM. *$P = 0.02$ vs. first day in the whey protein group. †$P = 0.03$ vs. casein group (between-groups comparison). I thank Elsevier and the journal *Nutrition* for permission to use this figure.[33]

prescribed for critically ill patients. Heyland et al published a meta-analysis that included 11 randomized trials; it aimed to detect a treatment effect of micronutrients and vitamins on clinically important outcomes in critically ill patients. One of the endpoints of this meta-analysis was to compare the effect of selenium, either alone or in association with other trace elements. Seven out of these 11 studies included selenium alone or in combination with other antioxidants as a component of the antioxidant strategy. Selenium supplementation (alone and in combination with other antioxidants) was associated only with a trend toward a lower mortality (RR 0.59, 95% CI 0.32, 1.08; $P = 0.09$). However, non-selenium therapy (vitamins and other trace elements) showed no effect on mortality.

CONCLUSION

Oxidative stress is a common feature among critically ill patients, especially the elderly. In this subset of hospitalized patients enteral therapy should start as early as possible, at least within 48 hours of admission. Evidence from randomized trials and meta-analysis suggests that formulations containing either whey protein or a mixture of vitamins and micronutrients are most effective for improving antioxidant defenses; such formulations therefore enhance the recovery of critically ill patients. Selenium is probably the most effective trace element to enhance protection from oxidative damage.

SUMMARY POINTS

- Oxidative stress affects a wide spectrum of critically ill elderly patients.
- Nutritional therapy should begin as early as possible in the ICU provided that the patient is hemodynamically stable, and enteral nutrition should be preferred over total parenteral nutrition.
- An enteral nutrition formula containing whey protein may have antioxidant properties owing to the rich concentration of cysteine, which is essential for the production of endogenous glutathione, one of the most important organic scavengers.
- An enteral nutrition formula containing vitamins such as E, C, and β-carotene may have a beneficial activity against oxidative stress.
- Selenium is probably the most effective micronutrient against oxidative stress in critically ill elderly patients.

References

1. Masoro EJ. Overview of caloric restriction and ageing. *Mech Ageing Dev* 2005;**126**:913–22.
2. Luevano-Contreras C, Chapman-Novakofski K. Dietary advanced glycation end products and aging. *Nutrients* 2010;**2**(12):1247–65.
3. Ahmed N. Advanced glycation endproducts – role in pathology of diabetic complications. *Diabetes Res Clin Pract* 2005;**67**:3–21.
4. Zhang Q, Ames JM, Smith RD, et al. A perspective on the Maillard reaction and the analysis of protein glycation by mass spectrometry: probing the pathogenesis of chronic disease. *J Proteome Res* 2009;**8**:754–69.
5. Uribarri J, Woodruff S, Goodman S, et al. Advanced glycation end products in foods and a practical guide to their reduction in the diet. *J Am Diet Assoc* 2010;**110**:911–6.
6. Guo WA, Davidson BA, Ottosen J, et al. Effect of high advanced glycation end-product diet on pulmonary inflammatory response and pulmonary function following gastric aspiration. *Shock* 2012;**38**(6):677–84.
7. Paddon-Jones D, Rasmussen BB. Dietary protein recommendations and the prevention of sarcopenia. *Curr Opin Clin Nutr Metab Care* 2009;**12**:86–90.
8. Yoo SH, Kim JS, Kwon SU, et al. Undernutrition as a predictor of poor clinical outcomes in acute ischemic stroke patients. *Arch Neurol* 2008;**65**:39–43.
9. de Aguilar-Nascimento JE, Bicudo-Salomao A, Portari-Filho PE. Optimal timing for the initiation of enteral and parenteral nutrition in critical medical and surgical conditions. *Nutrition* 2012;**28**(9):840–3.
10. McClave SA, Martindale RG, Vanek VW, et al. Guidelines for the provision and assessment of nutrition support therapy in the adult critically ill patient: Society of Critical Care Medicine (SCCM) and American Society for Parenteral and Enteral Nutrition (A.S.P.E.N.). *J Parenter Enteral Nutr* 2009;**33**:277–316.
11. Tyml K. Critical role for oxidative stress, platelets, and coagulation in capillary blood flow impairment in sepsis. *Microcirculation* 2011;**18**(2):152–62.
12. Brealey D, Brand M, Hargreaves I, et al. Association between mitochondrial dysfunction and severity and outcome of septic shock. *Lancet* 2002;**360**:219–23.
13. Lovat R, Preiser JC. Antioxidant therapy in intensive care. *Curr Opin Crit Care* 2003;**9**:266–70.
14. Brealey D, Brand M, Hargreaves I, et al. Association between mitochondrial dysfunction and severity and outcome of septic shock. *Lancet* 2002;**360**:219–23.
15. Luo M, Fernandez-Estivariz C, Jones DP, et al. Depletion of plasma antioxidants in surgical intensive care unit patients requiring parenteral feeding: effects of parenteral nutrition with or without alanyl-glutamine dipeptide supplementation. *Nutrition* 2008;**24**:37–44.
16. Flaring UB, Rooyackers OE, Hebert C, et al. Temporal changes in whole-blood and plasma glutathione in ICU patients with multiple organ failure. *Intensive Care Med* 2005;**31**:1072–8.
17. Young B, Ott L, Kasarskis E, et al. Zinc supplementation is associated with improved neurologic recovery rate and visceral protein levels of patients with severe closed head injury. *J Neurotrauma* 1996;**13**:25–34.
18. Berger MM, Spertini F, Shenkin A. Trace element supplementation modulates pulmonary infection rates after major burns: a double-blind, placebo-controlled trial. *Am J Clin Nutr* 1998;**68**:365–71.
19. Angstwurm MW, Schottdorf J, Schopohl J, Gaertner R. Selenium replacement in patients with severe systemic inflammatory response syndrome improves clinical outcome. *Crit Care Med* 1999;**27**:1807–13.
20. Crimi E, Liguori A, Condorelli M, et al. The beneficial effects of antioxidant supplementation in enteral feeding in critically ill patients: a prospective, randomized, double-blind, placebo-controlled trial. *Anesth Analg* 2004;**99**:857–63.
21. Angstwurm MW, Engelmann L, Zimmermann T, et al. Selenium in intensive care (SIC): results of a prospective randomized, placebo controlled, multiple-center study in patients with severe systemic inflammatory response syndrome, sepsis, and septic shock. *Crit Care Med* 2007;**35**:118–26.
22. Berger MM, Binnert C, Chiólero RL, et al. Trace element supplementation after major burns increases burned skin trace element concentrations and modulates local protein metabolism but not whole-body substrate metabolism. *Am J Clin Nutr* 2007;**85**:1301–6.
23. Valenta J, Brodska H, Drabek T, et al. High-dose selenium substitution in sepsis: a prospective, randomized clinical trial. *Intensive Care Med* 2011;**37**:808–15.
24. Manzanares W, Biestro A, Torre MH, et al. High dose selenium reduces ventilator associated pneumonia and illness severity in critically ill patients with systemic inflammation. *Intensive Care Med* 2011;**37**:1120–7.
25. Andrews PJ, Avenell A, Noble DW, et al. Scottish Intensive Care Glutamine or Selenium Evaluative Trials Group. Randomised trial of glutamine, selenium, or both, to supplement parenteral nutrition for critically ill patients. *BMJ* 2011;**17**:d1542.

26. Rieu I, Balage M, Sornet C, et al. Increased availability of leucine with leucine-rich whey proteins improves postprandial muscle protein synthesis in aging rats. *Nutrition* 2007;**23**:323–31.
27. Dangin M, Boirie Y, Guillet C, Beaufrère B. Influence of the protein digestion rate on protein turnover in young and elderly subjects. *J Nutr* 2002;**132**:3228S–33S.
28. Rieu I, Balage M, Sornet C, et al. Leucine supplementation improves muscle protein synthesis in elderly men independently of hyperaminoacidaemia. *J Physiol* 2006;**575**:305–11.
29. Middleton N, Jelen P, Bell G. Whole blood and mononuclear cell glutathione response to dietary whey protein supplementation in sedentary and trained male human subjects. *Int J Food Sci Nutr* 2004;**55**:131–41.
30. Yamaguchi M, Yoshida K, Uchida M. Novel functions of bovine milk-derived alpha-lactalbumin: anti-nociceptive and anti-inflammatory activity caused by inhibiting cyclooxygenase-2 and phospholipase A2. *Biol Pharm Bull* 2009;**32**:366–71.
31. Hays NP, Kim H, Wells AM, et al. Effects of whey and fortified collagen hydrolysate protein supplements on nitrogen balance and body composition in older women. *J Am Diet Assoc* 2009;**109**:1082–7.
32. Grobler L, Siegfried N, Visser ME, et al. Nutritional interventions for reducing morbidity and mortality in people with HIV. *Cochrane Database Syst Rev* 2013;**2**. CD004536.
33. de Aguilar-Nascimento JE, Prado Silveira BR, Dock-Nascimento DB. Early enteral nutrition with whey protein or casein in elderly patients with acute ischemic stroke: a double-blind randomized trial. *Nutrition* 2011;**27**(4):440–4.
34. El-Attar M, Said M, El-Assal G, et al. Serum trace element levels in COPD patient: the relation between trace element supplementation and period of mechanical ventilation in a randomized controlled trial. *Respirology* 2009;**14**:1180–7.
35. Manzanares W, Biestro A, Torre MH, et al. High dose selenium reduces ventilator associated pneumonia and illness severity in critically ill patients with systemic inflammation. *Intensive Care Med* 2011;**37**:1120–7.
36. Manzanares W, Dhaliwal R, Jiang X, et al. Antioxidant micronutrients in the critically ill: a systematic review and meta-analysis. *Crit Care* 2012;**16**(2):R66.

CHAPTER

10

Herbs and Spices in Aging

Suhaila Mohamed

Institute of BioScience, Universiti Putra Malaysia, Serdang, Selangor, Malaysia

List of Abbreviations

ABETA β-amyloid
AGEs advanced glycation endproducts
AMD age-related macular degeneration
AP-1 activator protein 1
APP amyloid precursor protein
ATP adenosine triphosphate
BCL2 gene encoding apoptosis regulator proteins
COX cyclooxygenase
DNA deoxyribonucleic acid
FGF fibroblast growth factor
GFAP glial fibrillary acidic protein
HIV human immunodeficiency virus
IL interleukin
iNOS inducible nitric oxide synthase
L-NAME L-N-arginine methyl ester used to induce nitric oxide (NO) deficiency in rats
LOX lipooxygenase
MDA malondialdehyde
ncRNAs noncoding RNAs
NFκB nuclear factor-kappaB
NO nitric oxide
RNA ribonucleic acid
RPE retinal pigment epithelial
SBP systolic blood pressure
SOD superoxide dismutase
TNF tumor necrosis factor
UPP ubiquitin–proteasome pathway
UVB ultraviolet B
VEGF vascular endothelial growth factor

INTRODUCTION

Antioxidative herbs and spices can help to slow down ailments related to aging in animal models by reducing tissue oxidative stress. Many herbs and spices are used as traditional medicines for infections, various injuries, repelling insects, toothache, ulcers, wounds, arthritis, rheumatism, sprains, asthma, allergies, bronchitis, colic, dyspepsia, nausea and minor infections. They have digestive, mild anesthetic, stimulant, mild irritant, vasodilative, antinausea, antibacterial, antifungal, antilisteric, antioxidative and anti-inflammatory properties.[1] Chronic oxidative stress often results in degenerative chronic complications and premature aging affecting the eyes, heart, liver, blood vessels, nerves and kidneys. Aging animals show increased lipid peroxidation activities and decreased antioxidant levels in the kidneys, skins, lenses, hearts and aortas.[1] Circulating lipid peroxides can initiate atherosclerosis under oxidative stress. Prolonged oxidative stress increases protein glycation and the pathologic advanced glycation endproducts (AGEs) in organ tissues; some herbs and spices contain potent antiglycation compounds. The AGEs occur on intra- and extracellular proteins, lipids and nucleic acids; they not only cause overproduction of free radicals, which are damaging to the cells, but also change intracellular signaling and gene expression – resulting in a release of pro-inflammatory molecules and free radicals that cause lesions in tissues and organs. Over 42 herbs and spices have antiglycation properties[2] and their antiglycation properties are highly correlated to their total content of polyphenols.[3] These antiglycating herbs and spices include *Allium sativum, Zingiber officinale, Thymus vulgaris, Petroselinum crispum, Murraya koenigii Spreng, Mentha piperita* L., *Curcuma longa* L., *Allium cepa* L., *Allium fistulosum* and *Coriandrum sativum* L.

The ubiquitin–proteasome pathway (UPP) selectively degrades aberrant proteins to appropriately remove potential cytotoxic, damaged or abnormal proteins linked to aging and degenerative diseases. Age, oxidative-stress-induced impairment or overburdening of the UPP hinders the removal of abnormal proteins from the tissues. Certain herbs and spices (examples: turmeric, cinnamon and herbs containing 3,5-dicaffeoyl quinic acid) help to prevent health degeneration by enhancing or preserving UPP function or reducing the UPP burden.

SPICES

Spice is defined as any aromatic or pungent plant substance used to flavor food. Some food seasoning spices help to improve plasma glucose, glucose

Aging
http://dx.doi.org/10.1016/B978-0-12-405933-7.00010-X

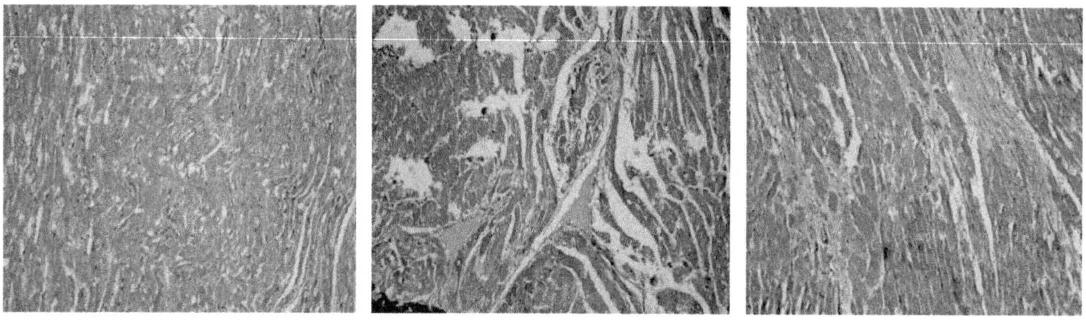

Normal rats: normal myocardial fibers in N and NC rats [H&E × 100].

Diabetic rats: vacuolation and fibrosis in D rats [H&E × 100].

Diabetes + cloves: DC rats showed reduced fibrosis [H&E × 100].

FIGURE 10.1 Histological evidence of the protective effects of cloves on oxidatively stressed diabetic rat's heart.[1] N, normal ats; NC, Normal rats with cloves in their diet; D, Diabetic rats; DC, Diabetic rats with cloves in their diet.

FIGURE 10.2 Histological evidence of the protective effects of antioxidative spice (clove) on oxidatively stressed diabetic rats' lenses. Reproduced with permission from reference 1.

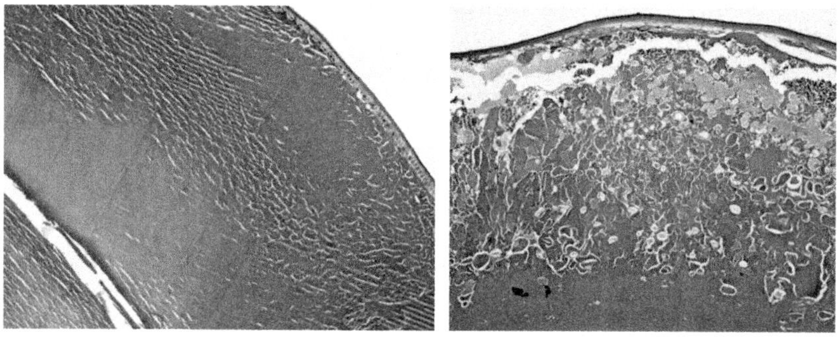

Normal lenses of normal rats and diabetic rats supplemented with clove.

Cataract in the lens of diabetic rats [H&E × 100].

metabolism and the lipid profile of hyperinsulinemic mammals. The modes of action include improving insulin-sensitivity and reducing oxidative stress in the tissues. Some culinary herbs and spices effectively inhibit protein glycation, and the most potent inhibitors reported were cloves, ground Jamaican allspice and cinnamon extracts.[1,4]

Clove

Clove, and some other culinary herbs and spice extracts, apparently have insulin-enhancing activity or cause increases in insulin sensitivity.[5] Consumption of cloves (equivalent to about 1g cloves/day for humans) gradually lowers fasting blood glucose levels.[1] Clove oil helps to modulate physiologic responses in aging rodents.[6,7] The eugenol and eugenyl acetate in cloves are antioxidative. Eugenol dose dependently binds to membranes, thus stabilizing them and protecting them against free radical attack.[8] Eugenol inhibits lipid oxidation and helps to limit structural changes to various tissues, such as the heart, kidney and liver.[1,9] Eugenol inhibits histamine release from mast cells to reduce hypersensitivity[10] besides having anti-anaphylactic and antispasmodic properties.[11] Cloves inhibit oxidative tissue damage and cataract formation in the eye lens of rats.[1] Cataractogenesis is caused by glucose diffusion into the lens, increased sorbitol levels, altered membrane permeability, loss of glutathione, decreased amino acid levels and decreased protein synthesis. Besides protecting the eye lens, dietary cloves also protect against damage to the heart (Figs 10.1 and 10.2).

Turmeric

Turmeric contains the antioxidative polyphenol curcumin (diferuloylmethane), which has anti-inflammatory, antiproliferative, anticancer and anti-angiogenic activities. Curcumin regulates the expression of genes related to apoptosis, tumor suppression, cell-cycle arrest, transcription factor, angiogenesis and metastasis.[12] Curcumin ameliorates various chronic illnesses, linked to free radical damage, affecting the eyes, lungs, liver, kidneys and the gastrointestinal and cardiovascular systems. Curcumin modulates signaling molecules, including inflammatory molecules, transcription factors, enzymes, protein kinases, protein reductases, carrier proteins, cell-survival proteins, drug resistance proteins, adhesion molecules, growth factors, receptors, cell-cycle regulatory proteins, chemokines, DNA, RNA and metal ions, important in oxidation, obesity, neurologic and psychiatric disorders and cancer.[12] Additionally, curcumin has antibacterial, chemopreventive, chemotherapeutic, antinociceptive, antiparasitic and antimalarial properties[13] and it blocks HIV replication. Curcumin is non-toxic in normal cells and kills only tumor cells.[14] Turmeric and curcumin prevented oxidative stress in diabetic mammals without changing blood sugar levels.[15] Turmeric and curcumin benefit acute and chronic diseases associated with enhanced inflammation (such as arteriosclerosis, cancer, respiratory, hepatic, pancreatic, intestinal and gastric diseases, neurodegeneration and degenerative diseases)[16] by inhibiting inflammation mediators such as NFKB, cyclooxygenase-2 (COX-2), lipooxygenase (LOX) and inducible nitric oxide synthase (iNOS). Curcuminoids, like many other polyphenols, have antioxidant effects greater than those of vitamins,[17] and they induce enzymes of the glutathione-linked detoxification pathways.[18] Phase I clinical trials have shown that turmeric is safe, even at high doses (12 g/day), in humans because of its poor bioavailability, poor absorption, rapid metabolism and rapid systemic elimination.[19]

Curcumin prevents degenerative disease by modulating (i) several cell-signaling pathways at multiple levels, (ii) cellular enzymes, e.g. cyclooxygenase and glutathione S-transferases, (iii) the immune system, (iv) angiogenesis, (v) cell–cell adhesion, (vi) gene transcription and (vii) apoptosis in preclinical cancer models. Despite the low systemic bioavailability following oral dosing, biologically active levels in the gastrointestinal tract have been reached.[20]

Turmeric is a promising herb for the prevention of age-related Alzheimer's disease. Curcumin significantly lowered oxidized proteins and proinflammatory cytokines in rodents' brains. Low doses of curcumin reduced the astrocytic marker (GFAP-glial fibrillary acidic protein), insoluble β-amyloid (ABETA), soluble ABETA and plaque burden by almost 50%, without reducing levels of amyloid precursor protein (APP) in the membrane fraction. These effects were not observed with a high dose of curcumin. Microgliosis was suppressed in neuronal layers but not near the plaques.[21]

Turmeric and curcumin inhibited the expression of vascular endothelial growth factor (VEGF) in rats,[22] and inhibited endothelial cell proliferation *in vivo*.[23] Hence, curcumin may be protective against various neovascularization-related diseases, stimulated by VEGF, such as cancer and obesity. Turmeric and curcumin prevented cataract development in rodents, even though it is not antihyperglycemic. Turmeric was more effective than curcumin in countering oxidative stress[15] when comparing the rates for changes in (i) lipid peroxidation, (ii) reduced glutathione, (iii) protein carbonyl content, (iv) antioxidant enzyme activities, (v) osmotic stress (assessed by polyol pathway enzymes), and (vi) aggregation or insolubilization of lens proteins induced by hyperglycemia.

Dietary curcumin is effective only in very low amounts (0.002%), but not at high levels (above 0.01%). Curcumin, at a low level (0.002%), inhibited oxidation, glycation, lipid peroxidation, AGE-fluorescence and protein aggregation. Under hyperglycemic conditions, higher levels of dietary curcumin (0.01%) showed the opposite effect of being pro-oxidative, enhancing AGE formation and protein aggregation. However, feeding curcumin to normal rats at up to a 0.01% level did not cause any changes in morphology or biochemical parameters[24] because of its poor bioavailability.

The mechanisms for curcumin protection against lipid peroxidation or galactose-induced oxidative stress are (i) through glutathione S-transferase isozyme induction[25] and (ii) by decreasing apoptosis in cells.[18] Dietary curcumin improved tissue morphology, physiology, gene expression, tissue structure and function in animal models of visual degeneration,[16,25–27] oxidative damage and Alzheimer disease.[21] Curcumin may benefit various degenerative diseases caused by protein trafficking defects such as diabetes, cancer and Alzheimer's disease. Treatment of COS-7 cells with curcumin results in dissociation of mutant protein aggregates and decreased endoplasmic reticulum stress.[27]

Excess free radical generation caused Ca^{2+} ATPase inactivation, leading to Ca^{2+} accumulation and calpain-mediated proteolysis. Pretreatment with curcumin, but not simultaneous or post-treatment, led to a decrease in oxidative stress and also rescued the (selenium-induced) increase in Ca^{2+} and inhibition of Ca^{2+} ATPase activity in the eye lens.[28] Curcumin mediates its effects through the inhibition of transcription factors (NFκB, AP-1), enzymes (COX-1, COX-2, LOX), cytokines (TNF, IL-1, IL-6) and the down-regulation of anti-apoptotic genes (BCL2, BCL2L1). These effects are helpful for various oxidative-stress -and inflammation-related degenerative diseases such as cancers, Alzheimer's disease, cardiovascular ailments, arthritis, alcohol-induced liver injury and multiple sclerosis. Curcumin inhibited the increase in tissue epithelial cells' pigments exclusively by inducing caspase-3/7-dependent but not via caspase-8-dependent cell death and necrosis.[29]

Cinnamon, Black Pepper, Cumin

Cinnamon extract helped safeguard against tissue changes by minimizing protein aggregation, preventing glycation and oxidative stress in fructose-fed rodents. Cumin, ginger, green tea, cinnamon and black pepper, at 1.0 mg/ml, inhibited the formation, *in vitro*, of advanced glycation endproducts (AGE).[2]

Piperine, from black pepper, effectively retarded memory impairment and neurodegeneration in the brain of an Alzheimer's disease animal model. Piperine inhibited lipid peroxidation as well as acetylcholinesterase, and had neurotrophic effects in the hippocampus.[30]

Ginger (*Zingiber officinalis*)

Ginger is known to be carminative, thermogenic, anti-inflammatory, antioxidative and stimulating. It is used traditionally for dyspepsia, gastroparesis, slow motility symptoms, constipation, nausea, coughs, colds or flu, vomiting, colic, pain relief, rheumatoid arthritis, osteoarthritis and joint and muscle injury. Dietary ginger has antiglycating properties, inhibits the polyol pathway, and is antihyperglycemic. Ginger significantly inhibited the formation of various AGE products, including carboxymethyl lysine.31 Dietary ginger delayed cataract onset and progression in rats. Dietary zerumbone from zerumbet ginger dose dependently prevented UVB-induced corneal damage or photokeratitis by inhibiting the accumulation of NFκB, iNOS, TNFα and MDA while increasing glutathione levels in mammals.26 Zerumbone, at 100 mg/kg after UVB exposure, reduced MDA (lipid peroxidation marker) levels, while increasing glutathione, glutathione reductase and SOD (superoxide dismutase) levels in the lens.26 Zerumbone is known for its protective and therapeutic effects against colorectal, breast, cervical, liver and other cancers linked to aging.32

HERBS

Herb is defined in the *English Oxford Dictionary* as 'any plant with leaves, seeds, or flowers used for food, flavoring, medicine, or perfume.'

Herbs Containing Phenolic Compounds and Catechins

Many leafy herbs, such as basil and Moringa leaf, contain beneficial phenolics such as catechins and flavonoids that are potent against various oxidative-stress-related degenerative diseases (Fig. 10.3).[31,33–35] The phenolic compounds often show beneficial effects on diabetes, hypertension, obesity, cancer, cognition and visual degeneration.

Oregano (*Origanum vulgare*)

Origanum majoricum, O. vulgare ssp. *hirtum* and *Poliomintha longiflora* have higher antioxidant ORAC (oxygen radical absorbance capacity) and phenolic contents than many other culinary and medicinal herbs. Oregano retards aging and degenerative diseases through antioxidant defense mechanisms.[36] These antioxidant mechanisms include preventing oxidation-induced DNA damage and enhancing the antioxidant response elements and enzymes.[37] Other herbs with very high ORAC values are *Catharanthus roseus, Thymus vulgaris, Hypericum perforatum* and *Artemisia annua*. Rosmarinic acid was among the most

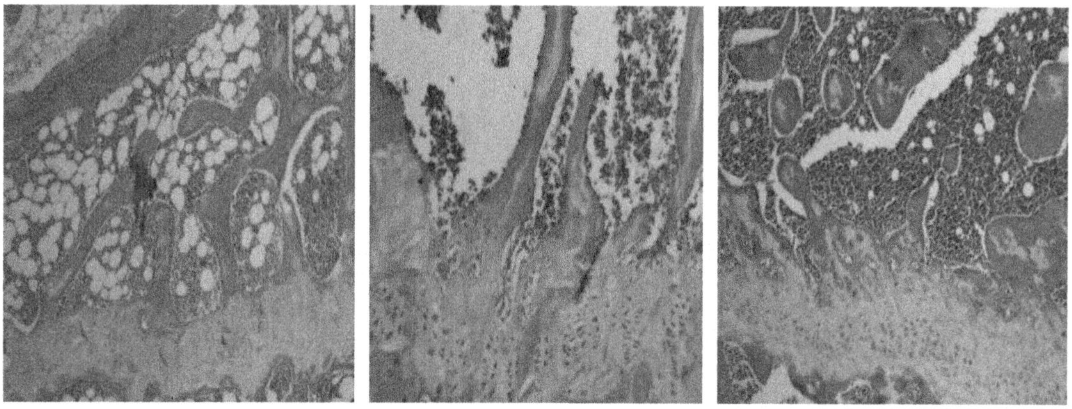

Normal rat's bone architecture.

Architecture of oxidatively stressed estrogen-deficient rat's bone.

Architecture of bone in an estrogen-deficient rat supplemented daily with 300 mg of catechin-rich botanical extract/kg body weight in drinking water.[52]

(H&E staining, 400 × magnification)

FIGURE 10.3 Femur bone architecture of rats after 3 months of estrogen deficiency-induced oxidative stress, with and without catechin-rich botanical extract supplementation.

predominant phenolic compounds in *Salvia officinalis, Thymus vulgaris, Origanum majoricum* and *P. longiflora*, whereas quercetin-3-O-rhamnosyl-(1→2)-rhamnosyl-(1→6)-glucoside and kaempferol-3-O-rhamnosyl-(1→2)-rhamnosyl-(1→6)-glucoside were predominant phenolic compounds in *Ginkgo biloba* leaves.[38]

Vitex negundo

Dietary antioxidative flavonoids from various herbs and botanical sources are protective against aging and degenerative ailments. They help to normalize blood sugar and the lens protein expression of cataract in animal models. The flavonoids from herbs such as *V. negundo* protect against selenite toxicity and cataractogenesis in enucleated rodent lenses by maintaining antioxidant status and calcium homeostasis, by protecting sulfhydryl groups and by decreasing oxidative stress in the lens.[39] In these studies, measures of antioxidant status included the activities of superoxide dismutase (SOD), catalase, $Ca^{(2+)}$ATPase, reduced glutathione (GSH) concentrations, protein sulfhydryl content, calpains, calcium and thiobarbituric acid-reactive substances (TBARS).

Moringa oleifera Leaves

Flavonoids from *Moringa oleifera* leaves were shown to help retard oxidative-stress-related aging ailments such as cataractogenesis by (i) enhancing the activities of antioxidant enzymes (superoxide dismutase and catalase), (ii) reducing the intensity of lipid peroxidation, (iii) inhibiting free radical generation in rats, (iv) improving reduced glutathione levels and (v) reducing protein oxidation.[34]

Brassica Flavonoids

Flavonoids from (especially) the brassicas prevented selenite-induced cataractogenesis in albino rat pups by (i) maintaining antioxidant status and ionic balance through the Ca^{2+} ATPase pump and (ii) inhibiting lipid peroxidation, calpain activation and protein insolubilization.[40] The antioxidant status was measured through superoxide dismutase (SOD), catalase activities, reduced glutathione (GSH) and the levels of lipid peroxidation products.

Ginkgo biloba

Ginkgo biloba is used for cognitive dysfunction, dementia and Alzheimer's disease. The flavonoid-rich leaves generally improve cerebral blood flow and memory. The ginkgolides inhibit platelet aggregation, while the bilobalide protects against neuronal death caused by global brain ischemia and excitotoxicity-induced damage. The mode of neuroprotection is via antioxidative, anti-amyloidogenic and anti-apoptotic effects.[41] The antioxidative effects of *Ginkgo biloba* are shown by its protective effects against membrane lipid peroxidation.[42] Lipid peroxidation in membranes was measured by the generation of lipid radicals (using electron paramagnetic resonance spectroscopy) and thiobarbituric acid-reactive substances (TBARS).[42] However, vitreous hemorrhage was reported with the use of *Ginkgo biloba* in a patient with age-related macular disease.[43]

Trigonella foenum-graecum Seeds (Fenugreek), Chili and Pennywort (*Centella asiatica*)

Alcoholic extract of fenugreek (*Trigonella foenum-graecum* Linn) seeds (2 g kg^{-1} day^{-1}) has beneficial effects on body weight, blood glucose and cataract development in aging humans.[44] The herbs and spices fenugreek, garlic, ginger, pennywort and red pepper are effective as hypocholesterolemics, and they help to retard oxidation and age-related cognitive decline and progression of metabolic syndrome.[45,46]

Saffron (*Crocus sativus* L.)

Saffron increases blood flow in certain tissues[47] affected by aging or chronic oxidative stress. Saffron consumption before tissue exposure to damaging light results in neuro-protective effects through the regulation of various known genes, including the genes encoding the antioxidative GPx3 and ncRNAs.[48] Saffron increases blood flow in the retina and choroid, improves memory and learning skills and has antihypertensive, anticonvulsant, antitussive, anti-genototoxic and cytotoxic effects, as well as anxiolytic, aphrodisiac, antioxidant, antidepressant, anti-nociceptive, anti-inflammatory and relaxant properties.[47] Saffron is reportedly anti-carcinogenic, antimutagenic and analgesic. It shows beneficial effects on opioid dependence, cardiovascular lipids, the respiratory system, gastric ulcer, the immune system, insulin resistance, tissue oxygenation and bronchodilation, and it has antibacterial effects.[47]

Saffron has antioxidants that are vision protective and help in the regulation of cell membrane fatty acid content, making vision cells more resilient. Saffron: (i) protected the photoreceptors from tissue stress, maintaining both morphology and function and regulated cell death, (ii) reduced the rate of photoreceptor death induced by bright continuous light (BCL) in animals, (iii) protected the outer nuclear layer (ONL) in a way similar to that of β-carotene and (iv) prevented the strong upregulation of FGF2 caused by BCL.[49] Saffron prevented vision loss with age, and reversed

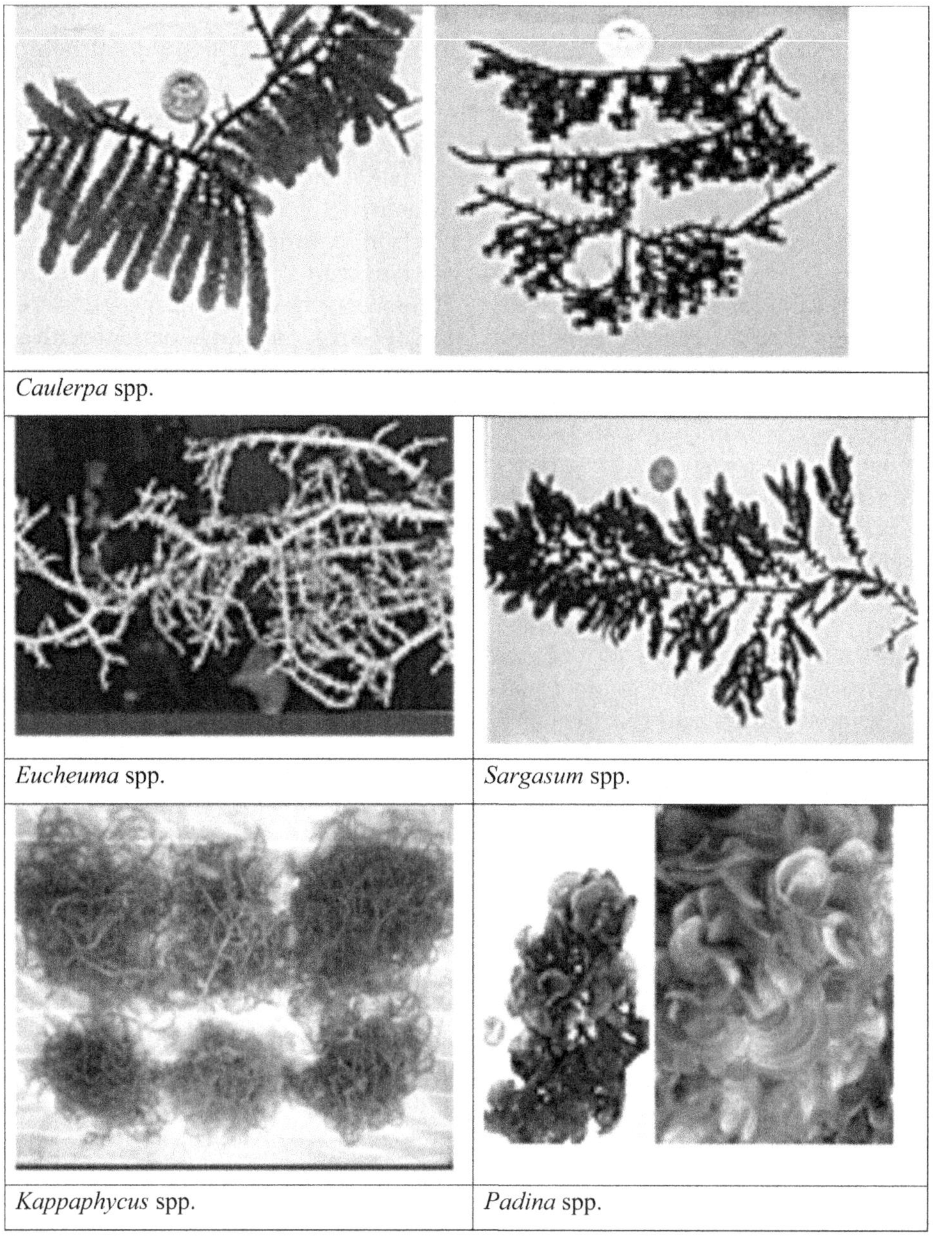

FIGURE 10.4 Examples of antioxidative seaweeds.

age-related macular degeneration (AMD) that often causes blindness in old age. Saffron protected photoreceptors from damage; this inhibited and possibly reversed AMD and *retinitis pigmentosa* (a genetic eye disease). AMD is incurable but can be retarded. When given a daily saffron pill for 3 months, pensioners with AMD – which gradually destroys sharp, central vision – reported improvements in their vision; the effect disappeared when the saffron supplement was stopped. Dietary saffron protected the eyes from damaging bright light and eye-related genetic diseases which can lead to blindness.[50]

SEAWEEDS

Edible seaweeds or sea vegetables may be considered as marine herbs that are rich in botanical antioxidants and beneficial for oxidative stress and health management during aging (Figs 10.4, 10.5 and 10.6). Seaweed compounds that include sulfated polysaccharides, phlorotannins, carotenoids (e.g. fucoxanthin), minerals, peptides and sulfolipids are beneficial against various degenerative metabolic diseases and their therapeutic modes of action and bioactive components have been well studied.[51]

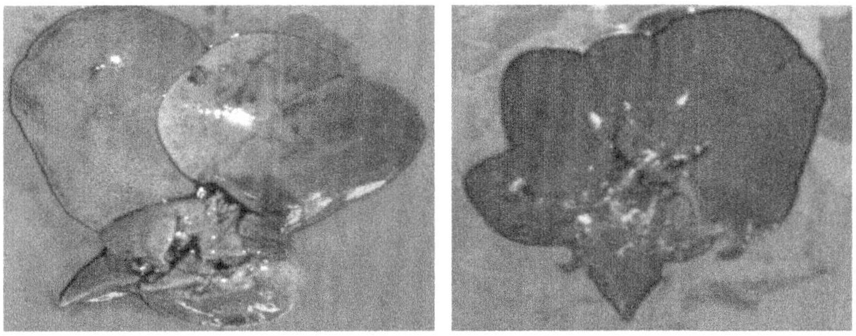

Liver of oxidative stress-induced rodents on a prolonged lipogenic diet. The liver appeared much paler than normal, soft, mottled and fatty.

Livers of rodents on a lipogenic diet + antioxidant herbs. The livers were slightly pale but were less fatty.[53]

FIGURE 10.5 The livers of oxidatively stressed rats on a prolonged lipogenic diet (with and without supplementation with antioxidative herbs).

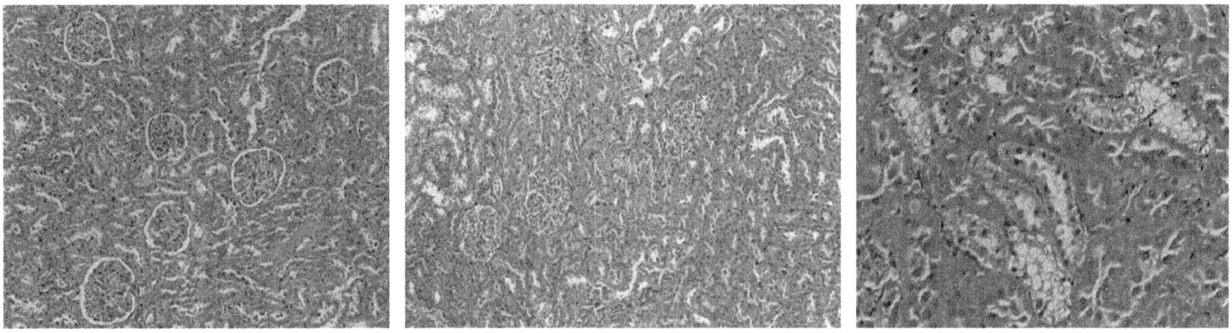

Normal rodent kidney showed normal glomeruli and tubules [H&E × 100].

Kidney of a rodent on a lipogenic diet showed mononuclear inflammatory cell infiltration and thickened glomerular basement membranes (glomerulopathy) [H&E × 100].

Kidney of a rodent on a lipogenic diet showed foamy tubular epithelial cells [H&E × 400].

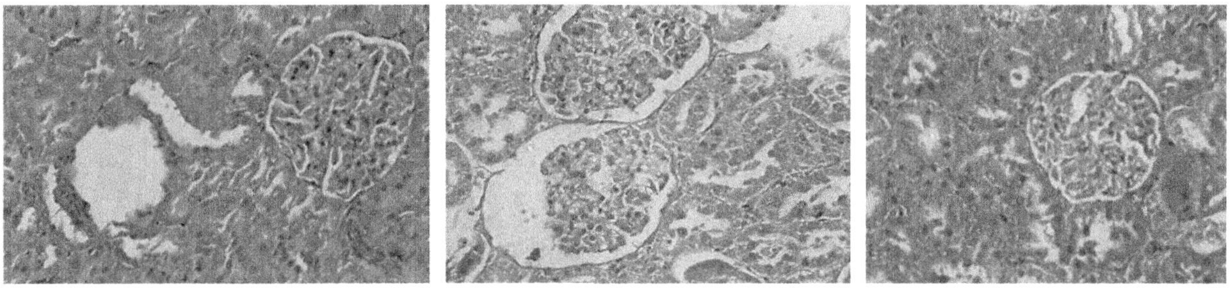

Kidney of a rodent on a lipogenic diet showed loss of whole glomerus and tubular dilation [H&E × 400].

Kidney of a rodent on a lipogenic diet showed mesangium expansion [H&E × 400].

Kidney of a rodent on a lipogenic diet + herbs showed normal glomerulus and tubule structures [H&E × 400].

FIGURE 10.6 Representative microscopic appearance of the kidney of oxidative stressed rats, induced by lipogenic diet (with and without antioxidative herb supplementation). Reproduced with permission from reference.[53]

SUMMARY POINTS

- The protective effects of certain herbs and spices against aging and oxidative stress are reviewed.
- The health conditions covered include cardiovascular ailments, cancer, obesity, diabetes, hypertension, dyslipidemia, oxidative stress, stroke, inflammation, allergy, thyroid problems, cognition, infection and tissue injuries.
- The herbs and spices discussed include cloves, saffron, turmeric, ginger, oregano, cinnamon, cumin, black pepper, *Vitex negundo*, *Moringa olefera* leaves,

fenugreek, onion, garlic, chili, *Ginkgo biloba* and seaweeds.
- Some bioactive components of the herbs and spices responsible for the effects are mentioned.
- Some modes of action of the herbs and spices (mostly antioxidant, antiglycating, biochemical signaling, and modulation of gene expression) are compiled.

References

1. Shukri R, Mohamed S, Mustapha NM. Cloves protect the heart, liver and lens of diabetic rats. *Food Chem* 2010;**122**(4):1116–21.
2. Elosta A, Ghous T, Ahmed N. Natural products as anti-glycation agents: possible therapeutic potential for diabetic complications. *Curr Diabetes Rev* 2012;**8**(2):92–108.
3. Ramkissoon JS, Mahomoodally MF, Ahmed N, Subratty AH. Antioxidant and anti-glycation activities correlates with phenolic composition of tropical medicinal herbs. *Asian Pacific J Trop Med* 2013;**6**(7):561–9.
4. Dearlove RP, Greenspan P, Hartle DK, et al. Inhibition of protein glycation by extracts of culinary herbs and spices. *J Medicinal Food* 2008;**11**(2):275–81.
5. Broadhurst CL, Polansky MM, Anderson RA. Insulin-like biological activity of culinary and medicinal plant aqueous extracts in vitro. *J Agric Food Chem* 2000;**48**(3):849–52.
6. Nangle MR, Gibson TM, Cotter MA, Cameron NE. Effects of eugenol on nerve and vascular dysfunction in streptozotocin-diabetic rats. *Planta Medica* 2006;**72**(6):494–500.
7. Zari TA, Al-Attar AM. Effects of ginger and clove oils on some physiological parameters in streptozotocin-diabetic and non-diabetic rats. *J Med Sci* 2007;**7**(2):267–75.
8. Kumaravelu P, Subramaniyam S, Dakshinamoorthy DP, Devaraj NS. The antioxidant effect of eugenol on CCl4-induced erythrocyte damage in rats. *J Nutritional Biochem* 1996;**7**(1):23–8.
9. Wei A, Shibamoto T. Antioxidant/lipoxygenase inhibitory activities and chemical compositions of selected essential oils. *J Agric Food Chem* 2010;**58**(12):7218–25.
10. Chaieb K, Hajlaoui H, Zmantar T, et al. The chemical composition and biological activity of clove essential oil, *Eugenia caryophyllata* (*Syzigium aromaticum* L. Myrtaceae): a short review. *Phytother Res* 2007;**21**(6):501–6.
11. Fujisawa S, Atsumi T, Satoh K, et al. Radical generation, radical-scavenging activity, and cytotoxicity of eugenol-related compounds. *In Vitro Mol Toxicol* 2000;**13**(4):269–79.
12. Sreenivasan S, Thirumalai K, Krishnakumar S. Expression profile of genes regulated by curcumin in y79 retinoblastoma cells. *Nutrition Cancer* 2012;**64**(4):607–16.
13. Gupta SC, Patchva S, Koh W, Aggarwal BB. Discovery of curcumin, a component of golden spice, and its miraculous biological activities. *Clin Exp Pharmacol Physiol* 2012;**39**(3):283–99.
14. Shehzad A, Lee YS. Curcumin: multiple molecular targets mediate multiple pharmacological actions – a review. *Drugs Future* 2010;**35**(2):113–9.
15. Suryanarayana P, Saraswat M, Mrudula T, et al. Curcumin and turmeric delay streptozotocin-induced diabetic cataract in rats. *Invest Ophthalmol Vis Sci* 2005;**46**(6):2092–9.
16. Chen M, Hu DN, Pan Z, et al. Curcumin protects against hyperosmoticity-induced IL-1beta elevation in human corneal epithelial cell via MAPK pathways. *Exp Eye Res* 2010;**90**(3):437–43.
17. Bengmark S. Curcumin, an atoxic antioxidant and natural NFkappaB, cyclooxygenase-2, lipooxygenase, and inducible nitric oxide synthase inhibitor: a shield against acute and chronic diseases. *JPEN J Parenter Enteral Nutr* 2006;**30**(1):45–51.
18. Pandya U, Saini MK, Jin GF, et al. Dietary curcumin prevents ocular toxicity of naphthalene in rats. *Toxicol Lett* 2000;**115**(3):195–204.
19. Anand P, Kunnumakkara AB, Newman RA, Aggarwal BB. Bioavailability of curcumin: problems and promises. *Mol Pharm* 2007;**4**(6):807–18.
20. Sharma RA, Gescher AJ, Steward WP. Curcumin: the story so far. *Eur J Cancer* 2005;**41**(13):1955–68.
21. Lim GP, Chu T, Yang F, et al. The curry spice curcumin reduces oxidative damage and amyloid pathology in an Alzheimer transgenic mouse. *J Neurosci* 2001;**21**(21):8370–7.
22. Mrudula T, Suryanarayana P, Srinivas PN, Reddy GB. Effect of curcumin on hyperglycemia-induced vascular endothelial growth factor expression in streptozotocin-induced diabetic rat retina. *Biochem Biophys Res Commun* 2007;**361**(2):528–32.
23. Rema M, Pradeepa R. Diabetic retinopathy: an Indian perspective. *Indian J Med Res* 2007;**125**(3):297–310.
24. Suryanarayana P, Krishnaswamy K, Reddy GB. Effect of curcumin on galactose-induced cataractogenesis in rats. *Mol Vis* 2003;**9**: 223–30.
25. Awasthi S, Pandya U, Singhal SS, Lin JT, et al. Curcumin–glutathione interactions and the role of human glutathione *S*-transferase P1-1. *Chem Biol Interact* 2000;**128**(1):19–38.
26. Chen BY, Lin DP, Wu CY, et al. Dietary zerumbone prevents mouse cornea from UVB-induced photokeratitis through inhibition of NF-kappaB, iNOS, and TNF-alpha expression and reduction of MDA accumulation. *Mol Vis* 2011;**17**:854–63.
27. Vasireddy V, Chavali VR, Joseph VT, et al. Rescue of photoreceptor degeneration by curcumin in transgenic rats with P23H rhodopsin mutation. *PLoS One* 2011;**6**(6). e21193.
28. Manikandan R, Thiagarajan R, Beulaja S, et al. Curcumin prevents free radical-mediated cataractogenesis through modulations in lens calcium. *Free Radic Biol Med* 2009;**48**(4):483–92.
29. Alex AF, Spitznas M, Tittel AP, et al. Inhibitory effect of epigallocatechin gallate (EGCG), resveratrol, and curcumin on proliferation of human retinal pigment epithelial cells in vitro. *Curr Eye Res* 2010;**35**(11):1021–33.
30. Chonpathompikunlert P, Wattanathorn J, Muchimapura S. Piperine, the main alkaloid of Thai black pepper, protects against neurodegeneration and cognitive impairment in animal model of cognitive deficit like condition of Alzheimer's disease. *Food Chem Toxicol* 2009;**48**(3):798–802.
31. Saraswat M, Suryanarayana P, Reddy PY, et al. Antiglycating potential of *Zingiber officinalis* and delay of diabetic cataract in rats. *Mol Vis* 2010;**16**:1525–37.
32. Huang YC, Chao KSC, Liao HF, Chen YJ. Targeting sonic hedgehog signaling by compounds and derivatives from natural products. *Evid-Based Complem Altern Med* 2013. 2013, article ID 748587.
33. El-Beshbishy HA, Bahashwan SA. Hypoglycemic effect of basil (*Ocimum basilicum*) aqueous extract is mediated through inhibition of Î±-glucosidase and Î±-amylase activities: an in vitro study. *Toxicol Indust Health* 2012;**28**(1):42–50.
34. Sasikala V, Rooban BN, Priya SG, et al. *Moringa oleifera* prevents selenite-induced cataractogenesis in rat pups. *J Ocul Pharmacol Ther* 2010 Oct;**26**(5):441–7.
35. Gooda Sahib N, Abdul Hamid A, Saari N, et al. Anti-pancreatic lipase and antioxidant activity of selected tropical herbs. *Int J Food Properties* 2012;**15**(3):569–78.
36. Dailami KN, Azadbakht M, Pharm ZR, Lashgari M. Prevention of selenite-induced cataractogenesis by *Origanum vulgare* extract. *Pak J Biol Sci* 2010;**13**(15):743–7.
37. Kozics K, Klusova V, Srančiková A, et al. Effects of *Salvia officinalis* and *Thymus vulgaris* on oxidant-induced DNA damage and antioxidant status in HepG2 cells. *Food Chem* 2013;**141**(3):2198–206.
38. Zheng W, Wang SY. Antioxidant activity and phenolic compounds in selected herbs. *J Agric Food Chem* 2001;**49**(11):5165–70.

39. Rooban BN, Lija Y, Biju PG, et al. *Vitex negundo* attenuates calpain activation and cataractogenesis in selenite models. *Exp Eye Res* 2009;**88**(3):575–82.
40. Vibin M, Siva Priya SG, Rooban B, et al. Broccoli regulates protein alterations and cataractogenesis in selenite models. *Curr Eye Res* 2010;**35**(2):99–107.
41. Iriti M, Vitalini S, Fico G, Faoro F. Neuroprotective herbs and foods from different traditional medicines and diets. *Molecules* 2010;**15**(5):3517–55.
42. Boveris AD, Galleano M, Puntarulo S. *In vivo* supplementation with *Ginkgo biloba* protects membranes against lipid peroxidation. *Phytother Res* 2007;**21**(8):735–40.
43. MacVie OP, Harney BA. Vitreous haemorrhage associated with *Ginkgo biloba* use in a patient with age related macular disease. *Br J Ophthalmol* 2005;**89**(10):1378–9.
44. Subhashini N, Thangathirupathi A, Lavanya N. Antioxidant activity of *Trigonella foenum graecum* using various *in vitro* and *ex vivo* models. *Int J Pharm Pharmaceut Sci* 2011;**3**(2):96–102.
45. Pittella F, Dutra RC, Junior DD, et al. Antioxidant and cytotoxic activities of *Centella asiatica* (L) Urb. *Int J Mol Sci* 2009;**10**(9):3713–21.
46. Kempaiah RK, Srinivasan K. Antioxidant status of red blood cells and liver in hypercholesterolemic rats fed hypolipidemic spices. *Int J Vitam Nutr Res* 2004;**74**(3):199–208.
47. Srivastava R, Ahmed H, Dixit RK, et al. *Crocus sativus* L.: a comprehensive review. *Pharmacogn Rev* 2010;**4**(8):200–8.
48. Natoli R, Zhu Y, Valter K, et al. Gene and noncoding RNA regulation underlying photoreceptor protection: microarray study of dietary antioxidant saffron and photobiomodulation in rat retina. *Mol Vis* 2010;**16**:1801–22.
49. Maccarone R, Di Marco S, Bisti S. Saffron supplement maintains morphology and function after exposure to damaging light in mammalian retina. *Invest Ophthalmol Vis Sci* 2008 Mar;**49**(3):1254–61.
50. Falsini B, Piccardi M, Minnella A, et al. Influence of saffron supplementation on retinal flicker sensitivity in early age-related macular degeneration. *Invest Ophthalmol Vis Sci* 2010 Dec;**51**(12):6118–24.
51. Mohamed S, Hashim SN, Rahman HA. Seaweeds: a sustainable functional food for complementary and alternative therapy. *Trends Food Science Technol* 2012;**23**(2):83–96.
52. Bakhsh A, Mustapha NM, Mohamed S. Catechin-rich oil palm leaf extract enhances bone calcium content of estrogen-deficient rats. *Nutrition* 2013;**29**(4):667–72.
53. Mohamed S, Matanjun P, Mustapha NM, et al. Edible seaweeds: a functional food with organ protective and other therapeutic applications. In: Pomin V, editor. *Seaweed: ecology, nutrient composition and medicinal uses*. New York: NovaScience Publishers; 2011.

CHAPTER

11

Coenzyme Q_{10} as an Antioxidant in the Elderly

Elena M. Yubero-Serrano, Antonio Garcia-Rios, Javier Delgado-Lista, Pablo Pérez-Martinez, Antonio Camargo, Francisco Perez-Jimenez, Jose Lopez-Miranda

Lipids and Atherosclerosis Unit, IMIBIC/Reina Sofia University Hospital/University of Córdoba, and CIBER Fisiopatologia Obesidad y Nutricion (CIBEROBN), Instituto de Salud Carlos III, Córdoba, Spain

List of Abbreviations

CKD cardiovascular kidney disease
CoQ_{10} coenzyme Q_{10}
DNA deoxyribonucleic acid
ESRD end-stage renal disease
FMD flow-mediated dilation
HD Huntington's disease
HPLC high-performance liquid chromatography
LDL low-density lipoprotein
mtDNA mitochondrial DNA
NMD nitroglycerin-mediated dilation
NO nitric oxide
OS oxidative stress
PD Parkinson's disease
PeD peritoneal dialysis
ROS reactive oxygen species

OXIDATIVE STRESS AND ANTIOXIDANT DEFENSE

Reactive oxygen species (ROS) are produced from molecular oxygen as a result of normal cellular metabolism and they have beneficial effects at low concentrations. However, when there is an overproduction of ROS it produces adverse modifications to cell components, such as lipids, proteins and DNA, inhibiting their normal function.[1] Organisms have developed a series of defense mechanisms against ROS (Fig. 11.1); the enzymatic antioxidant defenses include superoxide dismutase, catalase and glutathione peroxidase, while the non-enzymatic antioxidants include ascorbic acid, tocopherol, glutathione, coenzyme Q_{10} (CoQ_{10}) and others. Oxidative stress (OS) is defined as the imbalance between ROS and antioxidant defenses, in favor of ROS. Oxidative stress is a condition associated with chronic degenerative diseases, such as cancer, metabolic disease and cardiovascular diseases, and also the aging process.

AGING

In all species, aging is defined as a normal decline in survival with advancing age. Understanding the molecular and cellular mechanisms which underlie the aging process would enable a good strategy to be formulated for addressing the problems presented by the aging of the world's population. Many theories have been proposed to explain the aging events, although one of the most accepted ideas suggests that aging, as well as the associated degenerative diseases,[2] is produced by the deleterious, irreversible changes and macromolecular damage produced by ROS (Fig. 11.2).[3,4] Some modifications (mainly those related to DNA) are not completely repaired and thus accumulate, leading to cell death, malfunction of the organism and the 'aging phenotype'. Today, a version of this theory of aging – in which mitochondria are the source as well as the target of ROS – is one of the most popular theories of aging.[5,6] This theory postulates an increased sensitivity of mitochondrial DNA (mtDNA) to oxidative damage, leading to an accumulation of mutated mtDNA; this idea led to the concept of the 'vicious cycle': an initial ROS-induced impairment of mitochondria produces an increase in

Aging
http://dx.doi.org/10.1016/B978-0-12-405933-7.00011-1

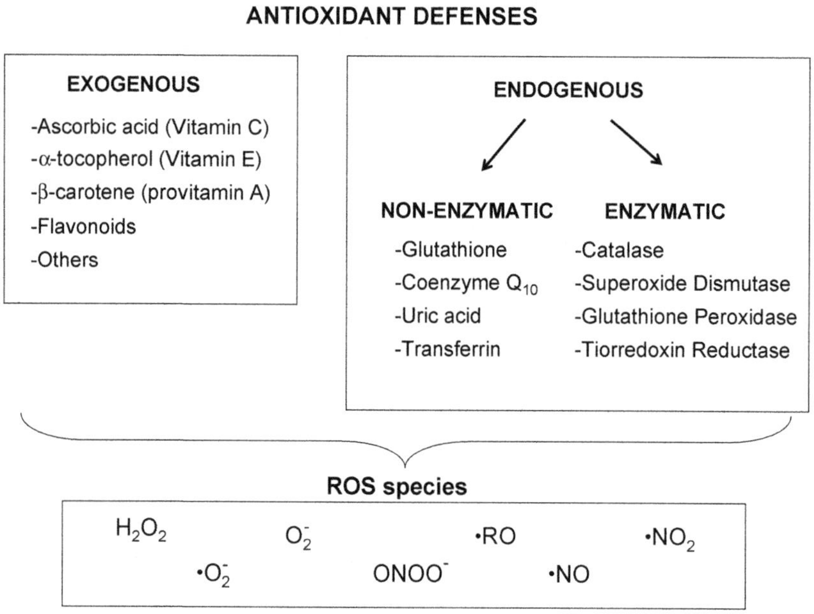

FIGURE 11.1 Enzymatic and non-enzymatic antioxidant defenses.

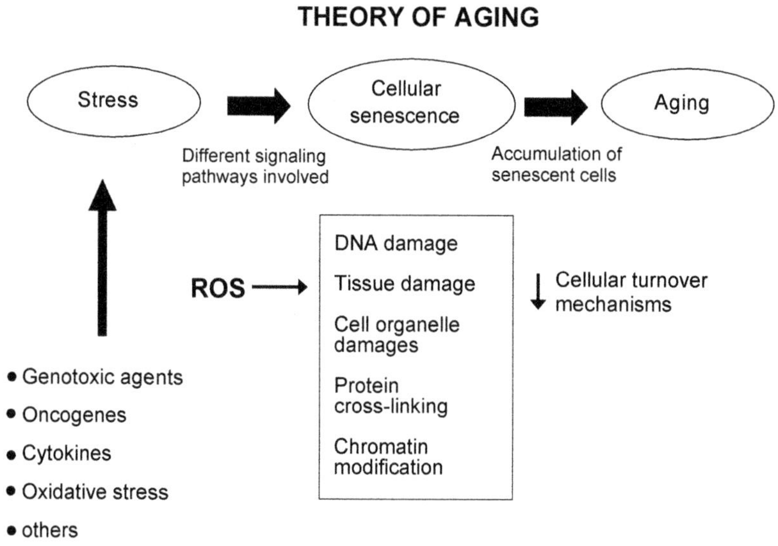

FIGURE 11.2 Theory of aging. One of the most accepted theories of aging is that aging is caused by the deleterious, irreversible changes and macromolecular damage produced by reactive oxygen species (ROS). Some modifications (mainly those related to DNA) are not completely repaired and thus accumulate, leading to cell death, organism malfunction, and the 'aging phenotype'.

oxidant production that, in turn, leads to mitochondrial damage.[7]

Strategies developed according to this concept and designed to delay and ameliorate the aging process have focused on extending the maximum lifespan and/or retarding the age-associated biologic changes.[8] Nutritional and pharmacologic interventions have been previously shown to extend lifespan in diverse model organisms, including *Saccharomyces cerevisiae, Drosophila melanogaster*, mice and rats, as well as monkeys. Some antioxidants, as well as interventions related to dietary fat, have proved to be useful as dietary anti-aging therapies.[9,10]

This chapter focuses on the most relevant biochemical characteristics of an important antioxidant, coenzyme Q_{10} (CoQ_{10}), including its main functions, levels and distribution in humans, and the therapeutic potential of CoQ_{10}, especially during aging and the associated degenerative diseases.

COENZYME Q_{10}

CoQ_{10} was identified by two different scientific groups as a component of the mitochondrial respiratory chain.[11] Festenstein et al (1955) isolated and characterized a substance which was named ubiquinone ('ubiquitous quinone') – reflecting the presence of this substance in all cells. Crane et al (1957) established that this compound functions as an electron carrier in the mitochondrial electron transport chain.[12,13]

The name CoQ_{10} also refers to its chemical structure, as the predominant human form of this molecule contains one

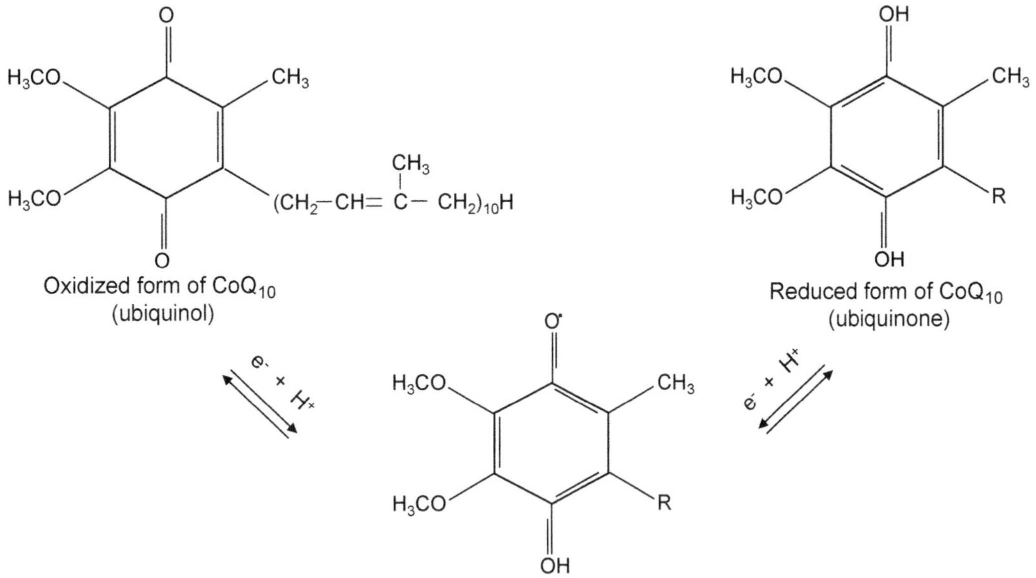

FIGURE 11.3 Chemical structures of the different forms of coenzyme Q_{10} (2,3-dimethoxy-5-methyl-6-decaprenyl-1,4-benzoquinone). Ubiquinone is reduced to ubiquinol through a semiquinone intermediate.

quinone group and 10 isoprenyl units. CoQ_{10} (2,3-dimethoxy-5-methyl-6-decaprenyl-1,4-benzoquinone) is a lipid-soluble and biologically active quinone, synthesized in the body from phenylalanine and mevalonic acid. It comprises a benzoquinone ring with an isoprenoid side chain; its structure is similar to that of vitamin E. The main chemical characteristic of CoQ_{10}, responsible for its various functions, is its ability to exist in three alternative redox states: the fully oxidized *ubiquinone* form, the *semiquinone* (free radical) form and the fully reduced *ubiquinol* form (Fig. 11.3).[14] CoQ_{10} is distributed in all membranes throughout the cell, but is present mainly in the inner membrane of the mitochondria in every cell in the human body.[15]

Coenzyme Q_{10} Functions

The main function of CoQ_{10} in cells is carried out in the mitochondrial inner membrane – where it acts as an essential cofactor, accepting electrons from several donors and transferring them to the cytochrome complex system (from respiratory complexes I and II to complex III). At the same time, CoQ_{10} transfers protons to the outside of the mitochondrial membrane.[15,16] The reduced form of CoQ_{10} (ubiquinol) is a potent antioxidant that protects the stability of the cell membranes (membrane phospholipids and mitochondrial membrane proteins) and the DNA from free-radical-induced oxidative damage. This property of CoQ_{10} is very important because it may be related to its role in extramitochondrial electron transfer in plasma membranes. CoQ_{10} has also been reported to protect LDL from oxidation.[17] In addition to direct antioxidant radical scavenging, CoQ_{10} is also capable of reducing and regenerating other antioxidants, such as tocopherol and ascorbate.[15,16,18] In these chemical reactions, the oxidized form (ubiquinone) is formed and is reduced to ubiquinol by, at least, three enzymes located in plasma and endomembranes: NADH/NADPH oxidoreductase (DT diaphorase), NADH cytochrome b5 reductase and NADPH coenzyme Q reductase.[19,20]

CoQ_{10} is considered to be an important antioxidant, more powerful than vitamin E, because it is the only antioxidant synthesized endogenously, regenerated by intracellular reducing mechanisms and present in relatively high concentrations.[21] Furthermore, recent data reveal that CoQ_{10} acts as a cofactor for the function of uncoupling proteins and as a modulator of the transition pore, and it affects the expression of genes involved in human cell signaling, metabolism and transport.[22,23]

Levels and Distribution of Coenzyme Q_{10} in Humans

In humans, the main form of the coenzyme is CoQ_{10}; it is produced in all cells and is present in all tissues in varying amounts. Human CoQ_{10} ranges from 8 μg/g in lung to 114 μg/g in heart. Small amounts (2–7%) of CoQ_9 are also found in all human tissues.

By employing rapid extraction, partition, and direct injection of the extract into HPLC columns, it is possible to estimate the degree of reduction most probably existing *in vivo*;[24] this approach indicates that the proportion of human CoQ_{10} in the reduced state is very high, with the exception of that in lung and brain. The distribution and redox state of CoQ_{10} in different human tissues have been described.[25] Arguably, as a general rule, tissues with high-energy requirements or metabolic activity, such as the heart, kidney, liver and muscle, contain relatively high concentrations of CoQ_{10}.[14] Data on the subcellular distribution of CoQ_{10} show high levels (40–50%) localized in the mitochondrial inner membrane, with smaller amounts in the other organelles (Golgi vesicles, endoplasmic reticulum or lysosomes) and also in the cytosol, reflecting a CoQ_{10} compartmentalization. The high concentration of CoQ_{10} in the mitochondria reflects its important role in the mitochondrial electron transport chain.

Tissue CoQ_{10} originates from various sources, such as endogenous synthesis as well as from food intake and from oral supplements. The concentrations of CoQ_{10} in humans depend on age, sex and race, and on the health of the individual. In a normal healthy young person, the total body content of CoQ_{10} has been estimated to be 0.5–1.5 g.[26] Decreased levels of CoQ_{10} are found in patients with cardiomyopathies, congestive heart failure and degenerative diseases, during aging, and in age-related diseases.[27,28]

Biosynthesis and Transport of CoQ_{10}

Intracellular synthesis in the human body, which depends on the mevalonate pathway, is the major source of CoQ_{10} – but not the only source of this ubiquinone. The mevalonate pathway is a sequence of cellular reactions leading to farnesyl pyrophosphate, the common substrate for the synthesis of cholesterol, dolichol, dolichyl phosphate and CoQ_{10}, and for protein prenylation (a post-translational modification necessary for the targeting and function of many proteins).[29] Cells synthesize CoQ_{10} *de novo*, starting with synthesis of the benzoquinone ring and the isoprenoid side chain, whose precursors are 4-hydroxybenzoate and acetyl-CoQ, respectively; synthesis of the isoprenoid side chain ends with farnesyl pyrophosphate. Then, both molecules are condensed by the enzyme polyprenil-4-hydroxybenzoate transferase (encoded by the COQ2 gene). It is known that, at least, six enzymes (encoded by COQ3-8) are implicated in the subsequent methylation, decarboxylation and hydroxylation reactions to synthesize CoQ_{10}.[18,30]

Independently of the locations of CoQ_{10} synthesis, or whether the CoQ_{10} source is through the diet, this ubiquinone is found in all subcellular compartments and, hence, there should be a transport system from the mitochondria to other membranes. Using *in vivo* labeling and cell fractionation, it was found that, in spinach leaves, CoQ_{10} is transported from the endoplasmic reticulum to other compartments through a vesicle-mediated process which involves the Golgi system.[31] The way in which CoQ_{10} is transported from the Golgi system to other cellular compartments is not known. CoQ_{10} transport also occurs in the opposite direction because exogenous CoQ_{10} is translocated through the plasma membrane, and further intracellular transport again requires an appropriate mechanism.

Uptake and Distribution of CoQ_{10}

CoQ_{10} is found naturally in dietary sources and is also used as a dietary supplement. It is present in a wide variety of foods from animal and vegetable sources. In animal sources, large amounts are present e.g. in chicken legs, heart, liver and herrings. In vegetable sources, it occurs e.g. in spinach, cauliflower and whole grains – but in a lower concentration compared with meat and fish (Table 11.1).

During the last decade, CoQ_{10} has been available as e.g. oil-based, softgel or powder-filled capsules and tablets in the context of dietary supplements. Many formulations have been developed to improve CoQ_{10} solubility

TABLE 11.1 Content of Coenzyme Q_{10} in Foods[a]

Food	CoQ_{10} Content (μg/g)
Meat	
Pork heart	203
Pork liver	3.1
Beef heart	41
Beef liver	19
Lamb leg	2.9
Chicken leg	17
Fish	
Trout	11
Herring	27
Tuna canned	0.3
Vegetables	
Spinach	2.3
Pea	0.1
Cauliflower	0.6
Fruits	
Orange	2.2
Strawberry	0.1
Apple	0.2
Cereals	
Bread (rye)	4.7
Bread (wheat)	2.1

[a]*Data based on reference.*[1]

in the human body. Comparisons between studies have indicated that CoQ_{10} bioavailability is influenced by the type of formulation, and that it is better to take CoQ_{10} with fatty foods.[25,29]

The lipophilic characteristics of CoQ_{10} are important for an understanding of its uptake and distribution following oral ingestion. The absorption of CoQ_{10} is enhanced in the presence of lipids. Exogenous CoQ_{10} is taken up from the intestine into chylomicrons and thence to the circulation – similar to the uptake of vitamin E – with a range of between 2 and 4% of the total uptake.[32] In the plasma, CoQ_{10} is carried mainly by lipoproteins, mostly in LDL particles, where it is predominantly in its reduced form.[32,33] The circulating concentrations of CoQ_{10} may be useful for assessing its status in the body and also for monitoring the response to CoQ_{10} supplementation.

CoQ_{10} and Aging

CoQ_{10} is synthesized in all cells in healthy individuals.[34] The CoQ_{10} levels increase in the first 20 years of life; however, the organism begins to lose its ability to synthesize CoQ_{10} during aging, and the coenzyme becomes deficient.[35–37] In addition to a decrease in its biosynthesis, other factors or situations may affect the levels or functions of CoQ_{10}, including an increase in degradation[38] or changes in membrane lipids which prevent the movement of this quinone – as occurs in different age-related diseases.[39] However, the changes in CoQ_{10} levels during aging are tissue- and organ-dependent. For example, old rats have increased levels of CoQ_{10} in mitochondria from the brain[40] but lower levels in mitochondria from skeletal muscle,[41]so it would be important to determine whether these tissue-dependent changes are related to a loss of function or antioxidant capability.

Supplementation with CoQ_{10} does not increase tissue levels above normal (except in liver and spleen) in young and healthy individuals, but in older animals, with decreased CoQ_{10}, supplemental CoQ_{10} can restore normal levels.[42,43] Low levels of CoQ_{10} have been related to the higher oxidative stress during aging, and in different related diseases. Oral CoQ_{10} supplementation could be a viable antioxidant strategy in many neurodegenerative disorders, diabetes, cancer and muscular and cardiovascular diseases, such as chronic heart failure and hypertension, where oxidative stress is implicated.

Therapeutic Uses of CoQ_{10} in Age-Related Diseases

The therapeutic use of CoQ_{10} is based upon its fundamental role in mitochondrial function and cellular bioenergetics. Animal data show that CoQ_{10} in large doses is taken up by all tissues, including the heart and brain mitochondria. This fact has implications for therapeutic applications in human diseases which involve oxidative stress, and there is evidence for its beneficial effects in cardiovascular and neurodegenerative diseases.[25,29,30]

CoQ_{10} and Cardiovascular Disease

Oxidative stress plays a central role in the pathogenesis of cardiovascular diseases, including congestive heart failure, hypertension and ischemic heart disease. Specifically, congestive heart failure, due to an energy depletion status in the mitochondria, has been strongly correlated with significantly low blood and tissue levels of CoQ_{10}. Generally, heart muscle cells have high CoQ_{10} levels due to the high energy requirements of this type of cell. However, in biopsy samples from human heart, a significant decrease in the CoQ_{10} content was found in cardiomyopathy, and this deficiency was correlated with the severity of disease.[44,45]

Several studies on the treatment of heart disease (congestive heart failure, cardiomyopathy and/or valvular heart disease) concluded that a CoQ_{10} supplement of between 50 and 300 mg/day appears to be the optimal dose; the supplement was safe and was well tolerated at doses of up to 1200 mg/day.[46] The majority of these clinical studies indicate that treatment with CoQ_{10} significantly improves heart muscle function – increasing ATP synthesis and enhancing myocardial contractility[44]—while producing no adverse effects or drug interactions.[47]

CoQ_{10} and Hypertension

The primary action of CoQ_{10} in clinical hypertension is vasodilatation; it acts directly on the endothelium and vascular smooth muscle, with the ability to counteract vasoconstriction and lower blood pressure without significant side effects. The health-giving effects of CoQ_{10} have also been investigated in several controlled intervention studies in humans designed to prevent high blood pressure;[48–50] in these studies, doses of CoQ_{10} were in the range of 100 mg to 200 mg/day. Among treated patients, decreases in systolic blood pressure ranged from 11 to 17 mmHg, and there was also a decrease of 8 mmHg in diastolic blood pressure. These results appear to demonstrate the role of CoQ_{10} as a hypotensive agent on its own or in combination with other conventional anti-hypertensive therapies.

CoQ_{10} and Endothelial Function

Endothelial dysfunction plays a key role in the development, progression, and clinical manifestations of atherosclerosis and cardiovascular diseases. The effect of oral CoQ_{10} supplementation on endothelial function in patients with coronary artery disease or diabetes mellitus, or in elderly people, has been investigated in many studies.[46,49–52]

The fact that endothelial function,[53,54] as measured by flow-mediated dilation (FMD) or nitroglycerin-mediated

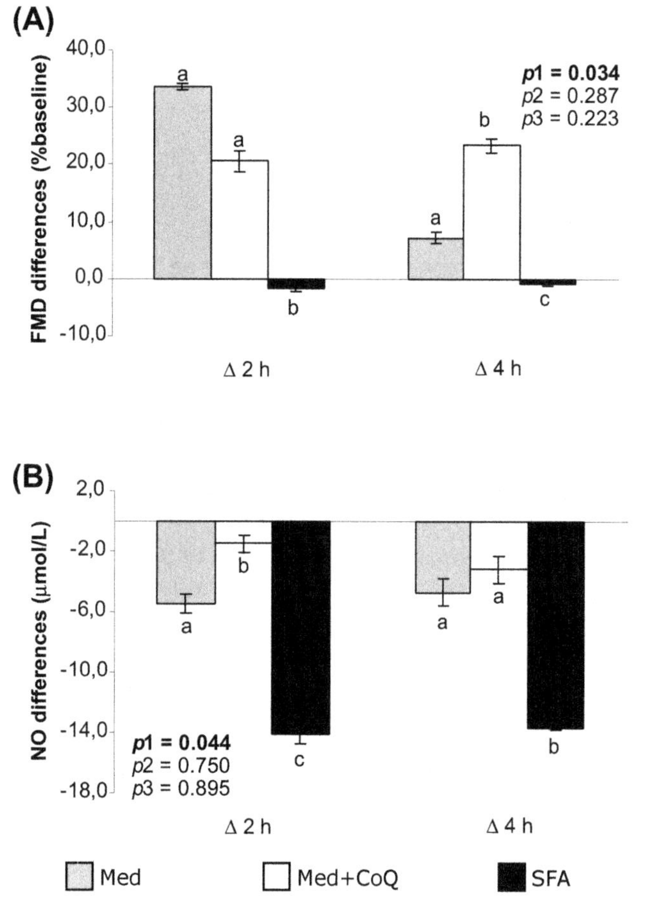

FIGURE 11.4 Effect of CoQ_{10} supplementation on endothelial dysfunction (measured as flow-mediated dilation – FMD) and nitric oxide (NO) levels. Differences in (A) FMD change and (B) NO levels in plasma after the consumption of three different diets, supplemented and not supplemented with CoQ_{10}. Published by Yubero-Serrano EM et al. Postprandial antioxidant effect of the Mediterranean diet supplemented with coenzyme Q_{10} in elderly men and women. Age (Dordr) 2011 33(4):579-590. Med: Mediterranean diet; Med+CoQ: Mediterranean diet supplemented with coenzyme Q_{10}; SFA: saturated rich diet. Differences in the parameter levels between time 0 and time 2 are shown as Δ 2h (value of time 2 minus value of time 0), and between time 0 and time 4 are shown as Δ 4h (value of time 4 minus value of time 0). Data were analyzed using ANOVA for repeated measures. All values represent the mean ± standard errors (SE). Bars with different superscript letters depict statistically significant differences ($p < 0.05$). $p1$: diet effect, $p2$: time effect, $p3$: diet × time interaction.

dilation (NMD), and the extracellular superoxide dismutase activity, improved in most of the subjects treated with CoQ_{10} could be due to its antioxidant activity and anti-inflammatory activity,[55] decreasing the rate of inactivation of nitric oxide (NO) to peroxynitrite by superoxide radicals, because these superoxide radicals may be reduced by CoQ_{10} under conditions of oxidative stress (Fig. 11.4, A and B).[50,51]

CoQ_{10} and Renal Failure

Oxidative stress is increased in patients with chronic kidney disease (CKD) and end-stage renal disease (ESRD).[56] In these patients, the balance between ROS and the antioxidant system has changed in favor of ROS. Only a few studies have evaluated the role of CoQ_{10} as a component of the antioxidant system in CKD patients. Lippa et al[57] determined the levels of CoQ_{10} in a group of 48 patients on chronic hemodialysis, in 15 uremic patients, and in a control group of healthy subjects. CoQ_{10} levels were significantly lower in CKD patients as compared with a healthy group. In a recent study, in which the authors investigated levels and associations of various markers of oxidative stress and CoQ_{10} in CKD, hemodialysis and peritoneal dialysis (PeD) patients, CoQ_{10} levels did not differ between CKD and hemodialysis groups, although other components of the antioxidant system were higher in patients undergoing PeD compared to CKD patients.[58] In relation to the effect of CoQ_{10} supplementation (120 mg/day) in patients with CKD, after 28 days of treatment, the number of patients on dialysis was significantly lower compared with the placebo group at the end of the study.[59]

CoQ_{10} and Neurodegenerative Diseases

A common characteristic of the neurodegenerative diseases is mitochondrial dysfunction with abnormal energy metabolism and an increase in cellular oxidative stress. As we have commented, CoQ_{10} levels in the brain and other tissues in humans and animals have been shown to decline with age. For that reason, it has been suggested that CoQ_{10}, as an antioxidant molecule, is involved in the mechanisms and development of neurodegenerative diseases such as Parkinson's disease (PD), Huntington's disease (HD) and other neurodegenerative disorders.[60,61] There is, therefore, a robust scientific rationale for testing this agent as a potential neuroprotective therapy.

There is evidence to support a role for mitochondrial dysfunction in PD because CoQ_{10} was significantly lower in mitochondria from Parkinsonian patients than in mitochondria from age- and sex-matched control subjects. Shults and colleagues demonstrated, in a randomized, placebo-controlled and double-blind study, that a dose of 1200 mg/day reduced the worsening of PD, with significant increases in plasma levels of CoQ_{10} and NADH-cytochrome *c* reductase activity. The greatest benefit was seen in everyday activities of the patients, such as dressing, bathing and feeding.[62] Another placebo-controlled, double-blind trial, conducted by Muller and colleagues, concluded that CoQ_{10} supplementation provided a significant mild benefit for PD symptoms compared with placebo.[63]

As in PD, there is substantial evidence linking impaired energy metabolism to HD, and a deficiency in CoQ_{10} is considered as part of the problem. CoQ_{10} doses, ranging from 600 to 1200 mg/day, tested in 10 HD patients during a 6-month open-label trial, did not bring about a significant effect on clinical scores.[64] However, treatment with CoQ_{10} resulted in significant decreases in cortical lactate concentrations, which reversed following withdrawal of therapy. This finding supports the predicted metabolic effect of oral CoQ_{10} in cerebral tissue, and is suggestive of an effect upon mitochondrial metabolism.[60]

CONCLUSIONS

Oxidative stress is caused by an overproduction of ROS, and shifts in the balance between oxidants and antioxidants in favor of the oxidants, producing damage to cell structures and consequently leading to various diseases and aging.

The importance of CoQ_{10} is not just as an agent for energy transduction in mitochondria. New findings show that CoQ_{10} is an important antioxidant with the unique feature of regenerating redox capacity. Biosynthesis in mitochondria and in the endoplasmic reticulum provides sufficient CoQ_{10} in normal individuals. Evidence for deficiency is based on genetic failure, and mainly, on aging and age-related diseases. The important properties of CoQ_{10} make this molecule an essential dietary supplement designed to delay and mitigate its deficiency in the body.

The beneficial effect of CoQ_{10} is reinforced because this antioxidant has an excellent safety record and is well tolerated in high doses for prolonged periods of time with few side effects.

However, more research is needed to determine the appropriate dose, effectiveness, and bioavailability of orally administered CoQ_{10}, especially in the elderly population. There is even the possibility of designing therapeutic agents that increase the endogenous synthesis of CoQ_{10}.

SUMMARY POINTS

- Reactive oxygen species (ROS) are produced as a result of normal cellular metabolism and they have beneficial effects at low concentrations; however, high levels of ROS produce adverse modifications to cell components, such as lipids, proteins and DNA, inhibiting their normal function.
- Oxidative stress is a condition associated with chronic degenerative diseases, such as cancer, metabolic and cardiovascular diseases, as well as with the aging process.
- Some antioxidants and interventions related to dietary fat have proved to be useful as dietary anti-aging therapies.
- In addition to its function as a cofactor in the mitochondrial electron transport chain, coenzyme Q_{10} is an important antioxidant with unique characteristics for regeneration of redox capacity.
- The therapeutic use of CoQ_{10} is based on its fundamental role in mitochondrial function and cellular bioenergetics.
- The therapeutic application of CoQ_{10} in humans has beneficial effects in conditions arising from oxidative stress, such as aging and age-related diseases (cardiovascular and neurodegenerative diseases and others).

References

1. Valko M, Rhodes CJ, Moncol J, et al. Free radicals, metals and antioxidants in oxidative stress-induced cancer. *Chemico-biological Interact* 2006 Mar 10;**160**(1):1–40.
2. Su Y, Sun H, Fang J, et al. Brain mitochondrial dysfunction in ovariectomized mice injected with D-galactose. *Neurochem Res* 2010;**35**(3):399–404.
3. Miquel J. An update on the oxygen stress-mitochondrial mutation theory of aging: genetic and evolutionary implications. *Exp Gerontol* 1998 Jan-Mar;**33**(1-2):113–26.
4. Sohal RS, Mockett RJ, Orr WC. Mechanisms of aging: an appraisal of the oxidative stress hypothesis. *Free Radic Biol Med* 2002 Sep 1;**33**(5):575–86.
5. Barja G. Mitochondrial oxygen consumption and reactive oxygen species production are independently modulated: implications for aging studies. *Rejuv Res* 2007 Jun;**10**(2):215–24.
6. Miquel J, Economos AC, Fleming J, Johnson Jr JE. Mitochondrial role in cell aging. *Exp Gerontol* 1980;**15**(6):575–91.
7. Gilmer LK, Ansari MA, Roberts KN, Scheff SW. Age-related changes in mitochondrial respiration and oxidative damage in the cerebral cortex of the Fischer 344 rat. *Mech Ageing Develop* 2010 Fep;**131**(2):133–43.
8. Lee CK, Pugh TD, Klopp RG, et al. The impact of α-lipoic acid, coenzyme Q10, and caloric restriction on life span and gene expression patterns in mice. *Free Radic Biol Med* 2004;**36**:1043–57.
9. Duntas LH. Resveratrol and its impact on aging and thyroid function. *J Endocrinol Invest* 2011 Nov;**34**(10):788–92.
10. Lopez-Dominguez JA, Khraiwesh H, Gonzalez-Reyes JA, et al. Dietary fat modifies mitochondrial and plasma membrane apoptotic signaling in skeletal muscle of calorie-restricted mice. *Age (Dordr)* 2012 Nov 20. PMID 23179253.
11. Crane FL, Hatefi Y, Lester RL, Widmer C. Isolation of a quinone from beef heart mitochondria. *Biochim Biophys Acta* 1957;**1989**(1000):362–3.
12. Festenstein GN, Heaton FW, Lowe JS, Morton RA. A constituent of the unsaponifiable portion of animal tissue lipids (lambda max. 272 m mu). *Biochem J* 1955 Apr;**59**(4):558–66.
13. Crane FL, Hatefi Y, Lester RL, Widmer C. Isolation of a quinone from beef heart mitochondria. *Biochim Biophys Acta* 1957 Jul;**25**(1):220–1.
14. Ernster L, Dallner G. Biochemical, physiological and medical aspects of ubiquinone function. *Biochim Biophys Acta* 1995 May 24;**1271**(1):195–204.
15. Crane FL. Biochemical functions of coenzyme Q10. *J Am Coll Nutr* 2001 Dec;**20**(6):591–8.
16. Crane FL, Navas P. The diversity of coenzyme Q function. *Mol Aspects Med* 1997;**18**(Suppl):S1–6.
17. Alleva R, Tomasetti M, Battino M, et al. The roles of coenzyme Q10 and vitamin E on the peroxidation of human low density lipoprotein subfractions. *Proc Natl Acad Sci USA* 1995 Sep 26;**92**(20):9388–91.

18. Turunen M, Olsson J, Dallner G. Metabolism and function of coenzyme Q. *Biochim Biophys Acta* 2004 Jan 28;**1660**(1-2):171–99.
19. Takahashi T, Okamoto T, Kishi T. Characterization of NADPH-dependent ubiquinone reductase activity in rat liver cytosol: effect of various factors on ubiquinone-reducing activity and discrimination from other quinone reductases. *J Biochem* 1996 Feb;**119**(2):256–63.
20. Villalba JM, Navas P. Plasma membrane redox system in the control of stress-induced apoptosis. *Antioxid Redox Signal* 2000;**2**(2):213–30.
21. Forsmark-Andree P, Dallner G, Ernster L. Endogenous ubiquinol prevents protein modification accompanying lipid peroxidation in beef heart submitochondrial particles. *Free Radic Biol Med* 1995 Dec;**19**(6):749–57.
22. Groneberg DA, Kindermann B, Althammer M, et al. Coenzyme Q10 affects expression of genes involved in cell signalling, metabolism and transport in human CaCo-2 cells. *Int J Biochem Cell Biol* 2005 Jun;**37**(6):1208–18.
23. Littarru GP, Tiano L. Bioenergetic and antioxidant properties of coenzyme Q10: recent developments. *Mol Biotechnol* 2007 Sep;**37**(1):31–7.
24. Yamashita S, Yamamoto Y. Simultaneous detection of ubiquinol and ubiquinone in human plasma as a marker of oxidative stress. *Analyt Biochem* 1997 Jul 15;**250**(1):66–73.
25. Bhagavan HN, Chopra RK. Coenzyme Q10: absorption, tissue uptake, metabolism and pharmacokinetics. *Free Radic Res* 2006 May;**40**(5):445–53.
26. Greenberg S, Frishman WH. Co-enzyme Q10: a new drug for cardiovascular disease. *J Clin Pharmacol* 1990 Jul;**30**(7):596–608.
27. Fotino AD, Thompson-Paul AM, Bazzano LA. Effect of coenzyme Q10 supplementation on heart failure: a meta-analysis. *Am J Clin Nutr* 2013;**97**:268–75.
28. Shetty RA, Forster MJ, Sumien N. Coenzyme Q(10) supplementation reverses age-related impairments in spatial learning and lowers protein oxidation. *Age (Dordr)* 2013;**35**(5):1821–34.
29. Villalba JM, Parrado C, Santos-Gonzalez M, Alcain FJ. Therapeutic use of coenzyme Q10 and coenzyme Q10-related compounds and formulations. *Expert Opin Investig Drugs* 2010 Apr;**19**(4):535–54.
30. Quinzii CM, DiMauro S, Hirano M. Human coenzyme Q10 deficiency. *Neurochem Res* 2007 Apr-May;**32**(4-5):723–7.
31. Wanke M, Dallner G, Swiezewska E. Subcellular localization of plastoquinone and ubiquinone synthesis in spinach cells. *Biochim Biophys Acta* 2000 Jan 15;**1463**(1):188–94.
32. Zhang Y, Aberg F, Appelkvist EL, et al. Uptake of dietary coenzyme Q supplement is limited in rats. *J Nutr* 1995 Mar;**125**(3):446–53.
33. Bhagavan HN, Chopra RK, Craft NE, et al. Assessment of coenzyme Q10 absorption using an in vitro digestion-Caco-2 cell model. *Int J Pharmaceut* 2007 Mar 21;**333**(1-2):112–7.
34. Schultz JR, Clarke CF. Functional roles of ubiquinone. In: Cardenas E, Packer L, editors. *Mitochondria, oxidants and ageing*. New York: M Dekker; 1999. pp. 95–118.
35. Blatt T, Littarru GP. Biochemical rationale and experimental data on the antiaging properties of CoQ(10) at skin level. *BioFactors (Oxford, England)* 2011 Sep-Oct;**37**(5):381–5.
36. Gutierrez-Mariscal FM, Perez-Martinez P, Delgado-Lista J, et al. Mediterranean diet supplemented with coenzyme Q10 induces postprandial changes in p53 in response to oxidative DNA damage in elderly subjects. *Age (Dordrecht, Netherlands)* 2011 Apr;**34**(2):389–403.
37. Ochoa JJ, Quiles JL, Lopez-Frias M, et al. Effect of lifelong coenzyme Q10 supplementation on age-related oxidative stress and mitochondrial function in liver and skeletal muscle of rats fed on a polyunsaturated fatty acid (PUFA)-rich diet. *J Gerontol* 2007 Nov;**62**(11):1211–8.
38. Nakamura T, Ohno T, Hamamura K, Sato T. Metabolism of coenzyme Q10: biliary and urinary excretion study in guinea pigs. *BioFactors (Oxford, England)* 1999;**9**(2-4):111–9.
39. Kagan VE, Nohl H, Quinn JP, Coenzyme Q. Its role in scavenging and generation of radicals in membranes. In: Cardenas E, Packer LM, editors. *Handbook of antioxidants*. New York: M Dekker; 1996. pp. 157–201.
40. Battino M, Svegliati Baroni S, Littarru GP, et al. Coenzyme Q homologs and vitamin E in synaptic and non-synaptic occipital cerebral cortex mitochondria in the ageing rat. *Mol Aspects Med* 1997;**18**(Suppl):S279–82.
41. Lass A, Kwong L, Sohal RS. Mitochondrial coenzyme Q content and aging. *BioFactors (Oxford, England)* 1999;**9**(2-4):199–205.
42. Beal MF. Coenzyme Q10 administration and its potential for treatment of neurodegenerative diseases. *BioFactors (Oxford, England)* 1999;**9**(2-4):261–6.
43. Rosenfeldt FL, Pepe S, Ou R, et al. Coenzyme Q10 improves the tolerance of the senescent myocardium to aerobic and ischemic stress: studies in rats and in human atrial tissue. *BioFactors (Oxford, England)* 1999;**9**(2-4):291–9.
44. Folkers K, Vadhanavikit S, Mortensen SA. Biochemical rationale and myocardial tissue data on the effective therapy of cardiomyopathy with coenzyme Q10. *Proc Natl Acad Sci USA* 1985 Feb;**82**(3):901–4.
45. Nobuyoshi M, Saito T, Takahira H, et al. Levels of coenzyme Q10 in biopsies of left ventricular muscle and influence of administration of coenzyme Q10. In: Folkers K, Yamamura Y, editors. *Biomedical and clinical aspects of coenzyme Q*;**vol. 4**. Amsterdam: Elsevier; 1984. pp. 221–9.
46. Gao L, Mao Q, Cao J, et al. Effects of coenzyme Q10 on vascular endothelial function in humans: a meta-analysis of randomized controlled trials. *Atherosclerosis* 2011 Apr;**221**(2):311–6.
47. Kaikkonen J, Tuomainen TP, Nyyssonen K, Salonen JT. Coenzyme Q10: absorption, antioxidative properties, determinants, and plasma levels. *Free Radic Res* 2002 Apr;**36**(4):389–97.
48. Young JM, Florkowski CM, Molyneux SL, et al. A randomized, double-blind, placebo-controlled crossover study of coenzyme Q10 therapy in hypertensive patients with the metabolic syndrome. *Am J Hypertens* 2012 Feb;**25**(2):261–70.
49. Yubero-Serrano EM, Gonzalez-Guardia L, Rangel-Zuniga O, et al. Postprandial antioxidant gene expression is modified by Mediterranean diet supplemented with coenzyme Q(10) in elderly men and women. *Age (Dordr)* 2013;**35**(1):159–70.
50. Yubero-Serrano EM, Delgado-Casado N, Delgado-Lista J, Perez-Martinez P, Tasset-Cuevas I, Santos-Gonzalez M, et al. Postprandial antioxidant effect of the Mediterranean diet supplemented with coenzyme Q10 in elderly men and women. *Age (Dordr)* 2011;**33**(4):579–90.
51. Tiano L, Belardinelli R, Carnevali P, et al. Effect of coenzyme Q10 administration on endothelial function and extracellular superoxide dismutase in patients with ischaemic heart disease: a double-blind, randomized controlled study. *Eur Heart J* 2007 Sep;**28**(18):2249–55.
52. Watts GF, Playford DA, Croft KD, et al. Coenzyme Q(10) improves endothelial dysfunction of the brachial artery in Type II diabetes mellitus. *Diabetologia* 2002 Mar;**45**(3):420–6.
53. Fuentes F, Lopez-Miranda J, Perez-Martinez P, et al. Chronic effects of a high-fat diet enriched with virgin olive oil and a low-fat diet enriched with alpha-linolenic acid on postprandial endothelial function in healthy men. *Br J Nutr* 2008 Jul;**100**(1):159–65.
54. Ruano J, Lopez-Miranda J, de la Torre R, et al. Intake of phenol-rich virgin olive oil improves the postprandial prothrombotic profile in hypercholesterolemic patients. *Am J Clin Nutr* 2007 Aug;**86**(2):341–6.
55. Yubero-Serrano EM, Gonzalez-Guardia L, Rangel-Zuniga O, et al. Mediterranean diet supplemented with coenzyme Q10 modifies the expression of proinflammatory and endoplasmic reticulum stress-related genes in elderly men and women. *J Gerontol A Biol Sci Med Sci* 2012 Jan;**67**(1):3–10.

56. Himmelfarb J, Hakim RM. Oxidative stress in uremia. *Curr Opin Nephrol Hypertens* 2003 Nov;**12**(6):593–8.
57. Lippa S, Colacicco L, Bondanini F, et al. Plasma levels of coenzyme Q(10), vitamin E and lipids in uremic patients on conservative therapy and hemodialysis treatment: some possible biochemical and clinical implications. *Clinica Chimica Acta* 2000 Feb 25;**292** (1-2):81–91.
58. Gokbel H, Atalay H, Okudan N, et al. Coenzyme Q10 and its relation with oxidant and antioxidant system markers in patients with end-stage renal disease. *Renal Failure* 2011;**33**(7):677–81.
59. Singh RB, Khanna HK, Niaz MA. Randomized, double-blind, placebo-controlled trial of coenzyme Q10 in chronic renal failure: discovery of a new role. *J Nutr Environ Med* 2000;**10**:281–8.
60. Koroshetz WJ, Jenkins BG, Rosen BR, Beal MF. Energy metabolism defects in Huntington's disease and effects of coenzyme Q10. *Ann Neurol* 1997 Feb;**41**(2):160–5.
61. Shults CW, Haas RH, Passov D, Beal MF. Coenzyme Q10 levels correlate with the activities of complexes I and II/III in mitochondria from parkinsonian and nonparkinsonian subjects. *Ann Neurol* 1997 Aug;**42**(2):261–4.
62. Shults CW, Oakes D, Kieburtz K, et al. Effects of coenzyme Q10 in early Parkinson disease: evidence of slowing of the functional decline. *Arch Neurol* 2002 Oct;**59**(10):1541–50.
63. Muller T, Buttner T, Gholipour AF, Kuhn W. Coenzyme Q10 supplementation provides mild symptomatic benefit in patients with Parkinson's disease. *Neurosci Lett* 2003 May 8;**341**(3):201–4.
64. Delanty N, Dichter MA. Oxidative injury in the nervous system. *Acta Neurologica Scandinavica* 1998 Sep;**98**(3):145–53.

CHAPTER

12

Vitamin C and Physical Performance in the Elderly

Kyoko Saito, Erika Hosoi

Research Team for Promoting the Independence of the Elderly, Tokyo Metropolitan Institute of Gerontology, Tokyo, Japan

Akihito Ishigami

Molecular Regulation of Aging, Tokyo Metropolitan Institute of Gerontology, Tokyo, Japan

Tetsuji Yokoyama

Department of Human Resources Development, National Institute of Public Health, Saitama, Japan

List of Abbreviations

ADL activities of daily living
AS antioxidant supplementation
DRI daily recommended intake
FFM fat-free mass
GPx-1 glutathione peroxidase
PGC-1α peroxisome proliferator-activated receptor gamma coactivator-1 alpha
QOL quality of life
ROS reactive oxygen species
RT resistance training
SOD superoxide dismutase
VO_{2max} maximal oxygen consumption

INTRODUCTION

In the elderly, a decline in physical function and performance, as well as impairments in activities of daily living (ADL), can lead to the deterioration of general health, dependence, and a decreased quality of life (QOL). Moreover, reduced physical function has been shown to lead to falls, fractures, and malnutrition, all of which have been related to morbidity and mortality in the elderly.[1,2]

Previous cross-sectional studies have found a decreased QOL and health status in under-nourished (low albumin) elderly people. Elderly people with vitamin C deficiency have been observed to experience a significant decline in physical function, shown by low muscle mass, weak grip strength, and slow walking speed.[3–5] Recent research has suggested that an intake of protein and vitamins within the diet can indeed have an effect on physical performance and function.[6–8] Understanding the effects that vitamins may have on physical performance in the elderly is crucial, as several studies have shown that deficiencies in substances with antioxidant properties – such as vitamin E, β-carotene, retinol, and vitamin C – may play a role within the aging process.

While most mammalian species are capable of synthesizing vitamin C (ascorbic acid), humans cannot.[9] The dangers of vitamin C deficiency were well known, as early sailors and travelers suffered from scurvy. Although scurvy is a result of extreme vitamin C deficiency, the intake of this antioxidant is vital for humans, and it must be ingested through the diet by the intake of fruits and vegetables. Epidemiologic studies show that diets high in fruits and vegetables are associated with a lower risk of cardiovascular disease, stroke, and cancer, and with increased longevity.[10–12]

Exercise increases oxygen free radicals and lipid peroxidation, which vitamin C is known to counter.[7,13,14] It is well known that old skeletal muscle may be more susceptible than young muscle to exercise-induced injury. Several researchers have investigated the effects of various antioxidants on physical performance and muscle

Aging
http://dx.doi.org/10.1016/B978-0-12-405933-7.00012-3

strength; however, these studies have focused on athletic performance, and young healthy people. The role of antioxidants, and specifically vitamin C, may be very different in the physical performance of athletes compared with the elderly. Older adults have significantly lower plasma vitamin C levels compared with young adults,[15–17] yet the literature lacks studies investigating the effects of vitamin C alone on physical performance in the elderly population, and the literature available is inconsistent and inconclusive.

This chapter aims to review and discuss the effects of exercise-induced oxidative stress and the water-soluble antioxidant vitamin C on physical performance in the elderly. The discussions focus on epidemiologic studies and the available animal and human intervention studies, as well as the mechanisms of vitamin C in relation to physical performance.

OXIDATIVE STRESS AND EXERCISE

Regular physical exercise is vital for the health of the elderly. Physical activity can reduce age-related reductions in muscle mass, promote strength for the maintenance of independence and QOL, and help protect against chronic diseases, including cardiovascular disease, cancer, and diabetes.[18,19] Acute, strenuous exercise, however, is known to be associated with the increased production of reactive oxygen species (ROS). The overproduction of ROS leads to oxidative stress and damage to biologic components such as nucleic acids, proteins, and lipids.[13,14,18] Ji[20] summarized that exercise-induced ROS and oxidative stress are greater in old age; in the review, Ji points out – based on rat studies – that older animals generate greater ROS in the heart and skeletal muscle from a relatively small amount of work. This seems problematic because oxidative stress and oxygen-derived free radicals play a major role in the aging process, as hypothesized by the free radical theory of aging.[18] Moreover, studies have suggested that oxidative damage plays an important role in the age-related decline of physical function.[21]

The source of free radicals during exercise has been assumed to be a result of skeletal muscle and cardiac muscle contraction, and in general, the mitochondria within skeletal muscles have been cited as a predominant source of ROS.[19] Specifically, researchers have reported that complexes I and III of the electron transport chain are the primary sources of superoxide production in mitochondria. An elaborate review by Powers and Jackson states that the increased ROS production via skeletal muscle contraction is directly related to the increase in oxygen consumption due to mitochondrial activity, which implies a 50- or 100-fold increase in superoxide generation by skeletal muscle during aerobic contractions.[19]

In terms of oxidative stress effects on physical performance, redox disturbances in skeletal muscle have been shown to reduce force production. ROS effects mainly involve changes in submaximal forces, as one mouse study pointed out that ROS reduce myofibrillar Ca^{2+} sensitivity in fatiguing skeletal muscle. Lamb and Westerblad[22] claimed that while several studies do not show significant effects of ROS on muscle fatigue, numerous studies have reported the integral role of ROS in force production. Several studies have shown that antioxidative ROS scavengers, such as *N*-acetylcysteine, can delay muscle fatigue during submaximal contractions, while antioxidant scavengers do not seem to have an effect in delaying fatigue during maximal muscular contractions.[19] Table 12.1 summarizes the results of human studies indicating that *N*-acetylcysteine delays fatigue during submaximal exercise. The literature suggests that the age-related physical decline might be related to oxidative damage perpetrated by free radicals.[23] Many animal studies have been conducted, as the collection of muscle biopsies during exercise is quite invasive,

TABLE 12.1 A Summary of Investigations Focused on the Effects of Exercise with *N*-Acetylcysteine Treatments on Various Dependent Measures. These Studies Show That Exercise with *N*-Acetylcysteine Improved Performance in the Populations Studied[a,b]

Mode of Exercise	Subject Pool	NAC Treatment	Exercise-Dependent Measure	Performance Improvement	References
Cycling to fatigue	Adult male endurance athletes	Multiple iv doses before and during exercise	Time to fatigue	+24%	McKenna et al[47]
Breathing against inspiratory load	Adult men	Single iv dose preexecise	Time to task failure	+50%	Travaline et al[48]
Repetitive handgrip exercise	Adult men and women	Single iv dose preexercise	Time to task failure	+15%	Matuszczak et al[49]
Repeated electrical stimulation of limb muscle	Adult men	Single iv dose preexercise	Force decline during 30 min of contractions	+15%	Reid et al[50]

[a]*NAC* = N-*acetylcysteine; iv = intravenous.*

[b]*From Powers SK & Jackson M.J. Exercise-induced oxidative stress: cellular mechanisms and impact on muscle force production, Table 3, in Physiological Reviews, 2008, 88(4), 1243-1276, with kind permission from the American Physiological Society.*

particularly for the elderly. Nevertheless, the presence of ROS in skeletal muscle of the elderly, in addition to age-related reductions in muscle mass and strength, can substantially decrease physical performance and function.

Although acute bouts of exercise increase ROS levels, exercise training or chronic exercise results in lower lipid peroxidation in the elderly.[24] With continuous training, skeletal and cardiac muscle can adapt, and while still unproven, such training is expected to reduce ROS generation in trained muscle as compared with untrained muscle in old age.[19–21] Although continuous submaximal exercise training may be feasible in healthy older adults, motivating the elderly to exercise continuously is a difficult feat. Further, frail older adults, and others with chronic diseases or disabilities may be incapable of exercise training, and may require other interventions in order to reduce the effects of oxidative stress. The relationship between oxidative stress and physical activity has been widely researched, yet it is still poorly understood. Studies focusing on physical activity and performance specifically in the elderly are particularly scarce.

OXIDATIVE STRESS, VITAMIN C SUPPLEMENTATION, AND PHYSICAL PERFORMANCE

There are a number of mechanistic hypotheses regarding the potential benefits of antioxidant vitamins.[13,25,26] Vitamin C, vitamin E, β-carotene, and retinol are important antioxidants that are not synthesized by humans and, therefore, are mainly supplied via dietary intake. Vitamin C is a water-soluble micronutrient required for multiple biologic functions, and it is also a cofactor for several enzymes participating in the post-translational hydroxylation of collagen, the biosynthesis of carnitine, the conversion of the neurotransmitter dopamine to norepinephrine, peptide amidation, and tyrosine metabolism.[27] In addition, vitamin C is an important regulator of iron uptake. It reduces ferric Fe^{3+} to ferrous Fe^{2+} ions, thus promoting dietary non-heme iron absorption from the gastrointestinal tract, and stabilizes iron-binding proteins. Most animals are able to synthesize vitamin C from glucose, but humans lack the last enzyme involved in the synthesis of vitamin C (gulonolactone oxidase) and so require the presence of this vitamin in the diet. The prolonged deprivation of vitamin C generates defects in the post-translational modification of collagen that cause scurvy and eventually death.[28]

The role of vitamin C as an antioxidant – reducing the effects of exercise-induced ROS, including muscle damage, immune dysfunction, and fatigue – are widely known and established.[29] Because exercise increases the generation of oxygen free radicals and lipid peroxidation,[7] and strenuous physical performance can increase oxygen consumption by 10- to 15-fold over the resting state to meet the energy demands, this may result in muscle injury.[14] Vitamin C may have potential protective effects against such exercise-induced muscle damage.[7,13,14] Supplementation with vitamin C may also decrease the elevated lipid peroxidation induced by exercise.[7,30,31] McGinley et al summarized that the rationale behind vitamin C supplementation after a bout of exercise lies in the location of vitamin C in the plasma. Increasing the plasma levels of vitamin C, by using a supplement, may potentially reduce the inflammatory phase of muscle damage caused by increases in ROS during peak oxidant production.[29] Vitamin C supplementation prior to exercise seeks to elevate plasma vitamin C levels – as well as levels in water-soluble compartments, including cytosol, mitochondrial matrix, and extracellular fluids – to increase bioavailability to active tissue, and hence, to counteract the increase in ROS during exercise. The results in the literature are, however, inconsistent. This may be because of differences in supplementation strategies, exercise protocols, and subject populations.

The molecular mechanism of the relationship between vitamin C supplementation and physical performance is unclear. Although much focus has been placed on the damage created by increased ROS levels, it is also known that ROS can activate signaling pathways involved in phenotypic adaptations.[32] There is growing evidence supporting the role for ROS in exercise-induced skeletal muscle mitochondrial biogenesis.[33,34] Higher mitochondrial content results in greater oxygen extraction by the exercising muscle, hence increasing the capacity for sustained work. There has been speculation suggesting that vitamin C completely diminishes the increase in several markers of exercise-induced skeletal muscle mitochondrial biogenesis, and may even prevent increases in the expression of antioxidant enzymes such as superoxide dismutase (SOD) and glutathione peroxidase (GPx-1).[33] However, some studies contradict this speculation, reporting that vitamin C does not prevent the increase in skeletal muscle metabolic signaling and activation of mitochondrial biogenesis.[35,36] The 'Intervention Study' section of this chapter will further discuss the available findings and proposed mechanisms.

EPIDEMIOLOGY STUDY

Investigation into the effects of oxidative stress and free radicals in relation to aging has been of interest for several decades. An early publication by Denham Harman[37] proposed that free radicals are causally related to aging. The study of the effects of antioxidants on physical performance is relatively newer, and interest in this area of research has been growing. However, whether or not antioxidant intake has beneficial effects

TABLE 12.2 Summary of Epidemiology Studies Examining the Association Between Vitamin C and Physical Performance in the Elderly. Physical Performance Was Measured by Functional Fitness Tests Commonly Used in the Elderly Population

Name of Study and Year of Publication	Country	Subjects	Age Mean ±SD	FFQ or Plasma Level	Physical Performance	Results
InCHIANTI study 2004	Italy	Male and female 986 (55.3% female)	75.3 ± 0.2	FFQ	Summary physical performance score (walking speed, chair-standing, standing balance test) Knee extension strength	Daily dietary intake of vitamin C was significantly correlated with knee extension strength (adjusted: β = 0.383, SE = 0.132, P = 0.02). The daily intake of vitamin C was significantly associated with performance test (adjusted: β = 0.0.029, SE = 0.014, P = 0.04).
Health examination of successful aging 2011	Japan	Female 655	75.7 ± 4.1	Plasma level	Hand grip strength One leg standing with eyes open Walking speed	Plasma vitamin C level was significantly correlated with handgrip strength (adjusted P for trend = 0.0004) and ability to stand on one leg with eyes open (adjusted P for trend = 0.049).
Hertfordshire Cohort Study 2011	UK	Male and female 628 (44.6% female)	M: 67.8 ± 2.5 F: 68.1 ± 2.5	FFQ	Short physical performance battery (walking speed, chair-standing, standing balance test)	Greater intake of vitamin C was associated with shorter chair-rise times (P = 0.039) in women.

FFQ = food frequency questionnaire.

on physical performance is still controversial. Some studies have shown improvements while others have not.[18,38,39] Most studies have been conducted on athletes and young adults, and research on antioxidant use for physical performance in the elderly is quite scarce. There are few epidemiologic studies attempting to clarify the relationship between vitamin C in physical performance in the elderly in large cohorts (summarized in Table 12.2).

The Invecchiare in Chianti (InCHIANTI) study explored the correlation of plasma antioxidant concentrations and physical performance in Italian elderly persons. A total of 986 elderly men and women over the age of 65 were assessed. Each subject completed the European Prospective Investigation into Cancer and Nutrition questionnaire to determine daily nutritional intakes of antioxidants such as vitamin C, vitamin E, β-carotene, and retinol. Physical performance measures included knee extension strength, walking speed, ability to rise from a chair, and standing balance. The results from this cross-sectional study revealed, as depicted in Table 12.3, that daily dietary intakes of vitamin C were significantly correlated with knee extension strength (β = 0.383, SE = 0.132, P = 0.02) and physical performance score (β = 0.0.029, SE = 0.014, P = 0.04) in the elderly. This study showed significant positive correlations between most antioxidants, especially vitamin C, and higher skeletal muscular strength and physical performance scores in the elderly population studied. The theoretical basis for the beneficial effects of antioxidants on performance

TABLE 12.3 Adjusted Regression Coefficients for Knee Extension Strength and Summary Physical Performance Scores per Antioxidants SD Increase (Log) in Plasma and in Dietary Intakes of Oxidants[a]

	Knee Extension Strength (kg)		Physical Performance Score	
	Regression Coefficient (SE)	P	Regression Coefficient (SE)	P
Plasma				
α-Tocopherol (μmol/L)	0.566 (0.193)	0.003	0.044 (0.017)	0.008
γ-Tocopherol (μmol/L)	0.327 (0.165)	0.04	0.004 (0.015)	0.80
Daily intake				
Vitamin C (mg)	0.383 (0.162)	0.02	0.029 (0.014)	0.04
Vitamin E (mg)	0.277 (0.204)	0.17	0.013 (0.018)	0.47
β-Carotene (μg)	0.311 (0.159)	0.05	0.015 (0.014)	0.29
Retinol (μg)	0.012 (0.147)	0.94	−0.010 (0.013)	0.31

[a]The coefficients were adjusted for age, sex, site, Mini-Mental State Examination score, smoking, education, BMI, physical activity, total cholesterol, triacylglycerol (log value), hypertension, coronary heart disease, stroke, congestive heart failure, number of medications taken, and daily dietary energy intake. The α-tocopherol SD = 0.278, γ-tocopherol SD = 0.441, daily vitamin C intake (log) SD = 0.449, daily vitamin E intake (log) SD = 0.323, daily β-carotene intake (log) SD = 0.508, and daily retinol intake (log) SD = 0.910. From Cesari M et al, Antioxidants and physical performance in elderly persons: the Invecchiare in Chianti (InCHIANTI) study, Table 3, in American Journal of Clinical Nutrition, 2004, 79(2), 289-294, with kind permission from the American Society for Nutrition.

TABLE 12.4 Relationship Between Plasma Vitamin C Concentration and Physical Performance Adjusted for Potential Confounder[a]

	Quartile of Plasma Vitamin C Level				
	Q1	Q2	Q3	Q4	
Physical Performance	Mean ± SE	Mean ± SE	Mean ± SE	Mean ± SE	*P* for Trend
Handgrip strength (kg), N	154	159	154	152	
Age-adjusted	17.70 ± 0.34	18.75 ± 0.33	18.75 ± 0.34	19.60 ±0.34	0.0001
Multivariate-adjusted[b]	17.83 ± 0.34	18.83 ± 0.32	18.89 ± 0.33	19.60 ± 0.33	0.0004
One leg standing with eyes open[c] (sec), N	162	163	164	161	
Age-adjusted	31.44 ± 1.71	33.98 ± 1.70	37.70 ± 1.70	37.89 ± 1.71	0.003
Multivariate-adjusted[b]	33.39 ± 1.74	34.08 ± 1.67	37.63 ± 1.67	37.50 ± 1.70	0.049
Usual walking speed (m/sec), N	146	154	145	147	
Age-adjusted	1.13 ± 0.02	1.19 ± 0.02	1.23 ± 0.02	1.21 ± 0.02	0.008
Multivariate-adjusted[b]	1.18 ± 0.02	1.19 ± 0.02	1.22 ± 0.02	1.21 ± 0.02	0.23
Maximal walking speed (m/sec), N	146	154	154	147	
Age-adjusted	1.70 ± 0.03	1.76 ± 0.03	1.82 ± 0.03	1.76 ± 0.03	0.15
Multivariate-adjusted[b]	1.76 ± 0.03	1.77 ± 0.03	1.80 ± 0.03	1.75 ± 0.03	0.94

[a]*Values are least squares mean and standard error (SE) adjusted for the factor(s) by ANCOVA. Q1 to Q4: 1st to 4th quartile groups of plasma vitamin C concentration, respectively.*
[b]*Adjusted for age, Body Mass Index, percent body fat, hypertension, diabetes mellitus, and fruit intake.*
[c]*Length of time standing on one leg with eyes open. From Saito K et al, A significant relationship between plasma vitamin C concentration and physical performance among Japanese elderly women, Table 3, in Journal of Gerontology: Medical Sciences, 2012, 67A(3), 295-301, by permission of Oxford University Press.*

enhancement is unclear; however, the authors suggest that an adequate antioxidant intake is needed to maintain healthy muscular activity.

A different study, by Saito et al,[40] also found a significant relationship between plasma vitamin C concentrations and physical performance (Table 12.2). This particular cross-sectional study specifically investigated the relationship between plasma vitamin C and physical performance among Japanese elderly women. Physical performance and plasma concentrations of vitamin C were assessed in 655 older women between the ages of 70 and 84, who did not take vitamin supplementation. Each participant completed lifestyle assessments providing information regarding their diet (fruit and vegetable intake) and other factors such as chronic diseases. Physical performance measures included handgrip strength, one leg standing time with eyes open, and usual and maximum walking speeds. The geometric mean (geometric standard deviation) of plasma vitamin C concentration was 8.9 (1.5) μg/mL. After adjusting for confounding factors, plasma vitamin C concentration was positively correlated with handgrip strength (*P* for trend = 0.0004) and length of time standing on one leg with eyes open (*P* for trend = 0.049), as shown in Table 12.4. The results did not show any significant associations between usual and maximum walking speed and plasma vitamin C levels. The authors suggested that vitamin C may have effects on relatively simple strength and balance functions in the elderly. Interestingly in this study, while the data were not shown in the manuscript, a subanalysis comparing supplement users and non-users (n = 238) was performed (Fig. 12.1). Figure 12.1 illustrates the relationship between plasma vitamin C concentration and handgrip strength in vitamin C supplement users and non-users. The data revealed that, after adjusting for selected factors by multiple stepwise linear regression, the relationship between plasma vitamin C concentration and handgrip strength was seen only in non-users of vitamin C supplement, but not in users (*P* = 0.0016 difference in slope between users and non-users). No differences in slope were observed between users and non-users for the other physical performance measures. The authors briefly mentioned that these results implied that vitamin C supplementation did not have beneficial effects on physical performance and muscle strength, supporting claims made by other researchers. The effects of vitamin C supplementation on physical performance, however, are still unclear, and results from studies have been equivocal. Regardless, the authors concluded that a significant relationship between plasma vitamin C concentration and physical performance was observed, although further study is warranted.

Martin et al[41] published the results from the Hertfordshire Cohort study investigating the relationship between diet and physical performance in community-dwelling older men and women (Table 12.2). Diet was assessed

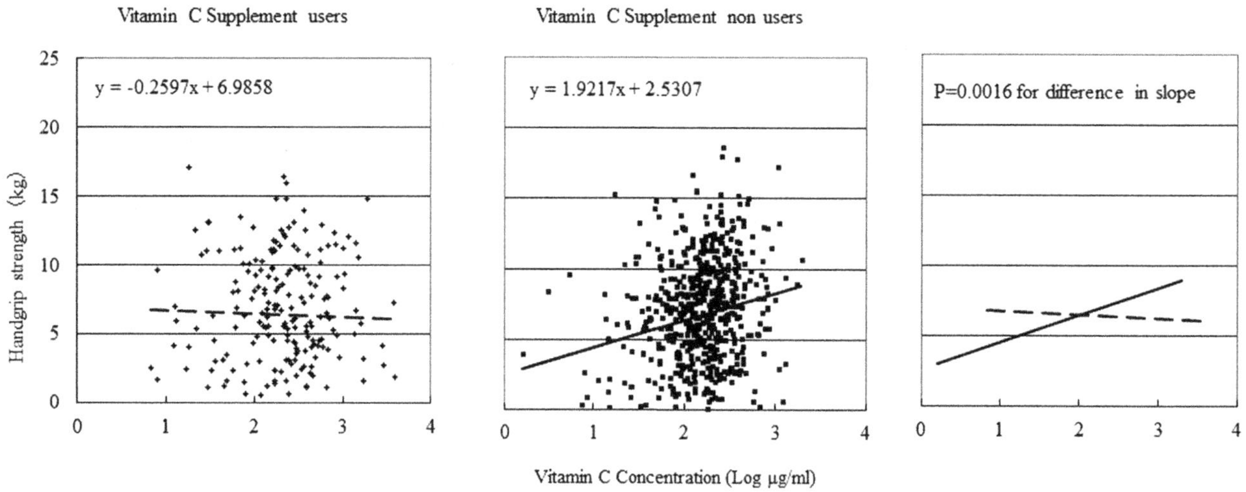

FIGURE 12.1 The relationship between plasma vitamin C concentration and handgrip strength in vitamin C supplement users and non-users. Values were adjusted for age, BMI, albumin, supplement use, exercise habit, and vegetable intake by multiple liner model. Unpublished.

in a total of 628 men and women between the ages of 63 and 73, using a food frequency questionnaire including 129 foods and food groups to assess frequency of consumption during a 3-month period. A short physical performance battery was completed by each participant, including the time taken to walk 3 m at usual speed, five sit–stand chair rises, and one-leg standing balance. The results of this cohort study showed that greater intakes of β-carotene and vitamin C were associated with shorter chair-rise times ($P = 0.003$ and $P = 0.039$, respectively). Mechanisms were not, however, discussed in this study.

The studies reviewed in this section imply a possible relationship between vitamin C and physical performance in the elderly. However, the three studies introduced were cross-sectional studies and, therefore, do not provide cause–effect relationships. Longitudinal follow-up studies and controlled clinical trials are necessary to confirm the role of plasma or food intake vitamin C and physical performance in elderly people. The evaluation of both dietary intake estimates and biologic measures is the ideal combination.

INTERVENTION STUDY

Few intervention studies, if any, have examined the effects of vitamin C alone on physical performance in the elderly. Vitamin C is almost always administered in congruence with other minerals, such as vitamin E, especially since α-tocopherol, a form of vitamin E, and vitamin C act in synergy to regulate oxidative stress, as they function in lipid and aqueous compartments, respectively.[29] This has led to interventional studies examining the effects of these two vitamins combined, as opposed to vitamin C alone. Interventional research investigating the effects of vitamin C on physical performance in elderly populations is even rarer. In fact, no such study was found for review within this chapter. However, there are several interesting intervention studies that depict the current understanding of antioxidant supplementation, including vitamin C, on physical performance.

One study by Bobeuf et al investigated the effects of resistance training combined with antioxidant supplementation on the antioxidant/pro-oxidant balance and on body composition, including muscle mass, in an elderly population.[42] Fifty-seven men and women (ages 59–73) were randomized into four groups: placebo, resistance training (RT), antioxidant supplementation (AS), and AS+RT. The AS was a combination of vitamin C (1000 mg) and vitamin E (400 IU), which was taken daily for 6 months. The RT group underwent training 3 days a week for 1 hour, during a 6-month period. Mean strength gains of the eight exercises performed in the intervention were analyzed. This study did not find significant improvements in the antioxidant/pro-oxidant profile; however, significant changes in body composition and strength gains were observed (Table 12.5). Both the RT-only group and AS+RT group significantly increased appendicular FFM, and significant strength gains were observed. The results showed that AS did not maximize strength gains during RT. Important to note is that this study was conducted on a very small sample size, and with the numerous analyses performed, there was an increased risk of type I error.

The effect of vitamin C supplementation on physical performance is often investigated in healthy young adults. Experimental results regarding vitamin C supplementation are equivocal, but earlier studies show a lack of beneficial effects on physical performance.[13] A review performed by Braakhuis[43] showed that vitamin C doses of over 1 gram per day actually impaired sports performance in the young. The review suggested that an inhibition of mitochondrial growth is a mechanism by which

TABLE 12.5 Physical Characteristics of Subjects by Group Before and After the RT and/or AS Intervention (Mean ± SD)

	Placebo (n = 12)		RT (n = 17)		AS (n = 14)		AS+RT (n = 14)	
Variables	**Baseline**	**6 Months**	**Baseline**	**6 Months**	**Baseline**	**6 Months**	**Baseline**	**6 Months**
Total FM (kg)	18.9 ± 6.6	19.5 ± 6.8†	23.5 ± 6.8	23.6 ± 7.5	21.3 ± 6.4	21.7 ± 5.7	22.2 ± 8.0	21.7 ± 7.9
Abdominal FM (kg)†	9.3 ± 3.7	10.1 ± 4.0†	12.3 ± 4.2[a]	12.6 ± 4.8	10.0 ± 3.3	10.3 ± 3.0	10.6 ± 3.6	10.6 ± 3.8
Total FM (%)	29.3 ± 9.9	30.1 ± 10.0†	33.6 ± 9.3	33.4 ± 9.6	32.8 ± 8.6	33.2 ± 7.9	32.9 ± 12.3	31.9 ± 12.4
Abdominal FM (%)	30.8 ± 10.0	32.3 ± 10.0†	35.9 ± 9.3	36.1 ± 9.8	33.5 ± 8.1	34.0 ± 7.0	34.4 ± 11.1	33.3 ± 11.6
VFM (cm^2)†	63.3 ± 21.5	63.5 ± 21.4	79.6 ± 23.8	80.0 ± 23.9	58.5 ± 25.0[b]	58.6 ± 25.1	62.8 ± 28.5	63.2 ± 28.6
Appendicular FFM (kg)*	21.5 ± 4.9	21.3 ± 5.4	21.8 ± 5.2	22.1 ± 5.4†	20.8 ± 6.2	20.7 ± 6.1	22.1 ± 6.5	22.6 ± 6.7†
FFM/FM ratio	2.8 ± 1.3	2.6 ± 1.3	2.2 ± 1.0	2.3 ± 1.3	2.2 ± 0.9	2.1 ± 0.7	2.6 ± 1.9	2.8 ± 2.3
Muscle mass (kg)*	30.2 ± 9.4	27.9 ± 8.3	27.8 ± 8.0	29.6 ± 8.1	30.7 ± 13.5	26.0 ± 8.1†	29.8 ± 10.1	30.0 ± 10.4
Strength gain (%)†				+65.00 ± 25.61†				+78.17 ± 61.46†

[a]*Different from group 1, P ≤ 0.05*
[b]*Different from group 2, P ≤ 0.05*
**Significant treatment effect; P ≤ 0.01*
†*Significantly different from baseline by paired t-test; P ≤ 0.05; RT = resistance training; AS = antioxidant supplementation; FM = fat mass; VFM = visceral fat mass; FFM = fat-free mass; strength gain (%) = mean of strength gain in percentage of the 8 exercises; SD = standard deviation. From Bobeuf F, Combined effect of antioxidant supplementation and resistance training on oxidative stress markers, muscle and body composition in an elderly population, Table 3, in Journal of Nutrition, Health and Aging, 2011, 15(10), 883-889, with kind permission from Springer Science and Business Media.*

vitamin C can impair performance, as mitochondria are crucial for aerobic energy metabolism. Braakhuis also mentioned that vitamin C may reduce exercise-induced blood flow, hence reducing exercise capacity and performance.[43] Researchers often perform animal studies to examine in detail the mechanistic effects of vitamin C on physical performance. In fact, Gomez-Cabrera et al found that the administration of vitamin C significantly hampered endurance capacity in young male rats, did not improve VO_{2max} associated with training in male humans aged 27–36 years, and lowered training efficiency.[33] This particular study investigated the effects of vitamin C on training efficiency in both rats and humans. In this double-blind randomized study conducted by Gomez-Cabrera et al,[33] 14 men aged between 27 and 36 years trained for 8 weeks. Five of the men were provided with daily 1 gram vitamin C supplementation. In the animal study, 24 male rats were exercised and treated with a daily dose of vitamin C (0.24 mg/cm^2 body surface area). The results revealed that vitamin C significantly hindered endurance capacity ($P = 0.014$) in rats. The authors suggested that vitamin C may reduce the exercise-induced expression of key transcription factors such as peroxisome proliferator-activated receptor co-activator 1, nuclear respiratory factor 1, and mitochondrial transcription factor A, involved in mitochondrial biogenesis. Furthermore, vitamin C hindered the expression of antioxidant enzymes SOD and GPx-1 that were activated with exercise, as well as cytochrome C, which is a marker of mitochondrial content. The study concluded that vitamin C decreases training efficiency in young male humans and rats by preventing cellular adaptations to exercise.

The literature also presents results contradictory to those found above. One rat study by Higashida et al[36] showed that very large doses of antioxidant vitamins did not prevent the exercise-induced adaptive responses of muscle mitochondria. This study investigated the short- and long-term effects of vitamin C and vitamin E administration on adaptive increases of mitochondrial enzymes and insulin-stimulated glucose transport. The authors reported results that disagree with those of Gomez-Cabrera summarized above, as they found that antioxidant supplementation did not affect the exercise-induced increases in SOD, PGC-1α, nor a number of mitochondrial marker proteins. The results showed that vitamin C did not have inhibitory effects on the adaptive responses of VO_{2max}, muscle citrate synthase, hydroxyacyl-CoA dehydrogenase activities, or insulin action in young healthy rats. Higashida et al concluded that a combined antioxidant supplementation of vitamin C and E did not have negative effects on the exercise-induced adaptation of skeletal muscle or mitochondrial content in sedentary animals.

Interestingly, one study by Roberts et al[44] also found that vitamin C did not hinder the effects of training. This study aimed to test the hypothesis that antioxidants can attenuate improvements in exercise performance resulting from high-intensity interval training in young healthy humans. The study was conducted on 18 recreationally active young males, who either consumed 1 gram of vitamin C or a placebo on weekdays, prior to the training sessions. The participants trained using a high-intensity interval training protocol on a motorized treadmill four times per week for 4 weeks. Subjects' VO_{2max}, 10-km time, and intermittent exercise performance were assessed. This double-blind placebo trial found that vitamin C supplementation did not impair the effects of training on oxygen consumption, running economy, or intermittent exercise performance, as there were no significant differences between the vitamin C and placebo groups. Interpretation of the results should be done with caution as the results of this study were based on a very small sample size. Furthermore, the authors point out that the data presented may be specific to the training mode, duration, and the population of young, active males studied.

In light of the research summarized in this section, it is clear that further study of vitamin C supplementation and its effects on physical performance is needed, especially in the elderly. As Roberts et al[44] discussed, the effects found in previous reports may be specific to the population studied. Because earlier investigations have found that the elderly have lower concentrations of plasma vitamin C, compared with young adults, due to impairments in absorption,[15–17] it is difficult to predict whether large doses of vitamin C would hinder or improve physical performance in the elderly. Nevertheless, an adequate dietary intake of vitamin C is likely to be required for normal muscle function,[43,45] and regular physical activity in the elderly is recommended to reverse age-related body composition modifications (i.e. decreasing lean mass and increasing fat mass).[18]

CONCLUSIONS

The relationship between vitamins, minerals, aging, and physical performance has been an area of interest in recent years. Epidemiologic studies have shown significant associations between antioxidants such as vitamin C, vitamin E, β-carotene, and retinol, and cancer, heart disease, stroke, and arteriosclerosis.[10–12,46] There is a gap in the literature, however, where studies investigating the effects of vitamin C on physical performance in the elderly are limited and scarce. While low vitamin C is reported to be a risk factor for aging, very few human epidemiologic studies are available.

Cross-sectional studies reveal that vitamin C may have an effect on muscle strength and performance. Greater plasma vitamin C concentration has been associated with grip strength and balance measures.[40] Similarly, a positive relationship between vitamin C intake and skeletal muscular strength, performance score,[6] and chair-rise times has been reported.[41] Nevertheless, the specific effects and mechanisms of the effects of vitamin C on physical performance remain unclear. Saito et al[40] found that greater plasma vitamin C concentration was associated with certain areas of physical performance, including grip strength and balance, yet no effects were observed in walking speed. Vitamin C appears to have beneficial effects on muscle strength, although the results of previous studies are inconclusive.

Further research into plasma vitamin C concentration, vitamin C intake, and physical performance is required to confirm the results of previous studies, especially as maintaining physical function among the elderly is crucial for daily activities. Because older adults have impaired absorption of vitamin C, elderly individuals should aim to fulfill the DRI established in their respective nations. Regardless of the contradicting evidence published in the literature, an adequate intake of vitamin C is required to maintain healthy muscular activity. Although high, non-physiologic doses of supplemental vitamin C cannot be recommended for healthy individuals, a high intake of fruits and vegetables is vital to sustain healthy antioxidant status in the elderly.

SUMMARY POINTS

- The underlying mechanism behind the relationship between vitamin C and physical performance is still unclear, and future research is necessary, especially regarding the effects that vitamin C may have on mitochondrial biogenesis.
- Cross-sectional epidemiologic studies reveal positive associations between plasma vitamin C levels and muscle strength and performance.
- There are inconsistencies and contradictions in both animal and human intervention studies regarding the effects of vitamin C supplementation on physical performance; however, adequate vitamin C intake is required for healthy muscle function.
- Further longitudinal investigations as well as intervention studies are needed to confirm the effects of vitamin C on physical performance in the elderly.
- While large doses of vitamin C do not seem to have beneficial effects on physical performance in the elderly, diets high in fruits and vegetables are considered healthy and beneficial.

References

1. Fried LP. Epidemiology of aging. *Epidemiol Rev* 2000;**22**:95–106.
2. Bartali B, Salvini S, Turrini A, et al. Age and disability affect dietary intake. *J Nutr* 2003;**133**:2868–73.
3. Kwon J, Suzuki T, Yoshida H, et al. Concomitant lower serum albumin and vitamin D levels are associated with decreased objective physical performance among Japanese community-dwelling elderly. *Gerontology* 2007;**53**:322–8.
4. Kwon J, Suzuki T, Yoshida H, et al. Association between change in bone mineral density and decline in usual walking speed in elderly community-dwelling Japanese women during 2 years of follow-up. *J Am Geriatr Soc* 2007;**55**:240–4.

5. Suzuki T, Kwon J, Kim H, et al. Low serum 25-hydroxy vitamin D levels associated with falls among Japanese community-dwelling elderly. *J Bone Miner Res* 2008;**23**:1309–17.
6. Cesari M, Pahor M, Bartali B, et al. Antioxidants and physical performance in elderly persons: the Invecchiare in Chianti (InCHIANTI) study. *Am J Clin Nutr* 2004;**79**:289–94.
7. Evans WJ. Vitamin E, vitamin C, and exercise. *Am J Clin Nutr* 2000;**72**:647S–52S.
8. Houston DK, Cesari M, Ferrucci L, et al. Association between vitamin D status and physical performance: the InCHIANTI study. *J Gerontol A Biol Sci Med Sci* 2007;**62**:440–6.
9. Traber MG, Stevens JF. Vitamins C and E: beneficial effects from a mechanistic perspective. *Free Radic Biol Med* 2011;**51**:1000–13.
10. Ness AR, Powles JW. Fruit and vegetables, and cardiovascular disease: a review. *Int J Epidemiol* 1997;**26**:1–13.
11. Genkinger JM, Platz EA, Hoffman SC, et al. Fruit, vegetable, and antioxidant intake and all-cause, cancer, and cardiovascular disease mortality in a community-dwelling population in Washington County, Maryland. *Am J Epidemiol* 2004;**160**:1223–33.
12. Yokoyama T, Date C, Kokubo Y, et al. Serum vitamin C concentration was inversely associated with subsequent 20-year incidence of stroke in a Japanese rural community. The Shibata study. *Stroke* 2000;**31**:2287–94.
13. Clarkson PM, Thompson HS. Antioxidants: what role do they play in physical activity and health? *Am J Clin Nutr* 2000;**72**: 637S–46S.
14. Clarkson PM. Antioxidants and physical performance. *Crit Rev Food Sci Nutr* 1995;**35**:131–41.
15. Blanchard J, Conrad KA, Mead RA, Garry PJ. Vitamin C disposition in young and elderly men. *Am J Clin Nutr* 1990;**51**:837–45.
16. Fletcher AE, Breeze E, Shetty PS. Antioxidant vitamins and mortality in older persons: findings from the nutrition add-on study to the Medical Research Council Trial of Assessment and Management of Older People in the Community. *Am J Clin Nutr* 2003;**78**:999–1010.
17. Jacob RA, Otradovec CL, Russell RM, et al. Vitamin C status and nutrient interactions in a healthy elderly population. *Am J Clin Nutr* 1988;**48**:1436–42.
18. Fusco D, Colloca G, Lo Monaco MR, Cesari M. Effects of antioxidant supplementation on the aging process. *Clin Interv Aging* 2007;**2**:377–87.
19. Powers SK, Jackson MJ. Exercise-induced oxidative stress: cellular mechanisms and impact on muscle force production. *Physiol Rev* 2008;**88**:1243–76.
20. Ji LL. Exercise at old age: does it increase or alleviate oxidative stress? *Ann N Y Acad Sci* 2001;**928**:236–47.
21. Pansarasa O, Castagna L, Colombi B, et al. Age and sex differences in human skeletal muscle: role of reactive oxygen species. *Free Radic Res* 2000;**33**:287–93.
22. Westerblad H, Allen DG. Emerging roles of ROS/RNS in muscle function and fatigue. *Antioxid Redox Signal* 2011;**15**:2487–99.
23. De la Fuente M. Effects of antioxidants on immune system ageing. *Eur J Clin Nutr* 2002;**56**(Suppl 3):S5–8.
24. Leeuwenburgh C, Fiebig R, Chandwaney R, Ji LL. Aging and exercise training in skeletal muscle: responses of glutathione and antioxidant enzyme systems. *Am J Physiol* 1994;**267**:R439–45.
25. Carr AC, Frei B. Toward a new recommended dietary allowance for vitamin C based on antioxidant and health effects in humans. *Am J Clin Nutr* 1999;**69**:1086–107.
26. Nishikimi M. Oxidation of ascorbic acid with superoxide anion generated by the xanthine-xanthine oxidase system. *Biochem Biophys Res Commun* 1975;**63**:463–8.
27. Duarte TL, Lunec J. Review: When is an antioxidant not an antioxidant? A review of novel actions and reactions of vitamin C. *Free Radic Res* 2005;**39**:671–86.
28. Halliwell B. Vitamin C and genomic stability. *Mutat Res* 2001;**475**:29–35.
29. McGinley C, Shafat A, Donnelly AE. Does antioxidant vitamin supplementation protect against muscle damage? *Sports Med* 2009;**39**:1011–132.
30. Davies KJ, Quintanilha AT, Brooks GA, Packer L. Free radicals and tissue damage produced by exercise. *Biochem Biophys Res Commun* 1982;**107**:1198–205.
31. Gee DL, Tappel AL. The effect of exhaustive exercise on expired pentane as a measure of in vivo lipid peroxidation in the rat. *Life Sci* 1981;**28**:2425–9.
32. Hood DA. Mechanisms of exercise-induced mitochondrial biogenesis in skeletal muscle. *Appl Physiol Nutr Metab* 2009;**34**:465–72.
33. Gomez-Cabrera MC, Domenech E, Romagnoli M, et al. Oral administration of vitamin C decreases muscle mitochondrial biogenesis and hampers training-induced adaptations in endurance performance. *Am J Clin Nutr* 2008;**87**:142–9.
34. Ristow M, Zarse K, Oberbach A, et al. Antioxidants prevent health-promoting effects of physical exercise in humans. *Proc Natl Acad Sci USA* 2009;**106**:8665–70.
35. Wadley GD, McConell GK. High-dose antioxidant vitamin C supplementation does not prevent acute exercise-induced increases in markers of skeletal muscle mitochondrial biogenesis in rats. *J Appl Physiol* 2010;**108**:1719–26.
36. Higashida K, Kim SH, Higuchi M, et al. Normal adaptations to exercise despite protection against oxidative stress. *Am J Physiol Endocrinol Metab* 2011;**301**:E779–84.
37. Harman D. Atherosclerosis: possible ill-effects of the use of highly unsaturated fats to lower serum-cholesterol levels. *Lancet* 1957;**273**:1116–7.
38. Gao X, Bermudez OI, Tucker KL. Plasma C-reactive protein and homocysteine concentrations are related to frequent fruit and vegetable intake in Hispanic and non-Hispanic white elders. *J Nutr* 2004;**134**:913–8.
39. Wijnen MH, Coolen SA, Vader HL, et al. Antioxidants reduce oxidative stress in claudicants. *J Surg Res* 2001;**96**:183–7.
40. Saito K, Yokoyama T, Yoshida H, et al. A significant relationship between plasma vitamin C concentration and physical performance among Japanese elderly women. *J Gerontol A Biol Sci Med Sci* 2012;**67**:295–301.
41. Martin H, Aihie Sayer A, Jameson K, et al. Does diet influence physical performance in community-dwelling older people? Findings from the Hertfordshire Cohort Study. *Age Ageing* 2011;**40**:181–6.
42. Bobeuf F, Labonte M, Dionne IJ, Khalil A. Combined effect of antioxidant supplementation and resistance training on oxidative stress markers, muscle and body composition in an elderly population. *J Nutr Health Aging* 2011;**15**:883–9.
43. Braakhuis AJ. Effect of vitamin C supplements on physical performance. *Curr Sports Med Rep* 2012;**11**:180–4.
44. Roberts LA, Beattie K, Close GL, Morton JP. Vitamin C consumption does not impair training-induced improvements in exercise performance. *Int J Sports Physiol Perform* 2011;**6**:58–69.
45. Carr AC, Bozonet SM, Pullar JM, et al. Human skeletal muscle ascorbate is highly responsive to changes in vitamin C intake and plasma concentrations. *Am J Clin Nutr* 2013;**97**:800–807.
46. Riccioni G, D'Orazio N, Salvatore C, et al. Carotenoids and vitamins C and E in the prevention of cardiovascular disease. *Int J Vitam Nutr Res* 2012;**82**:15–26.
47. McKenna MJ, Medved I, Goodman CA, et al. *N*-acetylcysteine attenuates the decline in muscle Na^+, K^+-pump activity and delays fatigue during prolonged exercise in humans. *J Physiol* 2006;**576**:279–88.
48. Travaline JM, Sudarshan S, Roy BG, et al. Effect of N-acetylcysteine on human diaphragm strength and fatigability. *Am J Respir Crit Care Med* 1997;**156**:1567–71.
49. Matuszczak Y, Farid M, Jones J, et al. Effects of N-acetylcysteine on glutathione oxidation and fatigue during handgrip exercise. *Muscle Nerve* 2005;**32**:633–8.
50. Reid MB, Stokic DS, Koch SM, et al. *N*-acetylcysteine inhibits muscle fatigue in humans. *J Clin Invest* 1994;**94**:2468–74.

CHAPTER

13

Tryptophan and Melatonin-Enriched Foodstuffs to Improve Antioxidant Status in Aging

M. Garrido, A.B. Rodríguez, M.P. Terrón

Department of Physiology (Neuroimmunophysiology and Chrononutrition Research Group), Faculty of Science, University of Extremadura, Badajoz, Spain

List of Abreviations

aMT6-s 6-sulfatoxymelatonin
BBB blood–brain barrier
GPx glutathione peroxide
H_2O_2 hydrogen peroxide
$O_2^{\cdot-}$ singlet oxygen
OH^- hydroxyl radical
$ONOO^-$ peroxynitrite anion
ROO^- peroxyl radical
ROS reactive oxygen species
SOD superoxide dismutase

INTRODUCTION

Cells in humans and other organisms are constantly exposed to a variety of oxidizing agents, some of which are necessary for life. These agents may be present in air, food and water, or they may be produced by metabolic activities within cells. Thus, despite the fact that oxygen is an essential element for the survival of aerobic organisms, the use of oxygen during normal metabolism generates reactive oxygen species (ROS), some of which are highly toxic and deleterious to cells and tissues. To protect cells from the damage caused by ROS and related reactants, organisms have evolved several antioxidant defense mechanisms to rapidly and efficiently remove ROS from the environment. Antioxidant defense systems may be generally classified into indirect antioxidant enzymes and low-molecular-weight molecules which act as scavengers of free radicals.[1] When the equilibrium between free radicals (oxidants) and the antioxidant defense systems becomes out of balance – in favor of oxidants – the result is *oxidative stress;* this is involved in the development of inflammatory, neurodegenerative and autoimmune diseases.[2]

The key factor is to maintain a balance between oxidants and antioxidants to sustain optimal physiologic conditions in the body. Antioxidants are found as dietary components (vitamins, phenolic compounds, flavonoids or carotenoids, among others) in fruits, vegetables and natural beverages. Interestingly, it is known that a balanced diet that is naturally rich in antioxidants protects against oxidative stress and also induces protective effects against several pathologies.[3]

As aging proceeds, the efficiency of antioxidant defense systems becomes lower, and the ability to remove deleterious reactive oxygen species and free radicals decreases. Consumption of fruits and vegetables containing large amounts of antioxidant compounds has been associated with the balance of free radical/antioxidant status, which helps to minimize oxidative stress and reduce the risk of chronic diseases. Because several thousand bioactive compounds have been identified in foodstuffs all over the world, there is an extensive scientific literature relating to the health-promoting effects of adopting a healthy diet. Scientists estimate that there may be as many as 10,000 different phytochemicals with potential biologic activities. In this context, phenolic compounds are phytochemicals broadly distributed in the plant kingdom, being the most abundant secondary metabolites in plants. Thus, numerous studies have reported improvement in antioxidant status and/or protection against oxidative stress in short-term intervention studies with various polyphenol-rich foodstuffs, including fruit juices, red wines, chocolates, and fruits such as strawberries.[4] The association between polyphenol intake, or the consumption of polyphenol-rich foods, and the incidence of cardiovascular diseases has been examined in several epidemiologic studies, and it was found that consumption of a polyphenol-rich diet was

Aging
http://dx.doi.org/10.1016/B978-0-12-405933-7.00013-5

associated with a lower risk of myocardial infarction in both case–control and cohort studies.[5]

Although polyphenols represent the archetype of the health-promoting effects of the Mediterranean diet, the pineal indole melatonin is currently revolutionizing the nutrition field by virtue of its biologic activities and its excellent bioavailability (Fig. 13.1). In fact, in contrast to other classic antioxidants, this molecule exhibits amphiphilic, hydrophilic and lipophilic properties, allowing it to penetrate all cells, fluids and intracellular compartments.[6] Melatonin has a widespread spectrum of physiologic effects, including, but not limited to, chronobiologic, immunomodulatory, neuroendocrine, antioxidant and free-radical-scavenging activities.[7] During the past decade, numerous publications have appeared concerning the role of melatonin in aging, demonstrating the potent gerontoprotective actions exerted by the indoleamine.[8] Thus, the beneficial effects derived from the administration of low doses of melatonin have been tested in rheumatoid arthritis (10 mg/day),[9] primary essential hypertension (5 mg/day)[10] and type 2 diabetes (5 mg/day)[11] in elderly patients. In this vein, numerous studies have also reported the beneficial effects of the ingestion of melatonin-enriched foodstuffs on antioxidant capacity.[12–15] Moreover, the inclusion of melatonin in the diet may reduce oxidation produced by lipids and other molecules present in the stomach, preserving the integrity of the gastrointestinal mucosa and protecting against the formation of gastric ulcers caused by stress or different drugs.[16] Similarly, this indole may be effective in limiting oxidative processes in serum and tissues because melatonin can freely cross cell membranes and it distributes to any aqueous compartment, including the cytosol, nucleus and mitochondria.[6]

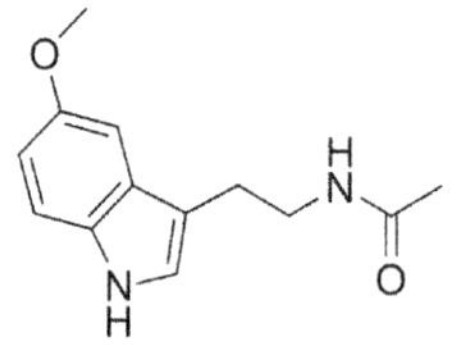

FIGURE 13.1 Chemical structure of the indole melatonin.

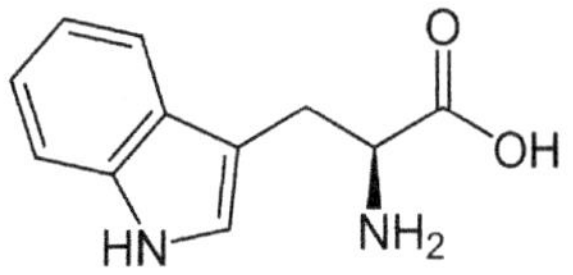

FIGURE 13.2 Chemical structure of the amino acid tryptophan.

Tryptophan is an indispensable amino acid that has to be supplied by dietary protein (Fig. 13.2). It has proved to be efficient as an antioxidant and as an anti-inflammatory and immune system modulator.[17] This amino acid is metabolized to serotonin and melatonin through the methoxyindole pathway (Fig. 13.3) and participates in the regulation of circadian rhythms. It has been reported that age reduces the transport of tryptophan across the blood–brain barrier (BBB).[18] Also, tryptophan hydroxylase, which catalyzes the rate-limiting step in the biogenesis of both serotonin and melatonin in cells, declines with aging due to oxidation by reactive oxygen species and alterations in the phosphorylation cascade that modulates enzyme activity.[19] In this sense, several reports have demonstrated that tryptophan administration was able to reverse age-related changes in the circulating levels of melatonin and serotonin in both humans and animals.[20,21]

DIETARY TRYPTOPHAN AND MELATONIN: SOURCES OF HEALTH

The presence of tryptophan in the plant kingdom is well known because it is classified as an essential amino acid owing to its nutritional properties. However, the

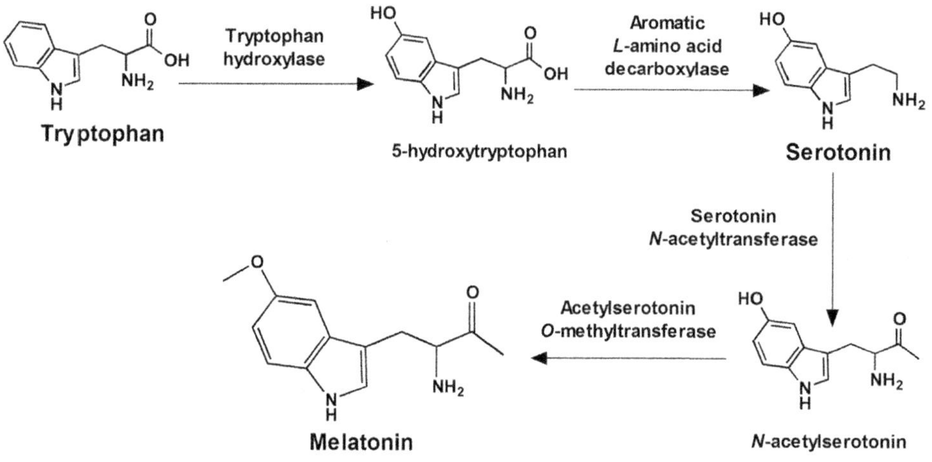

FIGURE 13.3 Methoxyindole pathway of tryptophan metabolism.

concept of phytomelatonin (plant melatonin) is relatively novel, despite the fact that this molecule is evolutionarily conserved and has the same chemical structure in both plants and animals. Melatonin has been linked to a wide variety of functions in organisms ranging from plants to humans. Indeed, the functional versatility of this indole amine has surprised even the most ardent scientists working on health sciences. Thus, during the last decade, research in this field has focused on the physiologic effect derived from the consumption of foods rich in melatonin and its precursor, the amino acid tryptophan.

Importance of Dietary Tryptophan

As well as serving as a building block for proteins, tryptophan is a critical nutrient for the synthesis of serotonin in the brain; this is required for mood alleviation and melatonin synthesis. Although several foods, including cheese, meat, fruits and vegetables, contain serotonin, the serotonin present in these foods is not easily accessible to the central nervous system owing to the existence of the BBB. However, the serotonin precursor, tryptophan, can easily cross the BBB and be utilized for the synthesis of serotonin and, subsequently, of melatonin.[18] Dietary tryptophan may affect biologic activities where serotonin and melatonin are also involved. For instance, a disturbed brain serotoninergic function, caused by inadequate availability of tryptophan, is recognized as a contributing factor in affective disorders, anxiety, aggression, stress, eating disorders and other conditions.[17] Similarly, there is a close link between dietary tryptophan and melatonin levels because oscillations in tryptophan concentration in maternal milk parallel oscillations in infant 6-sulfatoxymelatonin (aMT6-s).[22]

Evidence is now accumulating on the detrimental consequences of tryptophan deficiency in the diet during viral, bacterial and parasitic intracellular infections. In this respect, consuming foodstuffs rich in tryptophan may help the organism to maintain homeostasis. Mahan and Escott[23] listed an extensive inventory of foodstuffs containing this essential amino acid, including cereals, fish, seafood, meat, vegetables, nuts, eggs and fruits. Currently, the list of tryptophan-enriched food is broadening enormously due to the development of more accurate and efficient techniques for the detection and quantification of the amino acid (Table 13.1).

Numerous studies carried out in animals have corroborated the benefits of supplementation of tryptophan in the diet, reporting improvements in the sleep–wake cycle and in the oxidative and inflammatory pathways.[20,24] These effects have also been proved in humans because diets enriched with tryptophan itself, or foodstuffs containing high levels of this amino acid, were able to both consolidate sleep in newborns[22,25] and improve sleep and antioxidant capacity levels in young, middle-aged, and elderly individuals.[15,26]

Relevance of Phytomelatonin

The first complete publications showing that melatonin exists in plants were independently provided by Dubbels et al[27] and Hattori et al.[28] Since then, the number of articles published in relation to melatonin in plants has increased greatly in recent years. Thus, melatonin has been identified in the roots, stems, leaves, flowers and seeds of various plants.[29,30]

One important function of melatonin in plants may be the scavenging of free radicals, thereby protecting plants against oxidative stress and reducing damage to macromolecules in a manner similar to that in animals. This hypothesis is based on the discovery that cold stress significantly increases the production of melatonin in plants.[31] Furthermore, the identification of high levels of melatonin in the water hyacinth (*Eichhornia crassipes*), which is extremely resistant to pollution, has

TABLE 13.1 Tryptophan Levels in Plants

Common Name	Scientific Name	[Tryptophan]
Rice[a]	*Oryza sativa*	75 mg/100 g (fresh weight)
Banana[a]	*Musa paradisiaca*	9 mg/100 g (fresh weight)
Sweet cherry[b]	*Prunus avium*	3.6-8.2 mg/100 g (fresh weight)
White asparagus[a]	*Asparagus albus*	27 mg/100 g (fresh weight)
Strawberry[a]	*Fragaria vesca*	5 mg/100 g (fresh weight)
Chickpea[a]	*Cicer arietinum*	185 mg/100 g (fresh weight)
Lentil[a]	*Lens culinaris*	232 mg/100 g (fresh weight)
Corn[a]	*Zea mays*	67 mg/100 g (fresh weight)
Mandarin[a]	*Citrus reticulata*	2 mg/100 g (fresh weight)
Apple[a]	*Malus domestica*	1 mg/100 g (fresh weight)
Pear[a]	*Pyrus communis*	2 mg/100 g (fresh weight)

[a]*USDA National Nutrient Database for Standard Reference. http://ndb.nal.usda.gov/*
[b]*Cubero J et al (2010). Assays of the amino acid tryptophan in cherries by HPLC fluorescence. Food Anal Methods 3, 36–39.*

revealed the possible protective role of this indole in the plant kingdom. This species tolerates remarkably polluted environments and, therefore, is used in phytoremediation.[32] The high production of melatonin in this plant is believed to play a key role against environmental aggression generated by reactive oxygen species. In this respect, numerous studies have shown the protective effect exerted by melatonin against adverse environmental factors and different stressors when it is applied exogenously to plants.[33]

Melatonin has been reported in important Mediterranean commodities, including grape, apple, rice, almond and tomato, as well as in other common foodstuffs (Table 13.2) and natural beverages (Table 13.3). Interestingly, in animals and humans, an efficient uptake and bioavailability of melatonin from food sources has been demonstrated. Hattori et al[28] conducted the first animal study, which demonstrated that melatonin concentration increased from about 10 to 35 pg/mL 1.5 hours after chickens consumed melatonin-rich chick food composed of corn, milo, beans and rice (3 ng/g melatonin). Reiter et al[12] also reported higher serum melatonin concentrations and antioxidant capacity after walnuts were fed to Sprague Dawley rats. Likewise, in humans, the consumption of fruit rich in melatonin – particularly pineapple, banana and orange – produced an elevation in both melatonin circulating levels[34] and aMT6-s concentrations.[35]

The significance of high levels of melatonin in popular beverages could also prove beneficial for some types of pathology, including neurodegenerative diseases (Alzheimer's and Parkinson's disease), heart disease, metabolic disorders, tumors and accidental nuclear radiation.[33] Beer is commonly considered an example of a functional drink with varied components that may contribute to its overall therapeutic characteristics. Recently, Maldonado et al[36] have demonstrated that melatonin present in beer does contribute to the antioxidant capability of human serum; they concluded that moderate beer consumption may protect organisms from overall oxidative stress. Another interesting study emphasizing the therapeutic effects of melatonin present in red wine was published by Lamont et al.[37] These authors observed that the melatonin concentration measured in red wine (75 ng/L), when infused into isolated rat hearts subjected to ischemia/reperfusion injury, significantly reduced heart infarct size from 69% to 25%. The protective effect of melatonin on heart injury was equivalent to, or better than, that provided by resveratrol, which is considered to be a major beneficial ingredient found in red wine; this was particularly remarkable because the resveratrol concentration tested in this study (2.3 mg/L) was almost 300,000-fold higher than that of melatonin.

TABLE 13.2 Melatonin Levels in Different Plants

Common Name	Scientific Name	[Melatonin]
Almond[a]	*Prunus amygdalus*	In seed 39.9 ng/g (dry weight)
White lupin[b]	*Lupinus albus*	In root 16.2–18.4 ng/g (fresh weight)
Rice[28]	*Oryza sativa*	In seed 1 ng/g (fresh weight)
Oat[28]	*Avena sativa*	In seed 1.8 ng/g (fresh weight)
Banana[c]	*Musa paradisiaca*	0.5 ng/g (fresh weight)
Coffee[d]	*Coffea canephora*	In seed 11.5×10^4 ng/g (fresh weight)
Tart cherry[53]	*Prunus cerasus*	2–15 ng/g (fresh weight)
Sweet cherry[54]	*Prunus avium*	0–20 ng/100 g (fresh weight)
Sunflower[a]	*Helianthus annuus*	In seed 29 ng/g (dry weight)
Corn[28]	*Zea mays*	In seed 1.3 ng/g (fresh weight)
Apple[28]	*Malus domestica*	4.8×10^{-2} ng/g (fresh weight)
White mustard[a]	*Brassica hirta*	In seed 189 ng/g (dry weight)
Walnut[12]	*Juglans regia*	In seed 3.5 ng/g (fresh weight)
Tomato[e]	*Lycopersicon esculentum*	1.1 ng/g (fresh weight)
Grape vine[f]	*Vitis vinifera*	In grape peel 0.5×10^{-2}–0.9 ng/g (fresh weight)

[a] *Manchester LC et al (2000). High levels of melatonin in the seeds of edible plants: possible function in germ tissue protection. Life Sci 67, 3023–3029.*
[b] *Arnao MB & Hernández-Ruiz J (2007). Melatonin promotes adventitious and lateral root regeneration in etiolated hypocotyls of* Lupinus albus. *J Pineal Res 42, 147–152.*
[c] *Hardeland R & Pandi-Perumal SR (2005). Melatonin, a potent agent in antioxidative defense: actions as a natural food constituent, gastrointestinal factor, drug and prodrug. Nutr Metab 10, 2–22.*
[d] *Ramakrishna A et al (2011). Endogenous profiles of indoleamines: serotonin and melatonin in different tissues of* Coffea canephora *P ex Fr. as analyzed by HPLC and LC-MS-ESI. Acta Physiol Plant 34, 393–396.*
[e] *Pape C & Luening K (2006). Quantification of melatonin in phototrophic organisms. J Pineal Res 41, 157–165.*
[f] *Iriti M et al (2006). Melatonin content in grape: myth or panacea? J Sci Food Agric 86, 1432–1438.*

TABLE 13.3 Melatonin Levels in Plant-Derived Beverages

Beverage	[Melatonin]
Extra virgin olive oil[a]	0.05–0.12 ng/mL
Coffee[b]	60–78 ng/mL
Beer[36]	0.5×10^{-1}–1.9×10^{-1} ng/mL
Must[c]	1.1 ng/mL
Grappa[c]	0.3 ng/mL
Italian wine Albana[c]	0.6 ng/mL
Grape juice[c]	0.5 ng/mL

[a]*de la Puerta C et al (2007). Melatonin is a phytochemical in olive oil. Food Chem, 104, 609–612.*
[b]*Ramakrishna A et al (2012). Melatonin and serotonin profiles in beans of coffee species. J Pineal Res 52, 470–476.*
[c]*Mercolini L et al (2012). Content of melatonin and other antioxidants in grape-related foodstuffs: measurement using a MEPS-HPLC-F method. J Pineal Res 53, 21–28.*

CONSUMPTION OF FOODSTUFFS CONTAINING BIOACTIVE COMPOUNDS TO PROTECT AGAINST OXIDATIVE STRESS

Phytochemicals present in fruits and vegetables, e.g. phenols, anthocyanins and carotenoids, have a great potential to prevent or delay the development of disorders such as cancer, obesity and cardiovascular diseases. The current development of functional products with antioxidant effects is mainly focused on the production of phenolic-enriched products.[4] These phytochemicals can act as dietary therapeutics or nutraceuticals and, therefore, by virtue of their content, some plant foods and beverages can be considered as functional foods – products consumed as part of a normal diet that may provide health benefits beyond their basic nutritional functions.[38] Consumption of fruits and vegetables, foodstuffs rich in antioxidants, is an attractive strategy to reduce risk from chronic diseases. Increased dietary intake of vegetable-derived bioactive compounds may retard age-related decrements in immune function and prolong life-span.

The Mediterranean diet has long been known for contributing to a healthy life and preventing diseases associated with oxidative damage, such as coronary heart disease, cancer and degenerative diseases.[39] It is based on an array of plant secondary metabolites, such as carotenoids and indole amines, thereby supporting the hypothesis that the health benefits associated with the Mediterranean diet are due to the chemical diversity of plant foods. Endogenous and dietary antioxidants play an important role in moderating the damage associated with reactive oxygen species; unfortunately, antioxidant levels from both sources are decreased in the elderly. Micronutrient deficiency is common among elderly people, and this deficiency has been linked to the risk of chronic disease.[40] In advanced age, an exogenous supplementation of bioactive compounds may exert beneficial effects as potent antioxidants. Dietary components with high antioxidant activity have to receive particular attention because of their potential role in modulating oxidative stress associated with aging and chronic conditions.[41] Antioxidant vitamins and phytochemicals are important in maintaining antioxidant defenses against oxidative stress-related diseases, including cancer.

Tryptophan and Melatonin Against Oxidative Stress in Aging

In recent years research has been focused on the possible actions that newly identified plant bioactive compounds, i.e. melatonin and the related indolic compound serotonin, may exert on health. In particular, it has recently been suggested that the presence of melatonin in edible plants may improve human health by virtue of its biologic activities and its good bioavailability.[30] This is of importance owing to the numerous experimental data showing that melatonin may have utility in the treatment of several cardiovascular and neurodegenerative conditions.[42] Detoxification of free radicals, which are implicated in a number of pathophysiologic processes (including aging, inflammation, ischemia-reperfusion injury, atherosclerosis and cancer), is believed to be one of the underlying mechanisms in the health protection exerted by antioxidants contained in fruits and vegetables.[43] Melatonin is the main pineal hormone synthesized from tryptophan, predominantly at night. It is critical for the regulation of circadian and seasonal changes in various aspects of physiology and neuroendocrine function.[44] As age advances, various systemic changes occur in the rhythmic secretion of melatonin and serotonin. These changes are associated with a variety of chronopathologies and lead to a generalized deterioration in health.[7] Melatonin is a potent free radical scavenger and antioxidant[45,46] that not only scavenges especially highly toxic hydroxyl radicals, but also performs indirect antioxidant actions via its ability to stimulate antioxidant enzymes,[47] diminishing free radical formation at the mitochondrial level by reducing the leakage of electrons from the electron transport chain.[42] Melatonin has been shown, in a variety of studies, to possess anti-inflammatory, antioxidant and anticancer properties.[48] It is known to be a direct scavenger of the hydroxyl radical (OH^-)[49] and hydrogen peroxide (H_2O_2).[48] It also can scavenge the peroxynitrite anion ($ONOO^-$), nitric oxide,[50] and possibly the peroxyl radical (ROO^-) and singlet oxygen ($O_2^{\cdot-}$).[49] Melatonin also has immunoregulatory properties and is able to stimulate antioxidant enzymes; it is also able to inhibit the prooxidative enzyme nitric oxide synthase, and diminish free radical formation at the mitochondrial level, or synergize with other antioxidants to protect against oxidative stress.[51] The identification of melatonin, serotonin and the amino acid tryptophan in plants may have major implications for animal and human health, taking into account the potent antioxidant activity of these molecules in a variety of disorders and diseases where an exacerbated production of free radicals has been reported.[7] The consumption of plant materials that contain high levels of melatonin could alter blood levels of the indole and provide protection against oxidative damage. It is known that increased levels of circulating melatonin – produced directly by exogenous administration or indirectly by introducing it into the diet with vegetables rich in this compound – enhance the individual's antioxidant status.[15,52] As it was discovered in plants, there has been an ever-growing number of studies reporting the detection of melatonin in a variety of fruits,[29] including both tart and sweet cherry.[53,54] Cherry fruit also has a significant amount of important bioactive components such as anthocyanins, quercitin or carotenoids.[55] Several studies have reported the antioxidant effects *in vitro* of the aforementioned compounds.[56] Concerning *in vivo* effects, an increased activity of the antioxidant enzymes superoxide dismutase (SOD) and glutathione peroxidase (GPx), and a decrease in lipid peroxidation, were reported in mice fed with sour cherry juice. Garrido et al[26] also measured total anthocyanins in a Jerte Valley sweet cherry product. Anthocyanins show one of the strongest antioxidant activities among phenolic compounds. In fact, the anthocyanin content of cherries is compared to that of other plant foods in which evidence has suggested health-promoting effects related to their anthocyanin content.[56] Wang et al[57] demonstrated a positive linear relationship between the level of anthocyanins in cherries and the degree of protection from oxidative stress in neuronal cells. Even though levels of anthocyanins in tart cherries have been found to exceed those in sweet cherries and other fruits, sweet cherries, particularly Jerte Valley cherries, have been reported to present melatonin precursors – molecules with potent antioxidant activity – which may contribute to an increase in the potential biologic properties of these cherries. In this sense, Garrido et al[15,21,26] have evaluated the physiologic effects exerted by the consumption of both Jerte Valley cherries and a Jerte Valley cherry product. Thus, it has been shown that the ingestion of cherries, either fresh or as a cherry-based product, improved the antioxidant status in humans and elevated the aMT6-s levels.[15,21,26] Similarly, González-Flores et al[13,14] have concluded that the ingestion of plums and grapes contributed to increased urinary antioxidant capacity, as well as increased aMT6-s levels, in individuals, thereby suggesting the presence of melatonin or its precursors in these fruits. Finally, Bravo et al[58] observed a significant increase in urinary total antioxidant capacity after the intake of tryptophan-enriched cereals.

Because melatonin is found in plants, research has been conducted to evaluate whether supplementing the diet with vegetables rich in the indole amine exerts positive effects on health.[12]

Taking into account that the beneficial properties of fruits and vegetables come from phytonutrients,[59] the consumption of foodstuffs rich not only in phenolic compounds but also in melatonin and its precursors may be a natural approach to prevent or combat important diseases related to oxidative stress.

SUMMARY POINTS

- Antioxidants are found as dietary components in fruits, vegetables and natural beverages.
- Melatonin acts as a potent free radical scavenger and antioxidant.
- Consuming Jerte Valley cherries as well as plums or grapes, fruits rich in tryptophan and/or melatonin, elevates antioxidant capacity and 6-sulfatoxymelatonin levels in young to elderly individuals.
- A tryptophan-enriched diet improves oxidative and inflammatory pathways, and is able to reverse age-related changes in the circulating levels of melatonin.
- Functional foods are products consumed as part of a normal diet that may provide health benefits beyond their basic nutritional functions.

Acknowledgments

This investigation was supported by Gobierno de Extremadura (Re: GRU10003).

References

1. Reiter RJ, Paredes SD, Manchester LC, Tan DX. Reducing oxidative/nitrosative stress: a newly-discovered genre for melatonin. *Crit Rev Biochem Mol Biol* 2009;**44**:175–200.
2. Pham-Huy LA, He H, Pham-Huy C. Free radicals, antioxidants in disease and health. *Int J Biomed Sci* 2008;**4**:89–96.
3. Visioli F, Galli C. The role of antioxidants in the Mediterranean diet. *Lipids* 2001;**36**:S49–52.
4. Dai J, Mumper RJ. Plant phenolics: extraction, analysis and their antioxidant and anticancer properties. *Molecules* 2010;**15**:7313–52.
5. Peters U, Poole C, Arab L. Does tea affect cardiovascular disease? A meta-analysis. *Am J Epidemiol* 2001;**154**:495–503.
6. Menéndez-Peláez A, Poeggeler B, Reiter RJ, et al. Nuclear localization of melatonin in different mammalian tissues: immunocytochemical and radioimmunoassay evidence. *J Cell Biochem* 1993;**53**:373–82.
7. Poeggeler B. Melatonin, aging, and age-related diseases: perspectives for prevention, intervention, and therapy. *Endocrine* 2005;**27**:201–12.
8. Reiter RJ, Tan DX, Mayo JC, et al. Melatonin, longevity and health in the aged: an assessment. *Free Radic Res* 2002;**36**:1323–9.
9. Forrest CM, Mackay GM, Stoy N, et al. Inflammatory status and kynurenine metabolism in rheumatoid arthritis is treated with melatonin. *Br J Clin Pharmacol* 2007;**64**:517–26.

10. Kedziora-Kornatowska K, Szewczyk-Golec K, Czuczejko J, et al. Antioxidative effects of melatonin administration in elderly primary essential hypertension patients. *J Pineal Res* 2008;**45**:312–7.
11. Kedziora-Kornatowska K, Szewczyk-Golec K, Kozakiewicz M, et al. Melatonin improves oxidative stress parameters measured in the blood of elderly type 2 diabetic patients. *J Pineal Res* 2009;**46**:333–7.
12. Reiter RJ, Manchester LC, Tan DX. Melatonin in walnuts: influence on levels of melatonin and total antioxidant capacity of blood. *Nutrition* 2005;**21**:920–4.
13. González-Flores D, Velardo B, Garrido M, et al. Ingestion of Japanese plums (*Prunus salicina* Lindl. Cv. Crimson globe) increases the urinary 6-sulfatoxymelatonin and total antioxidant capacity levels in young, middle-aged, and elderly humans: nutritional and functional characterization of their content. *J Food Nutr Res* 2011;**50**:229–36.
14. González-Flores D, Gamero E, Garrido M, et al. Urinary 6-sulfatoxymelatonin and total antioxidant capacity increase after the intake of a grape juice cv. Tempranillo stabilized with HHP. *Food Funct* 2012;**3**:34–39.
15. Garrido M, Paredes SD, Cubero J, et al. Jerte Valley cherry-enriched diets improve nocturnal rest and increase 6-sulfatoxymelatonin and total antioxidant capacity in the urine of middle-aged and elderly humans. *J Gerontol A Biol Sci Med Sci* 2010;**65**:909–14.
16. Bubenik GA. Gastrointestinal melatonin: localization, function, and clinical relevance. *Dig Dis Sci* 2002;**47**:2336–48.
17. Le Floc'h N, Otten W, Merlot E. Tryptophan metabolism, from nutrition to potential therapeutic applications. *Amino Acids* 2011;**41**:1195–205.
18. Porter RJ, Mulder RT, Joyce PR, Luty SE. Trytophan and tyrosine availability and response to antidepressant in major depression. *J Affect Disord* 2005;**86**:129–34.
19. Hussain AM, Mitra AK. Effect of reactive oxygen species on the metabolism of tryptophan in rat brain: influence of age. *Mol Cell Biochem* 2004;**258**:145–53.
20. Delgado J, Terrón MP, Garrido M, et al. A cherry nutraceutical modulates melatonin, serotonin, corticosterone, and total antioxidant capacity levels: effect on ageing and chronotype. *J Appl Biomed* 2012;**10**:109–17.
21. Garrido M, González-Gómez D, Lozano M, et al. A Jerte Valley cherry product provides beneficial effects on sleep quality. Influence on aging. *J Nutr Health Aging* 2012;**17**:553–60.
22. Cubero J, Valero V, Sánchez J, et al. The circadian rhythm of tryptophan in breast milk affects the rhythms of 6-sulfatoxymelatonin and sleep in newborn. *Neuro Endocrinol Lett* 2005;**26**:657–61.
23. Mahan K, Escott-Stump S. *Nutrición y dietoterapia de Krausse*. Mexico: McGraw-Hill Interamericana; 1998.
24. Delgado J, Terrón MP, Garrido M, et al. Diets enriched with a Jerte Valley cherry-based nutraceutical product reinforce nocturnal behaviour in young and old animals of nocturnal (*Rattus norvegicus*) and diurnal (*Streptopelia risoria*) chronotypes. *J Anim Physiol Anim Nutr (Berl)* 2013;**97**:137–9.
25. Cubero J, Narciso D, Terrón MP, et al. Chrononutrition applied to formula milks to consolidate infants' sleep/wake cycle. *Neuroendocrinol Lett* 2007;**28**:360–6.
26. Garrido M, González-Gómez D, Lozano M, et al. Characterization and trials of a Jerte Valley cherry product as a natural antioxidant-enriched supplement. *Italian J Food Sci* 2013;**25**:90–7.
27. Dubbels R, Reiter RJ, Klenke E, et al. Melatonin in edible plants identified by radioimmunoassay and by high performance liquid chromatography-mass spectrometry. *J Pineal Res* 1995;**18**:28–31.
28. Hattori A, Migitaka H, Iigo M, et al. Identification of melatonin in plants and its effects on plasma melatonin levels and binding to melatonin receptors in vertebrates. *Biochem Mol Biol Int* 1995;**35**:627–34.
29. Paredes SD, Korkmaz A, Manchester LC, et al. Phytomelatonin: a review. *J Exp Bot* 2009;**60**:57–9.
30. Iriti M, Varoni EM, Vitalini S. Melatonin in traditional Mediterranean diets. *J Pineal Res* 2010;**49**:101–5.
31. Tan DX, Manchester LC, Reiter RJ, Plummer BF. Cyclic 3-hydroxymelatonin: a melatonin metabolite generated as a result of hydroxyl radical scavenging. *Biol Signals Recept* 1999;**8**:70–4.
32. Agunbiade FO, Olu-Owolabi BI, Adebowale KO. Phytoremediation potential of *Eichornia crassipes* in metal-contaminated coastal water. *Bioresour Technol* 2009;**100**:4521–6.
33. Tan DX, Hardeland R, Manchester LC, et al. Functional roles of melatonin in plants, and perspectives in nutritional and agricultural science. *J Exp Bot* 2012;**63**:577–97.
34. Sae-Teaw M, Johns J, Johns NP, Subongkot S. Serum melatonin levels and antioxidant capacities after consumption of pineapple, orange, or banana by healthy male volunteers. *J Pineal Res* 2012;**55**:58–64.
35. Johns NP, Johns J, Porasuphatana S, et al. Dietary intake of melatonin from tropical fruit altered urinary excretion of 6-sulfatoxymelatonin in healthy volunteers. *J Agric Food Chem* 2013;**61**:913–9.
36. Maldonado MD, Moreno H, Calvo JR. Melatonin present in beer contributes to increase the levels of melatonin and antioxidant capacity of the human serum. *Clin Nutr* 2009;**28**:188–91.
37. Lamont KT, Somers S, Lacerda L, et al. Is red wine a SAFE sip away from cardioprotection? Mechanisms involved in resveratrol- and melatonin-induced cardioprotection. *J Pineal Res* 2011;**50**:374–80.
38. Heber D. Vegetables, fruits and phytoestrogens in the prevention of diseases. *J Postgrad Med* 2004;**50**:145–9.
39. Alexandratos N. The Mediterranean diet in a world context. *Public Health Nutr* 2006;**9**:111–7.
40. Agarwal S, Rao AV. Carotenoids and chronic diseases. *Drug Metabol Drug Interact* 2000;**17**:189–210.
41. Meydani M. Nutrition interventions in aging and age-associated disease. *Ann N Y Acad Sci* 2001;**928**:226–35.
42. Reiter RJ, Paredes SD, Korkmaz A, et al. Melatonin in relation to the 'strong' and 'weak' versions of the free radical theory of aging. *Adv Med Sci* 2008;**53**:119–29.
43. Giugliano D, Esposito K. Mediterranean diet and cardiovascular health. *Ann N Y Acad Sci* 2005;**105**:253–60.
44. Pévet P, Bothorel B, Slotten H, Saboureau M. The chronobiotic properties of melatonin. *Cell Tissue Res* 2002;**309**:183–91.
45. Paredes SD, Terrón MP, Cubero J, et al. Tryptophan increases nocturnal rest and affects melatonin and serotonin serum levels in old ringdove. *Physiol Behav* 2007;**90**:576–82.
46. Paredes SD, Terrón MP, Marchena AM, et al. Tryptophan modulates cell viability, phagocytosis and oxidative metabolism in old ringdoves. *Basic Clin Pharmacol Toxicol* 2007;**101**:56–62.
47. Gitto E, Tan DX, Reiter RJ, et al. Individual and synergistic antioxidative actions of melatonin: studies with vitamin E, vitamin C, glutathione and desferrioxamine (desferoxamine) in rat liver homogenates. *J Pharm Pharmacol* 2001;**53**:1393–401.
48. Tan DX, Manchester LC, Reiter RJ, et al. Melatonin directly scavenges hydrogen peroxide: a potentially new metabolic pathway of melatonin biotransformation. *Free Radic Biol Med* 2000;**29**:1177–85.
49. Reiter RJ, Tan DX, Qi W, et al. Pharmacology and physiology of melatonin in the reduction of oxidative stress in vivo. *Biol Signals Recept* 2000;**9**:160–71.
50. Blanchard B, Pompon D, Ducrocq C. Nitrosation of melatonin by nitric oxide and peroxynitrite. *J Pineal Res* 2000;**29**:184–92.
51. Paredes SD, Reiter RJ. Melatonin: helping cells cope with oxidative disaster. *Cell Membr Free Radic Res* 2010;**2**:99–111.
52. Garrido M, Espino J, González-Gómez D, et al. A nutraceutical product based on Jerte Valley cherries improves sleep and augments the antioxidant status in humans. *e-SPEN* 2009;**4**:e321–3.
53. Burkhardt S, Tan DX, Manchester LC, et al. Detection and quantification of the antioxidant melatonin in Montmorency and Balaton tart cherries (*Prunus cerasus*). *J Agric Food Chem* 2001;**49**:4898–902.

54. González-Gómez D, Lozano M, Fernández-León MF, et al. Detection and quantification of melatonin and serotonin in eight sweet cherry cultivars (*Prunus avium* L.). *Eur Food Res Tech* 2009;**229**:223–9.
55. McCune LM, Kubota C, Stendell-Hollis NR, Thomson CA. Cherries and health: a review. *Crit Rev Food Sci Nutr* 2011;**51**:1–12.
56. Stevenson DE, Hurst RD. Polyphenolic phytochemicals – just antioxidants or much more? *Cell Mol Life Sci* 2007;**64**:2900–16.
57. Wang H, Nair MG, Strasburg GM, et al. Antioxidant polyphenols from tart cherries (*Prunus cerasus*). *J Agric Food Chem* 1999;**47**:840–4.
58. Bravo R, Matito S, Cubero J, et al. Tryptophan-enriched cereal intake improves nocturnal sleep, melatonin, serotonin, and total antioxidant capacity levels and mood in elderly humans. *AGE* 2013;**35**: 1277–85.
59. Milde J, Elstner EF, Grassmann J. Synergistic effects of phenolics and carotenoids on human low-density lipoprotein oxidation. *Mol Nutr Food Res* 2007;**51**:956–61.

CHAPTER

14

Protective Effects of Vitamin C on Age-Related Bone and Skin Phenotypes Caused by Intracellular Reactive Oxygen Species

Shuichi Shibuya

Department of Advanced Aging Medicine, Chiba University Graduate School of Medicine, Inohana, Chuo-ku, Chiba, Japan

Hidetoshi Nojiri

Department of Orthopaedics, Juntendo University Graduate School of Medicine, Bunkyo-ku, Tokyo, Japan

Daichi Morikawa

Department of Advanced Aging Medicine, Chiba University Graduate School of Medicine, Inohana, Chuo-ku, Chiba, Japan, and Department of Orthopaedics, Juntendo University Graduate School of Medicine, Bunkyo-ku, Tokyo, Japan

Hirofumi Koyama, Takahiko Shimizu

Department of Advanced Aging Medicine, Chiba University Graduate School of Medicine, Inohana, Chuo-ku, Chiba, Japan

List of Abbreviations

8-OHdG 8-hydroxy-2′-deoxyguanosine
AGEs advanced glycation end products
APPS L-ascorbyl 2-phosphate 6-palmitate trisodium salt
APS L-ascorbyl 2-phosphate trisodium salt
BMD bone mineral density
H_2O_2 hydrogen peroxide
μCT micro computed tomography
NAC *N*-acetylcysteine
O_2^- superoxide
OVX ovariectomy
ROS reactive oxygen species
SOD superoxide dismutase
SOD1 copper/zinc superoxide dismutase
UV ultraviolet
VC vitamin C
VCP-IS-2Na 2-O-L-ascorbyl phosphate

INTRODUCTION

Reactive oxygen species (ROS), including superoxide (O_2^-), hydrogen peroxide (H_2O_2) and the hydroxyl radical, are chemically reactive molecules that are generated as secondary products of cellular metabolism. Based on the free-radical theory of aging,[1] oxidative stress is thought to be a pivotal cause of age-related diseases in mammals. An imbalance between ROS production and antioxidant defense, called oxidative stress, causes damage to cell structures. To protect these structures from oxidative stress, cells possess multiple antioxidative systems, including endogenous antioxidants, such as vitamins C and E and glutathione, and enzymes such as superoxide dismutase (SOD), catalase and glutathione peroxidase.

Aging
http://dx.doi.org/10.1016/B978-0-12-405933-7.00014-7

TABLE 14.1 Aging-Like Phenotypes of *Sod1*$^{-/-}$ Mice

Bone	Osteopenia[16]
Skin	Skin atrophy[17,18]
	Skin inflammation[50]
Muscle	Skeletal muscle atrophy[8]
Liver	Fatty deposits[5]
	Hepatic cellular carcinoma[6]
Blood	Hemolytic anemia[7]
Ear	Hearing loss[13]
Ovary	Infertility[10,11]
	Luteal degeneration[12]
Brain	Acceleration of Alzheimer's disease[9]
Eye	Macular degeneration[4,59]
	Cataract[15]
	Lacrimal degeneration[14]

SODs play a central role in O_2^- metabolism in antioxidative systems *in vivo* due to their ability to catalyze the dismutation of cellular O_2^- into O_2 and H_2O_2. In mammals, there are three SOD isoforms: CuZn-SOD (*Sod1*), which exists in the cytoplasm, Mn-SOD (*Sod2*), which is distributed in the mitochondrial matrix, and extracellular SOD (*Sod3*), which is localized in extracellular fluids such as lymph, synovial fluid and plasma.[2,3] Many papers have reported that *Sod1*-deficient (*Sod1*$^{-/-}$) mice exhibit various aging-like phenotypes such as macular degeneration,[4] fatty liver,[5] hepatic carcinoma,[6] hemolytic anemia,[7] skeletal muscle atrophy,[8] acceleration of Alzheimer's disease,[9] infertility,[10,11] luteal degeneration,[12] hearing loss,[13] lacrimal degeneration[14] and cataracts[15] (Table 14.1). Recently, we demonstrated that *Sod1* deficiency also causes bone and skin atrophy associated with collagen malfunction[16–18] (Table 14.1). In human studies, it has been suggested that the accumulation of oxidative damage is responsible for age-associated bone fragility.[19–21] Because the erythrocyte SOD1 activity is significantly reduced in elderly human osteoporotic females,[20] it seems reasonable to hypothesize that a lack of *Sod1* would accelerate age-related pathologic changes in human bones. These data imply that excess cytoplasmic ROS are an important determinant of age-related changes in tissues.

In general, antioxidant treatment reduces tissue damage induced by oxidative stress in cells and tissues by neutralizing ROS. Vitamin C (VC), the most famous antioxidant, has been largely used in cosmetics and supplements. It is known that VC is also an essential cofactor for post-translational modifications in collagen formation and the synthesis of hydroxyproline and hydroxylysine, which are required for the formation of stable triple helices.[22] Therefore, VC is thought to play a physiologic role in collagen-rich bone and skin tissues. We herein review the protective effects of VC on age-related bone and skin phenotypes caused by intracellular ROS in mice.

SOD1 DEFICIENCY INDUCES BONE LOSS

We assessed the bone phenotypes of *Sod1*$^{-/-}$ mice as a mouse model in age-related bone disease. When X-ray irradiation was conducted, the aged *Sod1*$^{-/-}$ mice displayed a higher degree of kyphoscoliosis and decreased bone mass compared to that observed in the *Sod1*$^{+/+}$ mice (Fig. 14. 1A). The bone mineral density (BMD) of the whole body and femur was significantly decreased in the *Sod1*$^{-/-}$ mice compared to that observed in the *Sod1*$^{+/+}$ mice (Fig. 14.1B). A micro computed tomography (μCT) analysis also revealed a marked decrease in both cortical and cancellous bone volume in the *Sod1*$^{-/-}$ mice.[16] Notably, increased accumulation of ROS was also markedly observed in *Sod1*$^{-/-}$ osteoblasts (Fig. 14.2), leading to decreased cell proliferation and increased apoptosis. Furthermore, the *Sod1*$^{-/-}$ mice exhibited transcriptional alterations in genes associated with osteogenesis, such as *Col1a1*, *Alp* and *Nfatc1*. These results suggest that accumulated oxidative stress due to ROS production induces osteopenia accompanied by downregulation of collagen synthesis (Fig. 14.3). Interestingly, *Sod1* deficiency in osteoclasts did not induce ROS production or cellular abnormalities. These results indicate that *Sod1* plays a pivotal role in the viability and function of osteoblasts but not osteoclasts (Fig. 14.3). In addition, the *Sod1*$^{-/-}$ mice exhibited aging-like changes in collagen cross-linking associated with increased advanced glycation end products (AGEs) such as pentosidine in bone.[16] Accumulation of carboxymethyllysine, an AGE, was also observed in the retinas of the *Sod1*$^{-/-}$ mice.[4] These results indicate that excess cellular O_2^- may induce AGE formation in various tissues, including bone and retinas.

VITAMIN C PREVENTS BONE LOSS IN *SOD1*-DEFICIENT MICE

To determine whether oxidative damage plays a causative role in the pathogenesis of age-associated bone fragility, we investigated the effects of VC on bone fragility in *Sod1*$^{-/-}$ mice. When the *Sod1*$^{-/-}$ mice were administered 1% VC in drinking water, weakened bone stiffness, reduced BMD and decreased bone weight were completely reversed in the *Sod1*$^{-/-}$ mice following VC administration (Fig. 14.3). The *in vitro* experiments revealed that VC treatment completely diminished ROS production in osteoblasts isolated from the *Sod1*$^{-/-}$ mice

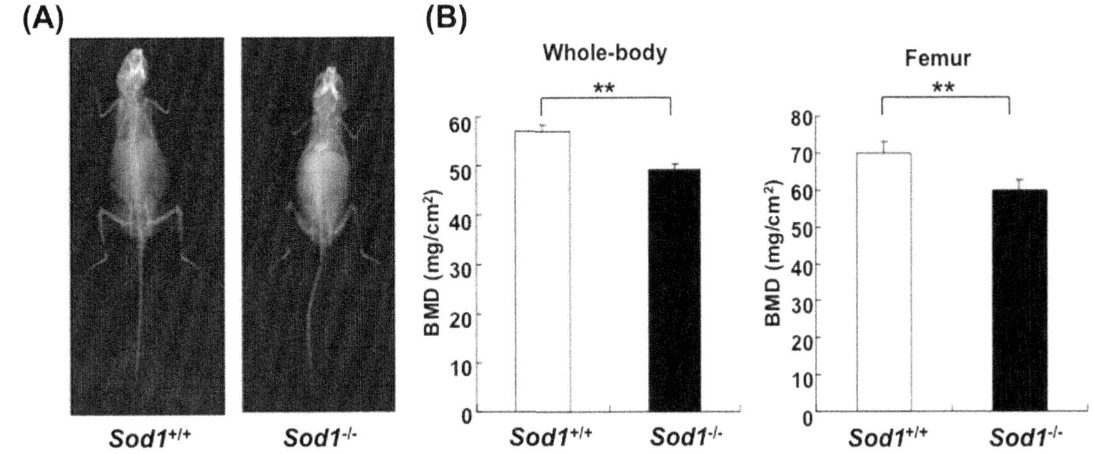

FIGURE 14.1 Decreased bone mass in *Sod1*$^{-/-}$ mice. (A) *Sod1*$^{+/+}$ and *Sod1*$^{-/-}$ mice were analyzed using X-rays at 69 weeks of age. (B) The BMD of the whole body and femur as assessed by dual-energy X-ray absorptiometry of the *Sod1*$^{-/-}$ males was compared with that of their *Sod*$^{+/+}$ littermates at 12 weeks of age (n = 5 for each). The data indicate the mean ± SD; $^{**}p < 0.01$.

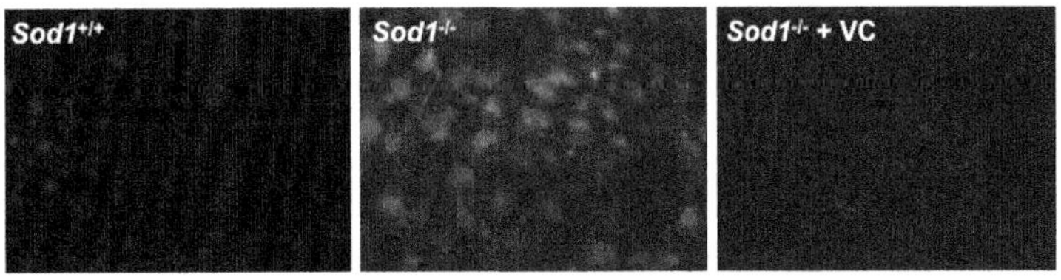

FIGURE 14.2 Vitamin C treatment reduces ROS generation in *Sod1*$^{-/-}$ osteoblasts. Osteoblasts were isolated enzymatically from the calvaria of P2 fetuses using sequential digestion with collagenase and cultured at the same density, as described previously.[58] For measurement of intracellular ROS, the cultured cells were stained with dichlorodihydrofluorescein diacetate dye (CM-H_2DCFDA). The intensity of the oxidized fluorescent DCF in the cells was observed using fluorescence microscopy.

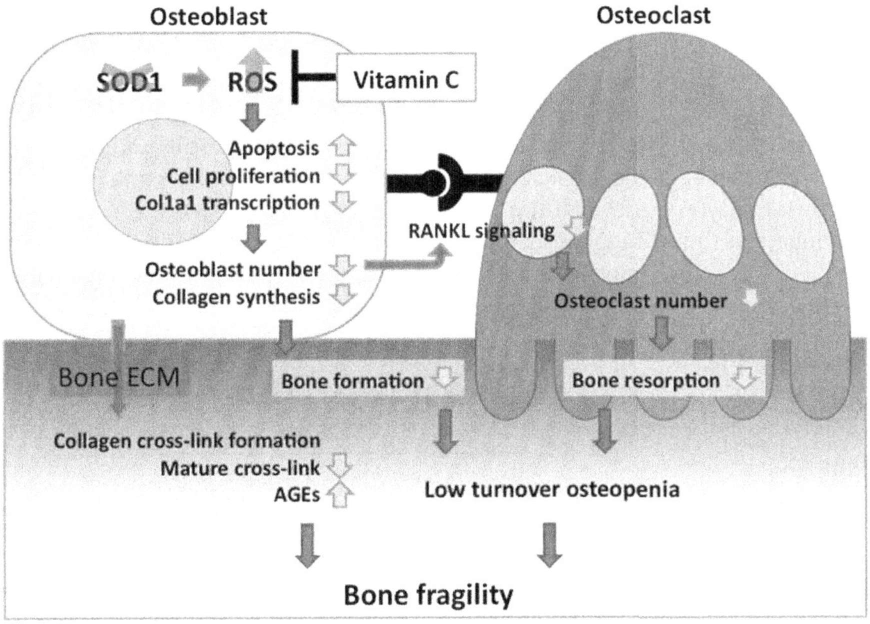

FIGURE 14.3 Schematic diagram of bone fragility induced by *Sod1* deficiency. Increased ROS production induced by *Sod1* deficiency in osteoblasts causes increased apoptosis and decreased cell proliferation. Fewer osteoblasts result in downregulation of collagen synthesis and RANKL signaling, leading to decreased numbers of osteoclasts. Both decreased bone formation and resorption cause low-turnover osteopenia. Furthermore, increased ROS production induces pathologic cross-linking (AGEs) of collagen, leading to bone fragility.[16]

(Fig. 14.2), indicating that the antioxidative actions of VC reverse the bone phenotypes of *Sod1*$^{-/-}$ mice both *in vivo* and *in vitro*. VC also accelerates the synthesis and accumulation of matrix collagen, which is involved in the differentiation of osteoblastic cells.[23,24] VC treatment may directly activate collagen formation, leading to the reversal of bone strength loss in *Sod1*$^{-/-}$ mice. VC therefore may represent a candidate for the treatment and prevention of bone loss in patients with oxidative stress or VC insufficiency. Recently, Massip et al implied that VC supplementation improved premature aging caused by mutations in WRN, a RecQ-like DNA helicase in mice.[25] These data suggest that VC therapy can be applied to treat age-related and premature aging disorders.

VITAMIN C IMPROVES BONE LOSS INDUCED BY ESTROGEN DEFICIENCY

In elderly females, postmenopausal osteoporosis occurs with a high prevalence and induces substantial morbidity and mortality. Estrogen insufficiency in females induces bone loss through an increased osteoclastic function resulting in postmenopausal osteoporosis.[26] In animal models, ovariectomy (OVX) has been used to estimate the effects of estrogen deficiency on osteoporosis. Indeed, in one study, the OVX mice exhibited significantly decreased BMD compared to the sham-operated group. In addition, estrogen exerts protective effects against oxidative stress in multiple tissues, including adipocytes, endothelial cells and neurons.[27–30] In bone marrow, the physiologic level of estrogen maintains thiol antioxidants, suggesting that ROS are a critical factor for postmenopausal osteoporosis.[31] In this context, in order to reverse OVX-induced osteoporosis using antioxidants, antioxidant experiments were conducted. Lean et al reported that intraperitioneal treatment with VC or *N*-acetylcysteine (NAC) improves bone loss induced by OVX in rats.[31] Recently, Zhu et al reported that oral administration of VC also prevents OVX-induced bone loss in mice.[32] These data suggest that antioxidant treatment, including VC, can ameliorate postmenopausal osteoporosis. We expect that novel strategies for the treatment of postmenopausal osteoporosis involving antioxidants will be developed in the future.

MECHANICAL UNLOADING INDUCES ROS PRODUCTION AND BONE LOSS

Mechanical loading plays an important role in the maintenance of the musculoskeletal function. Reduced mechanical loading causes bone and skeletal muscle atrophy under several conditions, including bed rest, paralysis and space flight.[33–36] Following bed rest for 35 days, mechanical unloading decreases bone mass by 1–3% in young people.[37] Such reductions are equivalent to the amount of bone loss induced each year by aging in elderly people.[38] Furthermore, mechanical loading treatment is required for 4-fold periods in order to recover the bone loss that occurs under conditions of bed rest.[39,40] In this context, preventing bone loss induced by mechanical unloading is indispensable for maintaining the human quality of life, especially among the elderly. In a human study, Smith et al reported that unloading increases the levels of urinary 8-hydroxy-2′-deoxyguanosine (8-OHdG) during long-duration space flight.[41] In addition, Rai et al revealed that head-down bed rest, a valuable ground-based model of mechanical unloading conditions, increases the serum levels of 8-OHdG and malondialdehyde,[42] thus indicating that mechanical unloading induces systemic oxidative stress via ROS production in humans.

We recently discovered that mechanical stimulation regulates ROS production in cells. When we suspended the tails of mice for 2 weeks, the BMD of the hind limb was significantly reduced as a model of mechanical unloading-induced bone loss. Under these conditions, we measured the ROS levels in bone marrow cells and sera. The results showed that mechanical unloading significantly increases the level of ROS in bone marrow cells as well as the serum levels of oxidative stress markers (Morikawa et al unpublished results). Interestingly, hind limb unloading also specifically upregulates the expression of *Sod1* but not that of any other antioxidant genes (*Sod2*, *Cat* or *Gpx1*), suggesting that *Sod1* suppresses ROS production and prevents mechanical unloading-induced bone loss. Next, we suspended the tails of $Sod1^{-/-}$ mice for 2 weeks to investigate the physiologic role of *Sod1* in mechanical unloading. Dual-energy X-ray absorptiometry and a μCT analysis revealed that loss of *Sod1* exacerbates bone loss via reduced osteoblastic abilities during mechanical unloading (Morikawa et al unpublished results). Interestingly, we found that the administration of VC significantly prevented bone loss in wild-type mice during unloading. These results indicate that mechanical stimulation, in part, regulates bone mass via intracellular ROS generation and expression of *Sod1*, suggesting that activating *Sod1* or antioxidant treatment may be a preventive strategy for ameliorating mechanical unloading-induced bone loss.

SKIN ATROPHY IN *SOD1*-DEFICIENT MICE

The skin is the largest and outermost organ of the body and it plays the role of protective barrier against external damage. In turn, the skin itself is constantly exposed to external stress and the progression of skin aging. Skin aging manifests as two main symptoms: photoaging induced by environmental stress, such as sunlight, which leads to skin hypertrophy, and intrinsic aging induced by chronologic or intrinsic factors, which leads to skin atrophy.[43] There are many studies of the former using ultraviolet (UV) radiation (such as UV-A or

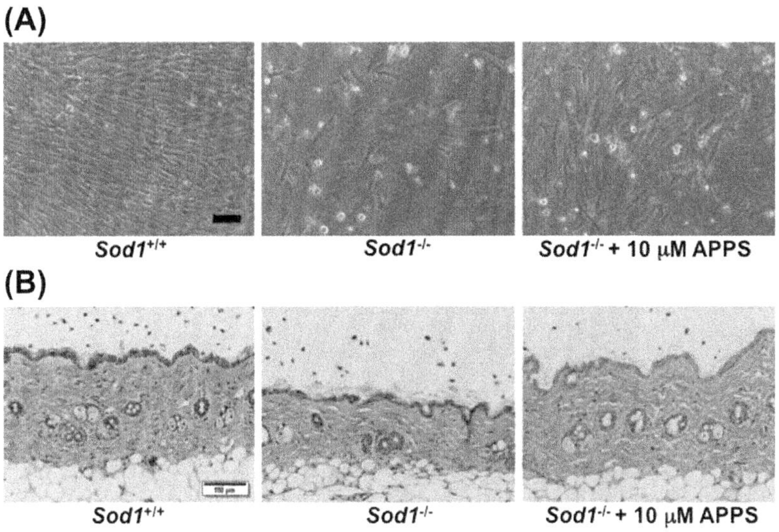

FIGURE 14.4 *Sod1* deficiency induces skin atrophy. (A) Dermal fibroblasts were isolated enzymatically from the skin of P4-6 fetuses using treatment with collagenase, as described previously.[18] Fibroblasts were cultured at the same density and treated with 10 μM of APPS. The scale bar represents 100 μm. (B) The hematoxylin and eosin staining of the back skin of *Sod1*$^{+/+}$ and *Sod1*$^{-/-}$ mice treated daily with 1% APPS for 4 weeks. The scale bar represents 100 μm.

UV-B).[44] The principal cause of photoaging is not only UV-induced direct DNA damage, but also UV-induced ROS production.[45,46] Studies on the intrinsic aging process have been reported using representative senescence models such as Klotho mice,[47] p53 mutant mice[48] and Ku86 knockout mice.[49] These model mice exhibit common phenotypes of premature aging, skin atrophy and loss of elasticity.

We have previously reported that *Sod1*$^{-/-}$ mice exhibit skin atrophy with degeneration of collagen and elastin.[17,18] When *Sod1*$^{-/-}$ dermal fibroblasts were cultured under 20% O_2 conditions, *Sod1*$^{-/-}$ fibroblasts showed significantly increased intracellular O_2^- with decreased viability and increased apoptotic cell death.[18] As expected, culture under hypoxic conditions (1% O_2) significantly suppressed intracellular O_2^- generation and improved cell viability, thus suggesting that O_2^- accumulation causes reduced viability of skin fibroblasts leading to skin atrophy in *Sod1*$^{-/-}$ mice[18] (Fig. 14.4A). Interestingly, elderly *Sod1*$^{-/-}$ mice frequently exhibit inflammation-like symptoms in the face, suggesting that atrophic changes induce inflammatory reactions in the skin of *Sod1*$^{-/-}$ mice (Shibuya, unpublished results). Indeed, Iuchi et al reported a significant delay in wound healing time in *Sod1*$^{-/-}$ mice.[50] These data suggest that redox imbalance in the skin caused by SOD1 depletion leads to skin inflammation and impaired repair processes.

A VITAMIN C DERIVATIVE IMPROVES SKIN ATROPHY IN *SOD1*-DEFICIENT MICE

Because vitamin C exhibits low stability and liposolubility, it does not easily absorb into the skin. For this reason, it has not been successfully used in skin-care materials. To increase the stability and liposolubility of VC, various VC derivatives have been developed for dermatologic application. Ascorbyl 2-phosphate 6-palmitate trisodium salt (APPS), a VC derivative, is more stable than VC and exhibits permeability through cell membranes due to having a conjugated long hydrophobic chain (Fig. 14.5). We tested the protective effects of APPS on skin atrophy in *Sod1*$^{-/-}$ mice.[17] When APPS was transdermally administered to the skin of *Sod1*$^{-/-}$ mice daily, the skin thinning of the *Sod1*$^{-/-}$ mice was completely reversed[17,18] (Fig. 14.4B). Histochemical analyses revealed that degeneration of collagen and elastin in *Sod1*$^{-/-}$ mice is reversed to almost normal levels by APPS treatment.[17] In contrast, oral administration of VC does not suppress dermatrophia in *Sod1*$^{-/-}$ mice, indicating that dietary VC intake might yield insufficient delivery to the skin (Shibuya, unpublished results). These data suggest that transdermal administration of APPS is more efficient for the treatment of skin atrophy compared to oral administration of VC. Additionally, to confirm the protective effects against skin atrophy in *Sod1*$^{-/-}$ mice, APPS was added to *Sod1*$^{-/-}$ fibroblasts *in vitro*. Pretreatment with APPS effectively suppressed O_2^- generation and increased cell numbers of *Sod1*$^{-/-}$ fibroblasts[18] (Fig. 14.4A). These findings indicate that APPS preserves the viability of *Sod1*$^{-/-}$ dermal fibroblasts by suppressing O_2^- production. Interestingly, when NAC is administered in the drinking water of *Sod1*$^{-/-}$ mice, the symptom onset of skin inflammation is significantly delayed.[50] These results indicate that both transdermal APPS and oral NAC treatment reverse the skin phenotypes of *Sod1*$^{-/-}$ mice.

Shibayama et al reported that disodium isostearyl 2-O-L-ascorbyl phosphate (VCP-IS-2Na) suppresses the loss of type I collagen induced by UV-A irradiation via increased permeability and conversion to VC *in vitro*.[51,52] Matsuda et al reported that VCP-IS-2Na inhibits melanogenesis in human melanoma cells and

Vitamin C (VC)

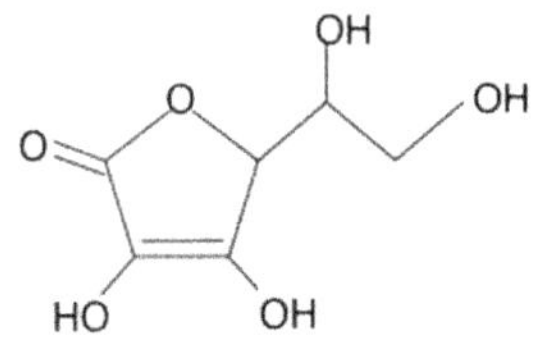

L-Ascorbyl 2-phosphate 6-palmitate trisodium salt (APPS)

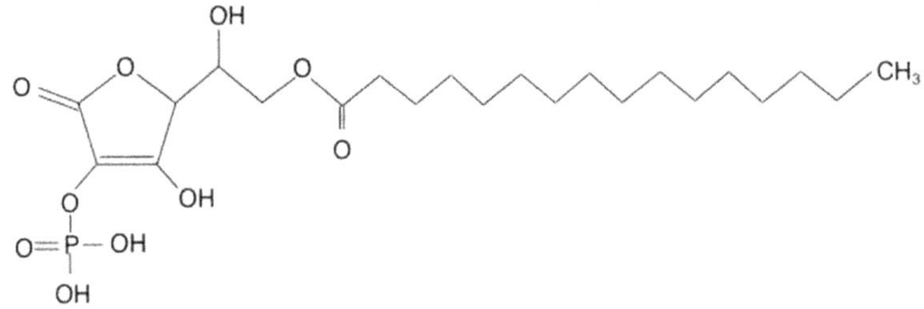

FIGURE 14.5 Structure of vitamin C and L-ascorbyl 2-phosphate 6-palmitate trisodium salt (APPS). Vitamin C exhibits low stability and liposolubility. APPS is additionally conjugated with a phosphate group and a long hydrophobic chain to enhance stability and liposolubility.

epidermal melanocytes.[53] In addition, Xiao et al reported that 2,3,5,6-O-tetra-2′-hexyldecanoyl L-ascorbic acid in liquid form exerts cytoprotective effects against UV-A ray-induced injury in human skin cells.[54] Saitoh et al reported that a fucoidan-VC complex suppresses tumor invasion through the basement membrane by decreasing oxidative stress and metalloproteinase activity in human fibrosarcoma cells.[55] It has been reported that L-ascorbyl 2-phosphate trisodium salt (APS) attenuates age-dependent telomere shortening in human skin keratinocytes by reducing intracellular oxidative stress.[56] Furthermore, APS prevents apoptotic cell death and DNA strand cleavage in murine epidermal keratinocytes.[57] *In vitro* experiments have shown that APS suppresses intracellular O_2^- generation and cell death in *Sod1*$^{-/-}$ fibroblasts (data not shown). In contrast, APS fails to induce any significant improvements in the skin thickness of *Sod1*$^{-/-}$ mice, suggesting that the permeability of VC derivatives is an essential factor for transdermal treatment.[18] These reports suggest that VC derivatives, including APPS, are also likely to behave more efficiently in terms of the pharmacologic actions of VC, due to their higher stability and permeability.

SUMMARY POINTS

- SOD catalyzes cellular O_2^- to O_2 and H_2O_2 and plays a central role in antioxidative systems.
- Cytoplasmic SOD1 loss leads to age-related diseases caused by intracellular oxidative stress.
- *Sod1*$^{-/-}$ mice exhibit distinct weakness in bone stiffness, decreased BMD and aging-like changes in collagen cross-linking induced by marked ROS production.
- Oral VC administration markedly restores the skeletal phenotypes of *Sod1*$^{-/-}$ mice.
- VC treatment improves bone loss induced by estrogen insufficiency and mechanical unloading.
- *Sod1*$^{-/-}$ mice exhibit skin thinning associated with decreased hydroxyproline content.
- *Sod1*$^{-/-}$ dermal fibroblasts exhibit increased O_2^- accumulation and lower cell viability.
- Transdermal APPS treatment completely normalizes skin atrophy in *Sod1*$^{-/-}$ mice.
- APPS treatment suppresses intracellular O_2^- generation and improves viability in *Sod1*$^{-/-}$ fibroblasts.
- Redox regulation by VC is a powerful anti-aging strategy for treating bone and skin tissues.

References

1. Harman D. Aging: a theory based on free radical and radiation chemistry. *J Gerontol* 1956;**11**:298–300.
2. Miao L, St Clair DK. Regulation of superoxide dismutase genes: implications in disease. *Free Radic Biol Med* 2009;**47**:344–56.
3. Fattman CL, Schaefer LM, Oury TD. Extracellular superoxide dismutase in biology and medicine. *Free Radic Biol Med* 2003;**35**:236–56.
4. Imamura Y, Noda S, Hashizume K, et al. Drusen, choroidal neovascularization, and retinal pigment epithelium dysfunction in SOD1-deficient mice: a model of age-related macular degeneration. *Proc Natl Acad Sci USA* 2006;**103**:11282–7.
5. Uchiyama S, Shimizu T, Shirasawa T. CuZn-SOD deficiency causes ApoB degradation and induces hepatic lipid accumulation by impaired lipoprotein secretion in mice. *J Biol chem* 2006;**281**:31713–9.
6. Elchuri S, Oberley TD, Qi W, et al. CuZnSOD deficiency leads to persistent and widespread oxidative damage and hepatocarcinogenesis later in life. *Oncogene* 2005;**24**:367–80.
7. Iuchi Y, Okada F, Onuma K, et al. Elevated oxidative stress in erythrocytes due to a SOD1 deficiency causes anaemia and triggers autoantibody production. *Biochem J* 2007;**402**:219–27.

8. Muller FL, Song W, Liu Y, et al. Absence of CuZn superoxide dismutase leads to elevated oxidative stress and acceleration of age-dependent skeletal muscle atrophy. *Free Radic Biol Med* 2006;**40**:1993–2004.
9. Murakami K, Murata N, Noda Y, et al. SOD1 (copper/zinc superoxide dismutase) deficiency drives amyloid beta protein oligomerization and memory loss in mouse model of Alzheimer disease. *J Biol Chem* 2011;**286**:44557–68.
10. Ho YS, Gargano M, Cao J, et al. Reduced fertility in female mice lacking copper-zinc superoxide dismutase. *J Biol Chem* 1998;**273**:7765–9.
11. Matzuk MM, Dionne L, Guo Q, et al. Ovarian function in superoxide dismutase 1 and 2 knockout mice. *Endocrinology* 1998;**139**:4008–11.
12. Noda Y, Ota K, Shirasawa T, Shimizu T. Copper/zinc superoxide dismutase insufficiency impairs progesterone secretion and fertility in female mice. *Biol Reprod* 2012;**86**:1–8.
13. Ohlemiller KK, McFadden SL, Ding DL, et al. Targeted deletion of the cytosolic Cu/Zn-superoxide dismutase gene (Sod1) increases susceptibility to noise-induced hearing loss. *Audiol Neurootol* 1999;**4**:237–46.
14. Kojima T, Wakamatsu TH, Dogru M, et al. Age-related dysfunction of the lacrimal gland and oxidative stress: evidence from the Cu,Zn-superoxide dismutase-1 (Sod1) knockout mice. *Am J Pathol* 2012;**180**:1879–96.
15. Behndig A, Karlsson K, Reaume AG, et al. In vitro photochemical cataract in mice lacking copper-zinc superoxide dismutase. *Free Radic Biol Med* 2001;**31**:738–44.
16. Nojiri H, Saita Y, Morikawa D, et al. Cytoplasmic superoxide causes bone fragility owing to low-turnover osteoporosis and impaired collagen cross-linking. *J Bone Miner Res* 2011;**26**:2682–94.
17. Murakami K, Inagaki J, Saito M, et al. *Handbook of diet, nutrition and the skin: protective effects of vitamin C derivatives on skin atrophy caused by Sod1 deficiency*. The Netherlands: Wageningen Academic; 2011. 351–364.
18. Shibuya S, Kinoshita K, Shimizu T. *Handbook of diet, nutrition and the skin: protective effects of vitamin C derivatives on skin atrophy caused by Sod1 deficiency*. The Netherlands: Wageningen Academic; 2011. 351–364.
19. Basu S, Michaelsson K, Olofsson H, et al. Association between oxidative stress and bone mineral density. *Biochem Biophys Res Commun* 2001;**288**:275–29.
20. Maggio D, Barabani M, Pierandrei M, et al. Marked decrease in plasma antioxidants in aged osteoporotic women: results of a cross-sectional study. *J Clin Endocrinol Metab* 2003;**88**:1523–7.
21. Ostman B, Michaelsson K, Helmersson J, et al. Oxidative stress and bone mineral density in elderly men: antioxidant activity of alpha-tocopherol. *Free Radic Biol Med* 2009;**47**:668–73.
22. Peterkofsky B. Ascorbate requirement for hydroxylation and secretion of procollagen: relationship to inhibition of collagen synthesis in scurvy. *Am J Clin Nutr* 1991;**54**:1135S–40S.
23. Harada S, Matsumoto T, Ogata E. Role of ascorbic acid in the regulation of proliferation in osteoblast-like MC3T3-E1 cells. *J Bone Miner Res* 1991;**6**:903–8.
24. Takeuchi Y, Nakayama K, Matsumoto T. Differentiation and cell surface expression of transforming growth factor-beta receptors are regulated by interaction with matrix collagen in murine osteoblastic cells. *J Biol Chem* 1996;**271**:3938–44.
25. Massip L, Garand C, Paquet ER, et al. Vitamin C restores healthy aging in a mouse model for Werner syndrome. *Faseb J* 2010;**24**:158–72.
26. Riggs BL, Khosla S, Melton 3rd LJ. Sex steroids and the construction and conservation of the adult skeleton. *Endocr Rev* 2002;**23**:279–302.
27. Sack MN, Rader DJ, Cannon 3rd RO. Oestrogen and inhibition of oxidation of low-density lipoproteins in postmenopausal women. *Lancet* 1994;**343**:269–70.
28. Sudoh N, Toba K, Akishita M, et al. Estrogen prevents oxidative stress-induced endothelial cell apoptosis in rats. *Circulation* 2001;**103**:724–9.
29. Arnal JF, Clamens S, Pechet C, et al. Ethinylestradiol does not enhance the expression of nitric oxide synthase in bovine endothelial cells but increases the release of bioactive nitric oxide by inhibiting superoxide anion production. *Proc Natl Acad Sci USA* 1996;**93**:4108–13.
30. Sawada H, Ibi M, Kihara T, et al. Mechanisms of antiapoptotic effects of estrogens in nigral dopaminergic neurons. *Faseb J* 2000;**14**:1202–14.
31. Lean JM, Davies JT, Fuller K, et al. A crucial role for thiol antioxidants in estrogen-deficiency bone loss. *J Clin Invest* 2003;**112**:915–23.
32. Zhu LL, Cao J, Sun M, et al. Vitamin C prevents hypogonadal bone loss. *PLoS One* 2012;**7**:e47058.
33. Tidball JG. Mechanical signal transduction in skeletal muscle growth and adaptation. *J Appl Physiol* 2005;**98**:1900–908.
34. Ikemoto M, Nikawa T, Takeda S, et al. Space shuttle flight (STS-90) enhances degradation of rat myosin heavy chain in association with activation of ubiquitin-proteasome pathway. *Faseb J* 2001;**15**:1279–81.
35. Rittweger J, Winwood K, Seynnes O, et al. Bone loss from the human distal tibia epiphysis during 24 days of unilateral lower limb suspension. *J Physiol* 2006;**577**:331–7.
36. Yang Li C, Majeska RJ, Laudier DM, et al. High-dose risedronate treatment partially preserves cancellous bone mass and microarchitecture during long-term disuse. *Bone* 2005;**37**:287–95.
37. Rittweger J, Simunic B, Bilancio G, et al. Bone loss in the lower leg during 35 days of bed rest is predominantly from the cortical compartment. *Bone* 2009;**44**:612–8.
38. Riggs BL, Khosla S, Melton 3rd LJ. A unitary model for involutional osteoporosis: estrogen deficiency causes both type I and type II osteoporosis in postmenopausal women and contributes to bone loss in aging men. *J Bone Miner Res* 1998;**13**:763–73.
39. Sibonga JD, Evans HJ, Sung HG, et al. Recovery of spaceflight-induced bone loss: bone mineral density after long-duration missions as fitted with an exponential function. *Bone* 2007;**41**:973–8.
40. Watanabe Y, Ohshima H, Mizuno K, et al. Intravenous pamidronate prevents femoral bone loss and renal stone formation during 90-day bed rest. *J Bone Miner Res* 2004;**19**:1771–8.
41. Smith SM, Zwart SR, Block G, et al. The nutritional status of astronauts is altered after long-term space flight aboard the International Space Station. *J Nutr* 2005;**135**:437–43.
42. Rai B, Kaur J, Catalina M, et al. Effect of simulated microgravity on salivary and serum oxidants, antioxidants, and periodontal status. *J Periodontol* 2011;**82**:1478–82.
43. Glogau RG. Physiologic and structural changes associated with aging skin. *Dermatol Clin* 1997;**15**:555–9.
44. Ichihashi M, Ueda M, Budiyanto A, et al. UV-induced skin damage. *Toxicology* 2003;**189**:21–39.
45. de Gruijl FR, van Kranen HJ, Mullenders LH. UV-induced DNA damage, repair, mutations and oncogenic pathways in skin cancer. *J Photochem Photobiol B* 2001;**63**:19–27.
46. Scharffetter-Kochanek K, Wlaschek M, Brenneisen P, et al. UV-induced reactive oxygen species in photocarcinogenesis and photoaging. *Biol Chem* 1997;**378**:1247–57.
47. Kuro-o M, Matsumura Y, Aizawa H, et al. Mutation of the mouse klotho gene leads to a syndrome resembling ageing. *Nature* 1997;**390**:45–51.
48. Tyner SD, Venkatachalam S, Choi J, et al. p53 mutant mice that display early ageing-associated phenotypes. *Nature* 2002;**415**:45–53.
49. Vogel H, Lim DS, Karsenty G, et al. Deletion of Ku86 causes early onset of senescence in mice. *Proc Natl Acad Sci USA* 1999;**96**:10770–5.
50. Iuchi Y, Roy D, Okada F, et al. Spontaneous skin damage and delayed wound healing in SOD1-deficient mice. *Mol Cell Biochem* 2010;**341**:181–94.
51. Shibayama H, Hisama M, Matsuda S, Ohtsuki M. Permeation and metabolism of a novel ascorbic acid derivative, disodium isostearyl 2-O-L-ascorbyl phosphate, in human living skin equivalent models. *Skin Pharmacol Physiol* 2008;**21**:235–43.

52. Shibayama H, Hisama M, Matsuda S, et al. Effect of a novel ascorbic derivative, disodium isostearyl 2-O-L-ascorbyl phosphate, on normal human dermal fibroblasts against reactive oxygen species. *Biosci Biotechnol Biochem* 2008;**72**:1015–22.
53. Matsuda S, Shibayama H, Hisama M, et al. Inhibitory effects of a novel ascorbic derivative, disodium isostearyl 2-O-L-ascorbyl phosphate on melanogenesis. *Chem Pharm Bull (Tokyo)* 2008;**56**:292–7.
54. Xiao L, Kaneyasu K, Saitoh Y, et al. Cytoprotective effects of the lipoidic-liquiform pro-vitamin C tetra-isopalmitoyl-ascorbate (VC-IP) against ultraviolet-A ray-induced injuries in human skin cells together with collagen retention, MMP inhibition and p53 gene repression. *J Cell Biochem* 2009;**106**:589–98.
55. Saitoh Y, Nagai Y, Miwa N. Fucoidan-vitamin C complex suppresses tumor invasion through the basement membrane, with scarce injuries to normal or tumor cells, via decreases in oxidative stress and matrix metalloproteinases. *Int J Oncol* 2009;**35**:1183–9.
56. Yokoo S, Furumoto K, Hiyama E, Miwa N. Slow-down of age-dependent telomere shortening is executed in human skin keratinocytes by hormesis-like-effects of trace hydrogen peroxide or by anti-oxidative effects of pro-vitamin C in common concurrently with reduction of intracellular oxidative stress. *J Cell Biochem* 2004;**93**:588–97.
57. Sugimoto M, Okugawa Y, Miwa N. Preventive effects of phosphorylated ascorbate on ultraviolet-B induced apoptotic cell death and DNA strand cleavage through enrichment of intracellular vitamin C in skin epidermal keratinocytes. *Free Radic Res* 2006;**40**:213–21.
58. Bellows CG, Aubin JE, Heersche JN, Antosz ME. Mineralized bone nodules formed *in vitro* from enzymatically released rat calvaria cell populations. *Calcif Tissue Int* 1986;**38**:143–54.
59. Hashizume K, Hirasawa M, Imamura Y, et al. Retinal dysfunction and progressive retinal cell death in SOD1-deficient mice. *Am J Pathol* 2008;**172**:1325–31.

CHAPTER

15

S-Equol, an Antioxidant Metabolite of Soy Daidzein, and Oxidative Stress in Aging: A Focus on Skin and on the Cardiovascular System

Richard L. Jackson, Jeffrey S. Greiwe, Richard J. Schwen

Ausio Pharmaceuticals, LLC, Cincinnati, Ohio, USA

List of Abbreviations

AREs antioxidant response elements
BH_4 tetrahydrobiopterin
CHD coronary heart disease
CIMT carotid intimal-media thickness
CRP C-reactive protein
CVD cardiovascular disease
DHEA dehydroepiandrosterone
eNOS endothelial nitric oxide synthase
ERβ estrogen receptor β
fMRI functional magnetic reasonance imaging
GPER G-protein-coupled estrogen receptor
HACE1 E3 ubiquitin-protein ligase
HO-1 heme oxygenase-1
3β-HSD 3β-hydroxysteroid dehydrogenase
HT hormone therapy
MMP-15 metalloperoxidase-15
NQO1 nitroquinone oxidoreductase
SNPs single-nucleotide polymorphisms
SOD superoxide dismutase
VMS vasomotor symptoms
WHI Women's Health Initiative
WHIMS Women's Health Initiative Memory Study

INTRODUCTION

Oxidative stress and cell senescence play an important role in the development of chronic diseases associated with aging. As is discussed elsewhere in this book, an extensive literature exists on reactive oxygen species (ROS) and their ability to modify a myriad of biologic molecules that affect aging. The body is well equipped to deal with oxidative stress if one consumes a healthy diet of fruits and vegetables. Fat-soluble vitamin E and water-soluble vitamin C provide the first line of antioxidant protection. Nonetheless, controlled clinical trials with dietary supplementation using these natural antioxidants have, in general, failed to produce any additional therapeutic benefit in the aging population, suggesting that other factors are also involved.

The oxidative stress associated with reproductive aging, as it relates to changes in hormonal balance, is another factor to be considered. The reduction of estrogen production in women at menopause, and decreased plasma testosterone in aging men, has a profound effect on physical function and age-associated chronic diseases. In men, testosterone deficiency is associated with impaired cognitive function.[1] Ota et al[2] developed a mouse model to understand this relationship and showed that testosterone regulates cerebral endothelial senescence through nitric oxide synthase (eNOS). These investigators proposed that age-associated vascular inflammation leads to a chronic stress response in endothelial cells, promoting senescence of adjacent neuronal cells. In women, the reduction in estrogen production with menopause is associated with dryness of the skin, reduced cognitive function, increased cardiovascular diseases (CVD) and numerous vasomotor symptoms (VMS), including hot flushes, depression and weight gain. For these reasons, hormone therapy (HT) has been used for decades by postmenopausal women for the treatment of VMS and general well-being. However, the results of the Women's Health Initiative (WHI) have dramatically reduced the use of this therapy. In the WHI study, subjects received either

Aging
http://dx.doi.org/10.1016/B978-0-12-405933-7.00015-9

conjugated equine estrogens (Premarin® 0.625 mg/d plus medroxyprogesterone acetate, 2.5 mg/day) or a placebo. After a mean 5.2 years of follow-up, the data and safety monitoring board recommended stopping the trial because of the higher incidence of invasive breast cancer in the treated versus the placebo group.[3] In addition, the women on HT had significantly increased risks of coronary heart disease (CHD), pulmonary embolism and stroke; HT reduced the observed hip and vertebral bone fracture rates by one third compared to placebo. Ancillary to WHI was the Women's Health Initiative Memory Study (WHIMS) that addressed the effects of HT on the incidence of dementia and cognitive function. Estrogen treatment alone or in combination with progestin resulted in increased risk of dementia.[4,5] As a consequence of WHI and WHIMS outcomes, there has been a dramatic decrease in the use of HT and, not surprisingly, an increase in the incidence of hip fractures.[6]

The findings from the WHI and WHIMS trials were unexpected because it had been assumed that estrogen replacement would show a positive effect. In retrospect, a major limitation in the clinical trials with HT was that the women were >65 years of age. It has been proposed that there is a 'window of opportunity' for HT that needs to be initiated within 10 years of menopause in order to have a clinical benefit.[7–9] In support of this proposal, recent findings from the Cache County Study have shown that women who used HT within 5 years of menopause had 30% less risk of Alzheimer's disease.[10]

Because of the potential risk associated with HT, many women are seeking alternative treatments for their menopausal symptoms. Isoflavones, also referred to as phyto-estrogens, contained in soy, have received much attention as an alternative approach to HT.[11] The major isoflavones in soy are the glycoside conjugates of genistein, daidzein and glycetein. In addition, certain individuals, mainly Asians, are able to convert daidzein to S-equol (Fig. 15.1) by gut bacteria.[12] Populations that consume sufficient soy and can carry out the biotransformation of daidzein to S-equol are referred to as 'equol producers'. These individuals have fewer menopausal symptoms, and a decreased incidence of prostate cancer, osteoporosis, neurodegenerative diseases and cardiovascular diseases.[13] In this chapter we focus the discussion on the role that isoflavones, particularly S-equol, may have in slowing the aging process in skin and in protecting against CVD by reducing oxidative stress.

PROPERTIES OF S-EQUOL

S-equol contains two phenoxyl groups and with this structural feature is capable of scavenging superoxide anions and hydroxyl radicals in solution. S-equol also has potent binding affinity (IC_{50} = 0.73 nM) to estrogen receptor β (ERβ) with lesser activity for ERα, the predominant estrogen receptor found in the breast and uterus; estrogen (17β-estradiol) binds to both receptors with equally high affinity (Kd = 0.15 nM).[12] It has been proposed that S-equol offers a treatment modality for menopausal symptoms in women.[14] Because S-equol is selective for ERβ it may not have the side effects associated with estrogen therapy. Equol has a chiral center at carbon 3 (Fig. 15.1) and can exist as either the S- or R-enantiomer. However, in man and animals, only S-equol is produced by the gut biotransformation of daidzein.[12] Several anaerobic, gram-positive bacteria have been shown to carry out the biotransformation,[13] as shown in Figure 15.2.

A major limitation with most of the animal studies to date is that racemic equol (± equol) has been used because it is commercially available. However, it should not be assumed that racemic equol, which contains 50% R-equol and 50% S-equol, has the same biologic activity as pure S-equol. After oral administration, S-equol is rapidly conjugated during first-pass metabolism.[15] As shown in Figure 15.3, the major metabolites are the 4′-glucuronide and 7-sulfate, with less than 1% of circulating S-equol

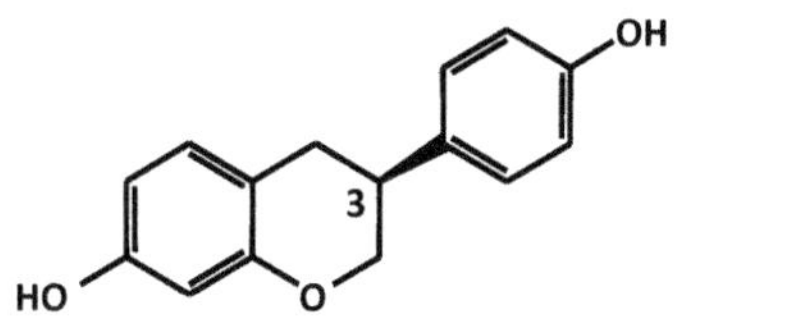

(S)-3,4-Dihydro-3-(4-hydroxyphenyl)-2H-1-benzopyran-7-ol

FIGURE 15.1 Structure of S-equol.

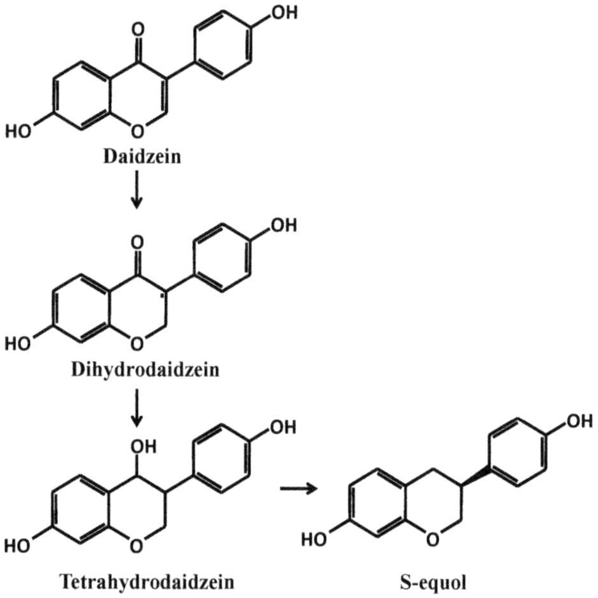

FIGURE 15.2 Biotransformation of daidzein to S-equol.

being unconjugated, the form that binds to ERβ. In most of the *in vitro* studies using racemic equol, micromolar concentrations were used. These concentrations are nonphysiologic, which compromises the interpretation of the results. The pharmacokinetics of S-equol after oral administration of oral doses of 10–160 mg of S-equol given twice a day has been reported.[16] At the 10 and 20 mg doses, the trough plasma levels of total (conjugated plus unconjugated) S-equol were approximately 100 ng/mL (Fig. 15.4), the same concentration in plasma of a Japanese person who is an equol producer and consumes 35 mg of daidzein per day or 60 g of soy protein.

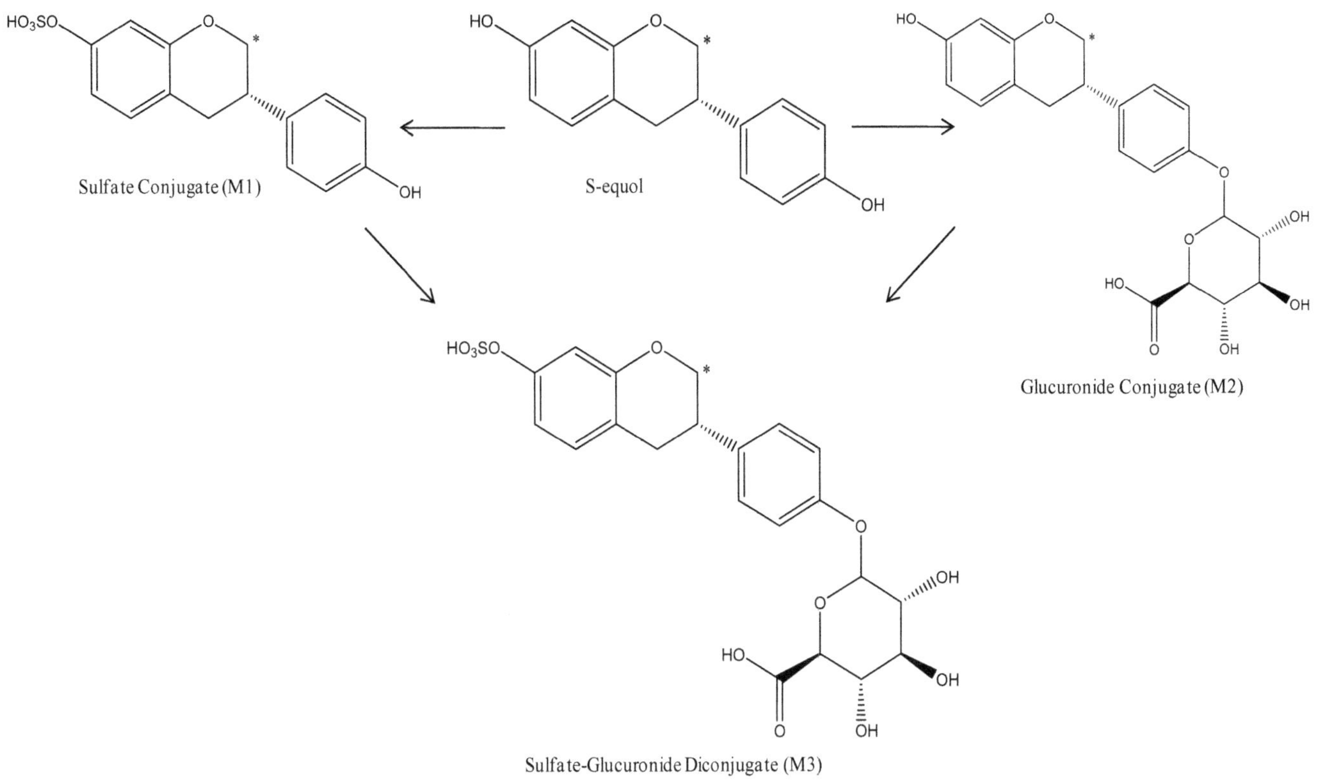

FIGURE 15.3 Metabolic pathways of [^{14}C]-S-equol in the rat, monkey and human.[15] Asterisks indicate the position of the [^{14}C]-label.

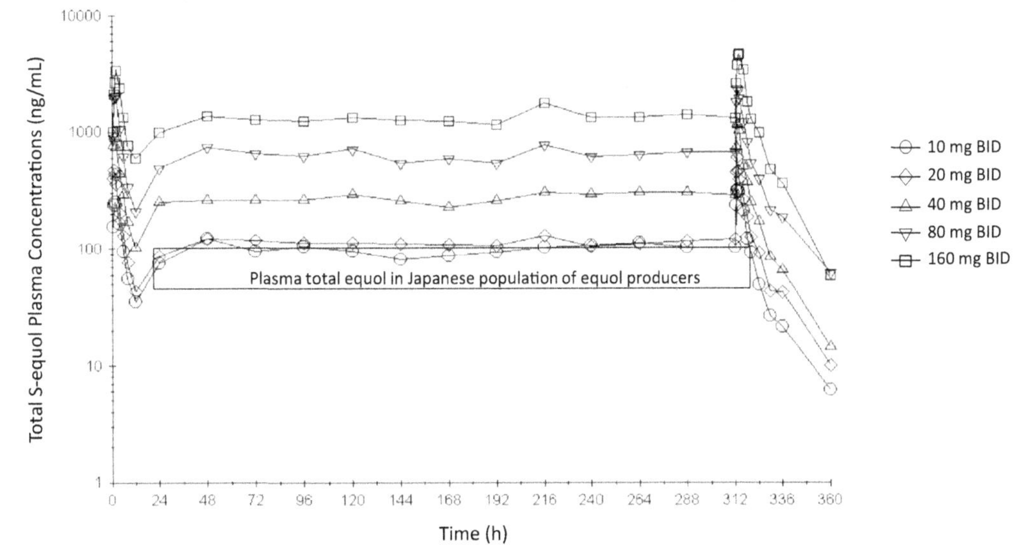

FIGURE 15.4 Plasma total S-equol dose pharmacokinetics.[16] Plasma total equol concentration appearance and disappearance curves were determined over 14 days of dosing. At the initial dose and at day 14, multiple blood samples were taken to determine $T_{1/2}$. The other values represent trough levels of S-equol. The bar shows the concentration of total S-equol in a population of Japanese people who are equol producers.

AGING SKIN

The greatest insult to skin is the long-term exposure to UV irradiation. The skin is well equipped to deal with UV-induced oxidative stress by antioxidant molecules, such as vitamin E and C, and by antioxidant enzymes, including superoxide dismutase (SOD) and catalase. The importance of SOD for the development of aging skin has been shown in mice with a genetic deficiency of SOD.[17] SOD deficiency accelerates the aging phenotype and cellular senescence by inducing DNA double-strand breaks leading to epidermal thinning. Likewise, in postmenopausal women with decreased estrogen production, and in aging men with decreased levels of testosterone, there is a decrease in the oxidative distress system, leading to aging skin.

It is well known that oral estrogen improves skin quality in postmenopausal women by increasing skin collagen content, elasticity and hydration.[18–20] The skin is able to synthesize testosterone and estrogen *de novo* and their concentrations decrease with aging. The conversion of dehydroepiandrosterone (DHEA) to testosterone in skin occurs through a two-step process: 3β-hydroxysteroid dehydrogenase (3β-HSD) catalyzes the conversion of DHEA to androstenedione which is then converted to testosterone by 17β-hydroxysteroid dehydrogenase; aromatization of testosterone by skin aromatase generates 17β-estradiol. Hang et al[21] have shown that young male skin (mean age 27.8 years) utilizes DHEA more efficiently for the synthesis of testosterone than does old skin (mean age 62.6 years), suggesting a possible explanation for aging skin in men.

ERβ is the predominant estrogen receptor in human skin, with high expression in the epidermis, dermal fibroblasts, blood vessels and hair follicles. The importance of ERβ in mediating the protective effect of estrogen against photoaging has been demonstrated by Markiewicz et al[22] using mice lacking ERα or ERβ, and by Chang et al[23] with estrogen receptor selective compounds. Markiewicz et al[22] assessed the role of skin estrogen receptors in ERα and ERβ knockout, aged female mice. In the ERα KO mouse, skin dermal thickness and collagen content were significantly increased as compared to wild-type mice; in the ERβ KO dermal thickness was slightly less than in wild-type mice, providing evidence for a role of ERβ in skin. The increase in skin collagen was reflected by a several-fold increase in collagen mRNA expression in cultured fibroblasts obtained from the ERα KO, whereas in the ERβ KO fibroblasts collagen mRNA decreased. To understand the mechanisms for the changes observed in these KO animals, Chang et al[22] measured the expression of matrix metalloproteinase-15 (MMP-15), an enzyme involved in collagen degradation. In the ERα KO mouse, mRNA levels of MMP-15 were significantly reduced, whereas in the ERβ KO they were increased. Since both estrogen receptors are present in skin, the changes in the ratio between ERα and ERβ may be an important factor in skin during the aging process.

Chang et al[23] determined the gene expression profile of UV-activated primary skin keratinocytes obtained from normal mice compared to keratinocytes obtained from ERβ heterozygote and ERβ KO mice. In wild-type mice, UV irradiation increased the expression of a number of inflammatory transcription factors and cytokines, including IL-1, IL-6 and NFκB. Topical application of the ERβ selective compound WAY-200070 significantly inhibited expression of these genes. The ER antagonist ICI-182,780 blocked expression, indicating that the reduction in gene expression occurred through a genomic mechanism of action. In UV-activated primary skin obtained from ERβ heterozygote animals, WAY-200070 inhibited gene expression of MMP-13 and MAPK-1 but in ERβ knockout animals the ERβ agonist had no effect, showing that ERβ is required for prevention of photoaging. Another ERβ selective compound, ERB-041, which is more selective for ERβ than WAY-200070, was also evaluated in an animal model of wrinkles. Five-week-old female albino hairless mice were exposed to UV irradiation three times per week for 6 weeks. After each irradiation, ERB-041 was applied topically to the dorsal area. ERB-041 applied topically to normal skin with no irradiation had no effect. The animals that were treated with ERB-041 followed by irradiation showed a significant reduction in wrinkles compared to control animals. Because ERB-041 is not an antioxidant, these findings suggest that its effects are mediated by a genomic ERβ-mediated mechanism and not by scavenging free radicals.

ISOFLAVONES AND AGING SKIN

Topically applied isoflavones have been shown to protect skin from UV damage.[24] Furthermore, racemic equol protects against UV-induced lipid peroxidation in mouse skin.[25] Daily application of racemic equol to skin following UV irradiation resulted in a significant reduction of epidermal skin thickness, a decrease in dermal mast cells, and a decrease in elastosis as compared to UV-irradiated control animals. Topical equol also reduced the amount of lipid peroxides present in the skin after irradiation. The only report to date showing that S-equol affects aging skin has been provided by Oyama et al.[26] A dietary supplement containing 5 mg S-equol was generated by bacterial fermentation of soy paste with *Lactococcus* 20-92. Oral administration of the supplement for 12 weeks in Japanese postmenopausal women resulted in a decrease in the area and depth of wrinkles (crows' feet) around the eyes. As with HT, these results show that S-equol given orally provides a systemic approach for retarding skin aging in women.

S-EQUOL'S MECHANISM OF REDUCING OXIDATIVE STRESS IN SKIN

Aging of the skin occurs naturally as a result of a decline in estrogen in women and in testosterone in men. In addition, the skin is subjected to exogenous factors, including UV- and chemical-induced damage. To understand the changes involved in protecting the skin from chronologic aging versus photoaging, Peres et al[27] used 8-, 18- and 24-week-old hairless mice and 8-week-old mice that were irradiated three times a week for 10 weeks. Total antioxidant capacity was significantly lower in both aged and UV-exposed skin. This reduced capacity was reflected by an increase in lipid peroxides. Interestingly, the content of lipid peroxides at 18 weeks was less than in the 8-week-old animals – but at 48 weeks there was a 5-fold increase, suggesting that skin can compensate for oxidative stress early on in the aging process, but that as the aging process continues there is uncompensated oxidative stress. In these studies, SOD activity was not changed with either chronologic or photoaging. However, catalase activity was significantly reduced, suggesting that elevated levels of H_2O_2 are associated with the aging process.

Nrf1 and Nrf2 are transcription factors that are critical for the regulation of antioxidant enzymes.[28] When a cell is exposed to UV light or ROS, Nrf2 is released from Keap 1, its cytoplasmic binding protein, causing translocation of Nrf2 to the nucleus. Nrf2 then binds to antioxidant response elements (ARE) that are present in the promoter region of the genes of antioxidant proteins and enzymes, e.g. glutathione-transferase, to promote transcription. One possible mechanism for S-equol's action is to increase the nuclear levels of Nrf2 by releasing Nrf2 from Keap 1 in the cytoplasma (Fig. 15.5). Another mechanism is that S-equol binds to estrogen-responsive elements (EREs) in the promoter region of the gene of Nrf2 or of antioxidant defense system genes to enhance their transcription. In this regard, Froyen & Steinberg[29] reported that racemic equol increased the expression of quinone reductase, a phase II xenobiotic metabolizing enzyme, by enhancing the binding of ERβ and Nrf2 to the quinone reductase gene. Racemic equol caused a 2-fold increase in quinone reductase mRNA and protein that was reflected by a >2-fold increase in the binding of ERβ and Nrf2 to the gene. Because S-equol binds to ERβ with high affinity, it is reasonable to speculate that ligand bound ERβ is a critical factor in the activation of transcription of antioxidant enzymes.

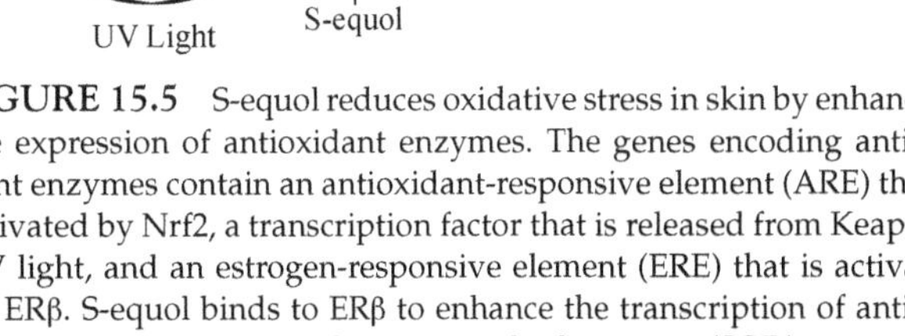

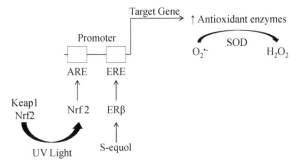

FIGURE 15.5 S-equol reduces oxidative stress in skin by enhancing the expression of antioxidant enzymes. The genes encoding antioxidant enzymes contain an antioxidant-responsive element (ARE) that is activated by Nrf2, a transcription factor that is released from Keap1 by UV light, and an estrogen-responsive element (ERE) that is activated by ERβ. S-equol binds to ERβ to enhance the transcription of antioxidant enzymes, for example, superoxide dismutase (SOD), to convert superoxide anion ($O_2^{\bullet-}$) to hydrogen peroxide (H_2O_2).

In conclusion, it is proposed that the mechanism by which S-equol protects skin from UV irradiation and chronologic aging in man is not through a pure chemical antioxidant mechanism but by enhancement of expression of enzymes that remove free radicals. In this regard, even under normal conditions S-equol may enhance the antioxidant defense system. The effects of orally administered racemic equol on the antioxidant defense system have been evaluated in the livers of normal mice.[30] After 1 week of oral administration there was a significant decrease in three markers of oxidative stress: liver concentrations of thiobarbituric acid reactive substances, protein carbonyls, and serum concentrations of 8-hydroxy-2-deoxyguanosine. Administration of racemic equol was associated with a significant increase in catalase and SOD protein and mRNA. Thus, in normal animals – those that have not been challenged by oxidative stress – the administration of equol further increases the antioxidant defense system and reduces markers of oxidative stress.

S-EQUOL AND CARDIOVASCULAR DISEASES

Jackman et al[31] have reviewed the experimental and clinical evidence for equol and the benefits of soy for the treatment of CVD. The WHO CARDIAC study was the first report to show a significant relationship between the mortality rate of coronary heart disease (CHD) and the consumption of soy products and isoflavones.[32] The amount of total isoflavones excreted in a 24-hour urine collection was used as an indicator of soy consumption. The age-adjusted mortality rate in individuals in Japan and China with >10 μmol isoflavones (approximately 30 mg) excreted per day had a much lower rate of CHD (<100 events per 100,000 people) than populations with <1 μmol/day (>500 events per 100,000 people). In these studies the equol producer status was not determined. However, because more than 60% of Japanese and Chinese people are equol producers, it could be assumed that a person with high levels of urinary isoflavones would have significant levels of S-equol.

Zhang et al[33] examined the associations of urinary isoflavones and equol with the risk of CHD in a nested case–control prospective study in Chinese adults in Shanghai. Participants reported no history of CHD or stroke at baseline. In contrast to the WHO CARDIAC study, there was no association between total urinary isoflavones and CHD. However, the urinary equol concentrations in women in the control group without CHD were significantly higher than those in subjects who had developed CHD; no association was found for men. For both men and women, urinary daidzein and genistein were not correlated with CHD.

The Women's Isoflavone Soy Health (WISH) trial is the first clinical study designed to determine the effect of an isoflavone supplement and equol producer status on atherosclerosis in a healthy population of postmenopausal women without pre-existing CVD.[34] Study participants were assigned to a placebo group or a treatment group that received a daily supplement containing 25 g of soy protein with 52 mg genistein and 36 mg daidzein. The primary end point in the study was the rate of change in the right distal carotid artery intima-media thickness (CIMT) as measured by high-resolution B-mode ultrasound images. The average follow up in this study was 2.8 years. The CIMT in the placebo group was 5.68 μm/year vs. 4.77 μm/year for the isoflavone supplemented group. Although the progression CIMT rate in the isoflavone-treated group was reduced by 16%, the decrease was not statistically different. However, women who had experienced menopause within 5 years of the study had a 68% lower CIMT progression rate compared to the placebo group, with a difference of −2.49 μm/year, $p < 0.05$. This finding is consistent with the results of WHI where it was found that older women who were 10 years past menopause had greater CVD, and it supports the 'window of opportunity' for isoflavones similar to HT.[7] Thirty nine of the subjects in the treated group were consistent equol producers (plasma equol >20 nmol/L, 5 ng/mL). In these subjects the CIMT was 4.56 μM (1.74–7.39 μM) for the equol producers versus 5.68 μM (4.2 –7.07 μM) for the non-producers. While the trend was positive there were no significant differences between the groups. In other findings, Cai et al[35] measured the equol producer status in a population of Chinese and reported a significantly lower CIMT in equol producers. No association was found between total dietary isoflavone intakes in the non-producers, suggesting that the benefit of soy in reducing atherosclerosis is dependent upon one's ability to transform daidzein to S-equol.

In a placebo-controlled clinical trial, Clerici et al[36] tested a novel soy germ pasta enriched with isoflavones. In this study the isoflavones were present only as aglycones; the glycosides were removed by the manufacturing process. Adults with hypercholesterolemia consumed a diet that included an 80 g serving per day of pasta containing 33 mg of aglycone isoflavones. After 4 weeks of consumption, the isoflavone-enriched pasta diet significantly lowered total-cholesterol and LDL-cholesterol; however, HDL-cholesterol and triglycerides were not changed. C-reactive protein (CRP), a measure of inflammation, was also significantly reduced in the isoflavone-enriched pasta diet group compared to the pasta diet without the isoflavones. Twenty of the subjects consuming the isoflavone-enriched soy pasta were equol producers and nine were not. The mean serum concentration in the equol producers was 25.5 ng/mL compared to 2.3 ng/mL in the non-producers. After the data were analyzed by equol producer status, only the equol producers showed a significant decrease in LDL-cholesterol and CRP. Wong et al[37] also reported that equol producers have lower levels of plasma CRP than equol non-producers. Hall et al[38] measured inflammatory markers of CVD in healthy postmenopausal women consuming an isoflavone-enriched (50 mg/day) cereal bar for 8 weeks. Consistent with findings discussed above, isoflavone consumption had a significant effect in lowering CRP levels in equol producers. Whether lower CRP levels cause a decrease in CVD or a decrease in CVD results in lower CRF is unknown.

One way CRP might reduce CVD is by affecting endothelial dysfunction, a major marker of cardiovascular risk. Several studies have shown that isoflavone supplements improve endothelial function, as determined by flow-mediated dilation (FMD) in the brachial artery. One of the first studies showing that isoflavones improve endothelial function was reported by Chan et al.[39] Chinese patients with primary or recurrent ischemic stroke were enrolled in this study; it should be noted that about one-half of the subjects also presented with diabetes mellitus. After 12 weeks' treatment, the group consuming 80 mg/day soy isoflavones showed a significant improvement in FMD as compared to the placebo group. Isoflavone consumption had no effect on blood pressure, plasma lipids or blood glucose. Serum SOD, 8-isoprostane and malondialdehyde, measures of oxidative stress, were also unchanged compared to the placebo group. The one CVD biomarker that was affected by isoflavone supplementation was CRP. The plasma level of CRP in the patients with a normal FMD was 2.8 ± 0.5 mg/dL compared to 4.4 ± 0.5 mg/dL with impaired FMD. This finding suggests that isoflavones mediate endothelial function by an anti-inflammatory mechanism. In these studies the equol producer status was not determined, but because the subjects were Chinese a large percentage of them would have conceivably been equol producers.

S-EQUOL'S MECHANISM FOR REDUCING OXIDATIVE STRESS IN THE CARDIOVASCULAR SYSTEM

Multiple mechanism(s) for equol's effect on the cardiovascular system have been proposed, as shown in Figure 15.6. Unconjugated S-equol contains two phenolic groups that function to scavenge free radicals in solution. With this structural feature, racemic equol has been shown to prevent the Cu^{2+}-induced oxidation of LDL in solution.[40] However, non-physiologic concentrations of equol were used in these experiments. Using L-buthionine (S, R)-sulfoximine to induce oxidative stress in fibroblasts from a patient with Fridreich's ataxia, a hereditary form of ataxia that results in the accumulation of ROS, Richardson & Simpkins[41] demonstrated that both S- and R-equol were equally cytoprotective. Becausde R-equol does not bind to ERβ with high affinity,[12] these authors proposed that R-equol is acting through an antioxidant mechanism as opposed to a genomic mechanism. This interpretation is consistent with the fact that concentrations of S- and R-equol of 100 nM or greater were required to attenuate cellular levels of ROS to prevent cell death.

Chung et al[42] examined the effects of racemic equol on the H_2O_2-induced death of bovine aortic endothelial cells. Pretreatment of cells with 0.5 μM equol for 30 minutes prior to incubation with H_2O_2 significantly inhibited cell death. The antioxidant effect of equol was shown by a decrease in ROS, cellular apoptotic bodies, and nuclear chromatin condensation. At the molecular level, equol pretreatment increased the phosphorylation of MAP kinase after the addition of H_2O_2 to cells. The MAP kinase inhibitor SB 203580 prevented the decrease

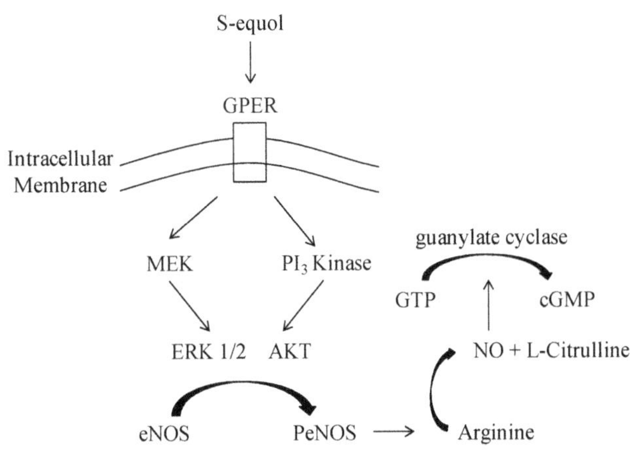

FIGURE 15.6 S-equol activates G protein-coupled estrogen receptor (GPER) to mediate rapid signaling in endothelial cells to induce vasodilation. Activation of GPER leads to enhanced PI_3 kinase activity, activation of MEK and AKT and phosphorylation of eNOS, generating increased amounts of NO and vasorelaxation by the activation of soluble guanylate cyclase and the production of cGMP.

in cell viability. These results suggest that equol needs to be present in the cell to prevent oxidative damage and that the MAP kinase signaling pathway is required for its activity. Furthermore, these results are consistent with a non-chemical antioxidant mechanism.

The importance of cell-associated equol in the prevention of oxidative damage has also been observed in the human macrophage cell line J774.[40] In these experiments, J774 cells were activated with lipopolysaccharide (LPS) and then the cells were incubated with freshly prepared LDL. In the absence of equol, LDL is oxidatively modified by macrophages to produce an electronegative, modified LDL^-. Pre-incubation of the J774 cells with equol reduced the amount of oxidized LDL from 4.1 to 2.0% after 8 hours; without pretreatment, equol had no effect on the amount of LDL^- formed. Equol-pretreated J774 cells also produced less superoxide ($O_2^{\bullet-}$). The $O_2^{\bullet-}$ formation rates were 2.01, 1.97 and 1.52 nmol $O_2^{\bullet-}$/min/mg protein for control cells, equol-supplemented control cells and equol pre-treated cells, respectively. The authors proposed that the equol in the pre-treated cells prevented the activation of the reduced nicotinamide adenine dinucleotide phosphate oxidase complex. With decreased $O_2^{\bullet-}$ production, the levels of nitric oxide (NO) increased, and the investigators suggested that the increased levels of NO prevent the oxidation of LDL. To summarize this mechanism for S-equol, LDL is oxidized by activated macrophages and, through a non-genomic mechanism, equol inhibits its oxidation.

Estrogen receptors are expressed in the vascular endothelial and smooth muscle cells and are activated by both ligand-dependent and ligand-independent mechanisms to regulate gene transcription and oxidative stress.[43] The level and transcriptional activity of ERα and ERβ change with age within the cardiovascular system, and the ratio between the two estrogen receptors could affect gene expression and the aging process. In addition, an orphan-G-protein-coupled estrogen receptor (GPER), formerly called GPR-30, is a membrane-bound estrogen receptor that rapidly activates signaling pathways, particularly in endothelial cells.[44,45] Joy et al[46] examined the molecular mechanisms involved in the rapid activation of eNOS and NO production by racemic equol in human aortic and umbilical vein endothelical cells in culture. The addition of equol to endothelium-intact rat aortic rings preconstricted with phenylephrine caused a rapid dose-dependent (100 nM to 10 μM) relaxation; equol had no effect on Ca^{2+} mobilization, cAMP levels or PGI_2 production. The rapid action of equol and the lack of effect of ICI 182,780, an estrogen receptor antagonist, indicate that ERα or ERβ, estrogen receptors that are present in these cells, is not involved. Joy et al[46] further showed the extracellular protein kinase ERK1/2 was rapidly phosphorylated by 1 nM equol, a physiologic concentration, and that the increase in activity was paralleled by an increase in intracellular cGMP. Because ERK1/2 is phosphorylated by MEK, the MEK1/2 inhibitor U0126 prevented the phosphorylation of ERK1/2 by equol (Fig. 15.6).

Another mechanism for activation of eNOS is by phosphorylation of AKT (Fig. 15.6). Equol increases AKT phosphorylation, and a PI3 kinase inhibitor prevents the phosphorylation.[46] With increased eNOS activity, the NO reacts with superoxide anions to form peroxynitrite. NO and peroxynitrite then, in turn, enhance the nuclear accumulation of Nrf 2 that binds to an ARE in target genes to enhance the transcription of the phase II antioxidant defense enzymes, such as SOD, catalase, glutathione-S-transferase, glutathione peroxidase, quinine reductase and heme oxygenase-1 (HO-1). eNOS requires the cofactor tetrahydrobiopterin (BH_4) for activity.[43] The levels of BH_4 are reduced with aging and, as a result, eNOS becomes uncoupled, generating superoxide and a lesser amount of NO. It is not known whether S-equol affects BH_4 levels with aging, but this is a potential mechanism to explain its activity in the cardiovascular system.

Activation of Nrf2 by ROS has been proposed to play an important role in vascular disease and aging.[47–50] During hypoxia/reoxygenation injury, myocardial cells in the heart attempt to compensate for oxidative stress by enhancing the expression of HO-1, SOD, glutathione-*S*-transferase and glutamate cysteine ligase amide, enzymes that represent the antioxidant defense system in cells. Transcription of these enzymes is upregulated by Nrf2.[51] Estrogen also increases the mRNA expression of these enzymes and is blocked by ICI 182,780, an estrogen receptor antagonist. Estrogen also promotes the translocation of cytoplasmic Nrf2 to the nucleus during hypoxic injury in heart myocytes. Thus, there are multiple mechanisms by which estrogen and S-equol reduce oxidative stress.

In coronary arteries, exposure of endothelial cells to protardim, a phyto-estrogen dietary supplement, increases the expression of HO-1, nitroquinone oxidoreductase (NQO1) and glutathione reductase.[52] Pre-exposure of endothelial cells with protardim prevented the H_2O_2-induced cell damage. Protardim also stimulated the expression and nuclear content of Nrf2. Donovan et al[52] concluded that the increase in antioxidant enzymes was mediated through an Nrf2 mechanism; silencing Nrf2 expression abrogated the expression of HO-1. Consistant with these *in vitro* results, Brydun et al[53] reported that there is an inverse relationship between coronary atherosclerosis and the amount of HO-1 mRNA, providing evidence that HO-1 is a key determinant of oxidative stress. Nrf2 binds to the ARE in the promoter of antioxidant genes to enhance transcription. Activation of the Nrf2-ARE pathway is essential for the induction of HO-1 and MnSOD.[54] Huang et al[54] examined the effect of hypoxia in H9c2 cells, a clonal line derived from embryonic rat heart. The cells were exposed to hypoxia followed by reoxygenation. Hypoxic preconditioning of these cells followed by oxygenation is known to upregulate the expression of HO-1 and MnSOD. In Nrf2 siRNA-transfected cells, Nrf2 expression was knocked down and HO-1 and MnSOD protein levels were abolished showing that Nrf2 was essential for the expression of these enzymes during the reoxygenation period. In support of this conclusion, the siRNA-transfected cells showed an increase in ROS.

It is well known that high concentrations of ROS result in cell and tissue damage. However, at low concentrations ROS are important second messagers of cellular activity by acting through intracellular signaling cascades. Rowlands et al[55] reported that the racemic equol-induced activation of eNOS in human umbilical vein endothelial cells was inhibited by pretreatment of the cells with SOD. Furthermore, the mitochrondrial Complex I inhibitor rotenone reduced the activation, providing evidence that $O_2^{\bullet-}$ was involved. Rotenone also blocked the acute phosphorylation of eNOS, AKT and ERK1/2, implicating $O_2^{\bullet-}$ in the kinase pathway. Equol stimulated the levels of intracellular cGMP that were blocked by rotenone and an inhibitor of eNOS, N-nitro-L-arginine ester. These surprising results in fetal endothelial cells suggest that $O_2^{\bullet-}$(ROS) acts as a second messenger to activate the kinase cascade.

IMPORTANCE OF S-EQUOL EXPOSURE EARLY IN LIFE

Food products containing soy have been consumed for centuries, but only within the last 10 years has it been known that daidzein, an isoflavone contained in soy, is converted to S-equol. Many clinical studies have been carried out with soy protein, and soy derived isoflavones, with mixed results. For most of these studies, the equol producer status was not determined, thus it is not surprising that many of the clinical findings were inconclusive. In those studies where plasma total equol levels were determined, a value >5 ng/mL provides a clinical benefit for a decrease in menopausal symptoms, a gain in bone mineral density and improvement in cardiovascular risk factors.[13] Consistent with an S-equol value >5 ng/mL, Hozawa et al[56] found that equol concentrations of 23.5 ng/mL or greater were associated with a lower risk of disability and death. In this study of Japanese people, almost one-half of the participants were equol producers (serum equol level $\geq$1 ng/mL). Serum levels of daidzein, genistein and glycitein did not correlate with disability and death, providing further evidence that the benefit of soy consumption is greater in equol producers.

Being an equol producer begins in early life, and most evidence shows that one becomes an equol producer in the first few years after birth. How one becomes an equol producer is not known, but if one does become an equol producer one is always an equol producer. This would imply that the anaerobic bacteria required for the biotransformation of daidzein are acquired soon after birth because the human fetus develops within a bacteria-free environment. At delivery, the neonate is exposed to bacteria provided by the mother primarily during and after the passage through the birth canal. The development of the newborn's microbiome has a long-lasting effect, and we have much to learn on the effects of bacterial imprinting on future health.

It is known that if a pregnant woman consumes soy phyto-estrogens they are transferred to the fetus. Furthermore, if the mother is an equol producer, the concentration of S-equol is proportional between maternal serum and fetal cord serum.[57] This relationship raises an important question as to the importance of fetal programming for enhancing the body's ability to deal with future oxidative stress. In other words, does S-equol exposure determine the response to soy later in life? Because Asian populations are mostly equol producers, the offspring would have been equol producers. However, today, fewer than 20% of Japanese children are equol producers, suggesting that they will be prone to future oxidative stress and chronic diseases, a trend that already appears to be occuring. An important question is why S-equol exposure would provide a different fetal imprint than estrogen. The fetal adrenal gland is the major source of estrogen production. Both ERα and ERβ are expressed in the fetal adrenal gland but little is known about the molecular events and the effects that S-equol might have on the imprint. For example, it is not known how the binding of S-equol to ERβ recruits gene enhancers and repressors to regulate expression *in utero*. The 'imprint' expressed *in utero* by S-equol may provide the defense to ward off oxidative stress later in life.

As discussed above, being an equol producer requires consumption of dietary daidzein and possession of appropriate gut bacteria to transform the phyto-estrogen into S-equol. In addition, genetic factors may also be important. Hong et al[58] carried out a genome-wide association study of single-nucleotide polymorphisms (SNPs) in a Korean population of equol producers. Of the 1391 subjects in the study, 70.1% were equol producers based on a serum total equol concentration of $\geq$68 ng/mL. Systolic and diastolic blood pressures were significantly lower in the equol producers. The most significant SNPs were located in a cluster of genes in the chromosome 6q 21 region encoding an Ankyrin repeat containing E3 ubiquitin-protein ligase (HACE1). Individuals with a minor allele and an A > C mutation were equol non-producers. Another SNP was strongly linked to the estrogen receptor activator protein 1 (AP-1). While these gene association studies do not explain the mechanism of S-equol's action they do provide evidence that genetic factors are involved. Hong et al[58] have suggested that the association of HACE1 and equol producers may involve host immune reponses. Expression profile analysis from a gene omnibus database showed that HACE1 expression rises in the mesenchymal layer of the embryonic small intestine and is important in intestinal immune responses. These investigators further proposed that estrogen receptors in the fetal mesenchymal layer bind S-equol to modulate the intestinal environment for maintaining the equol-producing bacteria at birth. This suggests that innate immune response genes are developmentally regulated by the mother's diet. Thus, therapeutic and nutritional interventions during the early years of life may affect immunity later and may be determined by the gut microtome. While this is a hypothesis, it does provide for the design of future experiments and offers the possibility that critical components of longevity may actually be established *in utero* and during the first years of one's life.

SUMMARY POINTS

- Estrogen receptor beta plays a key role in skin and in the cardiovascular system to reduce oxidative stress.
- S-equol is a selective estrogen receptor beta agonist produced from daidzein present in soy.
- The health benefits of soy may be determined by one's equol producer status.
- Although the S-equol molecule is a potent antioxidant, it decreases oxidative stress by enhancing the expression of antioxidant defense proteins and enzymes.
- In endothelial cells, S-equol binds to a membrane-associated estrogen receptor to rapidly activate the cell.

References

1. Holland J, Bandelow S, Hogervorst E. Testosterone levels and cognition in elderly men: a review. *Maturitas* 2011;**69**:322–37.
2. Ota H, Akishita M, Akiyoshi T, et al. Testosterone deficiency accelerates neuronal and vascular aging of SAMP8 mice: protective role of eNOS and SIRT1. *PLoS One* 2012. http://dx.doi.org/10.1371/journal/pone.0029598.
3. Rossouw JE. For the Writing Group for the Women's Health Initiative Investigators. Risks and benefits of estrogen plus progestin in healthy postmenopausal women: principal results from the Women's Health Initiative randomized controlled trial. *JAMA* 2002;**288**:321–33.
4. Coker LH, Espeland MA, Rapp SR, et al. Postmenopausal hormone therapy and cognitive outcomes: the Women's Health Initiative Memory Study (WHIMS). *J Steroid Biochem Mol Biol* 2010;**118**: 304–10.

5. Shumaker SA, Legault C, Kuller L, et al. Conjugated equine estrogens and incidence of probable dementia and mild cognitive impairment in postmenopausal women: Women's Health Initiative Memory Study. *JAMA* 2004;**291**:2947–58.
6. Karim R, Dell RM, Greene DF, Mack WJ, Gallagher JC, Hodis HN. Hip fracture in postmenopausal women after cessation of hormone therapy: results from a prospective study in a large health management organization. *Menopause* 2011;**18**:1172–7.
7. Hodis HN, Mack WJ. A 'window of opportunity': the reduction of coronary heart disease and total mortality with menopausal therapies is age - and time-dependent. *Brain Res* 2011;**1379**:244–52.
8. Rocca WA, Grossardt BR, Shuster LT. Oophorectomy, menopause, estrogen treatment, and cognitive aging: clinical evidence for a window of opportunity. *Brain Res* 2011;**1379**:188–98.
9. Sherwin BB. Estrogen therapy: is time of initiation critical for neuroprotection? *Nat Rev Endocrinol* 2009;**5**:620–7.
10. Shao H, Breitner JC, Whitmer RA, et al. Hormone therapy and Alzheimer disease dementia: new findings from the Cache County Study. *Neurology* 2012;**79**:1846–52.
11. Zhao L, Brinton RD. In search of estrogen alternative for the brain. In: Hogervorst E, Henderson VW, Gibbs RB, Brinton RD, editors. *Hormones, cognition and dementia: state of the art and emergent therapeutic strategies*. Cambridge: Cambridge University Press; 2009. pp. 93–100.
12. Setchell KDR, Clerici C, Lephart ED, et al. S-equol, a potent ligand for estrogen receptor β, is the exclusive enantiomeric form of the soy isoflavone metabolite produced by human intestinal bacterial flora. *Am J Clin Nutr* 2005;**81**:1072–9.
13. Jackson RL, Greiwe JS, Schwen RJ. Emerging evidence on the health benefits of S-equol, an estrogen receptor β agonist. *Nutr Rev* 2011;**69**:432–8.
14. Ishiwata N, Melby M, Mizuno S, Watanabe S. New equol supplement for relieving menopausal symptoms: randomized, placebo-controlled trial of Japanese women. *Menopause* 2009;**16**:141–8.
15. Schwen RJ, Nguyen L, Jackson RL. Elucidation of the metabolic pathway of S-equol in rat, monkey and man. *Food Chem Toxicol* 2012;**50**:2074–83.
16. Jackson RL, Greiwe JS, Desai PB, Schwen RJ. Single-dose and multi-dose 14 day pharmacokinetics studies of S-equol, a potent non hormonal, estrogen receptor β agonist being developed for the treatment of menopausal symptoms. *Menopause* 2011;**18**: 185–93.
17. Velarde MC, Flynn JM, Day NU, et al. Mitochondrial oxidative stress caused by Sod2 deficiency promotes cellular senescence and aging phenotypes in the skin. *Aging* 2012;**4**:3–12.
18. Wend K, Wend P, Krum SA. Tissue-specific effects of loss of estrogen during menopause and aging. *Front Endocrinol (Lausanne)* 2012;**3**:1–19.
19. Archer DF. Postmenopausal skin and estrogen. *Gynecol Endocrinol* 2012;**28**:2–6.
20. Jackson RL, Greiwe JS, Schwen RJ. Aging skin: estrogen receptor β agonists offer a new approach to change the outcome. *Exp Dermatol* 2011;**20**:879–82.
21. Hang M, Hamann T, Kulle AE, et al. Age and skin site related differences in steroid metabolism in male skin point to a key role of sebocytes in cutaneous hormone metabolism. *Dermatoendocrinol* 2012;**4**:58–64.
22. Markiewicz M, Znoyko S, Stawski L, et al. A role for estrogen receptor-α and estrogen receptor-β in collagen biosynthesis in mouse skin. *J Invest Dermatol* 2013;**133**:120–7.
23. Chang KC, Wang Y, Oh IG, et al. Estrogen receptor beta is a novel therapeutic target for photoaging. *Mol Pharmacol* 2010;**77**:744–50.
24. Lin JY, Tournas JA, Burch JA, et al. Topical isoflavones provide effective photoprotection to skin. *Photodermatol Photoimmunol Photomed* 2008;**24**:61–6.
25. Reeve VE, Widyarini S, Domanski D, et al. Protection against photoaging in the hairless mouse by the isoflavone equol. *Photochem Photobiol* 2005;**81**:1548–53.
26. Oyama A, Ueno T, Uchiyama S, et al. The effects of natural S-equol supplementation on skin aging in postmenopausal women: a pilot randomized placebo-controlled trial. *Menopause* 2012;**19**:202–10.
27. Peres PS, Terra VA, Guarnier FA, et al. Photoaging and chronological aging profile: understanding oxidation of the skin. *J Photochem Photobiol B* 2011;**103**:93–97.
28. Ohtsuji M, Katsuoka F, Kobayashi A, et al. Nrf1 and Nrf2 play distinct roles in activation of antioxidant response element-dependent genes. *J Biol Chem* 2008;**283**:33554–62.
29. Froyen EB, Steinberg FM. Soy isoflavones increase quinone reductase in hepa-1c1c7 cells via estrogen receptor beta and nuclear factor erythroid 2-related factor 2 binding to the antioxidant response element. *J Nutr Biochem* 2011;**22**:843–8.
30. Choi EJ. Evaluation of equol function on anti - or prooxidant status in vivo. *J Food Sci* 2009;**74**:H65–71.
31. Jackman KA, Woodman OL, Sobey CG. Isoflavones, equol and cardiovascular disease: pharmacological and therapeutic insights. *Curr Med Chem* 2007;**14**:2824–30.
32. Yamori Y. CARDIAC study and dietary intervention studies, Chapter 6. In: Sugano M, editor. *Soy in health and disease prevention*. New York: CRC Press; 2005. pp. 107–19.
33. Zhang X, Gao YT, Yang G, et al. Urinary isoflavonoids and risk of coronary heart disease. *Int J Epidemiol* 2012;**41**:1367–75.
34. Hodis HN, Mack WJ, Kono N, et al. Isoflavone soy protein supplementation and atherosclerosis progression in healthy postmenopausal women: a randomized controlled trial. *Stroke* 2011;**42**:3168–75.
35. Cai Y, Guo K, Chen C, et al. Soya isoflavone consumption in relation to carotid intima-media thickness in Chinese equol excretors aged 40-65 years. *Br J Nutr* 2012;**29**:1–7.
36. Clerici C, Setchell KD, Battezzati PM, et al. Pasta naturally enriched with isoflavone aglycons from soy germ reduces serum lipids and improves markers of cardiovascular risk. *J Nutr* 2007;**137**:2270–8.
37. Wong JM, Kendall CW, Marchie A, et al. Equol status and blood lipid profile in hyperlipidemia after consumption of diets containing soy foods. *Am J Clin Nutr* 2012;**95**:564–71.
38. Hall WL, Vafeiadou K, Hallund J, et al. Soy-isoflavone-enriched foods and inflammatory biomarkers of cardiovascular disease risk in postmenopausal women: interactions with genotype and equol production. *Am J Clin Nutr* 2005;**82**:1260–8.
39. Chan YH, Lau KK, Yiu KH, et al. Reduction of C-reactive protein with isoflavone supplement reverses endothelial dysfunction in patients with ischaemic stroke. *Eur Heart J* 2008;**29**:2800–807.
40. Hwang J, Wang J, Morazzoni P, et al. The phytoestrogen equol increases nitric oxide availability by inhibiting superoxide production: an antioxidant mechanism for cell-mediated LDL modification. *Free Radic Bio Med* 2003;**34**:1271–82.
41. Richardson TE, Simpkins JW. R- and S-Equol have equivalent cytoprotective effects in Friedreich's ataxia. *BMC Pharmacol Toxicol* 2012;**13**. http://dx.doi.org/10.1186/2050-6411-13-12.
42. Chung JE, Kim SY, Jo HH, et al. Antioxidant effects of equol on bovine aortic endothelial cells. *Biochem Biophys Res Comm* 2008;**375**:420–4.
43. Murphy E. Estrogen signaling and cardiovascular disease. *Circ Res* 2011;**109**:687–96.
44. Meyer MR, Prossnitz ER, Barton M. The G protein-coupled estrogen receptor GPER/GPR30 as a regulator of cardiovascular function. *Vascul Pharmacol* 2011;**55**:17–25.
45. Prossnitz ER, Barton M. The G-protein-coupled estrogen receptor GPER in health and disease. *Nat Rev Endocrinol* 2011;**7**:715–26.

46. Joy S, Siow RC, Rowlands DJ, et al. The isoflavone equol mediates rapid vascular relaxation: Ca2+-independent activation of endothelial nitric-oxide synthase/Hsp90 involving ERK1/2 and Akt phosphorylation in human endothelial cells. *J Biol Chem* 2006;**281**:27335–45.
47. Chapple SJ, Siow RC, Mann GE. Crosstalk between Nrf2 and the proteasome: therapeutic potential of Nrf2 inducers in vascular disease and aging. *Int J Biochem Cell Biol* 2012;**44**:1315–20.
48. Siow RC, Li FY, Rowlands DJ, et al. Cardiovascular targets for estrogens and phytoestrogens: transcriptional regulation of nitric oxide synthase and antioxidant defense genes. *Free Radic Biol Med* 2007;**42**:909–25.
49. Mann GE, Bonacasa B, Ishii T, Sio RCM. Targeting the redox sensitive Nrf2-Keap 1 defense pathway in cardiovascular disease: protection afforded by dietary isoflavones. *Curr Opin Pharmacol* 2009;**9**:139–45.
50. Siow RC, Mann GE. Dietary isoflavones and vascular protection: activation of cellular antioxidant defenses by SERMs or hormesis? *Mol Aspects Med* 2010;**31**:468–77.
51. Yu J, Zhao Y, Li B, et al. 17β-estradiol regulates the expression of antioxidant enzymes in myocardial cells by increasing Nrf2 translocation. *J Biochem Mol Toxicol* 2012;**26**:264–9.
52. Donovan EL, McCord JM, Reuland DJ, et al. Phytochemical activation of Nrf2 protects human coronary artery endothelial cells against an oxidative challenge. *Oxid Med Cell Longev* 2012. http://dx.doi.org/10.1155/2012/132931.
53. Brydun A, Watari Y, Yamamoto Y, et al. Reduced expression of heme oxygenase-1 in patients with coronary atherosclerosis. *Hypertens Res* 2007;**30**:341–8.
54. Huang XS, Chen HP, Yu HH, et al. Nrf2-dependent upregulation of antioxidative enzymes: a novel pathway for hypoxic preconditioning-mediated delayed cardioprotection. *Mol Cell Biochem* 2013. http://dx.doi.org/10.1007/s11010-013-1812-6.
55. Rowlands DJ, Chapple S, Siow RC, Mann GE. Equol-stimulated mitochondrial reactive oxygen species activate endothelial nitric oxide synthase and redox signaling in endothelial cells: roles for F-actin and GPR30. *Hypertension* 2011;**57**:833–40.
56. Hozawa A, Sugawara Y, Tomata Y, et al. Relationship between serum isoflavone levels and disability-free survival among community-dwelling elderly individuals: nested case-control study of the Tsurugaya Project. *J Gerontol A Biol Sci Med Sci* 2013;**68**:465–72.
57. Todaka E, Sakurai K, Fukata H, et al. Fetal exposure to phytoestrogens–the difference in phytoestrogen status between mother and fetus. *Environ Res* 2005;**99**:195–203.
58. Hong KW, Ko KP, Ahn Y, et al. Epidemiological profiles between equol producers and nonproducers: a genomewide association study of the equol-producing phenotype. *Genes Nutr* 2012;**7**:567–74.

CHAPTER

16

Magnesium, Oxidative Stress, and Aging Muscle

Mario Barbagallo, Ligia J. Dominguez

Geriatric Unit, Department of Internal Medicine DIBIMIS, University of Palermo, Italy

List of Abbreviations

ATP adenosine triphosphate
Ca calcium
cAMP cyclic adenosine monophosphate
CRP C-reactive protein
DNA deoxyribonucleic acid
IFN interferon
Ig immunoglobulin
IL interleukin
K potassium
MAPK mitogen-activated protein kinase
MDA malondialdehyde
Mg magnesium
MgT total serum Mg concentrations
NADPH nicotinamide adenine dinucleotide phosphate
NFκB activation of nuclear factor-kappaB
NHANES National Health and Nutrition
NMDA *N*-methyl-D-aspartate
PAI plasminogen activator inhibitor
PMN polymorphonuclear
PTH parathyroid hormone
RBCs red blood cells
RDA recommended dietary allowance
RNA ribonucleic acid
ROS reactive oxygen species
TNF tumor necrosis factor
VCAM vascular cell adhesion molecule

INTRODUCTION

In recent decades the clinical relevance and biologic significance of magnesium (Mg) have been extensively documented. However, the role of Mg in medicine started as far back as the 17th century covering a large span of the chemical and pharmacologic fields of knowledge. The recognition by Grew, in 1695, that magnesium sulfate is one of the essential constituents of Epsom salt, may be considered the entry of magnesium into medicine.[1]

The Mg ion, the second most abundant intracellular cation after potassium, plays essential roles in the structure and function of the human body; it is an essential cofactor in a wide variety of biologic processes, including protein synthesis, nucleic acid synthesis and stability, neuromuscular excitability and the conduction of neural impulses, stimulus–contraction coupling and muscle contraction.[2,3] In particular, it is now clear that the Mg ion, although not directly involved in the biochemical process of contraction, modulates vascular smooth muscle tone and contractility by affecting calcium ion concentrations and the availability of the calcium ion at critical sites. Magnesium serves to actively promote relaxation, offset calcium-related excitation–contraction coupling and decrease cellular responsiveness to depolarizing stimuli by different mechanisms: (i) stimulating Ca-dependent K channels; (ii) competitively inhibiting calcium (Ca) binding to calmodulin; (iii) stimulating both plasma membrane and sarcoplasmic reticulum Ca ATPases; and (iv) activating the membrane Na,K-ATPase pump.[2,4] Magnesium is a critical modulator of the tension with which the contractile apparatus of striated muscle responds to the prevailing ionized calcium concentration; the Mg complex with adenosine triphosphate (MgATP) is the substrate for the enzymatic reactions that underlie the sliding filament mechanism for myofibrillar contraction and relaxation. Magnesium also participates in many of the most vital oxidative, synthetic, and transport processes of the muscle cell. Consistent with the above, calcium-induced contraction in muscles is not only sensitive to changes in Mg concentration, but direct reduction of extracellular Mg raises smooth muscle Ca content, while conversely, elevations in Mg concentrations reciprocally lower calcium content in muscle.[5] Magnesium is an indispensable part of the activated MgATP complex, and it is required for adenosine triphosphate (ATP) synthesis in the mitochondria.[2,3] Cell signaling requires MgATP for the phosphorylation of proteins and the synthesis and activation of the cell-signaling molecule cyclic adenosine monophosphate (cAMP), involved in multiple biochemical processes. Magnesium is a necessary cofactor in over 300 enzymatic reactions; it is required for the activity of

Aging
http://dx.doi.org/10.1016/B978-0-12-405933-7.00016-0

TABLE 16.1 Physiological Role of Magnesium in the Body

ENZYME FUNCTION	
Enzyme Substrate • Kinases • ATPases/GTPases • Cyclases	*Direct Enzyme Activation* • Phosphofructokinase • Creatine kinase • 5-Phosphoribosyl-pyrophosphate synthetase • Adenylate cyclase • Na-K-ATPase
STRUCTURAL FUNCTION	
• Proteins • Polyribosomes • Nucleic acids	• Multiple enzyme complexes • Mitochondria
CALCIUM ANTAGONIST	
• Muscle contraction/relaxation • Neurotransmitter release • Action potential conduction in nodal tissue	
MEMBRANE FUNCTION	
• Cell adhesion • Transmembrane electrolyte flux	

TABLE 16.2 Characteristics of Ionic Magnesium

Element category	Alkaline earth metal
Atomic number	12
Atomic weight	24.305 g/mol
Valence	2
Normal serum	0.75-0.95 mmol/L 1.7-2.5 mg/dL
Total body content	24 g
Distribution in serum	– free ionized 70-80% – protein-bound 20-30% – complexed 1%

TABLE 16.3 Magnesium Equilibrium

- Main determinants are gastrointestinal absorption and renal excretion
- Healthy individuals need to ingest 0.2-0.4 mmol/kg of body weight/day to stay in balance
- Extracellular Mg is in equilibrium with that in the bone, kidneys, intestine, and other soft tissues
- Bone is the main reservoir of Mg
- Primary renal disorders cause hypomagnesemia by decreased tubular reabsorption of Mg
- Osmotic diuresis results in magnesium loss
- Drugs may cause magnesium wasting

all rate-limiting glycolytic enzymes, protein kinases, and, more generally, all ATP and phosphate transfer-associated enzymes. Magnesium may also bind the enzymes directly (i.e. RNA and DNA polymerases) and alter their structure.[2,3] Therefore, the availability of an adequate quantity of Mg may be considered a critical factor for normal cellular and body homeostasis and function (Table 16.1).

MAGNESIUM METABOLISM IN OLDER ADULTS

The adult human body contains approximately 24 g (1 mol) of Mg, of which about 65% resides in the mineral phase of bone, about 27% is found in muscle, and 6–7% is found in other cells. Extracellular Mg accounts for only 1% or so of total body Mg. The normal serum Mg concentration ranges from 0.75 to 0.95 mmol/L (1.7–2.5 mg/dL or 1.5–1.9 meq/L) and is tightly controlled and maintained in this range. In the serum, about 70–80% of Mg exists in the biologically active ionized (free) form, while the remainder is bound to circulating proteins (e.g. albumin) (20–30%) or complexed to anions (e.g. bicarbonate, phosphate) (1%). The cytosolic Mg concentration ranges between 0.5 and 1.0 mmol/L in various types of cell[2,6] (Tables 16.2 and 16.3). The magnesium status in the body is determined mainly by absorption through the gastrointestinal tract, the requirement of different tissues (i.e. skeletal and cardiac muscle uptake and usage), and renal excretion. The small intestine is the main site for Mg absorption. Healthy individuals need to ingest 0.15–0.2 mmol/kg/day to stay in balance. The kidney exerts the most predominant impact in controlling body Mg status. Diuretics, frequently used in older populations, may also modify renal Mg handling, reducing Mg reabsorption.[2,6]

Bone is the main storage location of Mg, which cannot be quickly exchanged with the Mg in extracellular fluids; more prompt requirements for Mg are satisfied from the Mg stored in the intracellular compartment. There is wide-ranging variability in Mg intake, absorption, conservation, and excretion. Alterations in Mg metabolism that have been associated with aging include a reduction in Mg intake and intestinal absorption, and an increase in Mg urinary and fecal excretion (Figure 16.1). Although no known hormonal factor is specifically involved in the regulation of Mg metabolism, several hormones are recognized to have an effect on Mg balance and transport. Among them, parathyroid hormone (PTH), calcitonin, catecholamines, and insulin have a major role.[2,6]

One of the main reasons why Mg metabolism has not become a greater focus of routine attention in clinical practice has been the difficulties in obtaining an easily available, accurate, and reproducible measurement of Mg status in the body. Total serum Mg concentrations (MgT) do not always reflect accurately the body Mg homeostasis; MgT has proved useful and it has been extensively utilized in epidemiologic studies, but it may not be helpful for the detection of a subclinical Mg deficit on an individual basis. MgT does not change with age, while intracellular free Mg tends to decrease with age.[6]

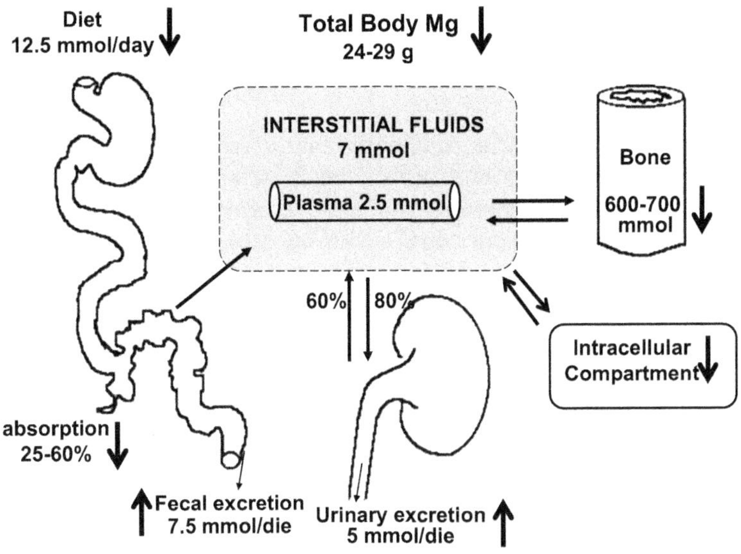

FIGURE 16.1 Magnesium homeostasis with age (arrows indicate possible sites of alteration with aging).

TABLE 16.4 Main Mechanisms of Magnesium Deficit with Aging

PRIMARY MAGNESIUM DEFICIT
• Inadequate Mg dietary intake • Reduced efficiency of Mg absorption (associated with reduced vitamin D levels) • Increased urinary excretion of Mg (associated with age-dependent reduction of kidney function and of Mg tubular reabsorption)
SECONDARY MAGNESIUM DEFICIENCY
• Associated with age-related diseases and comorbidities • Increased urinary Mg loss secondary to drugs (i.e. diuretics) frequently used in older persons

However, more precise and expensive techniques for measuring intracellular free Mg concentrations remain mainly at the research level, while available measurements of serum ionized free Mg, the active part of serum Mg, may still have some technical flaws.[2,6]

Aging represents a major risk factor for Mg deficit. The total body Mg content and intracellular Mg tend to decrease with age.[2] The most common mechanisms which may cause Mg deficits with aging are summarized in Table 16.4;[6] they include primary Mg deficit (inadequate Mg nutrient intake, reduced efficiency of Mg absorption, and increased urinary excretion of Mg), and secondary Mg deficiency (associated with drug use or pathologic conditions associated with aging).

MAGNESIUM, MUSCULAR PERFORMANCE, AND AGING MUSCLE

An adequate cellular Mg concentration seems necessary for maintaining optimal muscle performance and exercise tolerance. Magnesium status strongly affects muscle

TABLE 16.5 Mechanisms by Which Low Magnesium Status May Affect Muscle, Increasing Oxidative Stress

• Energetic metabolism (oxygen uptake and energy production) • Transmembrane transport • Muscle contraction and relaxation (by means of MgATP and the release of calcium)

performance, probably due to the key role of magnesium in energetic metabolism, transmembrane transport, and muscle contraction and relaxation.[7] Magnesium is involved in numerous processes that affect muscle function, including oxygen uptake and energy production. Myofibrillar contraction and relaxation is strictly connected to Mg and MgATP, as is the uptake and release of Ca from the sarcotubules. Thus, intracellular concentrations of Mg, even within narrow limits, can profoundly affect the contractile performance of the muscle cell (Table 16.5).

Magnesium supplementation (up to 8 mg/kg daily) was shown to enhance muscle strength (20% increase of peak knee-extension torque) in young individuals,[8] improve endurance exercise performance, and decrease oxygen use during submaximal exercise.[9] Exercise induces a redistribution of Mg in the body to accommodate metabolic needs, accounting for changes in Mg concentration in extracellular fluids, blood cellular components, myocytes, and adipocytes. It has been suggested that Mg deficiency may impair exercise performance and increase the oxidative stress linked to strenuous exercise.[10] Conversely, exercise increases urinary and sweat losses of magnesium, and thus the Mg requirement is increased in an individual performing exercise. A low Mg intake in athletes may result in a Mg-deficient status, and may thus reduce their performance. Magnesium supplementation, or increased dietary intake, is beneficial to physically active persons with a low or deficient Mg status to enhance strength and to improve exercise performance.[8,9] There is less evidence for Mg supplementation in physically active persons with an adequate Mg status for enhancing physical performance.[10]

A large portion of the energy used for physiologic functions in humans is produced by mitochondria through the movement of electrons in the respiratory chain. Magnesium in the mitochondria accounts for one third of total cellular Mg; it is present as a complex with ATP and as a component of membranes and nucleic acids. Magnesium is critical for basic mitochondrial functions, including ATP synthesis, electron transport chain complex subunits, and oxygen detoxification.[2,6] As discussed below, Mg seems fundamental for controlling oxidative stress and for the preservation of normally functioning muscle mitochondria. An inadequate availability of Mg may lead to reduced mitochondrial efficiency and increased production of reactive oxygen species (ROS), and subsequent structural and functional impairment of proteins that may lead to the decline in skeletal muscle mitochondrial function associated with aging in humans.[11]

In rats, Mg depletion is associated with structural damage to muscle cells, mitochondrial swelling and an altered ultrastructure that was shown to be associated with an increased production of ROS, lipid and protein damage, and impaired intracellular calcium homeostasis. Both Mg deficit and exercise contribute to the occurrence of oxidative stress.[12] Although the importance of Mg as a determinant of muscle performance in young athletes is well established, its role in maintaining muscle integrity and function in older adults is largely unknown.

Lukaski and Nielsen[7] examined the effects of different levels of dietary Mg on biochemical measures and physiologic responses in postmenopausal women (45 to 71 years old) during submaximal exercise, in relation to changes in erythrocyte and skeletal muscle Mg concentrations. The women consumed diets containing conventional foods with varying Mg content, totaling 112 mg/8.4 MJ (2000 kcal) supplemented with 200 mg of Mg daily for 35 days (control), then 112 mg/8.4 MJ for 93 days (depletion) followed by 112 mg/8.4 MJ supplemented with 200 mg of Mg daily for 49 days (repletion), in a depletion–repletion experiment. The concentration of Mg in erythrocytes, Mg retention, and skeletal muscle Mg content decreased when dietary Mg was restricted. Peak oxygen uptake, total and cumulative net oxygen uptake – determined by using indirect calorimetry – and peak heart rate increased during standardized submaximal work with restricted compared with adequate dietary Mg.[7] These findings indicate that dietary Mg depletion, in otherwise healthy women, results in a significant increase in energy needs and adversely affects cardiovascular function during sub-maximal work. These data, showing a decreased work economy and functional impairment in older women on restricted dietary Mg, extend the knowledge of magnesium status in young athletes to older women not participating in intense physical activity.

Older age is frequently characterized by sarcopenia, defined as a loss of skeletal muscle mass, quality, and function.[13] Sarcopenia of aging is almost a universal phenomenon, occurring in a wide range of species, from nematodes to flies, rodents, non-human primates, and humans. Muscle changes in humans start in the fourth decade of life. Sarcopenia in the aging population is a strong independent risk factor for disability and mortality. What gererally occurs with aging is a decrease in the rate of synthesis of several muscle proteins, specifically myosin heavy chain and mitochondrial proteins. The underlying causes of the reduction in mitochondrial activity and ATP production seem to be related to a decrease in mitochondrial DNA and in messenger RNA. Reduced ATP production has been suggested to be the basis of reduced muscle protein turnover, which requires energy[13] (Table 16.6). Magnesium depletion, because of the fundamental role of Mg in the MgATP complex, may play a role in this phenomenon, causing structural damage in muscle cells through increased oxidative stress (see below) and impaired intracellular calcium homeostasis. Both aerobic exercise and resistance exercise enhance muscle protein synthesis and mitochondrial activity. We do not know the role of Mg in aging-related muscle mitochondrial dysfunction, or the role of reduced physical activity. Because Mg status is strictly related to muscle ATP, and both Mg deficiency and sarcopenia tend to be more prevalent at older ages, we hypothesized that poor Mg status contributes to late life sarcopenia. Using data from the InCHIANTI study, a well-characterized representative sample of older men and women, we found a significant, independent, and strong relationship between circulating Mg and muscle performance; this was consistent across several muscle performance parameters for both men and women,[14] suggesting a role for Mg status in helping to maintain muscle function with age. As discussed below, in addition to the role of Mg in energetic metabolism, at least two other mechanisms may help to explain these findings: (i) increased oxidative stress in the presence of Mg deficiency, and (ii) the proinflammatory effect of Mg depletion.

TABLE 16.6 Effects of Aging on the Mitochondria by Which Oxidative Stress May Be Increased

- Decreased number
- Morphology modifications
- Increased DNA mutations
- Decreased biogenesis
- Decreased autophagy
- Increased apoptosis

MAGNESIUM, EXERCISE, AND OXIDATIVE STRESS

Studies on cultured cells, in experimental animal models, and in humans, have consistently shown that Mg deficiency is associated with increased oxidative stress and decreased antioxidant defense. Several studies have shown convincingly that Mg deficiency results in increased production of oxygen-derived free radicals in various tissues, increased free-radical-elicited oxidative tissue damage, increased production of superoxide anion by inflammatory cells, decreased antioxidant enzyme expression and activity, decreased cellular and tissue antioxidant levels, and increased oxygen peroxide production.[6,15-23] Magnesium deficiency in rats causes decreased hepatic glutathione, superoxide dismutase, and vitamin E, together with increased lipid peroxidation and malondialdehyde (MDA) levels secondary to upregulated nicotinamide adenine dinucleotide phosphate (NADPH) oxidase activity.[24]

In stroke-prone spontaneously hypertensive rats, Mg deficiency results in marked increases in oxidative stress, superoxide accumulation, mitogen-activated protein kinase (MAPK) activation, and the development of hypertension.[25] Magnesium has antioxidant capacities and may prevent oxygen radical formation by scavenging free radicals and by inhibiting xanthine oxidase and NADPH oxidase.[26] Magnesium supplementation may reduce oxidative stress. In experimental diabetes, a decreased intracellular Mg level and increased Mg urinary excretion were associated with increased plasma MDA and decreased expression of hepatic superoxide dismutase and glutathione S-transferase; all of these effects were corrected by Mg supplementation.[27]

We have shown that antioxidant capacity and the action to reduce oxidative stress of glutathione and vitamin E (measured as decreased reduced/oxidized glutathione ratio, increased lipohydroperoxides, and increased thiobarbituric acid-reactive substances) are associated with an increase in the intracellular concentration of Mg.[28,29]

MAGNESIUM, OXIDATIVE STRESS, AND THE AGING MUSCLE: THE ROLE OF INFLAMMATION

A state of chronic inflammation has been proposed as one of the main causes of frailty in older persons.[30] Poor Mg status may trigger the development of a pro-inflammatory state both by causing excessive production and release of IL-1β and TNF-α,[31] and by elevating circulating concentrations of pro-inflammatory neuropeptides that trigger activation of low-grade chronic inflammation.[32]

Several interventional studies in animal models of Mg deficiency have provided convincing evidence of the link between Mg, inflammation, and oxidative stress. Oxidative mitochondrial decay linked to aging may itself favor hypomagnesemia. Magnesium deficiency inhibits endothelial growth and migration, and it stimulates the synthesis of nitric oxide and some inflammatory markers, thus directly modulating microvascular functions.[33,34] Experimental studies in rats have shown that Mg deficiency induces a chronic impairment of redox status associated with inflammation, which could contribute to increased oxidized lipids, and may promote hypertension and vascular disorders,[35] confirming the link between oxidative stress and inflammation.

Several other experimental studies have demonstrated that Mg deprivation determines a pro-inflammatory state, confirmed by the elevated circulating plasma level of several markers of inflammation, including interleukin-6 (IL-6), IL-1β, tumor necrosis factor (TNF)-α, vascular cell adhesion molecule (VCAM), and plasminogen activator inhibitor (PAI)-1; other markers of inflammation, such as increased circulating inflammatory cells and increased hepatic production and release of acute phase proteins (i.e. complement, alpha2-macroglobulin, alpha1-acid glycoprotein, fibrinogen), have been reported as well.[33-39] A direct mechanistic link has been reported for the association of low Mg and increased production and secretion of TNF-α and IL-1β in cultured alveolar macrophages.[40]

Because Mg acts as a natural calcium antagonist, the molecular basis for the inflammatory response may also be the result of a modulation of the intracellular calcium concentration. Potential mechanisms include the priming of phagocytic cells, the opening of calcium channels, activation of *N*-methyl-D-aspartate (NMDA) receptors, and/or the activation of nuclear factor-kappaB (NFκB).[2,6]

Studies in humans have confirmed the clinical relevance of the link between low serum Mg levels, as well as inadequate dietary Mg, and low-grade systemic inflammation.[41-44] Data from the Women's Health Study have shown that Mg intake is inversely related to systemic inflammation, measured by serum C-reactive protein concentrations, and to the prevalence of the metabolic syndrome in adult women.[41] Likewise, using the 1999–2002 NHANES databases, King et al found that dietary Mg intake was inversely related to C-reactive protein levels. Among the 70% of the population not taking supplements, Mg intake below the RDA was significantly associated with a higher risk of having elevated C-reactive protein.[42] Other clinical studies have suggested that serum Mg levels are also inversely related to oxidative stress and inflammation markers, including C-reactive protein and TNF-α concentrations.[43,44]

MAGNESIUM, IMMUNE RESPONSES, AND OXIDATIVE STRESS

There is evidence that Mg plays a role in the immune response. Magnesium has a strong relationship with the immune system, in both non-specific and specific immune responses, while magnesium deficit has been shown to be related to impaired cellular and humoral immune function. Magnesium deficiency leads to immunopathologic changes that are related to the initiation of a sequential inflammatory response. Magnesium is a cofactor for immunoglobulin (Ig) synthesis, immune cell adherence, antibody-dependent cytolysis, IgM lymphocyte binding, macrophage response to lymphokines, and T helper-β cell adherence.[45,46] In addition, Mg deficiency seems to accelerate thymus involution. One of the most remarkable results regarding the effects of Mg deficiency on the organism is the higher level of apoptosis shown in thymuses from Mg-deficient rats as compared with controls.[47] Altered polymorphonuclear (PMN) cell number and function have been shown in rats fed a Mg-deficient diet for 8 days, together with the characteristic inflammatory response. In fact, an increased number of neutrophils, related to an increased activity of phagocytosis, has been found in Mg-deficient rats compared with control rats.[48] Clinical signs of inflammation, splenomegaly, and leukocytosis, have been reported as well in rats given a Mg-deficient diet. A reduced proportion of CD8-T cells has been shown under these conditions, which has been related to a decreased concentration of interferon (IFN)-γ in spleen homogenates.[49] Several changes in gene expression, including upregulation of TNF receptor 1 and IL-1 receptor type I, have also been demonstrated in rat thymocytes in early Mg deficiency.[21] A recent study discovered a key role for Mg in the T cell antigen receptor signaling pathway from the study of a novel primary immunodeficiency.[50] There are studies confirming the involvement of Mg in human cell apoptosis. Fas-induced β-cell apoptosis is Mg-dependent. Increased cytosolic free Mg levels are required for Fas molecule binding expression on the β-cell surface to initiate multiple signaling pathways that result in apoptotic cell death.[51] *In vitro* incubation of granulocytes in media with different Mg composition resulted in significant changes in chemotactic peptide-induced calcium transients.[52]

Because physical exercise may deplete Mg, many aspects of immune function can be depressed temporarily by either severe exercise or a longer period of excessive training.[53] Although the disturbance is usually quite transient, it has been suggested that it may be sufficient to allow clinical episodes of infection, particularly upper respiratory tract infections.[53] The risk may be higher in older persons who already have a reduced immune capacity and a consequent propensity to infections. However, regular and moderate exercise has been reported to improve the ability of the immune system to protect the host from infection.[53] Strenous exercise has been associated with an acute increase in oxidative stress free radical (ROS) production.[13] ROS may contribute to the development of muscle fatigue *in situ*, but there is still a lack of convincing direct evidence that ROS are able to impair exercise performance *in vivo* in humans. It remains unclear whether exercise-induced oxidative stress modifications may have clinical significance. Conversely, the antioxidant actions of Mg[26] may have a role in muscle protection in the aging population. Even though this needs to be proven by specific trials, it is likely that maintaining a good quality Mg homeostasis throughout life, and in particular in older persons, may help to protect from muscle performance decline associated with aging.

CONSEQUENCES OF MAGNESIUM IMBALANCE WITH AGE

The consequences of Mg imbalance in elderly people related to defective membrane function, chronic inflammation, increased oxidative stress, and immune dysfunction may include an increased vulnerability to age-related diseases and particularly to sarcopenia and frailty.

Several studies have reported alterations in cell physiology with senescence features during Mg deficiency in different cell types. Magnesium is an essential cofactor in cell proliferation and differentiation and in all steps of nucleotide excision repair. Magnesium deficiency-related alterations may include reduced oxidative stress defense, cell cycle progression, culture growth, cellular viability,[54-57] activation of proto-oncogenes (i.e. c-fos, c-jun), and expression of transcription factors (e.g. NFκB).[58] Magnesium deficiency may accelerate cellular senescence in cultured human fibroblasts. Continuous culture of primary fibroblasts in Mg-deficient media resulted in the loss of replicative capacity with accelerated expression of senescence-associated biomarkers. A marked decrease in the replicative lifespan was seen compared to fibroblast populations cultured in standard Mg media conditions. Human fibroblast populations cultured in Mg-deficient conditions also showed an increased senescence-associated β-galactosidase activity. Additionally, activation of cellular aging (p53 and pRb) pathways by Mg-deficient conditions also increased the expression of proteins associated with cellular senescence, including p16INK4a and p21WAF1. Telomere attrition was found to be accelerated in cellular populations from Mg-deficient cultures, suggesting that the long-term consequence of inadequate Mg availability in human fibroblast cultures is an accelerated cellular senescence.[59] Features of cellular senescence induced by low magnesium concentrations have also been reported in other cell types, e.g. endothelial cells.[60] Intracellular free Mg is a 'second messenger' for downstream events in apoptosis. There is increasing evidence from animal

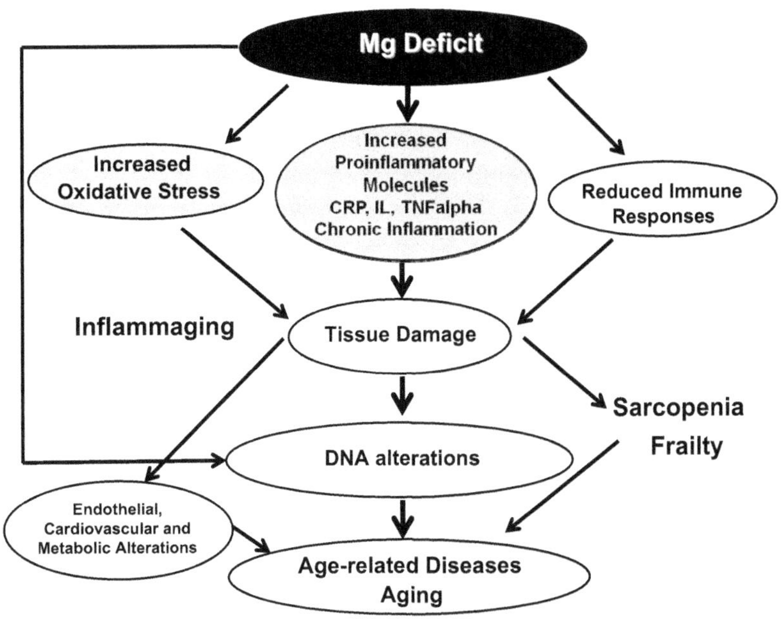

FIGURE 16.2 Overall hypothesis in which a chronic Mg deficit has been proposed as one of the physiopathologic links that may help to explain the interactions among inflammation, oxidative stress, and altered immune responses with sarcopenia of aging and with the aging process.

experiments and epidemiologic studies that Mg deficiency may decrease membrane integrity and membrane function, increasing the susceptibility to oxidative stress, cardiovascular disease, and accelerated aging. It is likely that many of these Mg effects may have relevance also to the aging of muscle and sarcopenia, as well as to many other age-related diseases.

CONCLUSIONS

Aging is very often associated with Mg inadequacy. Chronic Mg deficiency may result in excessive production of oxygen-derived free radicals. Oxidative stress and a chronic, low-grade inflammation have been proposed as an underlying condition linked to aging muscle, sarcopenia, and frailty. A chronic Mg deficit has been proposed as one of the physiopathologic mechanisms that may help to explain the interactions among inflammation and oxidative stress with sarcopenia, frailty, and a number of age-related diseases (Figs 16.2 and 16.3).

Magnesium status is strictly connected to muscle performance in older persons. A possible mechanism of this association is the effect of Mg concentrations on mitochondrial function in muscle, which may be particularly critical in the aging muscle.

Despite the physiologic importance of Mg, the multiple problems associated with its deficiency, and the ease of supplementation, inadequate Mg intake remains highly prevalent in various populations. Because Mg supplementation is inexpensive and in general well tolerated, it should be a key consideration in older subjects at particular risk for Mg deficiency. Although existing data confirm that the availability of an adequate quantity of Mg is a critical factor for normal cellular and body homeostasis, much remains to be done in this field to further clarify the potential role of Mg supplementation.

Even if the role of Mg supplementation as a possible intervention approach for delaying or preventing muscle aging deserves some consideration, a number of questions still need to be answered. What is the role of oxidative stress and inflammatory cytokines in mediating the adverse effect of Mg deficiency on muscle? Is low Mg a component of the frailty syndrome leading to sarcopenia of aging? Can Mg supplementation influence muscle strength/performance and cytokine concentrations in older persons? The possible role of Mg supplementation in aging-associated conditions remains unclear. At present, there are no data to support a potential role of dietary Mg supplementation as a possible health strategy in the aging population. Very few open and double-blind studies on the effects of the treatment of Mg deficiencies in geriatric populations have been done. The possibility that maintaining an optimal Mg balance throughout life might help in preventing or significantly retarding the oxidative stress and inflammation process, and the manifestations of chronic diseases, including sarcopenia and frailty, is a working hypothesis that needs to be tested in future prospective studies.

SUMMARY POINTS

- Mg is a key intracellular cation because it is an essential cofactor in a wide variety of biologic processes, including protein synthesis, nucleic acid synthesis and stability, neuromuscular excitability and conduction of neural impulses, stimulus–contraction coupling, and muscle contraction; it is also an antioxidant.

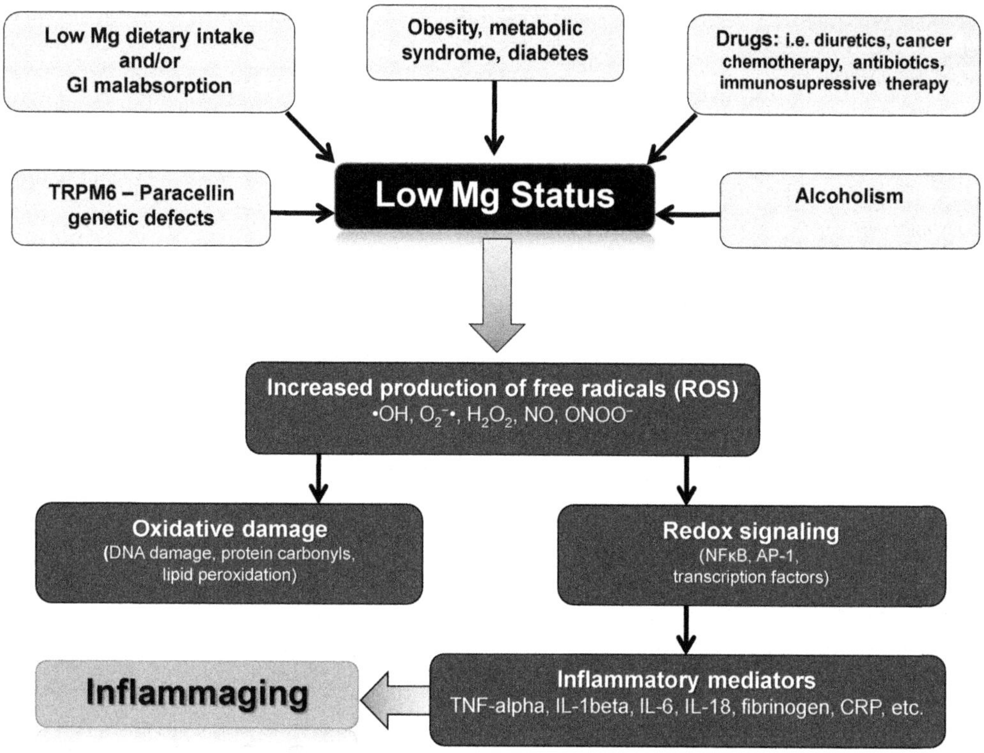

FIGURE 16.3 Low Mg status has different determinants that may converge in old age. Because Mg reduces the production of free radicals by the mitochondria as an antioxidant, its deficit may lead to the accumulation of oxidative damage and the release of inflammatory mediators. These events have been identified as 'inflammaging', that is, the low-grade chronic inflammation frequently seen in old age, which is associated with age-related conditions.

- Aging is frequently associated with a Mg deficit due to dietary reduced intake and/or absorption, increased renal wasting and/or reduced tubular reabsorption, age-related diseases, and/or drugs, leading to increased oxidative stress and chronic inflammation.
- Magnesium status affects muscle performance, probably due to the key role of magnesium in energetic metabolism, transmembrane transport, and muscle contraction and relaxation. Magnesium is integral to the function of ATP.
- Aging is associated with sarcopenia, defined as the loss of skeletal muscle mass, quality, and function. Sarcopenia of aging is a strong independent risk factor for disability and mortality.
- Oxidative stress and chronic, low-grade inflammation have been proposed as an underlying condition linked to aging muscle, sarcopeni, and frailty.
- Magnesium acts as an antioxidant against free radical damage of the mitochondria. Chronic Mg deficiency results in excessive production of oxygen-derived free radicals and low-grade inflammation.

References

1. Durlach J, Pages N, Bac P, et al. Magnesium research: from the beginnings to today. *Magnes Res* 2004;**17**:163–8.
2. Barbagallo M, Dominguez LJ. Magnesium metabolism in type 2 diabetes mellitus, metabolic syndrome and insulin resistance. *Arch Biochem Biophys* 2007;**458**:40–7.
3. Wolf FI, Cittadini A. Chemistry and biochemistry of magnesium. *Mol Aspects Med* 2003;**24**:3–9.
4. Altura BM, Altura BT. Magnesium ions and contractions on vascular smooth muscles: relationship to some vascular diseases. *Fed Proc* 1981;**40**:2672–9.
5. Turlapaty PDMV, Altura BM. Extracellular magnesium ions control calcium exchange and content of vascular smooth muscle. *Eur J Pharmacol* 1978;**52**:421–3.
6. Barbagallo M, Belvedere M, Dominguez LJ. Magnesium homeostasis and aging. *Magnes Res* 2009;**22**:235–46.
7. Lukaski HC, Nielsen FH. Dietary magnesium depletion affects metabolic responses during submaximal exercise in postmenopausal women. *J Nutr* 2003;**132**:930–5.
8. Brilla LR, Haley TF. Effect of magnesium supplementation on strength training in humans. *J Am Coll Nutr* 1992;**11**:326–9.
9. Brilla LR, Gunther KB. Effect of Mg supplementation on exercise time to exhaustion. *Med Exerc Nutr Health* 1995;**4**:230.
10. Nielsen FH, Lukaski HC. Update on the relationship between magnesium and exercise. *Magnes Res* 2006;**19**:180–9.
11. Short KR, Bigelow ML, Kahl J, et al. Decline in skeletal muscle mitochondrial function with aging in humans. *PNAS* 2005;**102**:5618–23.
12. Rock E, Astier C, Lab C, et al. Dietary magnesium deficiency in rats enhances free radical production in skeletal muscle. *J Nutr* 1995;**125**:1205–10.
13. Nair SK. Aging muscle. *Am J Clin Nutr* 2005;**81**:953–63.
14. Dominguez LJ, Barbagallo M, Lauretani F, et al. Magnesium and muscle performance in older persons: the InCHIANTI study. *Am J Clin Nutr* 2006;**84**:419–26.

15. Barbagallo M, Dominguez LJ. Magnesium and aging. *Curr Pharmaceut Design* 2010;**16**:832–9.
16. Dickens BF, Weglicki WB, Li YS, Mak IT. Magnesium deficiency in vitro enhances free radical-induced intracellular oxidation and cytotoxicity in endothelial cells. *FEBS Lett* 1992;**311**: 187–91.
17. Freedman AM, Mak IT, Stafford RE, et al. Erythrocytes from magnesium-deficient hamsters display an enhanced susceptibility to oxidative stress. *Am J Physiol* 1992;**262**:C1371–5.
18. Weglicki WB, Mak IT, Kramer JH, et al. Role of free radicals and substance P in magnesium deficiency. *Cardiovasc Res* 1996;**31**: 677–82.
19. Mazur A, Maier JA, Rock E, et al. Magnesium and the inflammatory response: potential physiopathological implications. *Arch Biochem Biophys* 2007;**458**:48–56.
20. Blache D, Devaux S, Joubert O, et al. Long-term moderate magnesium-deficient diet shows relationships between blood pressure, inflammation and oxidant stress defense in aging rats. *Free Rad Biol Med* 2006;**41**:277–84.
21. Petrault I, Zimowska W, Mathieu J, et al. Changes in gene expression in rat thymocytes identified by cDNA array support the occurrence of oxidative stress in early magnesium deficiency. *Biochim Biophys Acta* 2002;**1586**:92–8.
22. Rayssiguier Y, Durlach J, Gueux E, et al. Magnesium and ageing. I. Experimental data: importance of oxidative damage. *Magnes Res* 1993;**6**:369–78.
23. Yang Y, Wu Z, Chen Y, et al. Magnesium deficiency enhances hydrogen peroxide production and oxidative damage in chick embryo hepatocyte in vitro. *Biometals* 2006;**19**:71–81.
24. Calviello G, Ricci P, Lauro L, et al. Mg deficiency induces mineral content changes and oxidative stress in rats. *Biochem Mol Biol Int* 1994;**32**:903–11.
25. Touyz RM, Pu Q, He G, et al. Effects of low dietary magnesium intake on development of hypertension in stroke-prone spontaneously hypertensive rats: role of reactive oxygen species. *J Hypertens* 2002;**20**:2221–32.
26. Afanas'ev IB, Suslova TB, Cheremisina ZP, et al. Study of antioxidant properties of metal aspartates. *Analyst* 1995;**120**: 859–62.
27. Hans CP, Chaudhary DP, Bansal DD. Effect of magnesium supplementation on oxidative stress in alloxanic diabetic rats. *Magnes Res* 2003;**16**:13–9.
28. Barbagallo M, Dominguez LJ, Tagliamonte MR, et al. Effects of glutathione on red blood cell intracellular magnesium: relation to glucose metabolism. *Hypertension* 1999;**34**:76–82.
29. Barbagallo M, Dominguez LJ, Tagliamonte MR, et al. Effects of vitamin E and glutathione on glucose metabolism: role of magnesium. *Hypertension* 1999a;**34**:1002–6.
30. Ferrucci L, Guralnik JM. Inflammation, hormones, and body composition at a crossroad. *Am J Med* 2003;**115**:501–2.
31. Weglicki WB, Dickens BF, Wagner TL, et al. Immunoregulation by neuropeptides in magnesium deficiency: ex vivo effect of enhanced substance P production on circulation T lymphocytes from magnesium-deficient mice. *Magnes Res* 1996;**9**:3–11.
32. Kramer JH, Mak IT, Phillips TM, Weglicki WB. Dietary magnesium intake influences circulating pro-inflammatory neuropeptide levels and loss of myocardial tolerance to postischemic stress. *Exp Biol Med* 2003;**228**:665–73.
33. Maier JAM, Malpuech-Brugère C, Zimowska W, et al. Low magnesium promotes endothelial cell dysfunction: implications for atherosclerosis, inflammation and thrombosis. *Biochim Biophys Acta* 2004;**1689**:13–21.
34. Bernardini D, Nasulewicz A, Mazur A, Maier JAM. Magnesium and microvascular endothelial cells: a role in inflammation and angiogenesis. *Front Biosci* 2005;**10**:1177–82.
35. Barbagallo M, Dominguez LJ, Resnick LM. Magnesium metabolism in hypertension and type 2 diabetes mellitus. *Am J Clin Ther* 2007;**14**:375–85.
36. Malpuech-Brugere C, Nowacki W, Daveau M, et al. Inflammatory response following acute magnesium deficiency in the rat. *Biochim Biophys Acta* 2000;**1501**:91–8.
37. Weglicki WB, Phillips TM. Pathobiology of magnesium deficiency: a cytokine/neurogenic inflammation hypothesis. *Am J Physiol* 1992;**263**:R734–7.
38. Kurantsin-Mills J, Cassidy MM, Stafford RE, Weglicki WB. Marked alterations in circulating inflammatory cells during cardiomyopathy development in a magnesium-deficient rat model. *Br J Nutr* 1997;**78**:845–55.
39. Bussiere FI, Tridon A, Zimowska W, et al. Increase in complement component C3 is an early response to experimental magnesium deficiency in rats. *Life Sci* 2003;**73**:499–507.
40. Oono H, Nakagawa M, Miyamoto A, et al. Mechanisms underlying the enhanced elevation of IL-1beta and TNF-alpha mRNA levels following endotoxin challenge in rat alveolar macrophages cultured with low-Mg^{2+} medium. *Magnes Res* 2002;**15**:153–60.
41. Song Y, Ridker PM, Manson JE, et al. Magnesium intake, C-reactive protein, and the prevalence of metabolic syndrome in middle-aged and older U.S. women. *Diabetes Care* 2005;**28**:1438–44.
42. King DE, Mainous 3rd AC, Geesey ME, Woolson RF. Dietary magnesium and C-reactive protein levels. *J Am Coll Nutr* 2005;**24**:166–71.
43. Guerrero-Romero F, Rodriguez-Moran M. Hypomagnesemia, oxidative stress, inflammation, and metabolic syndrome. *Diabetes Metab Res Rev* 2006;**22**:471–6.
44. Rodriguez-Moran M, Guerrero-Romero F. Elevated concentrations of TNF-alpha are related to low serum magnesium levels in obese subjects. *Magnes Res* 2004;**17**:189–96.
45. Galland L. Magnesium and immune function: an overview. *Magnesium* 1988;**7**:290–9.
46. Tam M, Gomez S, Gonzalez-Gross M, Marcos M. Possible roles of magnesium on the immune system. *Eur J Clin Nutr* 2003;**57**:1193–7.
47. Malpuech-Brugere C, Nowacki W, Gueux E, et al. Accelerated thymus involution in magnesium-deficient rats is related to enhanced apoptosis and sensitivity to oxidative stress. *Br J Nutr* 1999;**81**:405–11.
48. Bussiere FI, Gueux E, Rock E, et al. Increased phagocytosis and production of reactive oxygen species by neutrophils during magnesium deficiency in rats and inhibition by high magnesium concentration. *Br J Nutr* 2002;**87**:107–13.
49. Malpuech-Brugere C, Kuryszko J, Nowacki W, et al. Early morphological and immunological alterations in the spleen during magnesium deficiency in the rat. *Magnesium Res* 1998;**11**:161–9.
50. Li FY, Chaigne-Delalande B, Kanellopoulou C, et al. Signaling role for Mg^{2+} revealed by immunodeficiency due to loss of MagT1. *Nature* 2011;**475**:471–6.
51. Chien MM, Zahradka KE, Newell MK, Freed JH. Fas-induced B cell apoptosis requires an increase in free cytosolic magnesium as an early event. *J Biol Chem* 1999;**274**:7059–66.
52. Mooren FC, Golf SW, Völker K. Effect of magnesium on granulocyte function and on the exercise inflammatory response. *Magnes Res* 2003;**16**:49–58.
53. Laires MJ, Monteiro C. Exercise, magnesium and immune function. *Magnes Res* 2008;**21**:92–6.
54. McKeehan WL, Ham RG. Calcium and magnesium ions and the regulation of multiplication in normal and transformed cells. *Nature* 1978;**275**:756–8.
55. Ames BN, Atamna H, Killilea DW. Mineral and vitamin deficiencies can accelerate the mitochondrial decay of aging. *Mol Aspects Med* 2005;**26**:363–78.
56. Hartwig A. Role of magnesium in genomic stability. *Mutat Res* 2001;**475**:113–21.

57. Sgambato A, Wolf FI, Faraglia B, Cittadini A. Magnesium depletion causes growth inhibition, reduced expression of cyclin D1, and increased expression of P27Kip1 in normal but not in transformed mammary epithelial cells. *J Cell Physiol* 1999;**180**:245–54.
58. Altura BM, Kostellow AB, Zhang A, et al. Expression of the nuclear factor-kappaB and proto-oncogenes c-fos and c-jun are induced by low extracellular Mg^{2+} in aortic and cerebral vascular smooth muscle cells: possible links to hypertension, atherogenesis, and stroke. *Am J Hypertens* 2003;**16**:701–7.
59. Killilea DW, Ames BN. Magnesium deficiency accelerates cellular senescence in cultured human fibroblasts. *Proc Natl Acad Sci USA* 2008;**105**:5768–73.
60. Ferre S, Mazur A, Maier JAM. Low magnesium induces senescent features in cultured human endothelial cells. *Magnes Res* 2007;**20**:66–71.

CHAPTER

17

Late-Life Depression and Antioxidant Supplements

Joanna Rybka

Department of Biochemistry, Collegium Medicum UMK in Bydgoszcz, Poland, and Life4Science Foundation, Bydgoszcz, Poland

Kornelia Kedziora-Kornatowska

Department and Clinic of Geriatrics, Collegium Medicum UMK in Bydgoszcz, Poland

Floor van Heesch

Division of Pharmacology, Utrecht Institute for Pharmaceutical Sciences (UIPS), Faculty of Science, Utrecht University, Utrecht, The Netherlands

Jozef Kedziora

Department of Biochemistry, Collegium Medicum UMK in Bydgoszcz, Poland

List of Abbreviations

ACTH adrenocorticotropic hormone
AD Alzheimer's disease
AGEs advanced glycation end-products
APP amyloid precursor protein
BBB blood–brain barrier
BDNF brain-derived neurotrophic factor
CaMKII calcium/calmodulin kinase II
CAT catalase
CNS central nervous system
COX2 cyclooxygenase
FEHP flavonoid-rich extract of *Hypericum perforatum* L.
GPx glutathione peroxidase
H_2O_2 hydrogen peroxide
iNOS inducible nitric oxide synthase
MAO-A monoamine oxidase A
MAPKs mitogen-activated protein kinases
NADPH oxidase nicotinamide adenine dinucleotide phosphate-oxidase
NF-κB nuclear factor kappa-light-chain-enhancer of activated B cells
NO nitric oxide
NOS nitric oxide synthase
Nrf2 nuclear factor (erythroid-derived 2)-like 2
PKC protein kinase C
RAGE receptor for advanced glycation products
RNS reactive nitrogen species
ROS reactive oxygen species
SJW St John's wort
SOD superoxide dismutase

INTRODUCTION

Late-life depression has been defined as the occurrence of depressive symptoms in people over the age of 65 (International Statistical Classification of Diseases and Related Health Problems, 10th Revision, World Health Organization). Depression is the most common mental health problem in the elderly. Prevalence studies suggest that 14–20% of elderly people living in the community experience depressive symptoms, and because of the demographic trend of an aging population, it is expected that the number of seniors suffering from depression will increase. There is also a body of evidence indicating that the prevalence of depressive symptoms increases with age; however, the most depressed people remain underdiagnosed and do not receive medical treatment for their psychiatric conditions.[1]

Aging
http://dx.doi.org/10.1016/B978-0-12-405933-7.00017-2

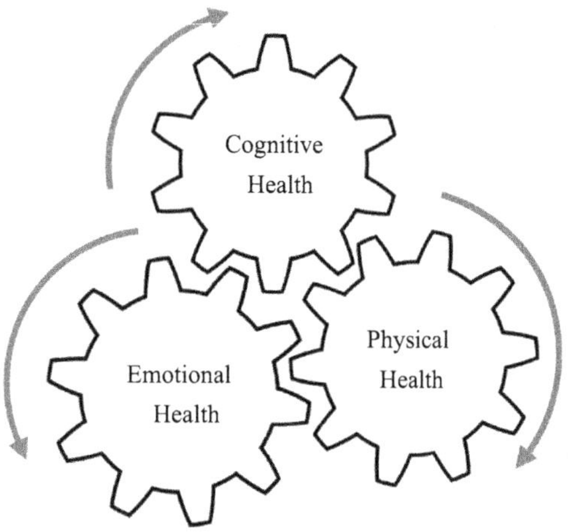

FIGURE 17.1 Triad of healthy aging.

Depression significantly affects health and shortens the life span owing to the increased risk of comorbidity with other chronic and life-threatening conditions, including other mental disorders, cardiovascular events, cancer, respiratory diseases, neuropathologies and metabolic diseases. Accordingly, the presence of one or more chronic physical diseases increases significantly the risk of comorbid depression. Comorbidity leads to worsening health compared with depression alone, also compared with chronic diseases alone, and compared with any combination of chronic diseases without depression.[2] Furthermore, individuals with late-onset depression (i.e. first depressive episode occurring at 60 years or older) exhibit more impairment in overall cognitive deficits and in memory and executive dysfunction than those with early-onset depression. Thus, healthy aging is maintained by the triad of physical, emotional and cognitive functions, and while all of these functions are required for good health, when any one of them worsens it negatively affects the others and the whole-body homeostasis (Fig. 17.1).

The pathobiology of aging is complex and includes many interrelated pathways and processes. One of the most recognized and explored is the free radical theory of aging which claims that aging is due at least partly to the accumulated damage from harmful reactive oxygen species (ROS).[3] Moreover, ROS play a role in a 'balance of power' and are involved in different molecular pathways controlling life span and directing the cell toward life or death.[4] Furthermore, inflammation, metabolic stress and circadian rhythm disturbances, which, together, have been implicated in aging, depression and other chronic diseases of later life, are related to the major physiologic redox systems, as presented in Figure 17.2.

ANTIOXIDANTS AND NEUROPSYCHOLOGIC FUNCTIONS IN THE ELDERLY: EVIDENCE FOR ANTIDEPRESSANT ACTIVITY

Late-Life Depression: Rationale for Antioxidants in Prevention and Treatment

Conceptualizations of the underlying neurobiology of major depression have changed their focus from dysfunctions of neurotransmission to dysfunctions of neurogenesis and neuroprotection. Stress, neuroinflammation, monoamine neurotransmitters, dysfunctional insulin regulation, oxidative stress, and alterations in neurotrophic factors possibly contribute to the development of depression.[5] Disruptions of the molecular mechanisms underlying the circadian clock system – including abnormalities in the expression of clock genes and hyperactivity of the hypothalamus–pituitary–adrenal axis, resulting in abnormal circadian rhythms – are yet another important feature of depression.[6] The data show that late-life depression may incorporate all of these factors. The origin of depression, specifically in later life, was presented by Blazer and Hybels, who revised the risk factors for late-life depression; the revised risk factors included biologic factors (genetic and heredity factors, neurotransmitter dysfunction, endocrine changes, vascular disorders and medical comorbidities), psychologic factors (personality attributes, neuroticism, cognitive distortions and the lack of emotional control and self-efficacy) and social factors (stressful life events, bereavement, chronic stress or strain, socio-economic disadvantage and impaired social support).[7]

Multiple comorbidities of physical, emotional and cognitive impairments in the elderly[8] – as well as a common pathologic mechanism in depression, cerebral aging and malignancies progressing with age – have implications for treatment methods in elderly patients. A classic pharmacologic approach provides single-disease solutions which result in multi-drug use. Polypharmacy increases the risk for adverse drug reactions, reduced drug efficiencies, drug–nutrient interactions, and deficiency in essential nutrients – further increasing the risk for health problems in elderly people.[9] The brain consumes an immense amount of energy relative to the rest of the body, and energy and nutrient intake influences brain chemistry and neural function. Disturbances in energy homeostasis have been linked to the pathobiology of several mental diseases, and so dietary management is becoming a realistic strategy to prevent and treat psychiatric disorders.[10]

There is a substantial amount of evidence showing the health benefits of dietary antioxidants and their

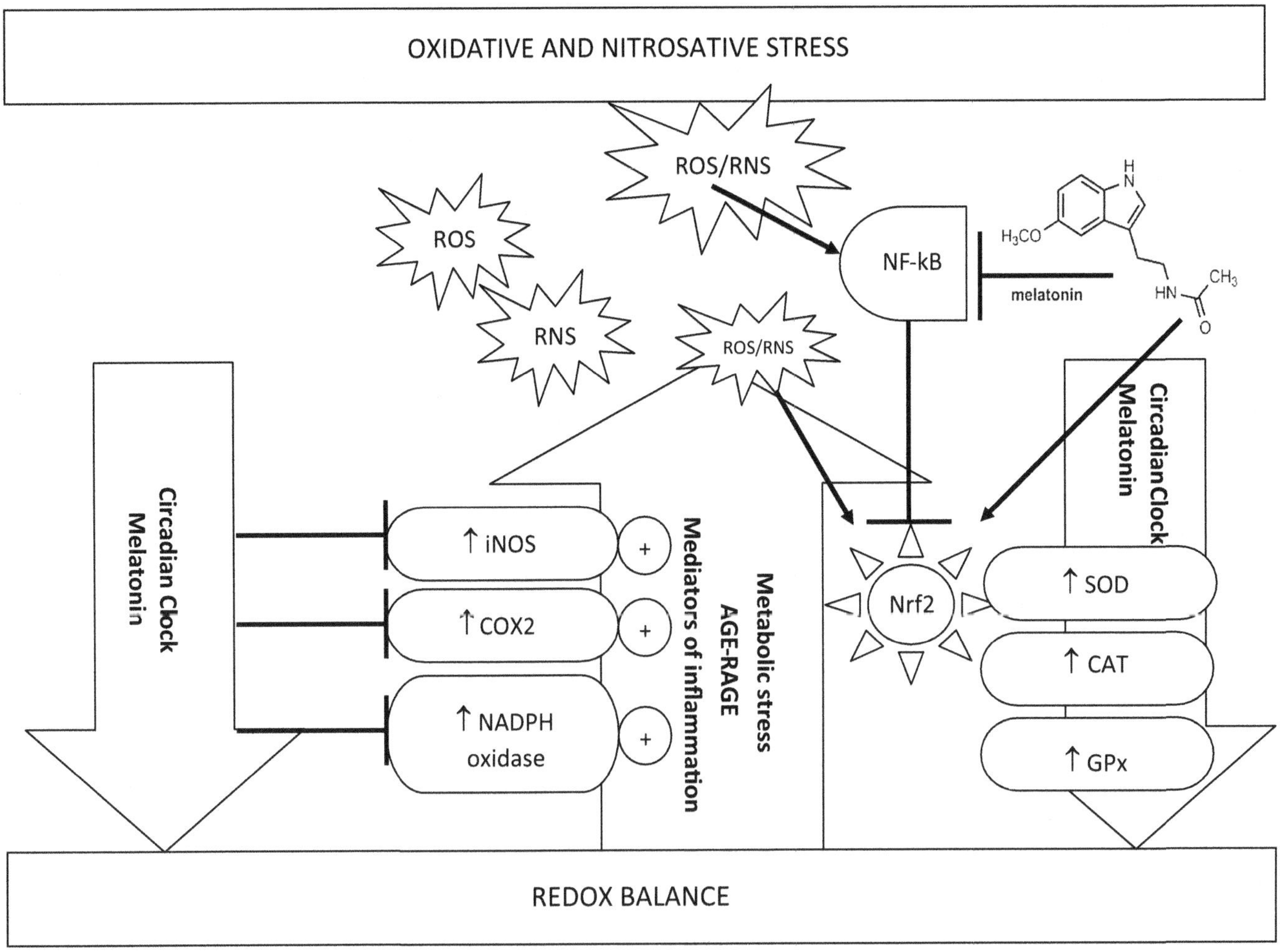

FIGURE 17.2 Physiologic pro- and antioxidants involved in biologic processes and molecular pathways associated with aging, depression and chronic disease. Aging, depression and other chronic diseases of late life are accompanied by inflammation, metabolic/glycemic dysregulation and disruption of the circadian clock system. Inflammatory mediators induce the cyclooxygenase (COX2) pathway, the production of prostaglandins and an increase in the secretion of cytokines, which can enhance the production of reactive oxygen species (ROS) and reactive nitrogen species (RNS). Another important source of ROS is the increased activity of NADPH oxidase in activated neutrophils. During inflammation, the inducible nitric oxide synthase pathway (iNOS) increases, resulting in overproduction of nitric oxide (NO). NO and its derivatives are able to damage cellular components such as proteins and DNA, and these effects are ascribed to nitrosative stress. Metabolic dysregulation, like inflammation, increases COX2, iNOS and NADPH oxidase pathways with the well recognized causative role of activated receptor for advanced glycation products (RAGE). RAGE has different ligands, including the products of non-enzymatic glycation and oxidation of proteins and lipids, the advanced glycation end-products (AGEs). NF-kappaB (NF-κB) is involved in transmission of inflammatory and RAGE signaling. NF-κB is a redox-sensitive transcription factor which induces intracellular events that concur to nuclear NF-E2-related factor 2 (Nrf2). Nrf2 coordinates the expression of genes required for free radical scavenging and maintenance of redox potential, which can help to reduce inflammatory potential and metabolic stress. One of the most potent physiologic antioxidants is melatonin, which contributes to appropriate internal circadian phasing – but also exerts direct radical scavenging effects and plays a role in the upregulation of antioxidant enzymes, including SOD, GPx and CAT as well as the downregulation of prooxidant enzymes, including COX2, iNOS and NADPH oxidase. These effects may also be mediated by the altered expression of Nrf2 and NF-κB. AGEs: advanced glycation end-products, CAT: catalase, COX2: cyclooxygenase, GPx: glutathione peroxidase, iNOS: inducible nitric oxide synthase, NADPH oxidase: nicotinamide adenine dinucleotide phosphate oxidase, NF-κB: nuclear factor kappa-light-chain-enhancer of activated B cells, NO: nitric oxide, Nrf2: nuclear factor (erythroid-derived 2)-like 2, RAGE: receptor for advanced glycation products, RNS: reactive nitrogen species, ROS: reactive oxygen species, SOD: superoxide dismutase.

potential to maintain whole-body homeostasis in the aging organism by improving physical, emotional and cognitive functions concurrently. At the same time, there are concerns that antioxidants, acting as any other treatment, may also have side effects and should be used with caution because there are no approved assessments and recommendations supporting the safety of antioxidant supplementation.[11] A study published in *Archives of General Psychiatry* found that people who follow a diet high in fruits, vegetables, nuts, whole grains, fish and unsaturated fat (common in olive and other plant oils) are up to 30% less likely to develop depression than those who typically consume meatier, dairy-heavy fare. The beneficial effects of certain food on mental health are ascribed, to a great extent, to antioxidant properties of dietary polyphenols.[12]

Antioxidant supplementations can decrease the vulnerability of the brain to increased oxidative stress, limiting associated cognitive and mental impairments during the aging process. Furthermore, classic antidepressants inhibit the cytochrome P450 enzyme-mediated metabolism of many medications, therefore co-administration of classic antidepressants may be associated with other altered drug activities. Safer alternatives are needed.[13] The effect of antidepressants on drug metabolism, and the deleterious effects of drug–drug interactions, can be especially dangerous in the elderly and in polymedicated patients; therefore, antioxidants – being the natural alternative to classic antidepressants – may provide additional benefits in late-life depression.

Antidepressant Actions of Selected Antioxidants

Antioxidant treatment has been shown to both prevent and treat depression, as demonstrated by *in vivo* studies in both humans and animals. There are also numerous experiments providing explanations of certain behavioral and physio-anatomic effects of antioxidants at the molecular level. Here, we overview the modulatory effects on depressive symptoms and cognitive function of the following antioxidants: hypericin, quercetin, *Ginkgo biloba*, epicatechin, resveratrol and curcumin. Whenever possible, we included a range of quality, randomized controlled trials, demonstrating the efficacy of antioxidants.

One of the best recognized natural remedies for depression – which has an effect similar to that of standard antidepressants – is an extract of St John's wort (*Hypericum perforatum*).[14] Preclinical data that are in favor of, or against, the use of *Hypericum perforatum* as an antidepressant have been revised by Crupi and colleagues.[15] A meta-analysis of 22 randomized controlled trials of St John's Wort (SJW) in depression has shown that SJW is significantly more effective than placebo (relative risk (RR) 1.98 (95% CI 1.49–2.62)) but not significantly different in efficacy from active antidepressants (RR 1.0 (0.90–1.11)).[16]

Many pharmacologic activities – including the antidepressant effect of *Hypericum* – appear to be attributable to hypericin and to the flavonoid constituents. The antioxidant properties of *Hypericum perforatum* were demonstrated by physicochemical examination,[17] and were confirmed in biologic systems.[18] A flavonoid-rich extract of *Hypericum perforatum* L. (FEHP) has been shown to inhibit the peroxidation of liposomes induced by both hydroxyl radical (generated by the iron–ascorbic acid system) and peroxyl radical; the antioxidant mechanism of FEHP was attributed to its free radical scavenging activity, its metal-chelation activity, and its ability to quench ROS.[17] In the rat brain, *H. perforatum* extract demonstrated antioxidant properties against elevated oxidative stress induced by an amnestic dose of scopolamine.[18]

The antidepressant effects of *H. perforatum* may involve the regulation of genes that control hypothalamic–pituitary–adrenal axis function and influence, at least in part, stress-induced effects on neuroplasticity and neurogenesis.[15] Possible mechanisms of the antidepressant action of *H. perforatum* are presented in Table 17.1.

Flavonoids are antioxidants with a profound effect on mental and general health. One of the major flavonoids is *quercetin*. Quercetin, which is present in some fruits and vegetables, has much stronger antioxidative and anticarcinogenic activities than the traditional antioxidants vitamin C and vitamin E.[19] The antidepressant capabilities of quercetin have been confirmed in laboratory animals based on behavioral studies and electropharmacogram data.[20]

The antidepressant and neuroprotective properties of quercetin have been related to the antioxidant properties of this flavonoid, which has been found to protect neurons from glutamate-induced oxidative damage by blocking hydrogen peroxide (H_2O_2) toxicity.[21] Possible mechanisms of antidepressant action of quercitin are presented in Table 17.1.

Flavonoids are also the major constituent of *Ginkgo biloba* extract, which is recognized as one of the most beneficial herbal treatments and one of the most potent antioxidants; it is also used as an antidepressant.[22] The antidepressant and anti-stress activities of *Ginkgo* have been demonstrated in a rodent model of depression in the behavioral despair test and learned helplessness.[23] The evidence from clinical studies has shown a positive effect of *Ginkgo biloba* on the subjective emotional well-being measured with Subjective Intensity Score Mood (SIS Mood) in elderly persons.[24] Meta-analysis also indicated that cognition, and the accompanying psychopathologic symptoms, are improved by treatment with *Ginkgo biloba* (Ginkgo) in Alzheimer's disease.[25]

The antidepressant action of Ginkgo has been related to its ability to increase levels of brain-derived neurotrophic factor (BDNF) in the hippocampus. Furthermore, 24-hour incubation of primary neuronal cells with Ginkgo extract resulted in glutamate-evoked activation of pCREB (c-AMP response element-binding).[26] Interestingly, increased pCREB may be an essential step in regulating the maturation and survival of newly generated neurons in the adult hippocampus.[27] Current evidence indicates that adult hippocampal neurogenesis may be preferentially involved in the regulation of emotions and

TABLE 17.1 Biologic Substances with Antioxidant Activity Known for Their Antidepressant Effect Together with Their Mechanism of Action

Antioxidant	Possible Mechanism of Antidepressant Action	References
Hypericin (St John's wort)	– antioxidative effects – modulation of adrenal axis – modulation of serotonergic neurotransmission – modulation of cholinergic neurotransmission	18,15,47
Quercetin	– antioxidative effects – modulation of serotonergic neurotransmission – MAO-A inhibition – modulation of BDNF-TrkB-PI3K/Akt pathways	21,48,49,50
Ginkgo biloba	– antioxidative effects – modulation of BDNF – glutamate-evoked activation of pCREB	26,51
Epicatechin	– antioxidative effects – modulation of calcium homeostasis – modulation of extracellular mitogen-activated protein kinases (MAPK) and protein kinase C (PKC) signaling pathways – modulation of cell death and cell survival genes and proteins – modulation of genes associated with mitochondrial function, such as Bcl-2 family members – modulation of amyloid precursor protein (APP) processing pathway – modulation of iron regulators and sensors encoding genes and proteins – suppressed production of proinflammatory cytokines – modulation of serotonergic neurotransmission – modulation of cholinergic neurotransmission – downregulation of markers of neurodegeneration in the hippocampus – reduction of serum corticosterone and ACTH levels in mice exposed to the FST (forced swimming test) – inhibition of the hypothalamic–pituitary–adrenal axis	52,53,30,54
Resveratrol	– antioxidative effects – modulation of serotonergic neurotransmission – modulation of noradrenergic neurotransmission – MAO-A inhibition	34
Curcumin	– antioxidative effects – reduction of serum corticosterone – modulation of glucocorticoid receptor mRNA – modulation of BDNF – modulation of serotonergic neurotransmission – modulation of noradrenergic neurotransmission – modulation of glutamatergic neurotransmission – protection of hippocampal neurons against corticosterone-induced toxicity – downregulation of calcium/calmodulin kinase II (CaMKII) – downregulation of glutamate receptor levels – MAO-A inhibition	37,55–59
Melatonin	– antioxidative effects – interactions with the hypothalamic–pituitary–adrenal axis and hormonal stress response – melatonin inhibits glucocorticoid receptor nuclear translocation – modulation of serotonergic neurotransmission	60–62

BDNF: Brain-derived neurotrophic factor, MAO-A: monoamine oxidase A, ACTH-adrenocorticotropic hormone.

memory, and may be required for some of the behavioral effects of antidepressants.[28] In fact, multiple cellular and molecular neuroprotective mechanisms – including attenuation of apoptosis and the direct inhibition of amyloid-beta aggregation – have been shown to underlie the neuroprotective effects of *Ginkgo biloba* extract.[29] Possible mechanisms of antidepressant action of *Ginkgo biloba* are presented in Table 17.1.

Green tea is another natural antioxidant with multiple health benefits; these benefits are attributed to polyphenolic compounds, among which epicatechin has gained much attention. Tea polyphenols act as antioxidants *in vitro* by scavenging ROS and RNS, and by chelating redox-active transition metal ions. They may also function indirectly as antioxidants through different pathways: inhibition of redox-sensitive transcription factors, nuclear factor-κB and activator protein-1; inhibition of 'prooxidant' enzymes (such as inducible nitric oxide synthase, lipoxygenases, cyclooxygenases and xanthine oxidase); and induction of phase II and antioxidant enzymes, such as glutathione *S*-transferases and superoxide dismutases.[30] There is a large body of literature confirming the broad spectrum of molecular effects of epicatechin, and many of these effects are directly related to neural function (Table 17.1). These may explain the positive effects of green tea on behavioral functions, including the antidepressive-like effect such as that observed in animals[31] and humans.[32] Possible mechanisms of the antidepressant action of epicatechin are presented in Table 17.1.

Resveratrol is another compound found to have an effective *in vitro* antioxidant and radical-scavenging activity.[33] It has been observed that, in mouse models, *trans*-resveratrol significantly decreases the immobility time in despair tests, suggesting antidepressant activity.[34] An antidepressant-like effect of *trans*-resveratrol has also been confirmed in chronic stress in a rat model.[35] In humans, beneficial neuroprotective effects of resveratrol have been observed following the consumption of red wine, which could attenuate clinical dementia produced by Alzheimer's disease (AD).[36] Possible mechanisms of antidepressant action of resveratrol are presented in Table 17.1.

Curcumin is a strong antioxidant that can intercept and neutralize potent prooxidants and carcinogens, both ROS (superoxide, peroxyl, hydroxyl radicals) and RNS (nitric oxide (NO), peroxynitrite).[37] Curcumin has been recognized for its antidepressant properties, and it has also been proposed for the treatment of neurodegenerative disease.[38] There is much evidence, from behavioral animal models, of depression showing positive effects from curcumin supplementation.[39] The potential of curcumin as an antidepressant is discussed by Kulkarni et al.[40]

Curcumin has therapeutic potential in the amelioration of a host of neurodegenerative ailments, as evidenced by its antioxidant, anti-inflammatory and anti-aggregation effects.[41] It has also been demonstrated that curcumin is capable of improving the blood–brain barrier function under ischemic conditions, and this beneficial effect might be reversed by inhibition of a heme oxygenase.[42] Possible mechanisms of antidepressant action of curcumin are presented in Table 17.1.

Melatonin is an important physiologic antioxidant with potential therapeutic applications in the elderly. Although melatoninergic compounds demonstrate antidepressant properties, and there are some data suggesting an antidepressant-like effect of melatonin in animal models of depression,[43] in major depressive disorders melatonin has no proven antidepressant action.[44] This suggests that supplementation with melatonin may not be a first-choice treatment in late-life depression. However, the taking of melatonin by elderly depressed patients may exert benefits on their psychologic and physical well-being related to circadian rhythms. Although melatonin does much more than help some people to sleep better, this particular effect of the pineal gland hormone should not be neglected, as obtaining sufficient amounts of quality sleep is an absolute necessity for good health. In the use of melatonin for treating disorders of the central nervous system (CNS), melatonin is defined by its ability to pass into the brain, as previously observed experimentally; however, the permeability of melatonin is lower than that of many drugs developed and targeted specifically to the brain.[45]

Some data have suggested that many other natural compounds have antidepressant activities, as demonstrated by their action on different neurotransmitter systems; a substantial body of these findings has been presented in a review discussing the potential of natural polyphenols in the management of major depression.[46] The data presented there suggest that many polyphenolic compounds are able to modulate brain neurotransmission, the main effector system of antidepressant treatment. These compounds are *naringenin, nobiletin, ferulic acid, ellagic acid, apigenin, rutin, fisetin, proanthocyjanidins, chlorogenic acid and amentoflavone*, to name but a few (Fig. 17.3).

As for any agent acting on the CNS, the therapeutic potential of antioxidants depends on their ability to cross the blood–brain barrier (BBB). According to the existing body of evidence, hypericin, quercetin, *Ginkgo biloba*, epicatechins, resveratrol and curcumin can all be delivered to the brain from the circulating blood (Table 17.2).

When antioxidants are used in depression, it may require additional monitoring to prevent the risk of fatal adverse events, and to ensure the best effectiveness of the treatment, because each patient has an individual redox threshold. Internal and external stressors contribute to pro- and antioxidant imbalance – which can worsen pathologies due to the loss of control of cell signaling and metabolism. Furthermore, elderly patients are more susceptible to medical complications owing to their greater difficulty in tolerating doses that are of therapeutic value. Consideration should also be given to possible drug–drug and drug–food interactions. The recognized side effects of antioxidants discussed in this chapter are presented in Table 17.3. Furthermore, many of these antioxidants also interact with multiple neurotransmitter systems which may limit concomitant use of antioxidants and antidepressants (Table 17.1).

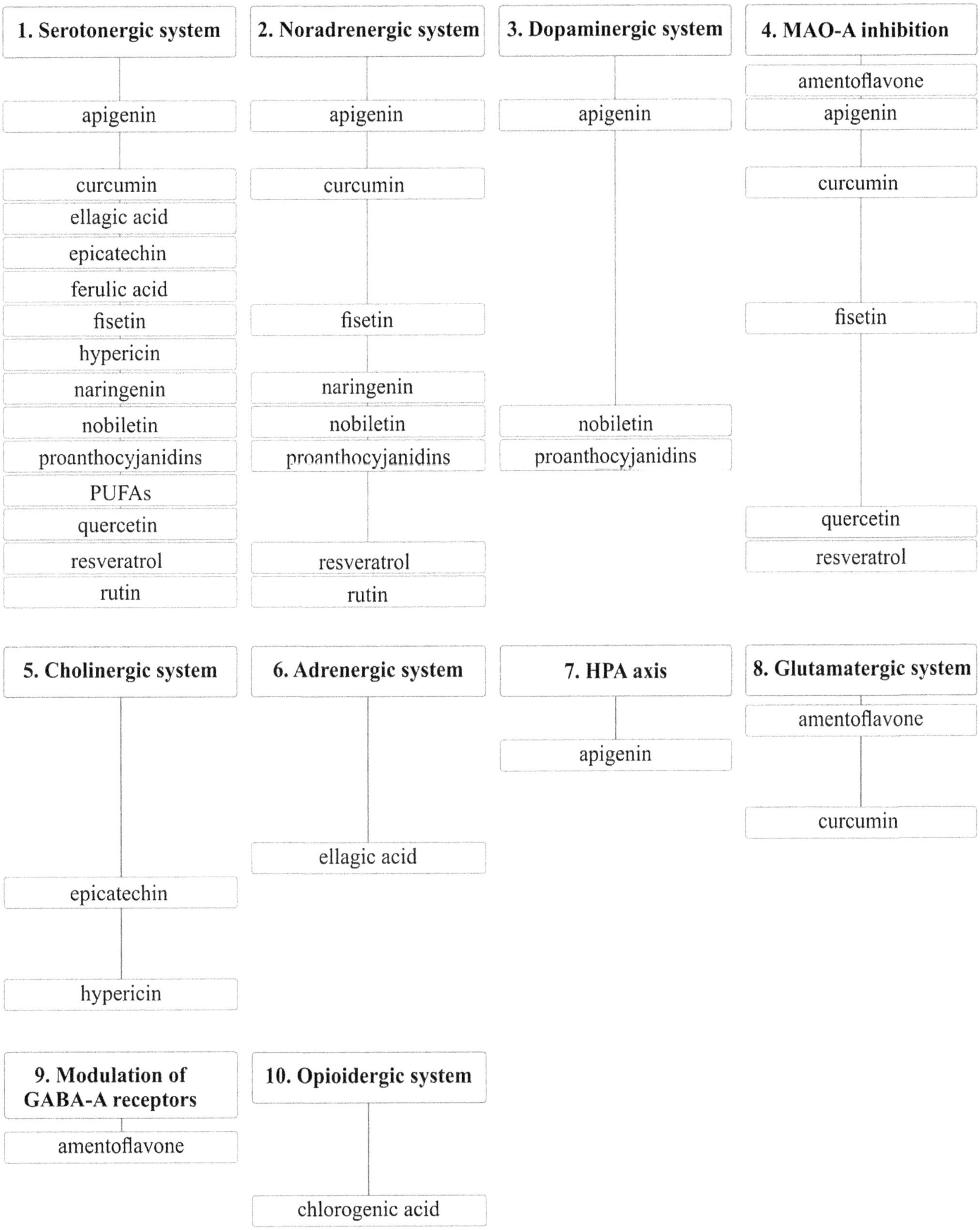

FIGURE 17.3 Antioxidants acting on the major neurotransmitter systems of depression.

TABLE 17.2 Pharmacokinetic Properties and Distribution Over the Blood–Brain Barrier of Biologic Substances with Antioxidant Activity Known for Their Antidepressant Effect

Hypericin (St John's wort)	Hypericin has been shown to be taken up by the brain in mice.[63]
Quercetin	Quercetin has been shown to permeate the BBB effectively.[64]
Ginkgo biloba	Significant levels of *Ginkgo biloba* terpene trilactones and flavonoids cross the blood–brain barrier and enter the CNS of rats after oral application of *Ginkgo biloba* extract.[65]
Epicatechins	Presence of epicatechin glucuronides in the brain after oral administration of epicatechin in rats.[66]
Resveratrol	Ability to cross the blood–brain barrier and exert protection against cerebral ischemic injury.[67]
Curcumin	Curcumin can cross the blood–brain barrier (BBB) to enter brain tissue.[68]
Melatonin	Melatonin passes to the brain; however, permeability of the BBB to melatonin is lower than that of many drugs developed and targeted specifically for the brain.[69]

BBB: blood–brain barrier, CNS: central nervous system.

TABLE 17.3 Potential Side Effects of Biologic Substances with Antioxidant Activity Known for Their Antidepressant Effect

Hypericin (St John's wort)	Interactions between St John's wort and certain prescribed medicines, including warfarin, ciclosporin, theophylline, digoxin, HIV protease inhibitors, anticonvulsants, triptans and oral contraceptives[70]
Quercetin	Interactions with liver metabolism[71]
Ginkgo biloba	Not found
Epicatechins	Not found
Resveratrol	Interactions with liver metabolism[72]
Curcumin	Alterations in glucose metabolism manifested by lowering of blood sugar caused by tumeric[74] Possible interactions of curcumin with anti-clotting and blood pressure medications[75]
Melatonin	Sleepiness and headaches[76,77]

SUMMARY POINTS

- Many dietary antioxidants exert antidepressant-like effects, as demonstrated in both animals and humans.
- The mechanism of antidepressant action involves different pathways related to oxidative stress, inflammation and neurotransmission.
- Antioxidants, when used in late-life depression, may also help to attenuate common comorbidities, including neurodegenerative and metabolic disorders.
- The rationale for using antioxidants in depressed elderly patients is dictated by many factors, including bioavailability, potential interactions and side effects.
- Antioxidant supplementation in depressed patients may not always exert an immediate and direct effect on disease symptomatology, but may have a long-term positive impact on the general health and the disease outcome.
- Considerable effort will be required to adapt antioxidants as agents to be used in the treatment of late-life depression.

References

1. Schoevers RA, Geerlings MI, Beekman AT, et al. Association of depression and gender with mortality in old age. Results from the Amsterdam Study of the Elderly (AMSTEL). *Br J Psychiatry* 2000;**177**:336–42.
2. Moussavi S, Chatterji S, Verdes E, et al. Depression, chronic diseases, and decrements in health: results from the World Health Surveys. *Lancet* 2007;**370**(9590):851–8.
3. Liochev SI. Reactive oxygen species and the free radical theory of aging. *Free Radic Biol Med* 2013;**60**:1–4.
4. Marchi S, Giorgi C, Suski JM, et al. Mitochondria-Ros crosstalk in the control of cell death and aging. *J Signal Transduct* 2012;**2012**:1–17.
5. Gardner A, Boles RG. Beyond the serotonin hypothesis: mitochondria, inflammation and neurodegeneration in major depression and affective spectrum disorders. *Prog Neuropsychopharmacol Biol Psychiatry* 2011;**35**(3):730–43.
6. Mendlewicz J. Disruption of the circadian timing systems: molecular mechanisms in mood disorders. *CNS Drugs* 2009;**23**(Suppl 2):15–26.
7. Blazer 2nd DG, Hybels CF. Origins of depression in later life. *Psychol Med* 2005;**35**(9):1241–52.
8. Vu T, Finch CF, Day L. Patterns of comorbidity in community-dwelling older people hospitalised for fall-related injury: a cluster analysis. *BMC Geriatrics* 2011;**11**(1):45.
9. Jyrkkä J, Mursu J, Enlund H, Lönnroos E. Polypharmacy and nutritional status in elderly people. *Curr Opin Clin Nutr Metab Care* 2012;**15**(1):1–6.
10. Gómez-Pinilla F. Brain foods: the effects of nutrients on brain function. *Nature Rev Neurosci* 2008;**9**(7):568–78.
11. Bast A, Haenen GRMM. Ten misconceptions about antioxidants. *Trends Pharmacol Sci* 2013;**34**(8):430–6.
12. Sánchez-Villegas A, Delgado-Rodríguez M, Alonso A, et al. Association of the Mediterranean dietary pattern with the incidence of depression: the Seguimiento Universidad de Navarra/University of Navarra follow-up (SUN) cohort. *Arch Gen Psychiatry* 2009;**66**(10):1090–8.
13. Otton SV, Ball SE, Cheung SW, et al. Venlafaxine oxidation in vitro is catalysed by CYP2D6. *Br J Clin Pharmacol* 1996;**41**(2):149–56.
14. Linde K. St John's wort for depression: meta-analysis of randomised controlled trials. *Br J Psychiatr* 2005;**186**(2):99–107.
15. Crupi R, Kareem Abusamra YA, Spina E, Calapai G. Preclinical data supporting/refuting the use of *Hypericum perforatum* in the treatment of depression. *CNS Neurol Disord Drug Targets* 2013;**12**(4):474–86.
16. Whiskey E, Werneke U, Taylor D. A systematic review and meta-analysis of *Hypericum perforatum* in depression: a comprehensive clinical review. *Int Clin Psychopharmacol* 2001;**16**(5):239–52.
17. Zou Y, Lu Y, Wei D. Antioxidant activity of a flavonoid-rich extract of *Hypericum perforatum* L. *in vitro*. *J Agric Food Chem* 2004;**52**(16):5032–9.

18. El-Sherbiny DA, Khalifa AE, Attia AS, Eldenshary EE-DS. *Hypericum perforatum* extract demonstrates antioxidant properties against elevated rat brain oxidative status induced by amnestic dose of scopolamine. *Pharmacol Biochem Behav* 2003;**76**(3-4):525–33.
19. Geetha T, Malhotra V, Chopra K, Kaur IP. Antimutagenic and antioxidant/prooxidant activity of quercetin. *Indian J Exp Biol* 2005;**43**(1):61–7.
20. Dimpfel W. Rat electropharmacograms of the flavonoids rutin and quercetin in comparison to those of moclobemide and clinically used reference drugs suggest antidepressive and/or neuroprotective action. *Phytomedicine* 2009;**16**(4):287–94.
21. Ishige K, Schubert D, Sagara Y. Flavonoids protect neuronal cells from oxidative stress by three distinct mechanisms. *Free Radic Biol Med* 2001;**30**(4):433–46.
22. Blumenthal M. Systematic reviews and meta-analyses support the efficacy of numerous popular herbs and phytomedicines. *Altern Ther Health Med* 2009;**15**(2):14–5.
23. Kalkunte SS, Singh AP, Chaves FC, et al. Antidepressant and antistress activity of GC-MS characterized lipophilic extracts of Ginkgo biloba leaves. *Phytother Res* 2007;**21**(11):1061–5.
24. Cieza A, Maier P, Pöppel E. The effect of ginkgo biloba on healthy elderly subjects. *Fortschr Med Orig* 2003;**121**(1):5–10.
25. Janssen IM, Sturtz S, Skipka G, et al. *Ginkgo biloba* in Alzheimer's disease: a systematic review. *Wien Med Wochenschr* 2010;**160** (21–22):539–46.
26. Hou Y, Aboukhatwa MA, Lei D-L, et al. Anti-depressant natural flavonols modulate BDNF and beta amyloid in neurons and hippocampus of double TgAD mice. *Neuropharmacology* 2010;**58**(6):911–20.
27. Jagasia R, Steib K, Englberger E, et al. GABA-cAMP response element-binding protein signaling regulates maturation and survival of newly generated neurons in the adult hippocampus. *J Neurosci* 2009;**29**(25):7966–77.
28. Sahay A, Hen R. Adult hippocampal neurogenesis in depression. *Nat Neurosci* 2007;**10**(9):1110–5.
29. Luo Y. Inhibition of amyloid-beta aggregation and caspase-3 activation by the *Ginkgo biloba* extract EGb761. *Proc Natl Acad Sci USA* 2002;**99**(19):12197–202.
30. Frei B, Higdon JV. Antioxidant activity of tea polyphenols in vivo: evidence from animal studies. *J Nutr* 2003;**133**(10):3275S–84S.
31. Singal A, Tirkey N, Chopra K. Reversal of LPS-induced immobility in mice by green tea polyphenols: possible COX-2 mechanism. *Phytother Res* 2004;**18**(9):723–8.
32. Pham NM, Nanri A, Kurotani K, et al. Green tea and coffee consumption is inversely associated with depressive symptoms in a Japanese working population. *Public Health Nutr* 2013; March; **4**:1–9. [Epub ahead of print].
33. Gülçin İ. Antioxidant properties of resveratrol: a structure–activity insight. *Innovative Food Science & Emerging Technologies* 2010;**11**(1):210–8.
34. Xu Y, Wang Z, You W, et al. Antidepressant-like effect of trans-resveratrol: involvement of serotonin and noradrenaline system. *Eur Neuropsychopharmacol* 2010;**20**(6):405–13.
35. Yu Y, Wang R, Chen C, et al. Antidepressant-like effect of trans-resveratrol in chronic stress model: behavioral and neurochemical evidences. *J Psychiatr Res* 2013;**47**(3):315–22.
36. Orgogozo JM, Dartigues JF, Lafont S, et al. Wine consumption and dementia in the elderly: a prospective community study in the Bordeaux area. *Rev Neurol (Paris)* 1997;**153**(3):185–92.
37. Jovanovic SV, Boone CW, Steenken S, et al. How curcumin works preferentially with water soluble antioxidants. *J Am Chem Soc* 2001;**123**(13):3064–8.
38. Darvesh AS, Carroll RT, Bishayee A, et al. Curcumin and neurodegenerative diseases: a perspective. *Expert Opin Investig Drugs* 2012;**21**(8):1123–40.
39. Xu Y, Ku B, Tie L, et al. Curcumin reverses the effects of chronic stress on behavior, the HPA axis, BDNF expression and phosphorylation of CREB. *Brain Res* 2006;**1122**(1):56–64.
40. Kulkarni S, Dhir A, Akula KK. Potentials of curcumin as an antidepressant. *Sci World J* 2009;**9**:1233–41.
41. Darvesh AS, Carroll RT, Bishayee A, et al. Curcumin and neurodegenerative diseases: a perspective. *Expert Opin Investig Drugs* 2012;**21**(8):1123–40.
42. Wang Y-F, Gu Y-T, Qin G-H, et al. Curcumin ameliorates the permeability of the blood–brain barrier during hypoxia by upregulating heme oxygenase-1 expression in brain microvascular endothelial cells. *J Mol Neurosci* 2013;**51**(2):344–51. [Epub ahead of print].
43. Raghavendra V, Kaur G, Kulkarni SK. Anti-depressant action of melatonin in chronic forced swimming-induced behavioral despair in mice, role of peripheral benzodiazepine receptor modulation. *Eur Neuropsychopharmacol* 2000;**10**(6):473–81.
44. Quera Salva MA, Hartley S, Barbot F, et al. Circadian rhythms, melatonin and depression. *Curr Pharm Des* 2011;**17**(15):1459–70.
45. Bars DL, Thivolle P, Vitte P, et al. PET and plasma pharmacokinetic studies after bolus intravenous administration of [11C]melatonin in humans. *Int J Radiation Applic Instrument Part B Nuclear Medicine and Biology* 1991;**18**(3):357–62.
46. Pathak L, Agrawal Y, Dhir A. Natural polyphenols in the management of major depression. *Expert Opin Investig Drugs* 2013;**22**(7):863–80.
47. Butterweck V, Böckers T, Korte B, et al. Long-term effects of St John's wort and hypericin on monoamine levels in rat hypothalamus and hippocampus. *Brain Res* 2002;**930**(1-2):21–9.
48. Kaur R, Chopra K, Singh D. Role of alpha2 receptors in quercetin-induced behavioral despair in mice. *J Med Food* 2007;**10**(1): 165–8.
49. Anjaneyulu M, Chopra K, Kaur I. Antidepressant activity of quercetin, a bioflavonoid, in streptozotocin-induced diabetic mice. *J Med Food* 2003;**6**(4):391–5.
50. Dixon Clarke SE, Ramsay RR. Dietary inhibitors of monoamine oxidase A. *J Neural Transm* 2011;**118**(7):1031–41.
51. Naik SR, Pilgaonkar VW, Panda VS. Evaluation of antioxidant activity of *Ginkgo biloba* phytosomes in rat brain. *Phytother Res* 2006;**20**(11):1013–6.
52. Weinreb O, Amit T, Mandel S, Youdim MBH. Neuroprotective molecular mechanisms of (−)-epigallocatechin-3-gallate: a reflective outcome of its antioxidant, iron chelating and neuritogenic properties. *Genes Nutr* 2009;**4**(4):283–96.
53. Zhu W-L, Shi H-S, Wei Y-M, et al. Green tea polyphenols produce antidepressant-like effects in adult mice. *Pharmacol Res* 2012;**65**(1):74–80.
54. Assunção M, Santos-Marques MJ, Carvalho F, Andrade JP. Green tea averts age-dependent decline of hippocampal signaling systems related to antioxidant defenses and survival. *Free Radic Biol Med* 2010;**48**(6):831–8.
55. Xu Y, Ku B, Tie L, et al. Curcumin reverses the effects of chronic stress on behavior, the HPA axis, BDNF expression and phosphorylation of CREB. *Brain Res* 2006;**1122**(1):56–64.
56. Sanmukhani J, Anovadiya A, Tripathi CB. Evaluation of antidepressant like activity of curcumin and its combination with fluoxetine and imipramine: an acute and chronic study. *Acta Pol Pharm* 2011;**68**(5):769–75.
57. Andrade JP, Assunção M. Protective effects of chronic green tea consumption on age-related neurodegeneration. *Curr Pharm Des* 2012;**18**(1):4–14.
58. Bhutani MK, Bishnoi M, Kulkarni SK. Anti-depressant like effect of curcumin and its combination with piperine in unpredictable chronic stress-induced behavioral, biochemical and neurochemical changes. *Pharmacol Biochem Behav* 2009;**92**(1):39–43.
59. Lin TY, Lu CW, Wang C-C, et al. Curcumin inhibits glutamate release in nerve terminals from rat prefrontal cortex: possible relevance to its antidepressant mechanism. *Prog Neuropsychopharmacol Biol Psychiatry* 2011;**35**(7):1785–93.

60. Vega-Naredo I, Poeggeler B, Sierra-Sánchez V, et al. Melatonin neutralizes neurotoxicity induced by quinolinic acid in brain tissue culture. *J Pineal Res* 2005;**39**(3):266–75.
61. Van Liempt S, Arends J, Cluitmans PJM, et al. Sympathetic activity and hypothalamo-pituitary-adrenal axis activity during sleep in post-traumatic stress disorder: a study assessing polysomnography with simultaneous blood sampling. *Psychoneuroendocrinology* 2013;**38**(1):155–65.
62. Domínguez-López S, Mahar I, Bambico FR, et al. Short-term effects of melatonin and pinealectomy on serotonergic neuronal activity across the light-dark cycle. *J Psychopharmacol (Oxford)* 2012;**26**(6):830–44.
63. Chung PS, Saxton RE, Paiva MB, et al. Hypericin uptake in rabbits and nude mice transplanted with human squamous cell carcinomas: study of a new sensitizer for laser phototherapy. *Laryngoscope* 1994;**104**(12):1471–6.
64. Ren S, Suo Q, Du W, et al. Quercetin permeability across blood–brain barrier and its effect on the viability of U251 cells. *Sichuan Da Xue Xue Bao Yi Xue Ban* 2010;**41**(5):751–4. 759.
65. Ude C, Schubert-Zsilavecz M, Wurglics M. *Ginkgo biloba* extracts: a review of the pharmacokinetics of the active ingredients. *Clin Pharmacokinet* 2013;**52**(9):727–49.
66. Abd El Mohsen MM, Kuhnle G, Rechner AR, et al. Uptake and metabolism of epicatechin and its access to the brain after oral ingestion. *Free Radic Biol Med* 2002;**33**(12):1693–702.
67. Wang Q, Xu J, Rottinghaus GE, et al. Resveratrol protects against global cerebral ischemic injury in gerbils. *Brain Res* 2002;**958**(2):439–47.
68. Tsai Y-M, Chien C-F, Lin L-C, Tsai T- H. Curcumin and its nanoformulation: the kinetics of tissue distribution and blood–brain barrier penetration. *Int J Pharm* 2011;**416**(1):331–8.
69. Johns J. Estimation of melatonin blood brain barrier permeability. *J Bioanalysis Biomed* 2011;**3**(3):64.
70. Barnes J, Anderson LA, Phillipson JD. St John's wort (*Hypericum perforatum* L.): a review of its chemistry, pharmacology and clinical properties. *J Pharm Pharmacol* 2001;**53**(5):583–600.
71. Raucy JL. Regulation of CYP3A4 expression in human hepatocytes by pharmaceuticals and natural products. *Drug Metab Dispos* 2003;**31**(5):533–9.
72. Chang TK, Chen J, Lee WB. Differential inhibition and inactivation of human CYP1 enzymes by trans-resveratrol: evidence for mechanism-based inactivation of CYP1A2. *J Pharmacol Exp Ther* 2001;**299**(3):874–82.
73. Piver B, Fer M, Vitrac X, et al. Involvement of cytochrome P450 1A2 in the biotransformation of trans-resveratrol in human liver microsomes. *Biochem Pharmacol* 2004;**68**(4):773–82.
74. Seo K-I, Choi M-S, Jung UJ, et al. Effect of curcumin supplementation on blood glucose, plasma insulin, and glucose homeostasis related enzyme activities in diabetic db/db mice. *Mol Nutr Food Res* 2008;**52**(9):995–1004.
75. Kim D-C, Ku S-K, Bae J- S. Anticoagulant activities of curcumin and its derivative. *BMB Rep* 2012;**45**(4):221–6.
76. Bauer ME, Muller GC, Correa BL, et al. Psychoneuroendocrine interventions aimed at attenuating immunosenescence: a review. *Biogerontology* 2013;**14**(1):9–20.
77. Burgess HJ. Melatonin: an adjunctive treatment for cardiometabolic disease? Sleep 2012;**35**(10):1319–20.

CHAPTER

18

Antioxidant and Anti-Inflammatory Role of Melatonin in Alzheimer's Neurodegeneration

Sergio A. Rosales-Corral

Centro de Investigacion Biomedica de Occidente, Instituto Mexicano Del Seguro Social, Guadalajara, Jalisco, Mexico, and University of Texas Health Science Center at San Antonio, Department of Cellular and Structural Biology, San Antonio, TX, USA

Russel J. Reiter, Dun-Xian Tan, Lucien C. Manchester

University of Texas Health Science Center at San Antonio, Department of Cellular and Structural Biology, San Antonio, TX, USA

Xiaoyan Liu

University of Texas Health Science Center at San Antonio, Department of Cellular and Structural Biology, San Antonio, TX, USA, and The Preclinical Medicine Institute of Beijing, University of Chinese Medicine, Chao Yang District, Beijing, China

List of Abbreviations

AANAT *N*-acetyltransferase
Aβ amyloid-beta
AD Alzheimer's disease
AFMK *N*-acetyl-*N*-formyl-5-methoxykynuramine
AMK *N*-acetyl-5-methoxykynuramine
APP amyloid precursor protein
C-3-OHM cyclic 3-hydroxymelatonin
CSF cerebrospinal fluid
ERK extracellular-signal-regulated kinase
ETC electron transport chain
GPx glutathione peroxidase
GR glutathione reductase
Grx glutaredoxin
GSH reduced glutathione
GSK-3β glycogen synthase kinase 3
GSSG glutathione disulfide
H_2O_2 hydrogen peroxide
HIOMT hydroxyindole-O-methyltransferase
5-HT 5-hydroxytryptamine
LO• alkoxyl radical
LOO• peroxyl
MAP mitogen-activated protein kinase
MIC mild cognitive impairment
NADPH nicotinamide adenine dinucleotide phosphate
NMDAR *N*-methyl-D-aspartate receptor
•NO nitric oxide
GSSG oxidized glutathione
O_2 diatomic oxygen
•OH hydroxyl radical
$^1O_2$ singlet oxygen
$O_2^{-\bullet}$ superoxide anion
$ONOO^-$ peroxynitrite
PI3K/Akt phosphoinositide 3/Akt kinase
PKA protein kinase
RAGE receptor for advanced glycation end products
RNS reactive nitrogen species
ROS reactive oxygen species
RZR/ROR retinoid related orphan nuclear receptor
SOD superoxide dismutase enzyme
SR scavenger receptor
TLR2 Toll-like receptor 2
TrkB tyrosine-related kinase B

INTRODUCTION

Oxidative stress in the brain contributes to neurodegeneration. There is a precise coordination among neurons, microglia and astrocytes to control free radicals and to maintain a balanced redox environment. This occurs in a complex net of signaling pathways and endogenous antioxidant systems in constant activity. Oxidative

Aging
http://dx.doi.org/10.1016/B978-0-12-405933-7.00018-4

pressures, on the other hand, derived from different sites, attack the abundant lipids in brain cells. Lipid peroxidation involves the propagation of oxidative chain reactions. Also, the persistence of the stimuli causing the oxidative stress does not cease, and a second factor worsens the situation, i.e. neuroinflammation. Neuroinflammatory events, accompanied by the activation of stress-activated pathways and endoplasmic reticulum stress, foment and feed the oxidative stress, which, in turn, promotes neuroinflammation. The well-formed vicious cycle prevents the return to a homeostatic balance (Fig. 18.1).

When or how the Alzheimer's disease (AD) pathology starts is not well defined. However, an important part of the investigations is focused on amyloid-beta (Aβ), a 40- to 43-amino-acid peptide, which has the ability to trap molecular oxygen and reduce it to H_2O_2 in the presence of iron.[1] Aβ can become oxidized on its methionine residue at position 35, forming a positively charged sulfuranyl radical ($MetS^+$). This cation radical, in turn, may react with allylic H atoms on unsaturated acyl chains of lipids.[2] Thus, Aβ itself might directly initiate free radical chain reactions.

The mechanisms involved in Alzheimer-related pathogenic events are highly complex and will be briefly analyzed in this chapter. However, in spite of the lack of a unified hypothesis or a consensus, oxidative stress seems to be connected with all the proposed hypotheses which try to explain this devastating disorder.

Since the first reports identifying oxidative stress as a pathogenic factor in AD, antioxidants have been proposed as therapeutics.[3] The natural history of the disease, in addition to its chronicity, which has presented almost insurmountable obstacles to realize the proper randomized control trials in humans, is complex, as is evaluating results in the long term. Reports about antioxidants in AD have also shown a series of failures in the methodologic design or they have not taken into account basic considerations, such as the low permeability of the blood–brain barrier to most of the antioxidants used, the role of vascular factors in AD pathogenesis, or the lack of previously established efficacy on oxidative stress biomarkers, etc. In fact, there has been no homogeneous scientific clinical approach to this problem.[4] Finally, outcome evaluations must be carefully chosen, and it is also necessary to consider what is feasible to expect when giving antioxidants to an AD patient, keeping in mind that the more advanced the disease the greater the loss of neurons, which are irreplaceable.

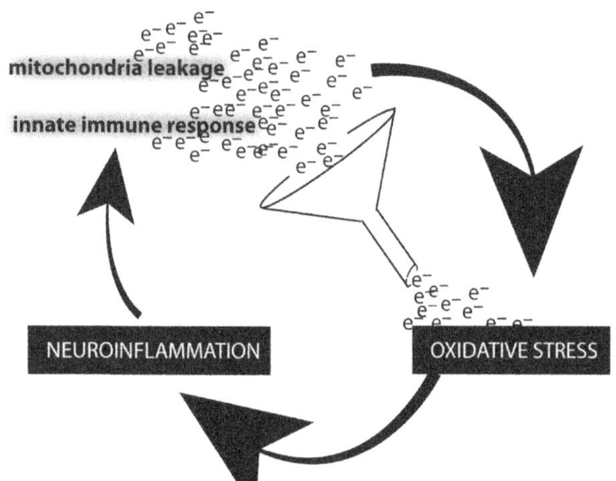

FIGURE 18.1 Oxidative stress and neuroinflammation feed each other in a vicious cycle.

On the other hand, clinical trials using melatonin in AD patients document its ability to reduce the mild cognitive impairment (MIC),[4] and its slowing down of the transition from MIC to Alzheimer's.[5] Also, melatonin may slow the progression of AD.[6] What is necessary is the assesment of many clinical or preclinical outcomes and some corrections to the methodologic design. However, experimental evidence is significant and suggests that melatonin is a tool to delay MIC or AD progression. By examining Alzheimer's most compelling hypotheses, it may be possible to find different checkpoints where melatonin may have a specific role as a free radical scavenger. As an antioxidant, melatonin may interfere with the Aβ peptide or it may modulate neuroinflammatory or stress-related endpoints.

MELATONIN: SYNTHESIS AND MECHANISMS OF ACTION

Melatonin (*N*-acetyl-5-methoxytryptamine), a tryptophan-derived indoleamine, is a highly conserved molecule whose origin can be traced back an estimated 2.5 billion years.[7] The essential amino acid tryptophan is hydroxylated by the enzyme tryptophan hydroxylase at the 5 carbon to form 5-hydroxytryptophan, which, in turn, is subsequently decarboxylated, giving rise to 5-hydroxytryptamine (5-HT, serotonin). This important neurotransmitter is acetylated by the enzyme arylalkylamine *N*-acetyl transferase (AANAT) to produce the immediate precursor of melatonin, *N*-acetylserotonin (NAS). AANAT is regulated on a circadian basis, but also by cAMP, which inhibits its proteasomal proteolysis.[8]

N-acetylserotonin (NAS) can be many times higher in concentration than melatonin. But it is also a neurotransmitter with an important role in activating the tyrosine-related kinase B (TrkB) receptor, which, in turn, stimulates neurotrophins such as BDNF (brain-derived neurotrophic factor), NT-4 (neurotrophin-4), and NT-3 (neurotrophin-3). NAS transduces the BDNF signal via Ras-ERK, PI3K, and PLCγ, which is related to neuroprotective and neurotrophic effects.[9] NAS is the melatonin precursor, but it also has an independent role in cell biology and it does not necessarily rate limit melatonin production.

Melatonin levels fluctuate in synchrony with the activity of the enzyme hydroxyindole-O-methyltransferase (HIOMT, also known as acetyl-serotonin-methyltransferase, or ASMT), which is indirectly controlled by light/darkness variations via a photoneural system [retina, the suprachiasmatic nuclei of the hypothalamus (SCN), and the superior cervical ganglia].[10] HIOMT O-methylates NAS to *N*-acetyl-5-methoxytryptamine (melatonin), and is the rate-limiting enzyme in pineal melatonin production[9] (Fig. 18.2).

Noradrenergic messages coming from nerve endings stimulated by suprachiasmatic nuclei and the superior cervical ganglia activate α1 and β1-adrenergic postsynaptic receptors in the pineal gland, leading to a 100-fold rise in the intracellular cAMP accumulation. cAMP accumulation increases the expression and activity of the tryptophan hydroxylase and AANAT messenger RNAs, events linked to the acetylation of serotonin, as mentioned before. During the day, AANAT protein is rapidly degraded via proteasomal proteolysis which is only terminated upon the nocturnal increase in cAMP levels. The following protagonist, HIOMT, has constitutively a constant high activity and it is not acutely (within hours) stimulated by using a beta-adrenergic agonist; rather, it is upregulated by chronic application of the agonist.[11] Thus, it seems to be another factor co-regulating HIOMT and this could be neuropeptide Y (NPY), with circadian variations of its content in the pineal gland. NPY originates mainly from the sympathetic nerve fibers, where it is co-localized with noradrenaline.

Melatonin production, following a circadian rhythm, is found even in primitive bacteria such as *Rhodospirillum rubrum*, which is phylogenetically linked to the origin of mitochondria. This could offer a clue about the ancient history of melatonin, which can be traced back 2.5 billion years.[7] Thus, as soon as the free radicals emerged during evolution, this indoleamine evolved as a free radical scavenger. For that reason, melatonin is found not only in primitive bacteria but in unicellular dinoflagellates and small aquatic metazoans, such as green algae. It is possible that in most organisms, if not all, melatonin

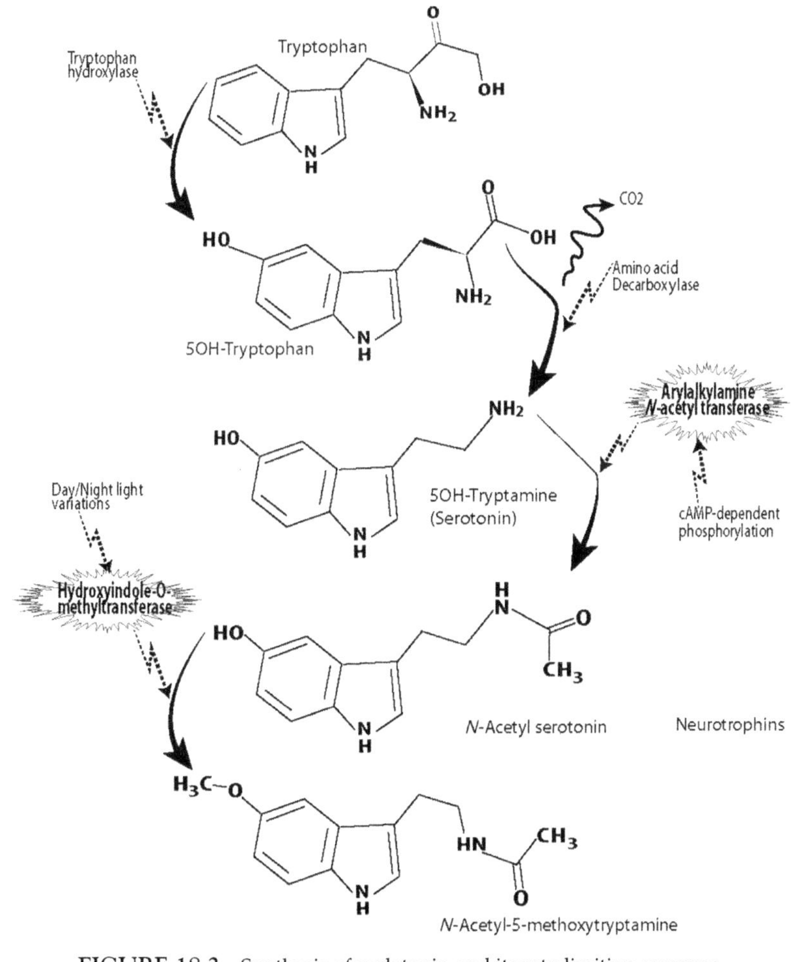

FIGURE 18.2 Synthesis of melatonin and its rate-limiting enzymes.

biosynthetic ability has been transferred from its original site of production, mitochondria and chloroplasts, to other compartments of the cell, for example, the cytosol. In the early stage of biosymbiosis, mitochondria and chloroplasts were the only sites of melatonin synthesis in cells.[12]

Apart from its presence in practically all living beings, plants and animals, melatonin has another feature also related to its ancient origin. Melatonin is an amphipathic antioxidant and immunomodulatory molecule, and it diffuses into cells freely. Phylogenetically, this characteristic could be an adaptive and cytoprotective mechanism particularly linked, again, to the mitochondrial overproduction of free radicals. Thus, even in the absence of receptors, melatonin crosses all the natural barriers both at cellular level and at organellar level, and this omnipresence is mostly related to its activity as an antioxidant and as a free radical scavenger.[13]

Melatonin also has its own receptors in cytoplasmic membranes. In brain, G protein-coupled melatonin receptors have been observed in the suprachiasmatic nuclei, the pars tuberalis, the pineal gland, and in hippocampal neurons. These receptors are related to circadian rhythms and immunomodulatory activities of melatonin. Via these receptors, melatonin is involved in several signal transduction pathways leading to modulatory effects on stress signals related to mitogen-activated protein kinases (MAP) and extracellular-signal-regulated kinases (ERK) with the activation of the phosphoinositide 3/Akt kinase (PI3K/Akt) pathway, which implies neuroprotective and cell survival roles for melatonin. Astrocytes also express MT1 and MT2 receptors, and both seem to be necessary to activate the Akt/PI3k signaling pathway. However, the expression or the specific activity of melatonin receptors remains to be fully clarified for astrocytes, microglia, and neurons under different physiologic and pathologic conditions. As an example, studies by Kaur et al[14] revealed that CR3 receptors, MHC antigens, and CD4 antigens on macrophages/microglia are upregulated following melatonin administration in postnatal rat brain; whether this is a receptor-mediated effect remains to be clarified (Fig. 18.3).

Melatonin may also regulate Ca^{2+} channels and Ca^{2+} signaling. Acting on its MT2 receptor, melatonin inhibits adenylyl cyclase, and by this means it decreases cAMP formation, blocking the cAMP-dependent protein kinase (PKA), which activates calcium-release channels. Additionally, melatonin may inhibit the mobilization of Ca^{2+} from the endoplasmic reticulum (ER) as well as the Ca^{2+} influx through voltage-sensitive channels, because of its great avidity for calmodulin, the calcium-binding messenger protein.[15] The implications of these and the above mentioned melatonin features may become important in more than one pathogenic process related to Alzheimer's disease.

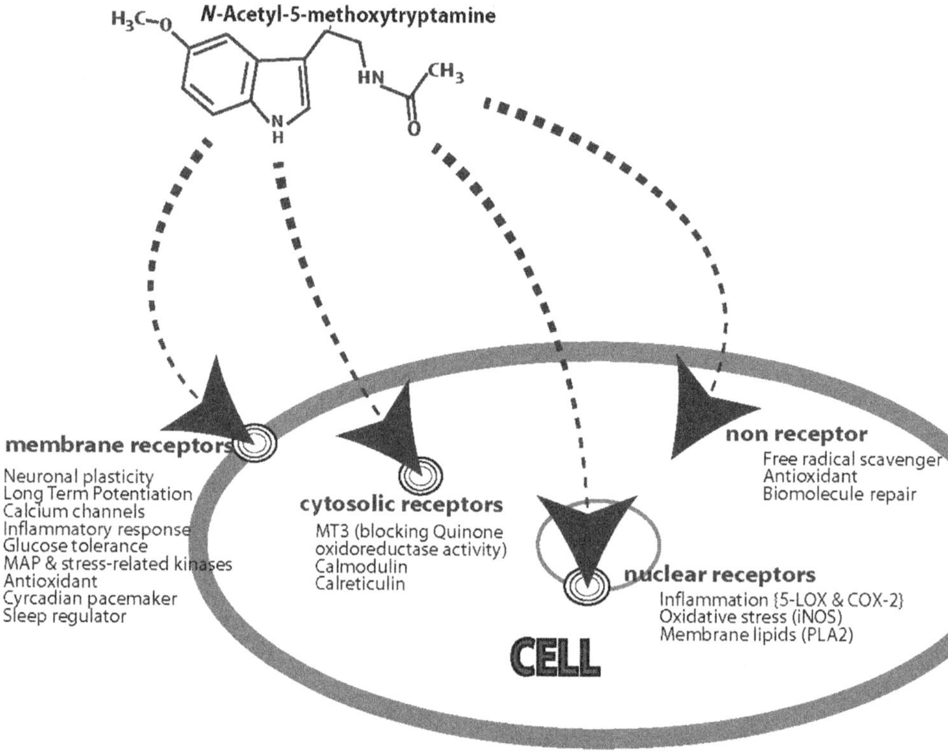

FIGURE 18.3 Different receptor- and non-receptor-dependent activities of melatonin.

Melatonin also has receptors and/or binding sites in the cell nucleus. The retinoid related orphan RZR/ROR receptors belong to a nuclear orphan hormone receptor family and they are specific for melatonin. RZR has been found in brain, pineal gland, retina, and spleen, whereas ROR alpha is more ubiquitously expressed, being particularly important in T- and B-lymphocytes, neutrophils, and monocytes. Melatonin binds and activates these receptors even at low nanomolar concentrations, and it has been documented that these are key events for transmission of photoperiodic information and regulation of seasonal reproductive cycles by melatonin. Additionally, melatonin regulates cell processes via nuclear signaling through RZR/ROR transcription factors, which, in turn, regulate key genes involved in: (a) neuroinflammatory pathways (5-lipoxygenase), (b) cell-cycle interruption, terminal differentiation, and cellular senescence ($p21^{WAF1/CIP1}$), (c) lipid metabolism (apolipoprotein A-1), (d) cell-cycle arrest and apoptosis (N-myc), and (e) Ca^{2+} channel modulatory functions in cerebellum (Purkinje cell protein 2).[16]

Due to these mechanisms, melatonin is a highly pleiotropic signaling molecule. What determines its activity, or how melatonin uses its receptors to mediate some of its action, remains to be fully clarified. Thanks to its ability to permeate through all compartments, trespassing all kind of barriers, melatonin has easy access to free radicals wherever they are formed.

HOW FREE RADICALS ARE FORMED

The normal, complete reduction of oxygen (O_2), as occurs in the production of energy by mitochondrial cytochrome oxidase, depends on the addition of four electrons: $O_2 + 4\,H^+ + 4e^- \rightarrow 2\,H_2O$.

O_2 has two unpaired electrons, each located in a different orbit but both having the same spin quantum number, which gives diatomic oxygen (O_2) a low reactivity. To oxidize a molecule of O_2 directly, it would need to accept a pair of electrons having spins opposite to those of the unpaired electrons in O_2 (spin restriction) (Fig. 18.4).

However, O_2 may accept electrons one at a time, being partially reduced to produce: (a) $O_2^{-\bullet}$ (superoxide anion radical), (b) H_2O_2 (hydrogen peroxide), and (c) $\bullet OH$ (hydroxyl radical). The main oxidative and nitrosative pathways are illustrated in Figure 18.4. The electron transport chain, in mitochondria, is responsible for most of these radicals.

$O_2^{-\bullet}$ undergoes dismutation to generate H_2O_2, and the subsequent addition of an electron produces hydroperoxyl radical ($HO_2^\bullet$). In this way, $O_2^{-\bullet}$ and its protonated form ($HO_2^\bullet$) exist in equilibrium:[17] $O_2^{-\bullet} + H^+ \sim HO_2^\bullet$.

Under tissue acidosis, a disequilibrium favoring $HO_2^\bullet$ is observed – as happens in certain medical conditions such as stroke or trauma. But $HO_2^\bullet$ is particularly soluble in lipids and is prone to oxidize tocopherol and polyunsaturated fatty acids in membranes. It is more reactive than $O_2^{-\bullet}$. Fortunately, it has a 4.7–4.8 pKa, thus only 0.25% of the $O_2^{-\bullet}$ generated under physiologic conditions (pH 7) becomes the hydroperoxyl radical.

$O_2^{-\bullet}$ also reacts with nitric oxide ($\bullet NO$, not toxic *per se* but particularly diffusible) to produce an oxidant and nitrating agent, peroxynitrite ($ONOO^-$), which is in equilibrium with peroxynitrous acid (ONOOH; pKa = 6.8). It reacts rapidly with carbon dioxide, forming carbonate and nitrogen dioxide, which, in turn, are one-electron oxidants. $ONOO^-$ is a powerful nitrosylating agent with a half-life around 10 ms and a rapid influence in surrounding molecules; 10 ms is a long half-life when compared with the other major free radicals (Table 18.1).

The partial reduction of O_2 may also generate H_2O_2, as mentioned above. H_2O_2 readily diffuses through cellular compartments and reacts avidly with Fe^{2+} or Cu^{1+} via either the Haber–Weiss or Fenton reactions. In this manner it is reduced to $\bullet OH$, which is highly reactive. Fortunately, several enzymes, such as catalase and glutathione peroxidases (GPx), intervene at this critical point of H_2O_2 to reduce the generation of

TABLE 18.1 Half-Life Time of Radicals

Radical	Half-Life Time
Hydroxyl ($\bullet OH$)	1 nanosec
Singlet oxygen ($^1O_2$)	1 microsec
Superoxide ($O_2^{-}\bullet$)	1 microsec
Alkoxyl (LO•)	1 microsec
Peroxyl (LOO•)	10 ms
Peroxynitrite ($ONOO^-$)	10 ms
Nitric oxide ($\bullet NO$)	2–5 sec

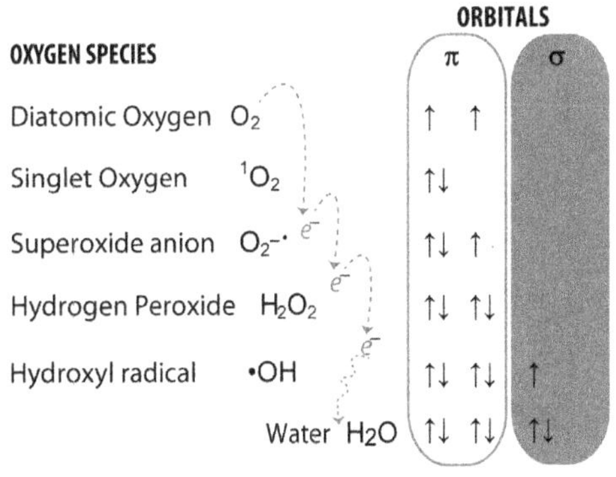

FIGURE 18.4 Oxygen species and spin restriction.

•OH to neutralize these free radicals by recycling them into redox systems. Glutathione peroxidases become remarkably important in brain tissue, where the reaction is 2 GSH + $H_2O_2 \rightarrow$ GSSG + 2 H_2. These enzymes neutralize free radicals and maintain a well tuned redox status[18] (Fig. 18.5).

The third possibility for the partial reduction of O_2 is related to the generation of •OH, where $H_2O_2 + Fe^{2+} \rightarrow Fe^{3+} + OH^- +$ •OH. Once formed, •OH reacts rapidly with any molecule within a few Ångstroms of where it is produced. In spite of its half-life – of the order of 1×10^{-9} seconds – this semi-reduced oxygen species readily damages nuclear and mitochondrial DNA, membrane lipids, and carbohydrates. Importantly, once •OH reacts with polyunsaturated fatty acids (PUFA) in membranes, and removes an allelic H^+, it initiates the chain reaction of lipid peroxidation: —CH_2^- + •OH → — •CH— + H_2O

This is a self-propagating phenomenon. Breakdown of lipids in cellular membranes – caused not only by •OH but also by singlet oxygen ($^1O_2$) and peroxynitrite $ONOO^-$ – generates the peroxyl radical (LOO•), which is responsible for propagating the process because it attacks a nearby PUFA. LOO• may take one H from the adjacent fatty acid to form a lipid hydroperoxide (LOOH), which re-initiates the chain reaction by reacting with the lipid alkyl radical (L•), as follows:

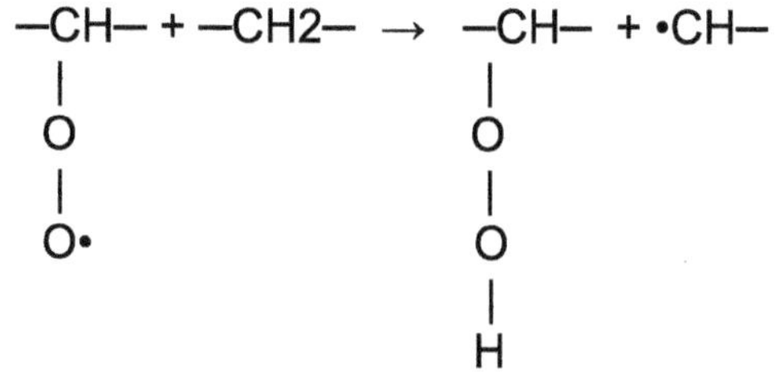

Moreover, LOOH may form an alkoxyl radical (LO•) in the presence of Fe^{2+} (Fenton reaction). There are also other enzymatic and non-enzymatic mechanisms that produce the LO•, but the Fenton reaction is the main mechanism. Both LOO• and LO• are responsible for the propagation of lipid oxidation in cell membranes.

$$\text{ROOH} + \text{Fe}^{2+}\text{complex} \rightarrow \text{Fe}^{3+}\text{complex} \rightarrow \text{RO}\bullet + \text{OH}^-$$

$$\text{ROOH} + \text{Fe}^{3+}\text{complex} \rightarrow \text{ROO}\bullet + \text{H}^+ + \text{Fe}^{2+}\text{complex}$$

Free radicals do not necessarily create oxidative stress. Oxidative stress is an inbalance between oxidative pressures and antioxidant forces. Whatever the origin or the level of the stimulus, changes in the redox environment are sensed by redox sensors; the number and availability of thiol-based redox sensors play a key role in modulating oxidative stress. Meanwhile, redox couples try to maintain a reduced physiologic environment. If that is not possible, oxidative stress becomes out of control and may eventually lead to disease. The redox status is defined as the ratio of interconvertible reduced↔oxidized forms of a molecule. It is maintained because of thiol-based sensors using redox modifications on the oxidizable -SH side-chain of cysteine (Cys-SH) to sense redox alterations. Glutaredoxin (Grx) is an oxidoreductase which catalyzes the reduction of disulfide bonds in proteins, converting reduced glutathione (GSH) to oxidized, glutathione disulfide (GSSG), whereas the enzyme glutathione reductase recycles GSSG back to GSH at the expense of NADPH (Fig. 18.5): GSSG + NADPH + $H^+ \rightarrow$ 2 GSH + $NADP^+$.

Deglutathionylation of protein thiols is also an important function of Grx. Also, the thioredoxin (Trx) system is another essential redox system for reducing oxidized proteins by cysteine thiol↔disulfide exchange. By acting

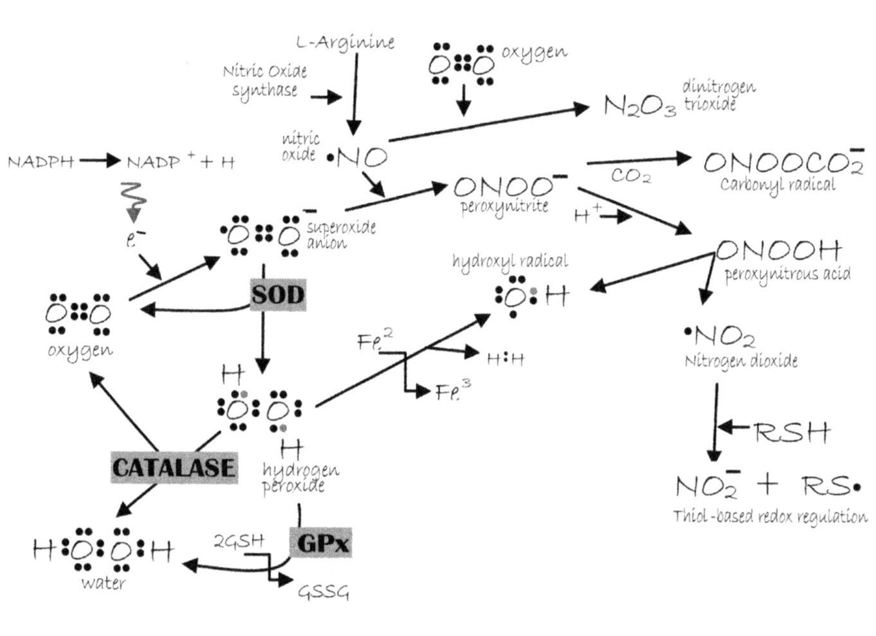

FIGURE 18.5 Main oxidative and nitrosative routes and the endogenous antioxidant enzymes.

as an electron donor, Trx reduces peroxidases and the ribonucleotide reductase. NADPH further reduces the oxidized Trx.

When oxidative stress augments the pressure on the intracellular environment, the effective reduction potential of redox pairs diminishes, and cell signaling shifts toward pro-inflammatory and pro-apoptotic signals. This creates a vicious cycle between oxidative stress and neuroinflammation. In addition, electrophilic compounds derived from the oxidative cascade react with key protein thiols and interfere with redox signaling.[18]

SOME OXIDATIVE-STRESS-RELATED FACTS ABOUT THE BRAIN

There is no organ as sensitive as the brain to changes in energetic metabolism. It consumes 20% of all the O_2 when a person is at rest, and 25% of total body glucose utilization. This consumption is 10 times the rate of the rest of the body per gram of tissue. Taking into account a consumption of 2400 kcal/24 hours for a typical adult, it means 100 Kcal/hour or 116.38 J/s in terms of energy. 20% of 116.38 J/s is 23.3 J/s or Watts, if expressed in terms of power. Because most neuronal energy is generated by oxidative metabolism, neurons critically depend on mitochondrial function, which explains the interconnection between neuronal activity and mitochondrial metabolic activity, and oxygen supply. The latter explains why oxygen consumption in the brain is 160 mmol/100 g·min, and why it consumes glucose at a rate of 31 μmol/100 g·min.[19]

An estimated 1–4% of the oxygen taken into cells, however, is prematurely and incompletely reduced, giving rise to reactive oxygen species (ROS).[17] A major source of oxidants, as a consequence of electron leakage, comes from mitochondria. In fact, mitochondrial dysfunction and associated oxidant stress have been linked to numerous complex diseases and aging. Oxidative stress produces membrane alterations particularly associated with lipid modifications.[20] In damaged mitochondrial membranes, the leak of electrons becomes predictable, feeding back to produce ROS and oxidative stress. In fact, Aβ-induced oxidative stress provokes significant alterations in cholesterol and fatty acids (their composition, disposition, and distribution) in mitochondrial membranes, as observed *in vivo*.[21] This is particularly significant because lipids may allow, facilitate, or even induce the amyloidogenic processing of the amyloid precursor protein.

Oxidative stress is thus linked to cellular membrane dysfunction. This phenomenon is due to damage to membrane lipids, which are among the most vulnerable cellular components to oxidative stress. It is well known that the brain has the highest concentration of fatty acids of any organ. The lipid content in brain white matter may reach 66%, 40% in gray matter, and more than 80% in isolated myelin from white matter.[22]

WHERE DO FREE RADICALS COME FROM IN THE ALZHEIMER'S DISEASE BRAIN?

Alzheimer's disease (AD) is a devastating disorder affecting around 35 million people worldwide. Ten years ago, there were 4.5 million persons with AD in the US population; however, according to the Alzheimer's Association, the number had increased to 5.3 million people in 2010.[23] It is estimated there will be 13.2 million people with this neurodegenerative disorder by 2050.[24]

There are several hypotheses to explain the pathogenesis of AD. The main protagonists are extracellular and intracellular aggregates of Aβ, although the cytoskeletal abnormalities due to tau protein aggregates are also relevant (however, aberrant aggregates of tau are present in most of the neurodegenerative pathologies with filamentous inclusions). Some of the most important hypotheses – including the loss of cholinergic neurotransmission, the amyloid cascade, the mitochondrial cascade, and the calcium or the insulin resistance hypotheses – all have a common factor, i.e. elevated oxidative stress. Even the very processing of Aβ brings with it oxidative stress.

Metabolic signs of oxidative stress in AD are always evident in the neocortex and hippocampus, and they appear to be related to alterations in synaptic density. Once brain peroxide metabolism increases, the AD brain shows elevated cerebral glucose-6-phosphate dehydrogenase activity, which is the first and rate-limiting enzyme of the pentose phosphate pathway, central to maintenance of the cytosolic pool of NADPH, and thus the cellular redox balance. Even the brain of preclinical AD individuals, with normal antemortem neuropsychologic test scores but abundant AD pathology at autopsy, may exhibit increased levels of 4-hydroxynonenal, which is the major product of lipid peroxidation, as well as acrolein, a powerful marker of oxidative damage to protein. Inferior parietal lobule samples from early AD patients compared to age-matched controls have been examined for proteomic identification of nitrated brain proteins; they revealed significant alterations in antioxidant defense proteins and energy metabolism enzymes, with all of them being directly or indirectly linked to AD pathology.

Amyloid-beta is a degradative peptide non-obligatory derivative from the catabolism of an integral membrane protein that normally plays an essential role in neural growth and repair, the amyloid protein precursor protein (APP); it emerges from an aberrant mechanism of cleavage. The APP chain of 695, 751, or 770 amino acids undergoes proteolysis. In the amyloidogenic pathway there is a concerted action of two secretases, the β-secretase, which cleaves the APP-N terminus, and the γ-secretase, which cleaves the APP-C terminus in the secondary transmembrane region. The resultant peptide of 39–43 amino acid residues, Aβ, is delivered to the extracellular milieu where it forms insoluble aggregates and becomes the major component of senile plaques. $A\beta_{1-40}$ and $A\beta_{1-42}$ are the most common Aβ isoforms. Aberrant $A\beta_{1-42}$ accumulation within distal neurites and synapses is directly associated with subcellular pathology and neurotransmitters, while $A\beta_{1-40}$ is the predominant form of the Aβ peptides but less prone to form fibrils. Aβ neurotoxic properties depend heavily on free radicals.[25] However, not only fibrillar Aβ but a variety of Aβ oligomers may cause cellular damage. Soluble oligomers – referred to as amorphous aggregates, micelles, protofibrils, prefibrillar aggregates, amyloid β-derived diffusible ligands (ADDLs), Aβ*56, globulomers, amylospheroids, toxic soluble Aβ, 'paranuclei', and annular protofibrils – appear within neuronal processes and synapses rather than within the extracellular space.[26] They are neurotoxic rather than amyloid fibrils found in amyloid plaques and may inhibit critical neuronal functions, including long-term potentiation, a classic experimental paradigm for memory and synaptic plasticity.

The overproduction of free radicals in the pathogeny of AD may come from the microglial respiratory burst in response to Aβ-induced neuroinflammatory events (Fig. 18.6). The microglial respiratory burst in AD may be a result of:

1. the interaction of Aβ with specialized receptors, such as scavenger receptors or the receptor for advanced glycation end products (RAGE);
2. the astrocyte/microglia intercommunication, by using cytokines or through Ca^{2+} waves;
3. detection of damage-associated molecular patterns (DAMPs) through their corresponding receptors, leading to the activation of the phagocytic-oxidase (PHOX).

PHOX is a membrane-bound enzyme complex responsible for the innate immune nicotinamide adenine dinucleotide phosphate (NADPH) response through the respiratory burst. The activation of the NADPH oxidase, probably both in neurons and in glia, links redox control and neuroinflammatory signaling pathways.

Aβ causes microglial proliferation mediated by PHOX, which is demonstrated by a marked translocation of the cytosolic factors p47phox and p67phox to the microglial membrane in the brain of patients with AD. This is correlated with proinflammatory factors, including TNF-α and IL-1β overproduction. The synergy between oxidative and nitrosative stress plus neuroinflammation may increase the excessive generation of $ONOO^-$ by 1,000,000-fold.[27] PHOX is a multicomponent enzyme system composed of two integral membrane proteins, p22phox and gp91phox, integrated as cytochrome b558, three essential cytosolic components, p47phox, p67phox, p40phox, and the above-mentioned GTPase Rac1, of the Rho family of small G proteins. In

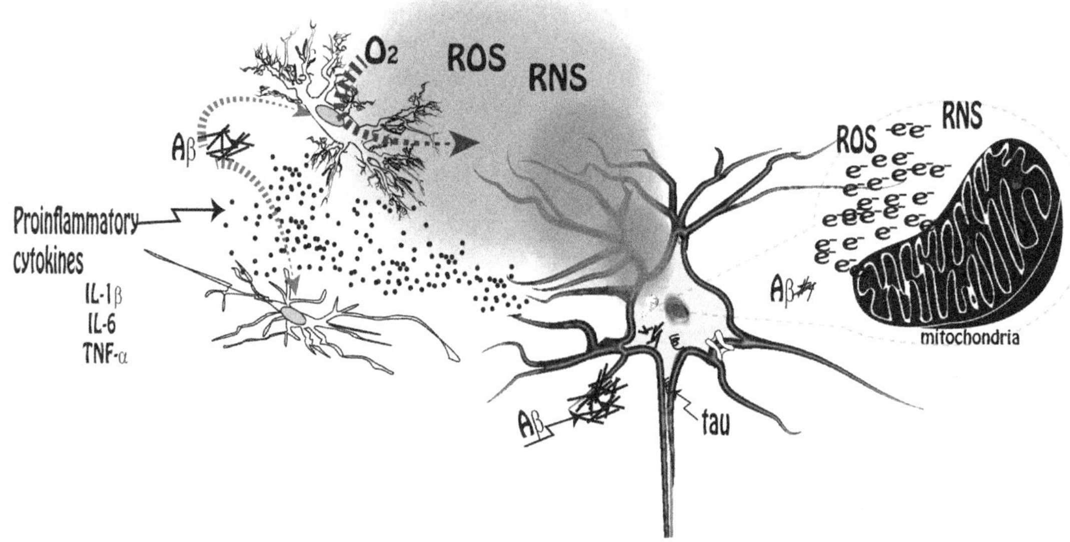

FIGURE 18.6 Amyloid-beta has the ability to generate oxidative stress (1) by itself, (2) by stimulating glia which cause oxidative stress directly as an innate immune response (PHOX activation) and indirectly through a neuroinflammatory response, and (3) by disturbing membrane structures and the electron transport chain in mitochondria, which becomes a main source of free radicals as a result of electron leakage.

general terms, the complex begins its integration when the cytosolic p47phox subunit becomes phosphorylated and transports the total cytosolic components to the docking site where they assemble to form flavocytochrome b558. GTP-bound Rac coordinates the translocation of the p47phox/p67phox/p40phox complex, and its dissociation from GTP permits the subsequent inactivation of the PHOX complex, a crucial step where the superoxide dismutase enzyme (SOD) plays a key role acting as a stabilizer of Rac. Once integrated, PHOX transfers electrons from NADPH to molecular oxygen, generating $O_2^{-\bullet}$. Because Aβ induces oxidative stress that is related to mitochondrial damage, a mechanism closely linked to apoptosis is established. Reciprocally, oxidative stress may induce intracellular accumulation of Aβ, enhancing the amyloidogenic pathway.[28]

$A\beta_{1-42}$ may initiate free radical chain reactions by itself. It has a critical methionine residue at position 35, which is highly hydrophobic and possesses a sulfur atom sensitive to oxidation (:S: → O=S: → O=S=O). If the lone pair of electrons on the S atom undergoes one-electron oxidation, it produces a positively charged sulfuranyl radical ($MetS^+$). In this manner, S–O bonded $MetS^+$ may initiate free radical chain reactions with allylic H atoms on unsaturated acyl chains of lipids generating the lipid hydroperoxide and propagating the chain reaction. Aβ can also directly trap molecular oxygen, reducing it to H_2O_2 in the presence of iron (Fenton reaction), as demonstrated by spectrochemistry in AD brain. Fe^{2+} ions are generated via a redox cycling of iron ($Fe^{2+} \leftrightarrow Fe^{3+}$), and in the presence of a metal chelator, such as clioquinol, Aβ neurotoxicity is reduced (Fig. 18.5). This matter is relevant because significant alterations in Cu, Zn, and Fe have been found in AD brain in those areas showing severe histopathologic alterations. In general, drugs that prevent oxidative stress include antioxidants, modifiers of the enzymes involved in ROS generation and metabolism, metal-chelating agents, and agents such as anti-inflammatory drugs that remove the stimulus for ROS generation.

Aβ peptides may activate microglia through (i) toll-like receptors 2 (TLR2), (ii) scavenger receptor (SR), (iii) receptor for advanced glycation end products (RAGE), (iv) a cell surface receptor complex, and (v) TNFR1, whose deletion, as observed in APP23 transgenic mice ($APP23/TNFR1^{(-/-)}$), may inhibit Aβ generation and diminish Aβ plaque formation in the brain. Aβ aggregates as foreign protein particles are recognized by TLRs, and these become important Aβ innate immune receptors, as demonstrated in antisense knockdown of TLR2 or using functional blocking antibodies against TLR2, which may suppress Aβ-induced expression of proinflammatory molecules and integrin markers in microglia. Even TLR4 could play a role, as demonstrated in mouse models homozygous for a destructive mutation of TLR4; these show significant increases in diffuse and fibrillar Aβ deposits. However, it is not clear whether TLR signaling pathways involve the clearance of Aβ deposits in the brain or whether they initiate a neuroinflammatory response responsible for the synaptic impairment observed in AD pathology.[29]

A cell surface receptor complex for fibrillar Aβ, linked to the small GTPase Rac1 and critical in signaling to PHOX, has been described (Fig. 18.6). This molecular complex mediates microglial activation through the stimulation of intracellular tyrosine kinase-based signaling cascades, and it is integrated by the B-class scavenger receptor CD36, the integrin-associated protein/CD47, and the a6b1-integrin.[30]

Other scavenger receptors, however, become prominent both in the oxidative response and in the neuroinflammatory response. The macrophage receptor with collagenous structure (MARCO), along with the chemotactic G-protein-coupled receptor formyl-peptide-receptor-like 1 (FPRL1), has been documented to be essential in the amyloid β-induced signal transduction in glial cells.[30]

Neurons, microglia, and endothelial cells, which surround the senile plaques in the AD brain, express higher levels of RAGE; this may trigger oxidative stress and NF-κB activation. The interaction of Aβ with RAGE may be a direct interaction, or it may involve damaged molecular patterns, such as the S100B protein. In primary cortical neurons, the transcription factor Sp1 mediates IL-1β induction by S100B without evidence of a role for NF-κB, whereas in microglia, S100B stimulates NF-κB or AP-1 transcriptional activity and upregulates Cox-2, IL-1β, IL-6, and TNF-α expression through RAGE engagement[31] (Fig. 18.6). The link between NF-κB and neurodegenerative disorders, particularly AD, is well-known. NF-κB may be activated from a variety of means, from the canonical pathway where the proinflammatory TNF-α, IL-1, and LPS exert their action in addition to DAMPS, to the non-canonical pathway where CD40 and lymphotoxin receptors activate a p52/relB complex. Moreover, there are other atypical pathways where genotoxic stress, hypoxia, UV light, H_2O_2, or the epidermal growth factor receptor 2, among others, may intervene.

In rat primary cultures of microglial cells and human neutrophils and monocytes, Aβ activates PHOX; this effect may be potentiated by the proinflammatory stimulus, such as interferon-gamma or TNF-α, but blocked by tyrosine kinase inhibitors.[32] Mediated by PHOX, oligomeric Aβ may induce ROS production, possibly through *N*-methyl-D-aspartate receptors (NMDAR), and these PHOX-related ROS, in turn, release the prostanoid precursor arachidonic acid through the activation of ERKs, which phosphorylate cytosolic phospholipase A2α.

It currently is debatable whether Aβ is a downstream product of the mitochondrial functional decline or whether Aβ-induced mitochondrial damage is an extension of the amyloid cascade hypothesis. The amyloid cascade, proposed 20 years ago, suggested that faulty metabolism of APP was the initiating event in AD pathogenesis, leading subsequently to the aggregation of Aβ, specifically $A\beta_{1-42}$. However, long before the appearance of extracellular Aβ deposits, they are detectable within mitochondria; in fact, it is possible that mitochondrial bioenergetic deficit precedes AD pathology.[33] Leakage of mitochondrial oxygen-free radicals (Fig. 18.7), loss of cellular energy charge, and enhanced opening of the mitochondrial membrane permeability transition pore (MPT) are serious consequences of the presence of Aβ, which has its own receptor inside mitochondria, the Aβ-binding alcohol dehydrogenase (ABAD) protein. Respiratory chain complexes III and IV become particularly affected by Aβ.

We have demonstrated *in vivo* that extracellular depositions of intracerebrally injected Aβ may cross the cytoplasmic membrane into the cytoplasm and, from there, Aβ attacks the mitochondrial membranes to appear inside the mitochondria at the end. Thirty-six hours following the intracerebral injection of fibrillar $A\beta_{1-42}$, brain tissue was obtained and subjected to conventional and transmission electron microscopic examination. Using a polyclonal antibody against Aβ for immunohistochemistry, extracellular deposits of this peptide accompanied by an intense microglial response were revealed, as expected. Ultrathin (70–90 nm) brain sections were incubated with the same anti-Aβ primary antibody, followed by incubation with a gold-labeled secondary antibody. The observation by electron microscopy revealed Aβ immunoreactivity inside mitochondria, accompanied by important swelling, rupture of the outer membrane, and dissolution of cristae. Aβ was found localized to the cristae of the inner membrane of mitochondria. The most prominent change was the peripheral vacuolization of the cristae (Fig. 18.8). Vacuolization in mitochondria was accompanied by electron-dense Aβ-immunoreactivity with inclusions being grouped and bound to the membranes, particularly to the inner cristae membranes. Rupture of the membranes resulted in discontinuity and formation of gaps. Mitochondria appeared swollen, with disorganized membrane structures, and the cristae were lost. In some localized areas, the intermembranous space was absent and the mitochondrion appeared like an irregularly enlarged, single-membrane sac. The incorporation of intracerebrally injected Aβ into mitochondria was related to mitochondrial free radical overproduction. Free radicals were measured by using CM-H2XRos, a potential-dependent probe; this evaluates the direct production of mitochondrial reactive oxygen species in cells. Once located within cells, the reduced CM-H2XRos is oxidized by ROS to a fluorescent mitochondrion-sensitive probe and sequestered in this organelle; thus, its oxidation is useful to detect both ROS (predominantly $O_2^{-\bullet}$) and possibly reactive nitrogen species (•NO and $ONOO^-$)[34] (Fig. 18.6).

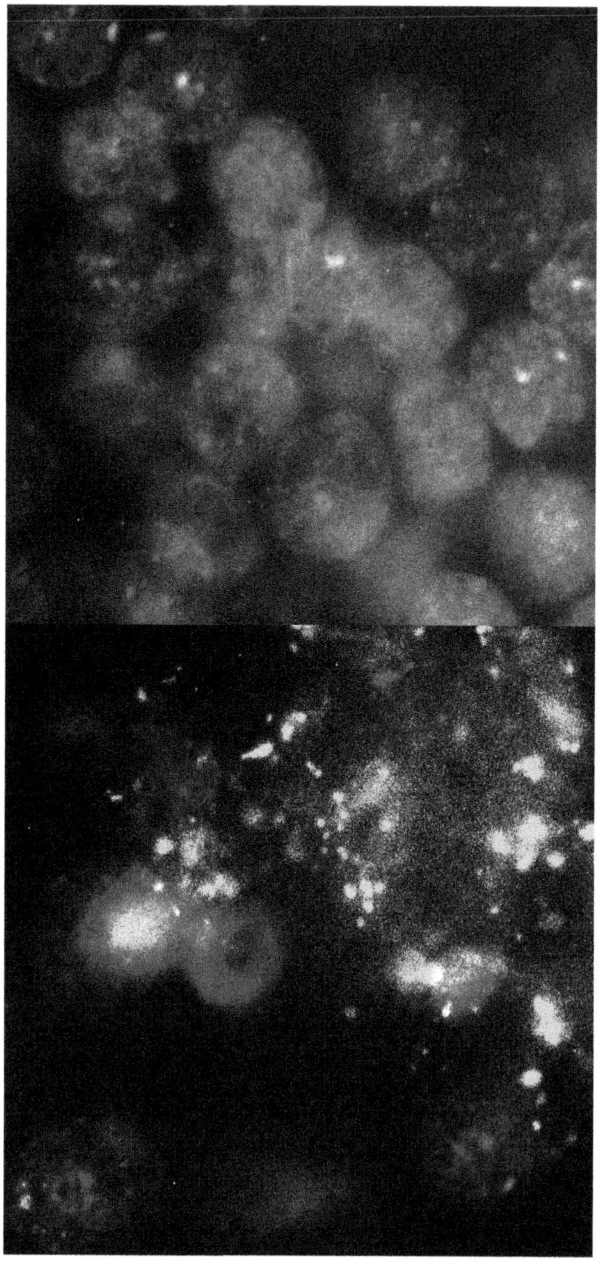

FIGURE 18.7 Red fluorescence identifies mitochondrial ROS by using CM-H2XRos, a specific marker which accumulates inside mitochondria because of the positive charge it acquires upon oxidation by ROS. The photomicrograph on top corresponds to the control group in which animals were injected with PBS into the hippocampus. The amount of mitochondrial free radicals is minimal [bright spots]. However, Aβ-injected brain, as shown in the lower panel, exhibits clusters of fluorescent mitochondria surrounding the nucleus. In fact, mitochondrial clusters are a feature of oxidative damage. The overproduction of ROS in Aβ-injected brain is highly significant, as revealed by multiple bright spots which correspond to the oxidation of CM-H2XRos.[21,34]

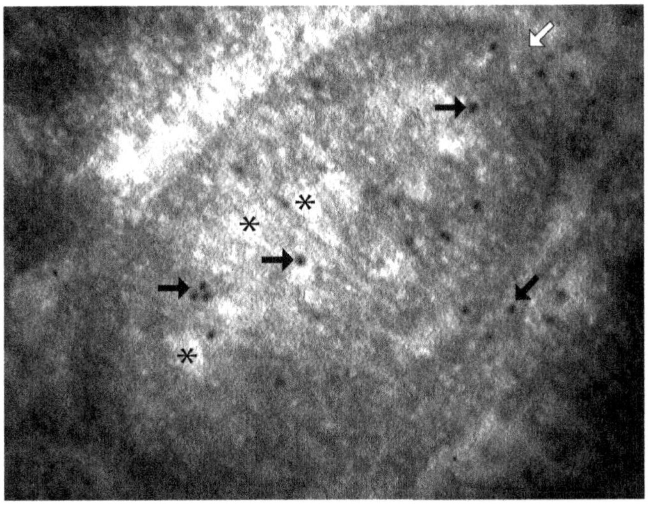

FIGURE 18.8 Aβ in mitochondria, particularly on cristae of the inner membrane accompanied by severe disruption of the organelle. Aβ immunoreactivity (black arrows) within mitochondria 36 hours following the injection of $fA\beta_{1-42}$ into hippocampus CA1 pyramidal neurons (39,000× magnification). Swollen, vacuolated (asterisks) mitochondria with broken cristae. Remnants appear intermixed with fine electron-dense dusty granules and a rupture of the external mitochondrial membranes is observed (white arrow).

Mitochondria represent a major source of oxidants as a consequence of electron leakage in AD brain. Electron leakage from the electron transport chain (ETC) to molecular oxygen leads to $O_2^{-\bullet}$, which combines at substantial rates with •NO to form $ONOO^-$, which affects Complex I, particularly vulnerable to S-nitrosation.[35] Once Complex I becomes damaged, the electron leakage is significantly increased, whereas mitochondrial functionality declines in such a manner that both the mitochondrial respiratory control ratio (RCR) and the hydrolysis activity of F_1F_o-ATPase decrease below their proper rates. RCR measures the ability of mitochondria to idle at a low rate yet respond to ADP by making ATP at a high rate in such a manner that it is feasible to infer the leaking of electron transfer without concomitant phosphorylation, or how much ATP-synthase is partially uncoupled from respiration; F_1F_o-ATPase is useful as a measure of the capacity to maintain an inner-membrane potential by coupling the energy of the electrochemical proton gradient with ATP synthesis. We have found that more extensive oxidative damage is related to a lower respiratory control ratio and a lower ATPase hydrolytic activity.[34]

Following the intracerebral injection of fAβ and its appearance inside mitochondria, a significant reduction in membrane fluidity (measured by estimating the excimer-to-monomer fluorescence intensity ratio (Ie/Im) of DPP) becomes evident. Membrane fluidity is a well-known pathogenic factor directly affecting the energetic coupling of Ca^{2+} pumping with the consequent energetic failure; this correlates with the degree of altered membrane lipid composition. We have reported how exogenous, intracerebrally injected Aβ forms deposits in the extracellular space. However, eventually Aβ peptides appear inside the cells, a phenomenon where axons seem to be the main entrance for Aβ, with a consequent demyelination as has been actually demonstrated in AD brain and AD transgenic mice brain.[36] Finally, in our *in vivo* experiments Aβ is found inside mitochondria where its presence was related to severe structural damage, particularly to mitochondrial membrane lipids, including an increment in cholesterol content, an increase in the cholesterol/total phospholipids ratio, a decrease in the phosphatidyl etanolamine ratio, and a significant rise in saturated fatty acids, loss of membrane fluidity, etc.[21] All of these changes were related to free radical overproduction due to enhanced electron leakage (Figs 18.6 and 18.7).

We speculate that extracellular Aβ paves its own pathway toward mitochondria through oxidative stress-induced alterations in membrane lipids. Once the cytosolic membrane is trespassed, Aβ reaches mitochondrial membranes and it becomes the key factor for mitochondrial failure.

HOW DOES MELATONIN SCAVENGE FREE RADICALS?

Melatonin directly may undergo at least five different transformations, all related to free radical scavenging[37] (Fig. 18.9).

1. Free radicals, such as •OH and LOO•, may abstract a single electron from melatonin, giving rise to a melatonyl cation radical which, in turn, could donate a second electron to $O_2^{-\bullet}$ to produce *N*-acetyl-*N*-formyl-5-methoxykynuramine (AFMK). Once melatonin donates an electron from its electron-rich aromatic indole ring it cannot undergo redox cycling because of the high potential (715 mV, as measured by cyclic voltametry). The melatonyl cation radical, on the other hand, is a resonance-stabilized nitrogen-centered radical which has low reactivity by itself. However, it can be converted by the catalase enzyme to *N*-acetyl-5-methoxykynuramine (AMK), which is also a radical scavenger (a small amount of AMK may be excreted in urine). The melatonyl radical may also donate a second electron directly to another •OH, giving cyclic 3-hydroxymelatonin (C-3-OHM).[38] Thus, each molecule of melatonin could scavenge two radicals. And its metabolites may scavenge numerous radicals.
2. Nitrosilation. Peroxynitrite reacts also with melatonin to form the melatoninyl radical cation. However, $ONOO^-$ may act upon the indole moiety as an oxidizing, a nitrating, and a nitrosating agent, according to the pH and CO_2 of the medium. Thus,

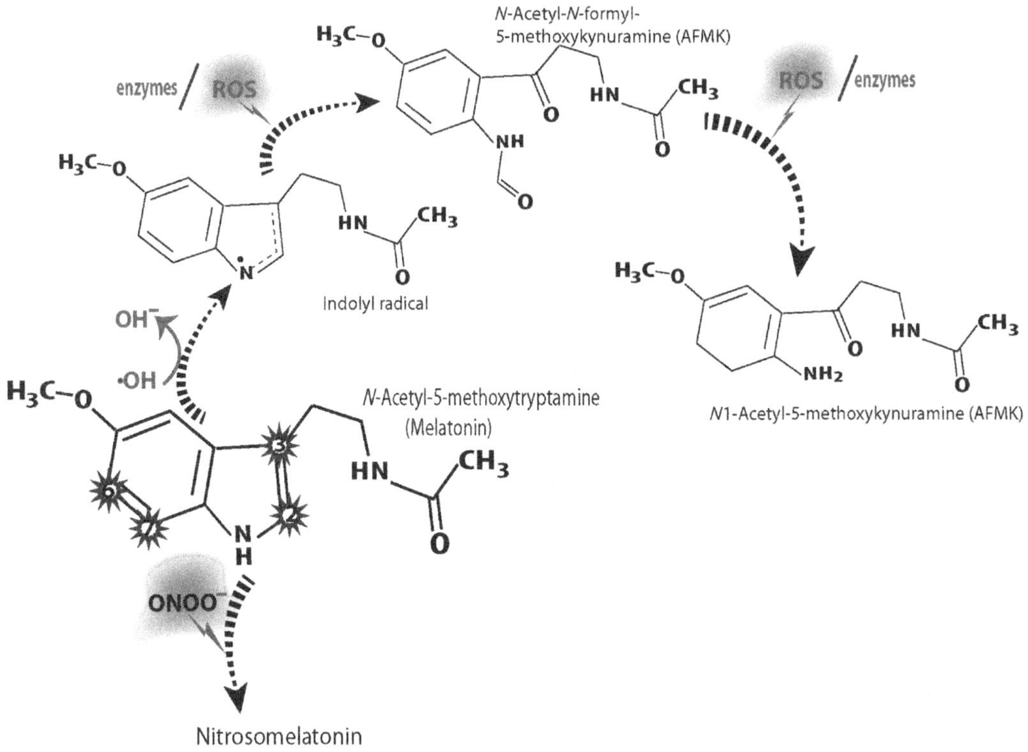

FIGURE 18.9 Free radical scavenging and antioxidant properties of melatonin.

within the pH range of 7.5–8 in the presence of CO_2 the formation of pyrroloindoles 4, indol-2-ones 2, and kynuramines 6 predominates. Nitrosation reactions are favored under physiologic, neutral pH in the absence of added bicarbonate. This latter seems to promote melatonin transformation and the oxidation routes toward pyrroloindoles and indol-2-ones as well as nitrations. Nitration of the indole ring is also feasible in a sequence where •OH is first oxidized and then coupled with $\bullet NO_2$. It involves the transient formation of melatonyl radical and the nitrosating radical ONOO• derived from $ONOO^-$ decay.[39] The resultant nitrosomelatonin may transnitrosate nucleophiles such as thiols and ascorbate.

3. 6-Hydroxylation, with O-demethylation (a relatively minor pathway resulting in 6-hydroxymelatonin (6-HMEL) and *N*-acetyl-5-hydroxytryptamine (*N*-acetylserotonin, NAS), which are excreted in urine as sulfate and glucuronide conjugates). These reactions are mediated by cytochromes P450.
4. 4-Hydroxylation. Small amounts, in transit toward AFMK and by arylamine formamidase or catalase to AMK. Direct scavenging of •OH mediates the immediate reduction of lipid peroxidation, protein oxidation, mitochondrial damage, and DNA damage.
5. 2-Hydroxylation. This is another non-enzymatic transformation of melatonin, mediated by the direct interaction with ROS and also observed following UV-B irradiation and possibly generated *in vivo* as a tautomer of C3-OHM. 2-OH-melatonin may be transformed to AFMK by the cells or directly excreted to urine.

Melatonin's metabolites AFMK and AMK also have the ability to scavenge ROS and RNS. This continuous protection exerted by melatonin and its metabolites is known as the free radical scavenging cascade and it helps to explain how melatonin differs from other conventional antioxidants.[3] Estimated in aqueous solution, the pKa values of AFMK and AMK can be 8.7 and 16.8, respectively; thus, they are expected to be mainly in their neutral form, at a physiologic pH of 7.4. Both react with •OH at diffusion-limited rates ($\sim 10^{10}$/M/s), regardless of the polarity of the environment, and the mechanisms of reaction are by radical adduct formation in a non-polar environment and by hydrogen transfer in aqueous solution.[40]

1. AFMK can be formed by enzymatic, pseudoenzymatic, and non-enzymatic metabolic pathways [i.e. melatonin may react directly with H_2O_2, generating AFMK]. It protects against lipid peroxidation and oxidative DNA damage by scavenging •OH. However, it is a poorer scavenger than AMK and melatonin. Nonetheless, AFMK is a ubiquitous molecule found in plants, unicellular algae, and metazoans, and its activity is

linked to the innate immune response in neutrophils and macrophages which are myeloperoxidase and $O_2^{-\bullet}$ dependent processes. Thus, AFMK may be a primary metabolite of melatonin in organisms under oxidative stress. C3-OHM has been found in a reaction of melatonin with 2,2′-azino-bis(3-ethylbenzthiazoline-6-sulphonic acid) (ABTS) cation radicals.[41] In this reaction, C3-OHM scavenges two ABTS cation radicals to form AFMK. When AFMK interacts with ABTS, this molecule has the ability to donate four electrons and cycles giving rise to indolinones. Finally, AFMK may also be deformylated, by arylamineformamidase, by hemoperoxidases, or by its interaction with ROS/RNS to form AMK.

2. Melatonin and AMK constitute an efficient team of scavengers acting on a wide variety of ROS under different polar and non-polar conditions either by using hydrogen transfer or by radical adduct formation, as mentioned above. AMK is the principal melatonin metabolite in the central nervous system via the kynuric pathway, but it is not the end, because AMK may react actively with RNS forming a stable product that does not easily re-donate NO.[42] It has been demonstrated that melatonin may prevent mitochondrial failure in models of sepsis through its ability to inhibit the expression and activity of both cytosolic (iNOS) and mitochondrial (i-mtNOS) inducible nitric oxide synthases, a function in which AMK has a key role by reducing oxidative/nitrosative stress and restoring Complex I activity.[43] Importantly, intramitochondrial accumulation of melatonin has been reported both under physiologic and pathologic conditions.[44]

MELATONIN STIMULATES ANTIOXIDANT SYSTEMS

Melatonin not only directly scavenges free radicals but it enhances antioxidant endogenous systems (Fig. 18.8).

1. Melatonin stimulates the activity of the enzyme GPx, which reduces intracellular levels of H_2O_2 and other hydroperoxides when they are used as substrates in the oxidation of glutathione. The general scheme for the glutathione system explains how GPx takes two electrons from the reactive oxygen intermediates to oxidize glutathione (GSH), transforming it into GSSG. The enzyme glutathione reductase (GR) reduces GSSG back to GSH again in a process in which NADPH intervenes in its role as an electron carrier. NADPH, in turn, comes from the pentose phosphate pathway generated during the transformation of glucose-6-phosphate to 6-phosphogluconolactone in the presence of the enzyme glucose-6-phosphate dehydrogenase (G6PDH).
The three enzymes GPx, GR, and G6PDH are all reportedly stimulated by melatonin, and the result is reflected in the GSH/GSSG ratio. For example, in experiments in brains under t-butyl hydroperoxide-induced mitochondrial oxidative stress, melatonin increased the activity of the GPx fourfold compared with the basal levels obtained for this enzyme. At the same time, melatonin decreased sigificantly the GSSG levels when practically all GSH had been oxidized to GSSG after the incubation with t-butyl hydroperoxide.[45] The effect was not achieved by using other antioxidants such as vitamin C or vitamin E.
There are significant data indicating that Aβ impairs endogenous antioxidant systems, particularly the GSH-GSSG system, and the process can be reproduced *in vivo*. Thus, following injection of Aβ into the hippocampus, an inverse correlation between GSH-Px activity and lipid peroxidation has been reported.[46] While the GSH-Px activity decreases due to the effect of Aβ, lipid peroxidation increases almost at the same rate.
2. Melatonin also stimulates mRNA levels of MnSOD (mitochondrial) and CuZuSOD (cytosolic) up to 35 and 51%, respectively.[47] This enzyme is responsible for the dismutation of $O_2^{-\bullet}$ leading to H_2O_2, which can be eliminated as water by the action of the enzymes catalase and GPx. It also reduces the likelihood of $O_2^{-\bullet}$ coupling with •NO to form the highly reactive $ONOO^-$. In brain, melatonin treatment may increase SOD activity in aged rats as well as preventing the reduction at the SOD/GPx and GR/GPx ratios.[48] It was reported that SOD enhanced activity in rat kidney, liver, and brain after a single melatonin injection.[49]
3. Melatonin may also restore catalase activity. For example, chronic melatonin treatment in the nigrostriatal dopamine system in the Zitter rat, which displays abnormal metabolism of $O_2^{-\bullet}$ leading to age-related degeneration of the dopaminergic system, attenuates the reduction in the expression of mRNA not only for SOD and GPx but also for catalase, providing a neuroprotective effect.[50] Catalase activity levels were significantly diminished in the oxidatively stressed brain induced by a high dose of adriamycin; however, the excessive lipoperoxidation was significantly reduced and the catalase activity was restored when animals received melatonin.[51] This feature gains importance because the CNS is highly prone to oxidation, and one

catalase molecule can convert millions of molecules of H_2O_2 to water and oxygen each second (one of the highest k_{cat}, or turnover number).

Thus, melatonin is effective as a free radical scavenger and as an antioxidant enhancer (Fig. 18.9). While these effects in the CNS have been observed primarily using pharmacologic doses of melatonin, in a small number of experiments melatonin has also been found to be physiologically relevant as an antioxidant as well.[17]

BREAKING THE CYCLE NEUROINFLAMMATION↔OXIDATIVE STRESS

Melatonin may regulate neuroinflammation through free radical control and modulation of important proinflammatory transcription factors and their signaling pathways while reducing glutamate excitotoxicity. This may be by inhibiting glutamate-induced ion currents or by controlling glutamate release. In this manner, melatonin interrupts the vicious cycle between oxidative stress and neuroinflammation (Fig. 18.10).

Several significant features have been reported, for example:

1. Melatonin prevents the Aβ-induced expression of NF-κB; specifically it inhibits p52 subunit binding to NF-κB as demonstrated by examining the expression of LPS-induced iNOS and COX-2.[52]
2. NF-κB DNA binding activity is inhibited by melatonin and the subsequent cascade of NF-κB-dependent proinflammatory cytokines, as demonstrated *in vivo* in animals receiving melatonin after the injection of Aβ directly into the hippocampus.[53]
3. The nuclear melatonin receptor RZR/ROR has been identified in the promoter region of 5-LOX, a key protagonist in neuroinflammation. By repressing the expression of 5-LOX mRNA in human B lymphocytes, melatonin may reduce the proinflammatory response via this receptor.[54] Furthermore, the transcriptional activation of RZR/RORa by melatonin occurs at a very low dose, i.e. in the nanomolar range.
4. Melatonin may also regulate the primarily microglia-guided neuroinflammatory response by regulating the overactivity of NF-κB, the amount of LPS-induced proinflammatory cytokines, or by preventing glycogen synthase kinase 3 (GSK-3β) activation and neuroinflammation in response to Aβ, as observed in astrocytes and microglial cells. Melatonin interferes with GSK-3β by activating and/or enhancing the activity of PKC, or by inducing Akt. Both PKC and Akt may turn off GSK-3 through phosphorylation. COX-2, related to APP synthesis in astrocytes, is also controlled by melatonin and its metabolites.[52]

MELATONIN: AN ANTI-AβAGENT

The above-mentioned interference with the GSK-3 activation could also interrupt APP synthesis, impairing in this manner the further release of Aβ peptides. The reason is that GSK-3 may interact directly with presenilins within the γ-secretase complex, required for the amyloidogenic APP processing. It has b activation, increases cerebral amyloidosis, and een demonstrated that insulin deficiency leads to enhancement of neural GSK-3α/β activation, increases cerebral amyloidosis, and exacerbates behavioral deficits, as observed in APP/PS1 transgenic mouse model of AD.

Additionally, by stopping GSK-3 activity, melatonin may have a role in reducing tau hyperphosphorylation.

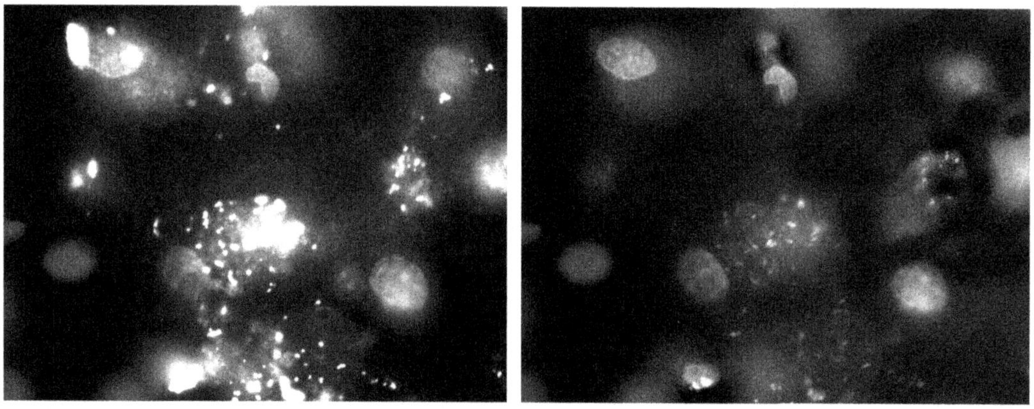

FIGURE 18.10 Fields at left and right are the same. Brain was injected with Aβ but animals received melatonin dissolved in their drinking fluid (Aβ+Mel group). Fluorescent green mitochondria [left] produce free radicals [right], as revealed by CM-H2XRos fluorescence. However, by comparing Aβ (Fig. 18.7) versus Aβ+Mel groups, according to the CM-H2XRos integrated optical density, i.e. $O_2^{-\bullet}$ mass, melatonin reduces very significantly the overproduction of ROS.[21,34]

Tau protein is the other key protagonist in AD pathology, responsible, once hyperphosphorylated, for paired helical filament (PHF) formation. GSK-3 phosphorylates tau; thus, GSK-3 regulates tau binding to microtubules, tau degradation, and tau aggregation.[55]

Furthermore, a direct interaction between melatonin and Aβ has been observed. This phenomenon is not related to the antioxidant properties of melatonin and involves the disruption of the His^+/Asp^- salt bridges in Aβ peptide, which are determinants of the formation and stabilization of β-sheet structures. Thus, 24 hours after incubation with melatonin, Pappolla et al[56] showed that the original β-sheet content of Aβ was significantly diminished in contrast to the increase in β-sheet content when Aβ was incubated alone.

Melatonin 5-methoxy group interacts with Aβ on its His-13 residue, which may be attributed to the higher binding energies in the 5-methoxyindole group, according to single-point energy calculations. By studying the hydrophobic nature of Aβ, this interaction with melatonin was corroborated to take place on the 29-40 Aβ-peptide segment, as revealed by electrospray ionization mass spectrometry (ESI-MS).[57]

MELATONIN PRODUCTION DECREASES WITH AGE

Unfortunately, melatonin levels decrease with age. This occurs in such a manner that children between 5 and 7 years of age have higher melatonin levels than at any point during life, while in young adults those levels are reduced to one-fourth.[58] Thereafter, serum melatonin concentrations in 144 persons aged 30–110 years showed a significant age-related decline.[59] This drop begins around the age of 60 but concentrations become significantly lower in subjects in their 70s and over 80 years of age. The circadian pattern of serum melatonin is also lost with increasing age, which leads to a weakened circadian system in elderly individuals.

In AD and other types of dementia, decreased levels of melatonin – exceeding those observed during normal aging – have been repeatedly reported.[60]

Melatonin levels in cerebral spinal fluid (CSF) from AD patients are reportedly significantly decreased with early neuropathologic AD-related changes in the temporal cortex. In aged patients, melatonin levels in the CSF have been found to be one-half of those in young control subjects, but in patients with AD, the CSF melatonin levels are only one-fifth of those in young subjects. In fact, it is possible to replicate hippocampal CA1 and CA3 pyramidal neuron loss in rats by merely removing the pineal gland (which lowers circulating melatonin levels) with this effect being reversed by melatonin replacement in the drinking water.[61] Also, constant light exposure, which decreases serum melatonin, shows that the loss of endogenous melatonin is enough to cause Alzheimer-like damage, such as memory deficits, tau hyperphosphorylation at multiple sites, and activation of GSK-3β and protein kinase A, as well as suppression of protein phosphatase-1 and prominent oxidative stress.[62]

CONCLUSIONS

There currently is no effective treatment for AD. Acetylcholinesterase inhibitors and memantine are the only currently FDA-approved drugs for AD. These drugs have only a modest trend favoring active treatment over placebo. Even worse, more than 100 years after the first clinical report of a case of AD, there is not yet a satisfactory hypothesis or a model capable of explaining or reproducing the pathogenic mechanisms of this devastating disease.

By examining the more plausible hypotheses, however, all of them eventually show that there is a common factor: oxidative stress. Oxidative stress and inflammation feed each other a vicious cycle which, eventually, kills the neurons. Melatonin is not only an outstanding free radical scavenger and antioxidant, as explained in this chapter, but it has a specific role as an anti-inflammatory agent. Thanks to its amphiphatic nature, melatonin neutralizes free radicals both in the aqueous cytosol and in lipidic membranes. It crosses all the natural barriers, acting even at nanomolar concentrations. Very importantly, scavenging of free radicals by melatonin is an irreversible process, which means that melatonin does not become a pro-oxidant agent.

Even though there is insufficient clinical evidence to support the effectiveness of melatonin by itself in managing the cognitive and noncognitive sequelae of people with dementia, there is a growing body of evidence indicating the potential role of melatonin as an effective adjuvant in AD management. There are molecular and physiologic bases that are worth analyzing, because melatonin may have an effective influence on several of the most prominent features which contribute to the cause of this disease.

References

1. Huang X, Atwood CS, Hartshorn MA, et al. The A beta peptide of Alzheimer's disease directly produces hydrogen peroxide through metal ion reduction. *Biochemistry* 1999;**38**(24):7609–16.
2. Pogocki D, Schoneich C. Redox properties of Met(35) in neurotoxic beta-amyloid peptide. A molecular modeling study. *Chem Res Toxicol* 2002;**15**(3):408–18.
3. Muller DP, Metcalfe T, Bowen DM. Vitamin E in brains of patients with Alzheimer's disease and Down's syndrome. *Lancet* 1986;**1**(8489):1093–4.

4. Mecocci P, Polidori MC. Antioxidant clinical trials in mild cognitive impairment and Alzheimer's disease. *Biochim Biophys Acta* 2012;**1822**(5):631–8.
5. Cardinali DP, Furio AM, Brusco LI. Clinical aspects of melatonin intervention in Alzheimer's disease progression. *Curr Neuropharmacol* 2010;**8**(3):218–27.
6. Brusco LI, Marquez M, Cardinali DP. Monozygotic twins with Alzheimer's disease treated with melatonin: case report. *J Pineal Res* 1998;**25**(4):260–3.
7. Tan DX, Hardeland R, Manchester LC, et al. The changing biological roles of melatonin during evolution: from an antioxidant to signals of darkness, sexual selection and fitness. *Biol Rev Camb Philos Soc* 2010;**85**(3):607–23.
8. Coon SL, Weller JL, Korf HW, et al. cAMP regulation of arylalkylamine *N*-acetyltransferase (AANAT, EC 2.3.1.87): a new cell line (1e7) provides evidence of intracellular AANAT activation. *J Biol Chem* 2001;**276**(26):24097–107.
9. Jang SW, Liu X, Pradoldej S, et al. *N*-Acetylserotonin activates Trkb receptor in a circadian rhythm. *Proc Natl Acad Sci USA* 2010;**107**(8):3876–81.
10. Stehle JH, Saade A, Rawashdeh O, et al. A survey of molecular details in the human pineal gland in the light of phylogeny, structure, function and chronobiological diseases. *J Pineal Res* 2011;**51**(1):17–43.
11. Ceinos RM, Chansard M, Revel F, et al. Analysis of adrenergic regulation of melatonin synthesis in Siberian hamster pineal emphasizes the role of HIOMT. *Neurosignals* 2004;**13**(6):308–17.
12. Tan DX, Manchester LC, Liu X, et al. Mitochondria and chloroplasts as the original sites of melatonin synthesis: a hypothesis related to melatonin's primary function and evolution in eukaryotes. *J Pineal Res* 2013;**54**(2):127–38.
13. Pappolla MA, Simovich MJ, Bryant-Thomas T, et al. The neuroprotective activities of melatonin against the Alzheimer beta-protein are not mediated by melatonin membrane receptors. *J Pineal Res* 2002;**32**(3):135–42.
14. Kaur C, Ling EA. Effects of melatonin on macrophages/microglia in postnatal rat brain. *J Pineal Res* 1999;**26**(3):158–68.
15. Pozo D, Reiter RJ, Calvo JR, Guerrero JM. Inhibition of cerebellar nitric oxide synthase and cyclic GMP production by melatonin via complex formation with calmodulin. *J Cell Biochem* 1997;**65**(3): 430–42.
16. Wiesenberg I, Missbach M, Carlberg C. The potential role of the transcription factor Rzr/Ror as a mediator of nuclear melatonin signaling. *Restor Neurol Neurosci* 1998;**12**(2-3):143–50.
17. Reiter RJ. Oxidative damage in the central nervous system: protection by melatonin. *Prog Neurobiol* 1998;**56**(3):359–84.
18. Rosales-Corral S, Reiter RJ, Tan DX, et al. Functional aspects of redox control during neuroinflammation. *Antioxid Redox Signal* 2010;**13**(2):193–247.
19. Andriezen WL. Brain energy metabolism: an integrated cellular perspective. In: Press R, editor. *Psychopharmacology – 4th generation of progress*. New York: Bloom, F.E. & Kupfer, D.J; 2000. p. 2002.
20. Axelsen PH, Komatsu H, Murray IV. Oxidative stress and cell membranes in the pathogenesis of Alzheimer's disease. *Physiology (Bethesda)* 2011;**26**(1):54–69.
21. Rosales-Corral SA, Lopez-Armas G, Cruz-Ramos J, et al. Alterations in lipid levels of mitochondrial membranes induced by amyloid-beta: a protective role of melatonin. *Int J Alzheimers Dis* 2012;**2012**:459806.
22. O'Brien JS, Sampson EL. Lipid composition of the normal human brain: gray matter, white matter, and myelin. *J Lipid Res* 1965;**6**(4):537–44.
23. Thies W, Bleiler L. Alzheimer's disease facts and figures. *Alzheimers Dement* 2011;**7**(2):208–44.
24. Hebert LE, Scherr PA, Bienias JL, et al. Alzheimer disease in the US population: prevalence estimates using the 2000 Census. *Arch Neurol* 2003;**60**(8):1119–22.
25. Rosales-Corral SA, Acuna-Castroviejo D, Coto-Montes A, et al. Alzheimer's disease: pathological mechanisms and the beneficial role of melatonin. *J Pineal Res* 2012;**52**(2):167–202.
26. Glabe CG. Structural classification of toxic amyloid oligomers. *J Biol Chem* 2008;**283**(44):29639–43.
27. Pacher P, Beckman JS, Liaudet L. Nitric oxide and peroxynitrite in health and disease. *Physiol Rev* 2007;**87**(1):315–424.
28. Shimohama S, Tanino H, Kawakami N, et al. Activation of NADPH oxidase in Alzheimer's disease brains. *Biochem Biophys Res Commun* 2000;**273**(1):5–9.
29. Salminen A, Suuronen T, Kaarniranta K. Rock, Pak, and Toll of synapses in Alzheimer's disease. *Biochem Biophys Res Commun* 2008;**371**(4):587–90.
30. Bamberger ME, Harris ME, McDonald DR, et al. A cell surface receptor complex for fibrillar beta-amyloid mediates microglial activation. *J Neurosci* 2003;**23**(7):2665–74.
31. Bianchi R, Giambanco I, Donato R. S100b/RAGE-dependent activation of microglia via Nf-Kappab and Ap-1 co-regulation of Cox-2 expression by S100b, IL-1beta and TNF-alpha. *Neurobiol Aging* 2010;**31**(4):665–77.
32. Bianca VD, Dusi S, Bianchini E, et al. Beta-amyloid activates the O-2 forming NADPH oxidase in microglia, monocytes, and neutrophils. A possible inflammatory mechanism of neuronal damage in Alzheimer's disease. *J Biol Chem* 1999;**274**(22):15493–9.
33. Caspersen C, Wang N, Yao J, et al. Mitochondrial abeta: a potential focal point for neuronal metabolic dysfunction in Alzheimer's disease. *FASEB J* 2005;**19**(14):2040–1.
34. Rosales-Corral S, Acuna-Castroviejo D, Tan DX, et al. Accumulation of exogenous amyloid-beta peptide in hippocampal mitochondria causes their dysfunction: a protective role for melatonin. *Oxid Med Cell Longev* 2012;**2012**:843649.
35. Dahm CC, Moore K, Murphy MP. Persistent S-nitrosation of Complex I and other mitochondrial membrane proteins by S-nitrosothiols but not nitric oxide or peroxynitrite: implications for the interaction of nitric oxide with mitochondria. *J Biol Chem* 2006;**281**(15):10056–65.
36. Lassmann H. Mechanisms of neurodegeneration shared between multiple sclerosis and Alzheimer's disease. *J Neural Transm* 2011;**118**(5):747–52.
37. Tan DX, Manchester LC, Terron MP, et al. One molecule, many derivatives: a never-ending interaction of melatonin with reactive oxygen and nitrogen species? *J Pineal Res* 2007;**42**(1):28–42.
38. Tan DX, Manchester LC, Reiter RJ, et al. A novel melatonin metabolite, cyclic 3-hydroxymelatonin: a biomarker of in vivo hydroxyl radical generation. *Biochem Biophys Res Commun* 1998;**253**(3):614–20.
39. Peyrot F, Martin M-T, Migault J, Ducrocq C. Reactivity of peroxynitrite with melatonin as a function of Ph and CO_2 content. *Eur J Organic Chem* 2003;**2003**(1):172–81.
40. Galano A, Tan DX, Reiter RJ. On the free radical scavenging activities of melatonin's metabolites, Afmk and Amk. *J Pineal Res* 2013;**54**(3):245–57.
41. Tan DX, Hardeland R, Manchester LC, et al. Mechanistic and comparative studies of melatonin and classic antioxidants in terms of their interactions with the Abts cation radical. *J Pineal Res* 2003;**34**(4):249–59.
42. Hardeland R, Backhaus C, Fadavi A. Reactions of the NO redox forms NO^+, *NO and HNO (protonated NO^-) with the melatonin metabolite N1-acetyl-5-methoxykynuramine. *J Pineal Res* 2007;**43**(4):382–8.
43. Tapias V, Escames G, Lopez LC, et al. Melatonin and its brain metabolite N(1)-acetyl-5-methoxykynuramine prevent mitochondrial nitric oxide synthase induction in Parkinsonian mice. *J Neurosci Res* 2009;**87**(13):3002–10.
44. Lopez A, Garcia JA, Escames G, et al. Melatonin protects the mitochondria from oxidative damage reducing oxygen consumption, membrane potential, and superoxide anion production. *J Pineal Res* 2009;**46**(2):188–98.

45. Martin M, Macias M, Escames G, et al. Melatonin but not vitamins C and E maintains glutathione homeostasis in t-butyl hydroperoxide-induced mitochondrial oxidative stress. *FASEB J* 2000;**14**(12):1677–9.
46. Rosales-Corral S, Tan DX, Reiter RJ, et al. Kinetics of the neuroinflammation-oxidative stress correlation in rat brain following the injection of fibrillar amyloid-β onto the hippocampus in vivo. *J Neuroimmunol* 2004;**150**(1):20–8.
47. Antolin I, Rodriguez C, Sainz RM, et al. Neurohormone melatonin prevents cell damage: effect on gene expression for antioxidant enzymes. *FASEB J* 1996;**10**(8):882–90.
48. Ozturk G, Akbulut KG, Guney S, Acuna-Castroviejo D. Age-related changes in the rat brain mitochondrial antioxidative enzyme ratios: modulation by melatonin. *Exp Gerontol* 2012;**47**(9):706–11.
49. Liu F, Ng TB. Effect of pineal indoles on activities of the antioxidant defense enzymes superoxide dismutase, catalase, and glutathione reductase, and levels of reduced and oxidized glutathione in rat tissues. *Biochem Cell Biol* 2000;**78**(4):447–53.
50. Hashimoto K, Ueda S, Ehara A, et al. Neuroprotective effects of melatonin on the nigrostriatal dopamine system in the Zitter rat. *Neurosci Lett* 2012;**506**(1):79–83.
51. Montilla P, Tunez I, Munoz MC, et al. Antioxidative effect of melatonin in rat brain oxidative stress induced by adriamycin. *Rev Esp Fisiol* 1997;**53**(3):301–5.
52. Deng WG, Tang ST, Tseng HP, Wu KK. Melatonin suppresses macrophage cyclooxygenase-2 and inducible nitric oxide synthase expression by inhibiting P52 acetylation and binding. *Blood* 2006;**108**(2):518–24.
53. Rosales-Corral S, Tan DX, Reiter RJ, et al. Orally administered melatonin reduces oxidative stress and proinflammatory cytokines induced by amyloid-beta peptide in rat brain: a comparative, in vivo study versus vitamin C and E. *J Pineal Res* 2003;**35**(2):80–4.
54. Steinhilber D, Brungs M, Werz O, et al. The nuclear receptor for melatonin represses 5-lipoxygenase gene expression in human B lymphocytes. *J Biol Chem* 1995;**270**(13):7037–40.
55. Hoppe JB, Frozza RL, Horn AP, et al. Amyloid-beta neurotoxicity in organotypic culture is attenuated by melatonin: involvement of Gsk-3beta, tau and neuroinflammation. *J Pineal Res* 2010;**48**(3):230–8.
56. Pappolla M, Bozner P, Soto C, et al. Inhibition of Alzheimer beta-fibrillogenesis by melatonin. *J Biol Chem* 1998;**273**(13):7185–8.
57. Skribanek Z, Balaspiri L, Mak M. Interaction between synthetic amyloid-beta-peptide (1-40) and its aggregation inhibitors studied by electrospray ionization mass spectrometry. *J Mass Spectrom* 2001;**36**(11):1226–9.
58. Waldhauser F, Weiszenbacher G, Frisch H, et al. Fall in nocturnal serum melatonin during prepuberty and pubescence. *Lancet* 1984;**1**(8373):362–5.
59. Zhao ZY, Xie Y, Fu YR, et al. Aging and the circadian rhythm of melatonin: a cross-sectional study of Chinese subjects 30-110 yr of age. *Chronobiol Int* 2002;**19**(6):1171–82.
60. Hardeland R. Neurobiology, pathophysiology, and treatment of melatonin deficiency and dysfunction. *Scientific World Journal* 2012;**2012**:640389.
61. De Butte M, Pappas BA. Pinealectomy causes hippocampal Ca1 and Ca3 cell loss: reversal by melatonin supplementation. *Neurobiol Aging* 2007;**28**(2):306–13.
62. Ling ZQ, Tian Q, Wang L, et al. Constant illumination induces Alzheimer-like damages with endoplasmic reticulum involvement and the protection of melatonin. *J Alzheimers Dis* 2009;**16**(2):287–300.

CHAPTER

19

Mitochondria-Targeted Antioxidants and Alzheimer's Disease

Vladimir P. Skulachev, Nikolay K. Isaev

Lomonosov Moscow State University, A.N. Belozersky Institute of Physico-Chemical Biology, Moscow, Russia

Nadezhda A. Kapay, Olga V. Popova, Elena V. Stelmashook

Department of Brain Research, Research Center of Neurology, Russian Academy of Medical Sciences, Pereulok Obukha 5, Moscow, Russia

Konstantin G. Lyamzaev

Lomonosov Moscow State University, A.N. Belozersky Institute of Physico-Chemical Biology, Moscow, Russia

Irina N. Scharonova

Department of Brain Research, Research Center of Neurology, Russian Academy of Medical Sciences, Pereulok Obukha 5, Moscow, Russia

Dmitry B. Zorov

Lomonosov Moscow State University, A.N. Belozersky Institute of Physico-Chemical Biology, Moscow, Russia

Vladimir G. Skrebitsky

Department of Brain Research, Research Center of Neurology, Russian Academy of Medical Sciences, Pereulok Obukha 5, Moscow, Russia

List of Abbreviations

Abeta (Aβ) β-amyloid peptide 1-42
AD Alzheimer's disease
APP amyloid precursor protein
BI brain ischemia
C_{12}TPP dodecyltriphenylphosphonium
GSK3 glycogen synthase kinase-3
HFS high frequency stimulation
LTP long-term post-tetanic potentiation of hippocampal pathway
MitoQ 10-(6′-ubiquinonyl)decyl-triphenylphosphonium
ROS reactive oxygen species
mROS mitochondrial reactive oxygen species
SkQ1 10-(6′-plastoquinonyl)decyltriphenylphosphonium
SkQR1 10-(6′-plastoquinonyl)decylrhodamine 19
TBI traumatic brain injury

INTRODUCTION

Available data suggest that human aging is accompanied by a decrease in brain volume and impairment in cognitive function rather than by an essential loss of brain neurons. These changes are caused by a shrinkage of neuronal bodies and atrophy of dendritic fields in neurophils, reducing the length of myelinated axons and lowering the number of synapses and causing changes in their structure.[1–3] These age-related effects are not critical for survival, but their increase promotes the onset of Alzheimer's disease (AD) associated with massive neuronal damage and disintegration of various

Aging
http://dx.doi.org/10.1016/B978-0-12-405933-7.00019-6

parts of the brain – leading to a complete block of its functions, and death. Alzheimer's disease is known to be accompanied by age-dependent deficits in cognition, and by essential memory and synapsis loss. Abeta and hyperphosphorylated tau protein were shown to trigger synaptic degeneration and mitochondrial dysfunction.[4] This occurs through a number of destructive signaling pathways, including excessive production of reactive oxygen species (ROS) by mitochondria.[5,6] The reason for the emergence of AD is not fully understood, but it has been demonstrated that traumatic brain injury (TBI) and brain ischemia (BI) increase the risk of onset of AD.[7,8] Recently it was shown that mitochondria-targeted antioxidants can reduce the toxic effects of Abeta[9–12] levels, which are elevated in the brains of AD patients. These compounds rescue hippocampal long-term potentiation, a recognized neuronal model of memory that has been impaired by Abeta. A set of data suggest that BI and TBI promote excessive production and aggregation of Abeta in the brain.[13,14] Moreover, the risk of AD onset is dramatically increased with age. Mitochondria-targeted antioxidants can reduce, and in some cases reverse, the development of a large group of typical traits of aging.[15–17] We suggest that mitochondria-targeted antioxidants have a high therapeutical potential for reducing the risk of AD onset and might alleviate the consequences of developed AD.

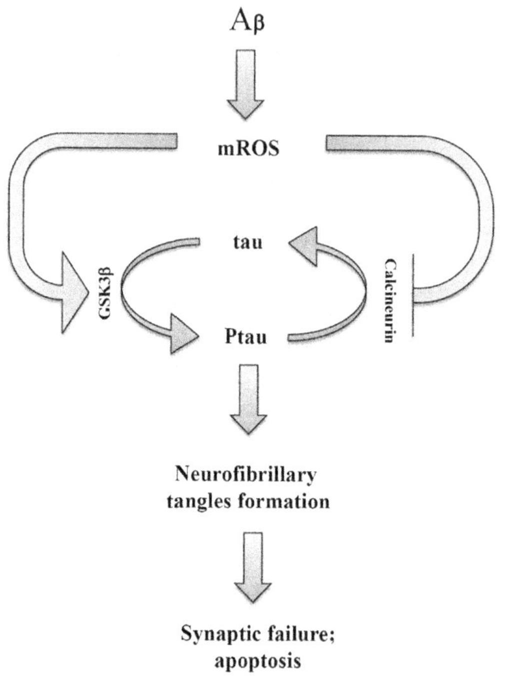

FIGURE 19.1 Mitochondrial reactive oxygen species (mROS) mediate the β-amyloid (Aβ)-induced damage to Alzheimer's disease neurons. *Modified from Lloret et al.*[11]

MITOCHONDRIAL REACTIVE OXYGEN SPECIES AS PROBABLE MEDIATORS OF THE ABETA-INDUCED DAMAGE TO ALZHEIMER'S DISEASE NEURONS

Alzheimer's disease is a neurodegenerative pathology characterized by age-dependent mitochondrial dysfunction.[5,6] Reddy and Reddy[5] described a vast set of pathologic changes in these organelles, namely, mutations in mtDNA, abnormal mitochondrial gene expression, decreased mitochondrial enzyme activities, binding of mutated amyloid precursor protein (APP) and β-amyloid peptide (Abeta) to mitochondrial membranes in neurons, mitochondrial fragmentation and decreased mitochondrial fusion, abnormal trafficking of mitochondria, and decreased bioenergetic functions of mitochondria. Abeta and hyperphosphorylated tau protein were shown to trigger mitochondrial dysfunction. This occurs through a number of pathways, including excessive production of reactive oxygen species (ROS), resulting in oxidative modification of mitochondrial proteins, which, in turn, impairs oxidative phosphorylation, apparently contributing to the onset and progression of the disease.[4] Vina and his colleagues[18] identified AD-inherent events occurring downstream of Abeta-induced mitochondrial ROS. They found that ROS upregulate expression of the regulator of the calcineurin 1 gene (RCAN1). The product of this gene inhibits the ability of calcineurin to operate as a phosphatase for phosphorylated tau protein. As a result, tau protein is hyperphosphorylated by glycogen synthase kinase-3 (GSK3). This effect is stimulated by an RCAN1-induced increase in tau-kinase activity of GSK3. Hyperphosphorylation of tau leads to neurofibrillary tangle formation, thus causing synaptic failure and apoptosis (Fig. 19.1).

MITOCHONDRIA-TARGETED ANTIOXIDANTS REDUCE THE TOXIC EFFECTS OF ABETA IN MODELS OF ALZHEIMER'S DISEASE

The use of antioxidants that can attenuate the effects of oxidative stress on the brain in AD has been discussed in many studies.[19–22] However, the effect of conventional antioxidants (vitamin E, alpha-lipoic acid, acetyl-L-carnitine, *N*-acetylcysteine) is rather nonspecific for cellular compartments. It is also worth noting that these antioxidants are efficient at rather high concentrations. For example, antioxidant *N*-acetylcysteine has been used with some efficacy in the treatment of AD, but its concentrations in this clinical trial were as high as 50 mg/kg/day (0.3 mmol/kg/day).[22] High concentrations of *N*-acetylcysteine and other classic antioxidants could have some dangerous side effects if used on a long-term basis. Potentially, this disadvantage might be overcome by using mitochondria-targeted antioxidants such as SkQ, MitoQ, MitoVitE, MitoPBN, MitoPeroxidase, glutathione choline esters, latripiridine, and Szeto-Schiller (SS) peptides.[15,23–25] It has been calculated that, under equilibrium, the SkQ1 concentration in the inner leaflet of the inner mitochondrial membrane is about 10^8 times higher than in the extracellular liquid.[15] Such a huge concentrative effect is due to (i) electrophoretic accumulation of SkQ1 supported by electric potentials on the plasma membrane and the inner mitochondrial membrane (collectively about 10^4 times), and (ii) a very high membrane/water distribution coefficient of SkQ1 (10^4 times more).[15] The SkQ-family of antioxidants show a wide concentration window between antioxidant and prooxidant activities inherent in quinones.[16] One more advantage is that SkQ is a rechargeable antioxidant. Its oxidized form, which appears during antioxidant activity, is immediately reduced by center *i* of the respiratory chain complex III, localized in the inner leaflet of the inner mitochondrial membrane.[16,26] This is why SkQs look like promising therapeutic tools to prevent ROS-induced oxidation of mitochondrial lipids and proteins *in vivo*.[15,25] Mitochondria-targeted drugs were suggested for the treatment of ROS-associated aging-related pathologies, e.g. AD.[9–12]

Submicromolar concentrations of Abeta peptide are known to impair synaptic transmission in glutamate synapses of the brain.[27] The electrical response of hippocampal slices to high-frequency stimulation, namely, long-term potentiation of synaptic transmission in the hippocampal pathway (Schaffer collaterals – CA1 field), can be used as a model of synaptic changes associated with learning and memory.[28] Using this model, we have shown that treatment with Abeta impairs long-term post-tetanic potentiation (LTP) induction in rat hippocampus slices. A single intraperitoneal injection of SkQ1 (Fig. 19.2) or SkQR1[10–12] 1 day before the preparation of the slice was found to abolish the inhibitory effect of Abeta on the LTP in the slice. Moreover, it was shown that SkQ1 abolishes the inhibitory effect of Abeta on LTP in the hippocampal slices pretreated with SkQ1 *in vitro* (Fig. 19.3).

Other researchers have established that the mitochondria-directed antioxidant MitoQ, which contains CoQ_{10} as an active part of the molecule, compensates the Abeta-induced impairment of long-term synaptic plasticity in the hippocampus.[9] It is notable that

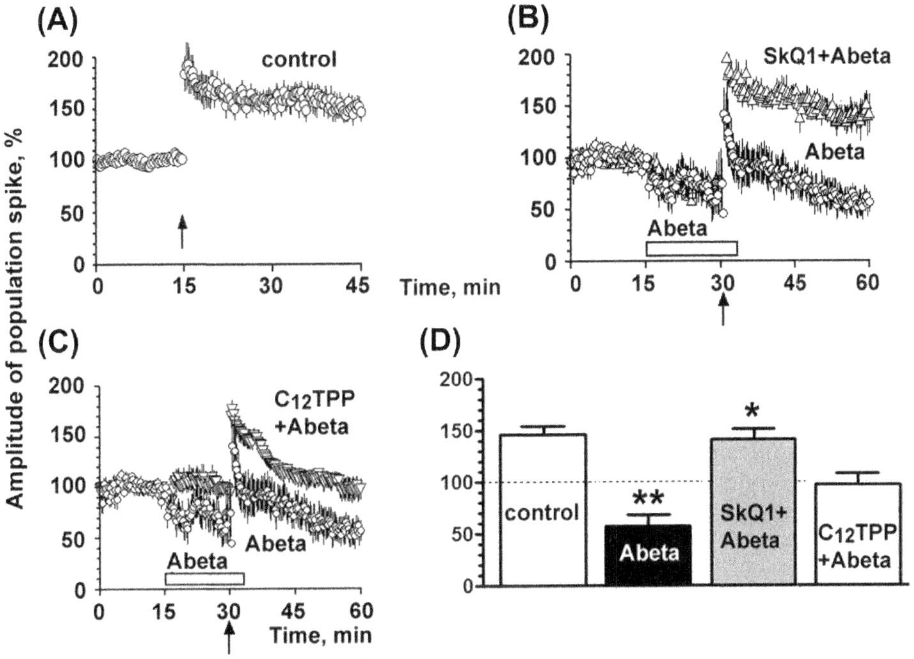

FIGURE 19.2 SkQ1, but not C_{12}TPP, abolishes the inhibitory effect of β-amyloid peptide 1-42 (Abeta) on LTP in the hippocampal slices from animals *in vivo* pretreated with SkQ1 or C_{12}TPP. The arrow indicates the start of high-frequency stimulation (HFS), and the box shows the duration of treatment of the slices with 200 nM Abeta. (A) Population spike (PS) in the hippocampal slices. The control animals. (B) PS in hippocampal slices treated with Abeta, obtained from the animals not treated with SkQ1 (circles) or from the animals *in vivo* pretreated with SkQ1 (single injection of 250 nmol SkQ1/kg a day before the experiment, triangles). (C) PS in hippocampal slices treated with 200 nM Abeta, obtained from the animals not treated with C_{12}TPP (circles) or from animals *in vivo* pretreated with 250 nmol C_{12}TPP/kg (triangles). (D) Mean amplitudes of population spike recorded 30 min after HFS. The number of slices in the control, Abeta, SkQ1+ Abeta, and C_{12}TPP+ Abeta groups were 8, 6, 4, and 4, respectively. $*p < 0.01$ as compared to Abeta; $**p < 0.01$ as compared to control. *From Kapay et al[12] with permission.*

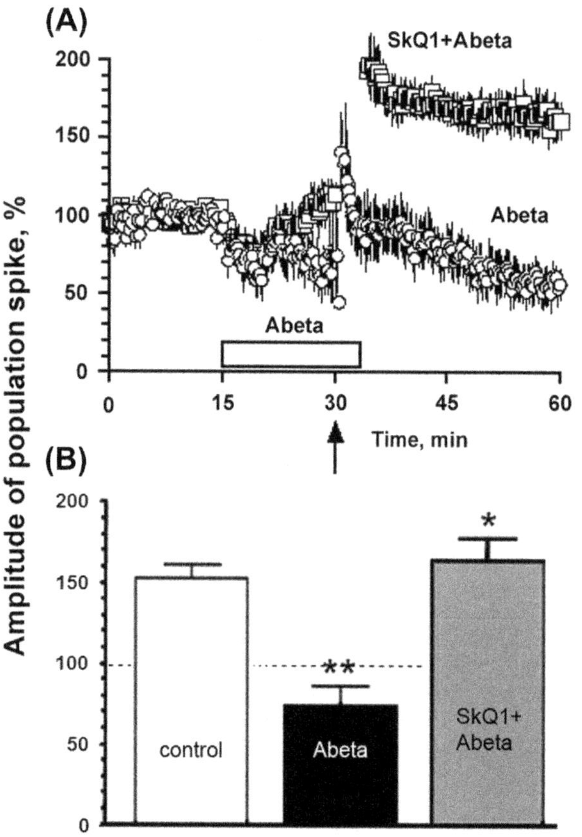

FIGURE 19.3 SkQ1 abolishes the inhibitory effect of β-amyloid peptide 1-42 (Abeta) on LTP in the hippocampal slices pretreated with SkQ1 *in vitro*. The arrow indicates the start of HFS, and the box shows the duration of treatment of the slices with 200 nM Abeta. (A) Population spike (PS) in hippocampal slices treated with 200 nM Abeta, not treated with SkQ1 (circles), or in hippocampal slices pretreated *in vitro* with 250 nM SkQ1 for 60 min (squares). (B) Mean amplitudes of PS recorded 30 min after HFS. The number of slices in the control, Abeta, and SkQ1+ Abeta groups were 8, 7, and 5, respectively. $*p < 0.01$ as compared to Abeta; $**p < 0.01$ as compared to control. *From Kapay et al[12] with permission.*

MitoQ lowered excessive mitochondrial ROS production induced by Abeta on hippocampal slices and N2a cells[9,24] and prevented a pathology similar to Alzheimer's disease in transgenic mice; this was revealed by (i) reduction of cognitive abilities, (ii) accumulation of Abeta, (iii) loss of synapses, and (iv) caspase activation in the brain.[25] Structurally, both MitoQ and SkQ1 contain two moieties, i.e. an antioxidative 'head' (quinone) and a cationic decyltriphenylphosphonium 'tail' responsible for their electrophoretic accumulation by a mitochondrion. To elucidate which part of the molecule is responsible for its activity, we explored the effect of C_{12}TPP, dodecyltriphenylphosphonium cation laking quinone moiety. We found that C_{12}TPP fails to abolish the Abeta effect (Fig. 19.2, B and D). Earlier, our group found that C_{12}TPP and its rhodamine-containing analog C_{12}R1 facilitate a decrease in mitochondrial membrane potential by free fatty acids.[29,30] Later, we showed that C_{12}R1 has some uncoupling (protonophorous) activity even without fatty acids.[31] The resulting mild uncoupling[32] was shown to reduce ROS generation in isolated mitochondria.[29,32] However, such an effect is apparently insufficient for complete prevention of the Abeta effect on hippocampus because such action was demonstrated in a statistically valid fashion by SkQ1 rather than by C_{12}TPP (Fig. 19.2). An advantage of SkQ substances is that they not only directly inhibit the production of mitochondrial ROS, but also – when administered *in vivo* – increase the resistance of cells to adverse effects by reducing the activity of proapoptotic protein GSK3β,[33,34] which is an important link in the cascade of toxic effects of Abeta (Fig. 19.1).

MITOCHONDRIA-TARGETED ANTIOXIDANTS IMPROVE NEUROLOGIC RECOVERY AFTER TRAUMATIC BRAIN INJURY OR STROKE AND REDUCE THE RISK OF DEVELOPING ALZHEIMER'S DISEASE

As already mentioned, TBI and BI increase the risk of onset of AD.[7,8] TBI can lead to the development of a long-term neurodegenerative process known as chronic traumatic encephalopathy. This pathology in many respects resembles AD.[35] The development of TBI and BI significantly depends on the increased mitochondrial production of ROS.[36,37] These active molecules can directly damage lipids, proteins, and nucleic acids in the cell. ROS also activate different molecular signaling pathways associated with cell death.[38] The increased production of ROS by mitochondria under ischemia and trauma is one of the most important pathogenetic factors in the mechanism of the neurodestruction, as well as under AD. AD-induced oxidative damage of proteins, lipids, or nucleic acids is limited to a defined group of neurons that degenerate when AD develops. The cytoplasmic nucleic acid damage may involve not only RNA but also mitochondrial DNA which – due to the absence of histones, proximity to the places of reactive oxygen generation, and poor repair mechanisms – is highly susceptible to oxidative damage.[39,40] So the oxidative damage of mitochondrial DNA appears to be more probable than that of nuclear DNA.[41] Some data suggest that BI and TBI promote excessive production and aggregation of Abeta in the brain.[13,14] Abeta was supposed to be an important factor in mediating initial pathogenic events in AD. Levels of total Abeta (Abeta-40 and Abeta-42) in the neocortical brain regions strongly correlate with cognitive decline by AD dementia.[42] Inhibition of glycogen synthase kinase-3, an effect which mediates the Abeta-induced damage of AD neurons, reduces spatial

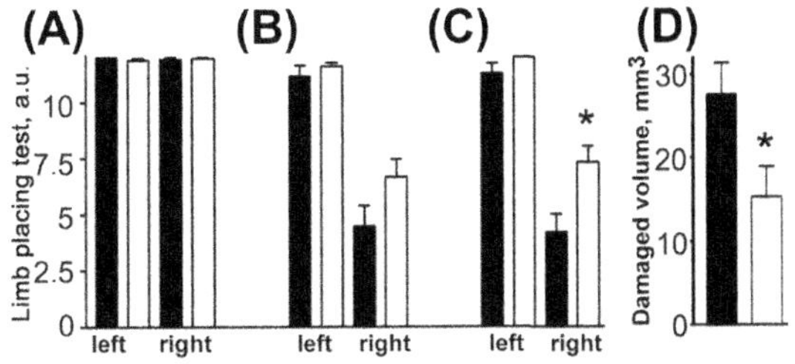

FIGURE 19.4 Effect of SkQR1 on the neurologic deficit and volume of brain damage caused by focal trauma of the sensorimotor cortex. (A) 'Limb placing test' in intact rats (before trauma), on the third (B) and on the seventh (C) day. (D) Volume of the brain-damaged area on the seventh day after the trauma. 100 nM SkQR1/kg was intraperitoneally injected 60 min after trauma and then once per day. $^*p < 0.05$ comparing the group of animals treated with SkQR1 (white columns) and not treated with SkQR1 (black columns). *From Isaev et al[44] with permission.*

learning deficits in mice with traumatic brain injury.[43] We recently found that intraperitoneal injection of SkQR1 after focal brain trauma or focal brain ischemia improves performance of a neurologic test and makes the damage to a cortical zone smaller[44,45] (Fig. 19.4).

Abrahamson et al[14] suggested that the same strategy as used for brain injury and stroke treatment can also be applied for reducing the damage caused by the accumulation of Abeta in order to delay or even prevent development of chronic Alzheimer's disease. Mitochondria-targeted antioxidants can potentially be used for these purposes.

THE GEROPROTECTIVE EFFECT OF SKQ1

As we pointed out above, the risk of AD onset increases dramatically with age, with Abeta possibly being involved in this effect. The amyloid-binding dye thioflavin T suppresses pathologic features of mutant metastable proteins and human β-amyloid-associated toxicity for adult *Caenorhabditis elegans*, slows down aging of this worm, and extends its lifespan.[46] Perhaps, mitochondria-targeted antioxidants might reduce the risk of AD with age because they were shown to decelerate some age-related pathologies in animal models. When studying the effects of SkQ1 on the behavior of rats in the cross maze and open field tests, it was observed that 1-year-old Wistar rats and rats with accelerated aging (OXYS strain) – if treated with SkQ1 – showed a significantly higher locomotive and exploratory activity when compared with the control groups. SkQ1 prevented age-related changes in rats, i.e. reduction in vision due to the development of retinopathy and cataracts; similar results were obtained for glaucoma, uveitis, conjunctivitis, and dry eye syndrome.[47,48] The geroprotective effect of SkQ1 is accompanied by inhibition, termination, and, in some cases, reversal of the development of a large group of

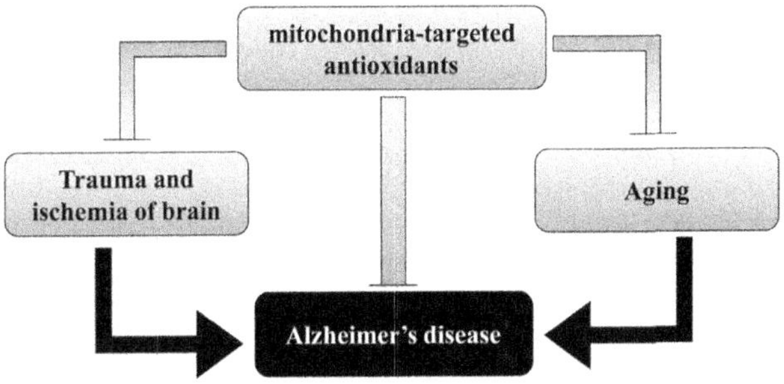

FIGURE 19.5 Therapeutic strategy in the use of mitochondria-targeted antioxidants for the treatment of Alzheimer's disease.

typical traits of aging, such as osteoporosis and lordokyphosis, canities, alopecia, loss of whiskers, slow wound healing, sarcopenia, extension of the left ventricle, and the disappearance of estrous cycles in females and of male libido.[48]

Summarizing the results presented above, we may conclude that mitochondria-targeted antioxidants seem promising for the treatment of ROS-associated aging-related pathologies, including Alzheimer's disease (Fig. 19.5).

SUMMARY POINTS

- Alzheimer's disease is a neurodegenerative pathology accompanied by age-related mitochondrial dysfunction and the production of mitochondrial reactive oxygen, triggered by β-amyloid and hyperphosphorylated tau protein.
- Traumatic brain injury and brain ischemia increase the risk of the development of Alzheimer's disease.
- Mitochondria-targeted antioxidants can reduce the toxic effects of β-amyloid *in vitro* (β-amyloid treatment of hippocampal slices) and *in vivo* (β-amyloid increase in brains due to traumatic injury and ischemia).
- It is assumed that mitochondria-targeted antioxidants have a high potential for reducing the risk of onset of Alzheimer's disease, and might alleviate the consequences of developed Alzheimer's disease.

Acknowledgment

This study was supported by the Research Institute of Mitoengineering; RFBR grants, 12-04-00025 and 11-04-00360.

References

1. Freeman SH, Kandel R, Cruz L, et al. Preservation of neuronal number despite age-related cortical brain atrophy in elderly subjects without Alzheimer disease. *J Neuropathol Exp Neurol* 2008;**67**:1205–12.
2. Fjell AM, Walhovd KB. Structural brain changes in aging: courses, causes and cognitive consequences. *Rev Neurosci* 2010;**21**:187–221.

3. Isaev NK, Stelmashook EV, Stelmashook NN, et al. Brain aging and mitochondria-targeted plastoquinone antioxidants of SkQ-type. *Biochem (Mosc)* 2013;**3** (in press).
4. Schmitt K, Grimm A, Kazmierczak A, et al. Insights into mitochondrial dysfunction: aging, amyloid-β, and tau-A deleterious trio. *Antioxid Redox Signal* 2012;**16**:1456–66.
5. Reddy PH, Reddy TP. Mitochondria as a therapeutic target for aging and neurodegenerative diseases. *Curr Alzheimer Res* 2011;**8**:393–409.
6. Chakrabarti S, Munshi S, Banerjee K, et al. Mitochondrial dysfunction during brain aging: role of oxidative stress and modulation by antioxidant supplementation. *Aging Dis* 2011;**2**:242–56.
7. Jellinger KA. Traumatic brain injury as a risk factor for Alzheimer's disease. *J Neurol Neurosurg Psychiatr* 2004;**75**:511–2.
8. Kalaria RN. The role of cerebral ischemia in Alzheimer's disease. *Neurobiol Aging* 2000;**21**:321–30.
9. Ma T, Hoeffer CA, Wong H, et al. Amyloid β-induced impairments in hippocampal synaptic plasticity are rescued by decreasing mitochondrial superoxide. *J Neurosci* 2011;**31**:5589–95.
10. Kapay NA, Isaev NK, Stelmashook EV, et al. *In vivo* injected mitochondria-targeted plastoquinone antioxidant SkQR1 prevents β-amyloid-induced decay of long-term potentiation in rat hippocampal slices. *Biochemistry (Mosc)* 2011;**76**:1367–70.
11. Skulachev VP. Mitochondria-targeted antioxidants as promising drugs for treatment of age-related brain diseases. *J Alzheimer's Dis* 2012;**28**:283–9.
12. Kapay NA, Popova OV, Isaev NK, et al. Mitochondria-targeted plastoquinone antioxidant SkQ1 prevents inhibiting effect of β-amyloid on long term potentiation in rat hippocampal slices while its analog lacking plastoquinone is much less effective. *J Alzheimer's Dis* 2012 (in press).
13. Pluta R, Furmaga-Jabłońska W, Maciejewski R, et al. Brain ischemia activates β- and γ-secretase cleavage of amyloid precursor protein: significance in sporadic Alzheimer's disease. *Mol Neurobiol* 2012. http://dx.doi.org/10.1007/s12035-012-8360-z.
14. Abrahamson EE, Ikonomovic MD, Dixon CE, DeKosky ST. Simvastatin therapy prevents brain trauma-induced increases in beta-amyloid peptide levels. *Ann Neurol* 2009;**66**:407–14.
15. Skulachev MV, Antonenko YN, Anisimov VN, et al. Mitochondrial-targeted plastoquinone derivates. Effect on senescence and acute age-related pathologies. *Curr Drug Targets* 2011;**12**:800–26.
16. Skulachev VP, Anisimov VN, Antonenko YN, et al. An attempt to prevent senescence: a mitochondrial approach. *Biochim Biophys Acta* 2009;**1787**:437–61.
17. Anisimov VN, Egorov MV, Krasilshchikova MS, et al. Effects of the mitochondria-targeted antioxidant SkQ1 on lifespan of rodents. *Aging (Albany NY)* 2011;**3**:1110–9.
18. Lloret A, Badia MC, Giraldo E, et al. Alzheimer's amyloid-β toxicity and tau hyperphosphorylation are linked via RCAN1. *J Alzheimer's Dis* 2011;**27**:701–9.
19. Muller WE, Eckert A, Kurz C, et al. Mitochondrial dysfunction: common final pathway in brain aging and Alzheimer's disease – therapeutic aspects. *Mol Neurobiol* 2010;**41**:159–71.
20. Palacios HH, Yendluri BB, Parvathaneni K, et al. Mitochondrion-specific antioxidants as drug treatments for Alzheimer's disease. *CNS Neurol Disord Drug Targets* 2011;**10**:149–62.
21. Facecchia K, Fochesato LA, Ray SD, et al. Oxidative toxicity in neurodegenerative diseases: role of mitochondrial dysfunction and therapeutic strategies. *J Toxicol* 2011. http://dx.doi.org/10.1155/2011/683728. 683728.
22. Adair JC, Knoefel JE, Morgan N. Controlled trial of N-acetylcysteine for patients with probable Alzheimer's disease. *Neurology* 2011;**57**:1515–7.
23. Calkins MJ, Manczak M, Reddy PH. Mitochondria-targeted antioxidant ss31 prevents amyloid beta-induced mitochondrial abnormalities and synaptic degeneration in Alzheimer's disease. *Pharmaceuticals (Basel)* 2012;**5**:1103–19.
24. Manczak M, Mao P, Calkins MJ, et al. Mitochondria-targeted antioxidants protect against amyloid-beta toxicity in Alzheimer's disease neurons. *J Alzheimers Dis* 2010;**20**:S609–31.
25. McManus MJ, Murphy MP, Franklin JL. The mitochondria-targeted antioxidant MitoQ prevents loss of spatial memory retention and early neuropathology in a transgenic mouse model of Alzheimer's disease. *J Neurosci* 2011;**31**:15703–15.
26. Antonenko YN, Avetisyan AV, Bakeeva LE, et al. Mitochondria-targeted plastoquinone derivatives as tools to interrupt execution of the aging program. 1. Cationic plastoquinone derivatives: synthesis and *in vitro* studies. *Biochem (Mosc)* 2008;**73**:1273–87.
27. Selkoe DJ. Alzheimer's disease is a synaptic failure. *Science* 2002;**298**:789–91.
28. Malenka RC, Nicoll RA. Long-term potentiation—a decade of progress? *Science* 1999;**285**:1870–4.
29. Severin FF, Severina II, Antonenko YN, et al. Penetrating cation/fatty acid anion pair as a mitochondria-targeted protonophore. *Proc Natl Acad Sci USA* 2010;**107**:663–8.
30. Plotnikov EY, Silachev DN, Jankauskas SS, et al. Mild uncoupling of respiration and phosphorylation as a mechanism providing nephro and neuroprotective effects of penetrating cations of the SkQ family. *Biochem (Mosc)* 2012;**77**:1029–37.
31. Antonenko YN, Avetisyan AV, Cherepanov DA, et al. Derivatives of rhodamine 19 as mild mitochondria-targeted cationic uncouplers. *J Biol Chem* 2011;**286**:17831–40.
32. Skulachev VP. Role of uncoupled and non-coupled oxidations in maintenance of safely low levels of oxygen and its one-electron reductants. *Q Rev Biophys* 1996;**29**:169–202.
33. Plotnikov EY, Chupyrkina AA, Jankauskas SS, et al. Mechanisms of nephroprotective effect of mitochondria-targeted antioxidants under rhabdomyolysis and ischemia/reperfusion. *Biochim Biophys Acta* 2011;**1812**:77–86.
34. Silachev DN, Isaev NK, Pevzner IB, et al. The mitochondria-targeted antioxidants and remote kidney preconditioning ameliorate brain damage through kidney-to-brain cross-talk. *PLoS One* 2012;**7**:e51553.
35. Blennow K, Hardy J, Zetterberg H. The neuropathology and neurobiology of traumatic brain injury. *Neuron* 2012;**76**:886–99.
36. Andriessen TM, Jacobs B, Vos PE. Clinical characteristics and pathophysiological mechanisms of focal and diffuse traumatic brain injury. *J Cell Mol Med* 2010;**14**:2381–92.
37. Zorov DB, Filburn CR, Klotz LO, et al. Reactive oxygen species (ROS)-induced ROS release: a new phenomenon accompanying induction of the mitochondrial permeability transition in cardiac myocytes. *J Exp Med* 2000;**192**:1001–14.
38. Niizuma K, Yoshioka H, Chen H, et al. Mitochondrial and apoptotic neuronal death signaling pathways in cerebral ischemia. *Biochim Biophys Acta* 2010;**1802**:92–9.
39. Beal MF. Aging, energy, and oxidative stress in neurodegenerative diseases. *Ann Neurol* 1995;**38**:357–66.
40. Nunomura A, Perry G, Pappolla MA, et al. RNA oxidation is a prominent feature of vulnerable neurons in Alzheimer's disease. *J Neurosci* 1999;**19**:1959–64.
41. Mecocci P, MacGarvey U, Beal MF. Oxidative damage to mitochondrial DNA is increased in Alzheimer's disease. *Ann Neurol* 1994;**36**:747–51.
42. Naslund J, Haroutunian V, Mohs R, et al. Correlation between elevated levels of amyloid beta-peptide in the brain and cognitive decline. *JAMA* 2000;**283**:1571–7.
43. Yu F, Zhang Y, Chuang DM. Lithium reduces BACE1 overexpression, beta amyloid accumulation, and spatial learning deficits in mice with traumatic brain injury. *J Neurotrauma* 2012;**29**:2342–51.
44. Isaev NK, Novikova SV, Stelmashook EV, et al. Mitochondria-targeted plastoquinone antioxidant SkQR1 decreases trauma-induced neurological deficit in rat. *Biochem (Mosc)* 2012;**77**:996–9.

45. Silachev DN, Pevzner IB, Zorova LD, et al. New generation of permeated cations as potential agents to rescue from ischemic stroke. *FEBS J* 2011;**278**:280.
46. Alavez S, Vantipalli MC, Zucker DJ, et al. Amyloid-binding compounds maintain protein homeostasis during ageing and extend lifespan. *Nature* 2011;**472**:226–9.
47. Stefanova NA, Fursova AZh, Kolosova NG. Behavioral effects induced by mitochondria-targeted antioxidant SkQ1 in Wistar and senescence-accelerated OXYS rats. *J Alzheimer's Dis* 2010;**21**:479–91.
48. Skulachev VP. What is 'phenoptosis' and how to fight it? *Biochem (Mosc) Phenoptosis* 2012;**77**:689–706.

CHAPTER

20

Downregulation of the Prooxidant Heart Failure Phenotype by Dietary and Nondietary Antioxidants

Victor Farah, Raza Askari, Shadwan Alsafwah, Dwight A. Dishmon, Syamal K. Bhattacharya, Karl T. Weber

Division of Cardiovascular Diseases, University of Tennessee Health Science Center, Memphis, TN, USA

List of Abbreviations

AAs African-Americans
ALDOST aldosterone-salt treatment
Amlod amlodipine
ANS adrenergic nervous system
CHF congestive heart failure
EF ejection fraction
mPTP membrane permeability transition pore
MSTE mitochondriocentric signal-transducer-effector
MT metallothionein
MTF metal-responsive transcription factor
NAC *N*-acetylcysteine
NADPH nicotinamide adenine dinucleotide phosphate
NO nitric oxide
PBMC peripheral blood mononuclear cells
PDTC pyrrolidine dithiocarbamate
PTH parathyroid hormone
RAAS renin–angiotensin–aldosterone system
SHPT secondary hyperparathyroidism
SSM subsarcolemmal mitochondria

INTRODUCTION

The congestive heart failure (CHF) syndrome, with its characteristic symptoms and signs, is now the leading admitting diagnosis at hospitals in the USA. Its origins are rooted in a salt-avid state induced by an inappropriate and deleterious homeostatic response – neurohormonal activation – evoked by kidneys underperfused by a failing heart. This maladaptive homeostatic response gone awry causes dyshomeostasis,[1] in which the retention of salt and water is mediated by effector hormones of the activated renin–angiotensin–aldosterone system (RAAS) and the adrenergic nervous system (ANS). The current medical management of patients with CHF therefore draws on pharmacologic regulation of these effector hormones by either disrupting their formation or receptor ligand binding.

Endocrine properties of RAAS hormones orchestrate a systemic illness whose features are referred to as a prooxidant and proinflammatory phenotype.[2–6] Its clinical manifestations include: (i) oxidative stress, where the rate of generation of reactive oxygen species overwhelms their rate of detoxification, largely rendered by Zn^{2+}-based endogenous antioxidant defenses; (ii) activated peripheral blood mononuclear cells (PBMC: lymphocytes and monocytes) releasing proinflammatory cytokines (e.g. TNF-α and IL-6); and (iii) a wasting of soft tissues and bone, termed cardiac cachexia.

In addition to this systemic illness there is a progressive pathologic remodeling of the myocardium, including the ongoing loss of cardiomyocytes via necrotic and apoptotic forms of cell death. Elevations in plasma troponins, biomarkers of myocyte necrosis, are found in patients hospitalized because of their CHF.[7–13] Each recurrent episode of decompensated failure is based on neurohormonal reactivation and can be punctuated with bouts of myocyte necrosis that account for progressive pathologic remodeling of myocardium, which is an arguably postmitotic organ and incapable of myocyte regeneration. A prominent feature of this remodeling is the appearance of foci of microscopic scars, footprints of cardiomyocyte necrosis. These scars are found widely scattered throughout both the right and left atria and

Aging
http://dx.doi.org/10.1016/B978-0-12-405933-7.00020-2

ventricles, irrespective of the etiologic origins of heart failure.[14–16] Contrary to the inflammatory cell and fibroblast responses which follow myocyte necrosis is the absence of tissue repair following the rapid engulfment of apoptotic cells by tissue macrophages. Hence, apoptosis leaves behind no such morphologic footprint.[17] In both the explanted failing human heart and postmortem tissue harvested from the failing myocardium – and irrespective of its etiologic origins – is widespread fibrosis recognized as the major component of pathologic cardiac remodeling. Microscopic scarring underscores the crucial pathophysiologic role of nonischemic cardiomyocyte necrosis.[14]

Mechanisms accounting for ongoing myocyte necrosis are therefore of fundamental importance. This turns attention to the pathophysiologic impact of oxidative stress and the role of catecholamines, angiotensin II, and aldosterone in promoting monovalent and divalent cation dyshomeostasis with compromised antioxidant defenses, and the susceptibility of cardiomyocytes to necrosis. Several micronutrients are deficient in patients with heart failure, including vitamin D, which may play a pathophysiologic role in the genesis of the prooxidant cardiomyopathic phenotype and its progressive nature.[18–21] Herein, we address pathophysiologic mechanisms leading to this prooxidant heart failure phenotype and the potential role of dietary and nondietary antioxidants in regulating this phenotype – salvaging cardiomyocytes, preventing myocardial scarring, and hence serving as cardioprotectants.

CONGESTIVE HEART FAILURE: A PROOXIDANT PHENOTYPE

Secondary Hyperparathyroidism and Intracellular Ca^{2+} Overloading

Increased circulating levels of parathyroid hormone (PTH) have been recognized as an integral feature of the aldosteronism seen with CHF.[22–33] As seen in Figure 20.1, there are marked urinary and fecal excretory losses of Ca^{2+} and Mg^{2+} that accompany aldosteronism and the resultant appearance of ionized hypocalcemia and hypomagnesemia, both of which are stimuli to the parathyroid glands' increased secretion of PTH; thus secondary hyperparathyroidism (SHPT) arises.[34–36] PTH-mediated excessive intracellular Ca^{2+} accumulation occurs via G protein-coupled receptor binding, which has been coined a Ca^{2+} paradox.[37] Collectively, the CHF syndrome therefore features catecholamine- and PTH-mediated excessive intracellular Ca^{2+} accumulation in cardiomyocytes leading to Ca^{2+} uptake and ensuing Ca^{2+} overloading of their mitochondria (see Fig. 20.2). This, in turn, provokes the induction of oxidative stress by these organelles, where their rate of reactive oxygen species generation overwhelms their rate of detoxification by endogenous intracellular antioxidant defenses. The ensuing opening of their inner membrane permeability pore leads to osmotic swelling and structural degeneration, with necrotic death of cardiomyocytes (see Fig. 20.2). Tissue repair with fibrous tissue formation, or scarring, follows. This replacement fibrosis preserves the structural integrity of this injured hollow muscular organ and constitutes a morphologic footprint of lost necrotic cells. Fibrosis, however, has adverse consequences: the addition of stiff fibrillar type I collagen in scar tissue compromises ventricular function in diastole and systole; and fibrosis, with its resident myofibroblasts, serves as substrate for arrhythmias.[38]

The catecholamine- and PTH-mediated Ca^{2+} overloading associated nonischemic necrosis of cardiomyocytes is accompanied by the release of intracellular contents (see Fig. 20.2), including troponin, a protein whose appearance in the circulation plays a crucial role in discerning myocardial injury. A modest increase in plasma troponins has been reported in diverse hyperadrenergic stressor states such as decompensated heart failure, sepsis, hemorrhagic shock, subarachnoid hemorrhage, trauma, gastrointestinal bleeding, or pulmonary embolus.[39] The levels to which plasma troponins rise, however, do not

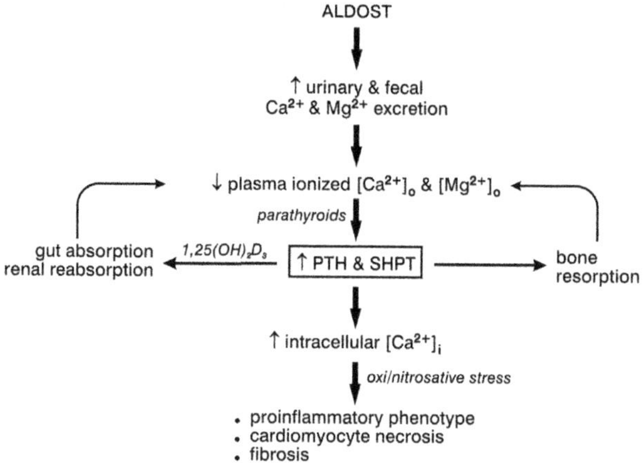

FIGURE 20.1 Excretory cation losses leading to secondary hyperparathyroidism. Inappropriate (relative to dietary Na^+) elevations in plasma aldosterone comparable to those found in human CHF accompany aldosterone/salt treatment (ALDOST) in rats. This chronic aldosteronism is responsible for increased urinary and fecal excretion of Ca^{2+} and Mg^{2+} leading to plasma ionized hypocalcemia and hypomagnesemia. Secondary hyperparathyroidism (SHPT) is invoked to restore extracellular Ca^{2+} and Mg^{2+} homeostasis via parathyroid hormone (PTH)-mediated bone resorption and increased absorption and reabsorption of these cations from the gut and kidneys, respectively, mediated by the steroid hormone $1,25(OH)_2D_3$ synthesized by the kidneys. Paradoxically, PTH is responsible for intracellular Ca^{2+} overloading, which leads to oxi/nitrosative stress and a proinflammatory phenotype with cardiomyocyte necrosis. Reprinted with permission from Kamalov G et al J Cardiovasc Pharmacol 2010;56:320–328.

reach those seen with the segmental loss of myocardium that accompanies ischemia with infarction.

Elevated PTH levels are found in patients hospitalized with decompensated heart failure and those awaiting cardiac transplantation,[22–25] where aldosteronism is expected. SHPT has recently been recognized as an independent predictor of CHF, the need for hospitalization, and cardiovascular mortality.[26–29] Moreover, PTH level is an independent risk factor for mortality and cardiovascular events in community-dwelling individuals.[30–32] SHPT induced by plasma ionized hypocalcemia is especially prevalent in African-Americans (AAs) with protracted (>4 weeks) decompensated biventricular failure, where chronic aldosteronism contributes to symptoms and signs of CHF.[23,33] SHPT is also associated with the prevalence of hypovitaminosis D in AAs, where the increased melanin content of dark skin serves as a natural sunscreen.[23] Accordingly, the prevalence of hypovitaminosis D, often of marked severity with plasma 25(OH)D <20 ng/mL, compromises Ca^{2+} homeostasis, predisposing AAs to ionized hypocalcemia and consequent SHPT.[23,40,41] Vitamin D deficiency is frequently found in Caucasians and Asians with heart failure whose effort intolerance predisposes to an indoors lifestyle.[26,27,42,43] Other factors, such as lactose intolerance with reduced dietary Ca^{2+}, which may be associated with compromised Ca^{2+} stores and contribute to the appearance of SHPT, especially in AAs with CHF, have been reviewed elsewhere.[3,21]

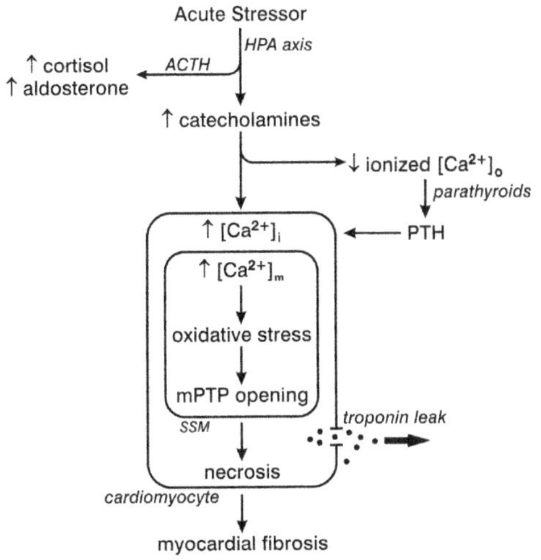

FIGURE 20.2 Mitochondriocentric signal-transducer-effector pathway to nonischemic cardiomyocyte necrosis. An acute stressor state, such as bodily injury, is accompanied by an activation of the hypothalamic–pituitary–adrenal (HPA) axis with elevated circulating catecholamines and adrenocorticotropin stimulating the release of cortisol and aldosterone from the adrenal glands. Epinephrine and norepinephrine are responsible for intracellular Ca^{2+} overloading with a subsequent fall in plasma ionized $[Ca^{2+}]_o$, which, in turn, provokes the parathyroid glands to release parathyroid hormone (PTH). It, too, is responsible for intracellular Ca^{2+} overloading. In cardiomyocytes this is accompanied by the induction of oxidative stress that leads to the opening of the mitochondrial permeability transition pore (mPTP) and osmotic injury of these organelles. The necrosis of a small number of cardiomyocytes follows, accompanied by the leak of troponins into the interstitial space and ultimate rise in plasma troponins, albeit a lesser rise than seen with acute myocardial infarction due to a critical reduction in blood flow to a segment of myocardium. Lost to necrosis, cardiac myocytes are replaced by fibrosis, or scarring, to preserve the structural integrity of myocardium. Adapted with permission from Whitted AD et al Am J Med Sci 2010;340:48–53.

Zinc Dyshomeostasis: A Scenario of Lost Antioxidants

A deficiency in antioxidant reserves also contributes to the imbalance in prooxidant:antioxidant equilibrium, leading to cardiomyocyte necrosis which accompanies neurohormonal activation of CHF.[44,45] Zinc is integral to antioxidant defenses, as well as Zn^{2+}-based matrix metalloproteinases involved in wound healing.[46] Upregulation of metallothionein, a Zn^{2+}-binding protein, occurs at sites of tissue injury, including the heart, where it promotes local sequestration of Zn^{2+} that participates in gene transcription and cell replication.[47,48] Zn^{2+} deficiency, evident with reduced tissue Zn^{2+} levels,[49,50] compromises these antioxidant reserves and healing after cardiomyocyte necrosis.[51–53]

In the aldosteronism of CHF, increased urinary and fecal losses of Zn^{2+} result in hypozincemia with simultaneous cellular and subcellular dyshomeostasis of Zn^{2+}.[48,49] The accompanying Zn^{2+} deficiency compromises the activity of Cu/Zn superoxide dismutase, an important endogenous antioxidant. Moreover, urinary Zn^{2+} excretion is increased in response to angiotensin-converting enzyme inhibitor or angiotensin receptor antagonist, commonly used in the management of CHF.[54,55] Serum and myocardial Zn^{2+} levels are reduced in patients with a dilated cardiomyopathy and individuals with arterial hypertension,[33,50,56–61] where underlying pathogenic mechanisms responsible for Zn^{2+} deficiency remain to be elucidated. A Zn^{2+}/Se^{2+} deficiency-associated cardiomyopathy has been reported in patients having malabsorption after intestinal bypass surgery.[61]

Intricate interactions between Zn^{2+} and Ca^{2+} have long been recognized.[46,62–64] The prooxidant effect involving intracellular Ca^{2+} overloading that accompanies elevations in either plasma catecholamines or PTH is intrinsically coupled to increased Zn^{2+} entry in cardiomyocytes acting as an antioxidant.[62,65–67] One of the modes of Zn^{2+} entry is known to occur via L-type Ca^{2+} channels. However, Zn^{2+} transporters activated by oxidative stress are the predominant source of intracellular Zn^{2+} entry (see Fig. 20.3). Increased cytosolic free $[Zn^{2+}]_i$ may be attained via its adaptive release from inactive intracellular Zn^{2+} bound to metallothionein (MT)-1 in response to acute or chronic cellular injury, and which can be induced by nitric oxide (NO) derived from endothelial or inducible NO synthases[68] or dietary sources, such as vegetables or $NaNO_3$.[69–72]

ANTIOXIDANTS DOWNREGULATE THE PROOXIDANT HEART FAILURE PHENOTYPE

Dietary Antioxidants

SHPT accounts for intracellular Ca^{2+} overloading of diverse cells, including PBMC, and the simultaneous appearance of oxidative stress in CHF. In rats with chronic aldosteronism, created by 4 weeks of aldosterone-salt treatment (ALDOST), we demonstrated that calcitriol ($1,25(OH)_2D_3$), together with a diet supplemented with Ca^{2+} and Mg^{2+}, would prevent SHPT and Ca^{2+} overloading of PBMC, and thereby abated oxidative stress in these cells. We found that this regimen prevented SHPT and abrogated PBMC intracellular Ca^{2+} overloading and their increased production of H_2O_2. These observations raise the intriguing prospect that acquired SHPT in patients with CHF could be managed by macro- and micronutrients with antioxidant properties.[73]

A dyshomeostasis of macro- and micronutrients, including vitamin D, together with oxidative stress, are common pathophysiologic features in AAs with CHF. Reductions in plasma 25(OH)D levels of moderate-to-marked severity (<20 ng/mL) may be accompanied by ionized hypocalcemia and consequent elevations in serum PTH in AAs.[23,33,56] The clinical management of hypovitaminosis D in AAs with CHF, however, had not been established. We therefore designed and tested a 14-week regimen: an initial 8 weeks of large-dose oral ergocalciferol (50,000 IU once weekly), followed by a 6-week maintenance phase of cholecalciferol (1400 IU daily), and $CaCO_3$ (1000 mg daily) supplement given throughout.[74] Fourteen AA patients having a dilated (idiopathic) cardiomyopathy with reduced ejection fraction (EF, <35%) were enrolled. We found the reduced 25(OH)D at entry (14.4±1.3 ng/mL) to be significantly improved in all patients at 8 weeks (30.7±3.2 ng/mL) and sustained at 14 weeks (30.9±2.8 ng/mL). Baseline serum PTH, abnormally elevated in five patients (104.8±8.2 pg/mL), was reduced at 8 and 14 weeks (74.4±18.3 and 73.8±13.0 pg/mL, respectively). Plasma 8-isoprostane at entry (136.1±8.8 pg/mL), a biomarker of lipid peroxidation, was also reduced at 14 weeks (117.8±7.8 pg/mL), together with significant improvement in EF (24.3±1.7% to 31.3±4.3%). Thus, our 14-week course of large-dose supplemental vitamin D and $CaCO_3$ led to healthy 25(OH)D levels in AAs with heart failure having vitamin D deficiency of moderate-to-marked severity. Albeit a small patient population, our findings suggest that this regimen may attenuate the accompanying SHPT and oxidative stress of CHF in AAs, and may also improve ventricular function.[74]

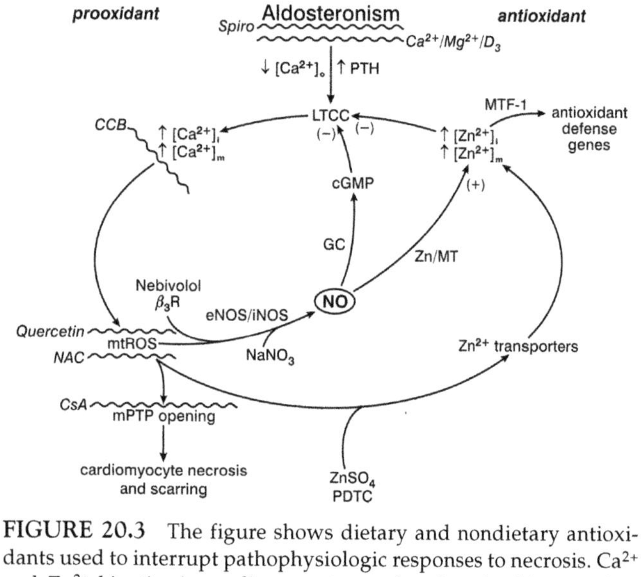

FIGURE 20.3 The figure shows dietary and nondietary antioxidants used to interrupt pathophysiologic responses to necrosis. Ca^{2+} and Zn^{2+} kinetics in cardiomyocytes under chronic aldosteronism leads to oxidative stress, upregulated antioxidant defenses, and ultimately cardiomyocyte necrosis. Dietary and nondietary antioxidants were tested (see Table 20.1) in downregulating the prooxidant heart failure phenotype in chronic aldosteronism. See text for details.

Increased urinary and fecal Zn excretion, as well as Zn translocation to injured tissues, including the heart, accompanies aldosteronism.[48,49] To assess temporal intracellular Zn kinetics in the heart in response to chronic aldosteronism, cytosolic free $[Zn^{2+}]_i$ and mitochondrial $[Zn^{2+}]_m$ in cardiomyocytes, the upregulation of MT-1 and biomarkers of oxidative stress were examined (see Fig. 20.1). Oxidative stress and cardiac pathology, in response to a $ZnSO_4$ supplement (40 mg/day), were also studied.[65] At 4 weeks ALDOST we found a rise in cardiac Zn, including increased $[Zn^{2+}]_i$ and mitochondrial $[Zn^{2+}]_m$, associated with increased tissue MT-1, 8-isoprostane, malondialdehyde, and $gp91^{phox}$, coupled with oxidative stress in plasma and urine. The $ZnSO_4$ supplement prevented hypozincemia, but not ionized hypocalcemia; however, it attenuated oxidative stress and microscopic scarring. Thus, the oxidative stress that appears in the heart is accompanied by increased tissue Zn^{2+} serving as an antioxidant. Cotreatment with $ZnSO_4$ attenuated cardiomyocyte necrosis. However, a *polynutrient supplement* with Ca^{2+}, Mg^{2+}, and Zn^{2+} appears to be necessary to mitigate the simultaneous dyshomeostasis of all three cations that accompanies aldosteronism, and presumed to be integral to the appearance of cardiac pathology.[65]

Elevation in $[Zn^{2+}]_i$ beyond that which accompanies aldosteronism could be achieved by a $ZnSO_4$ supplement.[65,67,75–80] Increased cytosolic free $[Zn^{2+}]_i$ activates its sensor, metal-responsive transcription factor (MTF)-1, which upon its translocation to the nucleus upregulates antioxidant defense genes (see Fig. 20.3).[66] These observations raise the therapeutic prospect that cation-modulating *nutriceutical supplementation* capable of favorably influencing the extra- and intracellular Ca^{2+} and Zn^{2+} equilibrium, as well as enhancing overall antioxidant capacity, could prove pivotal to combating mitochondria-based oxidative injury and cardiomyocyte necrosis, while simultaneously promoting Zn^{2+}-based endogenous cardioprotection.

We hypothesized the dyshomeostasis of intracellular Ca^{2+} and Zn^{2+} that alters the redox state of cardiac myocytes and mitochondria in chronic aldosteronism, analogous to CHF or low-renin hypertension, is intrinsically coupled with Ca^{2+} serving as a prooxidant and Zn^{2+} as antioxidant.[62] Accordingly, hearts were harvested from rats receiving 4 weeks ALDOST alone or cotreated with either spironolactone, an aldosterone receptor antagonist, or amlodipine (Amlod), an L-type Ca^{2+} channel blocker, and from age-/sex-matched untreated controls. In response to ALDOST, we reported increased $[Ca^{2+}]_i$ and $[Zn^{2+}]_i$, together with increased $[Ca^{2+}]_m$ and $[Zn^{2+}]_m$, each of which could be prevented by spironolactone and attenuated with Amlod; increased levels of 3-nitrotyrosine and 4-hydroxy-2-nonenal in cardiomyocytes, together with increased H_2O_2 production, malondialdehyde, and oxidized glutathione disulfide in mitochondria that were coincident with increased activities of Cu/Zn-superoxide dismutase and glutathione peroxidase; and increased expression of MT-1, Zn transporters (Zip1 and ZnT-1), and MTF-1, which were attenuated by spironolactone (see Fig. 20.3). Thus, an intrinsically coupled dyshomeostasis of intracellular Ca^{2+} and Zn^{2+} occurred in cardiac myocytes and mitochondria in rats receiving ALDOST, where they serve to alter their redox state via induction of oxidative stress and generation of antioxidant defenses.[62] The clinical relevance of therapeutic strategies that can uncouple these two pathophysiologically crucial divalent cations and modulate their equilibrium in favor of sustained antioxidant defenses is therefore highlighted.

We then investigated whether Ca^{2+} and Zn^{2+} dyshomeostasis and prooxidant:antioxidant dysequilibrium seen at 4 weeks ALDOST could be uncoupled in favor of antioxidants. Accordingly, we used cotreatment with a $ZnSO_4$ supplement, pyrrolidine dithiocarbamate (PDTC), a Zn^{2+} ionophore, or $ZnSO_4$ in combination with Amlod.[67] We reconfirmed an elevation in $[Ca^{2+}]_i$ coupled with $[Zn^{2+}]_i$ and increased mitochondrial H_2O_2 production with increased mitochondrial and cardiac 8-isoprostane levels with ALDOST. Cotreatment with the $ZnSO_4$ supplement alone, PDTC, or $ZnSO_4$+Amlod augmented the rise in cardiomyocyte $[Zn^{2+}]_i$ beyond that seen with ALDOST alone, while attenuating the rise in $[Ca^{2+}]_i$ which served to reduce oxidative stress. Thus, a coupled dyshomeostasis of intracellular Ca^{2+} and Zn^{2+} was found in cardiac myocytes and mitochondria at 4 weeks ALDOST, where prooxidants overwhelm antioxidant defenses. This intrinsically coupled Ca^{2+} and Zn^{2+} dyshomeostasis could be uncoupled in favor of antioxidant defenses (see Fig. 20.3) by selectively increasing free $[Zn^{2+}]_i$ and/or reducing $[Ca^{2+}]_i$ using cotreatment with $ZnSO_4$ as a *dietary supplement*, PDTC alone, or $ZnSO_4$+Amlod in combination.[67] Exogenous Zn^{2+} or treatment with pyrithione, another Zn^{2+} ionophore, has proven cardioprotective in promoting recovery from oxidative stress imposed by ischemia/reperfusion injury.[81–83]

Nondietary Antioxidants

Oxidative stress accounts for a proinflammatory/fibrogenic phenotype of both right and left ventricles in response to ALDOST. We hypothesized that cardioprotection would accompany antioxidant cotreatment (see Fig. 20.3). Uninephrectomized rats received ALDOST for 3, 4, or 5 weeks.[84] Other groups received this regimen in combination with an aldosterone receptor antagonist, spironolactone, or an antioxidant, either PDTC or *N*-acetylcysteine (NAC). Unoperated and untreated age- and sex-matched rats served as controls. At week 3 ALDOST and compared to controls, there was no evidence of oxidative stress or cardiac pathology. At 4 and 5 weeks, however, increased gp91^{phox} and 3-nitrotyrosine expression with persistent nicotinamide adenine dinucleotide phosphate (NADPH) oxidase activation were found in endothelial and inflammatory cells that appeared at sites of lost cardiomyocytes in both ventricles. Coincident in time and space with these events, there were increased mRNA expression of intercellular adhesion molecule-1, monocyte chemoattractant protein-1, and tumor necrosis factor-α. Macrophages, lymphocytes, and fibroblast-like cells were seen at each of these sites, together with an accumulation of fibrillar collagen, or fibrosis, as evidenced by a significant increase in ventricular collagen volume fraction. Co-treatment with spironolactone, PDTC, or NAC attenuated these molecular and cellular responses, as well as the appearance of fibrosis at sites of injury. Thus, chronic ALDOST is accompanied by a time-dependent sustained activation of NADPH oxidase with 3-nitrotyrosine generation and nuclear factor-κB activation expressed by endothelial and inflammatory cells. This leads to a proinflammatory/fibrogenic phenotype involving sites of injury found in both normotensive and hypertensive right and left ventricles. Spironolactone, PDTC, and NAC each attenuated these responses, suggesting that ALDOST with oxidative/nitrosative stress is responsible for the appearance of this proinflammatory/fibrogenic phenotype and that these agents serve as antioxidants in chronic aldosteronism (see Table 20.1).[84]

In this context, Noda et al[85] examined whether spironolactone provides additional benefit to olmesartan, an angiotensin II receptor blocker, in combating oxidative stress in postinfarct failing rat hearts induced

TABLE 20.1 Dietary and Nondietary Antioxidants That Were Tested in Downregulating the Prooxidant Heart Failure Phenotype in Chronic Aldosteronism

Dietary	Nondietary
Ca^{2+}/Mg^{2+}/vitamin D_3	*N*-acetylcysteine
$ZnSO_4$	Zn^{2+} ionophore
Polycation cocktail+D_3	Spironolactone
Flavonoid (Quercetin)	Amlodipine
$NaNo_3$	Cinacalcet
	Nebivolol
	Carvedilol

by coronary artery ligation. After 7 weeks monotherapy, improved left ventricular function and suppressed myocardial lipid peroxidation were seen, together with an attenuation of NADPH oxidase-dependent and mitochondrial superoxide production. Hence, this combination therapy had a synergistic attenuation of oxidative stress, leading to an improvement in cardiac function in postinfarct failing hearts.

Cardioprotection with the antioxidant NAC has been reported in rodents having streptozocin-induced diabetes,[86] after isoproterenol-induced cardiomyocyte necrosis,[87] or in response to aortic banding.[88] In a randomized controlled clinical trial, NAC provided cardioprotection in patients having surgical repair of their abdominal aneurysm.[89]

The mitochondriocentric signal-transducer-effector (MSTE) pathway (see Fig. 20.2), leading to necrotic cell death during ALDOST, includes intramitochondrial Ca^{2+} overloading, together with an induction of oxidative stress by these organelles and opening of their permeability transition pore (mPTP). We therefore hypothesized that mitochondria-targeted interventions would prove cardioprotective. Accordingly, rats receiving 4 weeks ALDOST were cotreated with either quercetin, a flavonoid with mitochondrial antioxidant properties,[90] or a well-known mPTP inhibitor, cyclosporine A, and compared to ALDOST alone or untreated, age-/sex-matched controls. At 4 weeks ALDOST we found a marked increase in mitochondrial H_2O_2 production and 8-isoprostane levels, an increased propensity for mPTP opening, and greater concentrations of mitochondrial free $[Ca^{2+}]_m$ and total tissue Ca^{2+}, coupled with a 5-fold rise in collagen volume fraction without any TUNEL-based evidence of cardiomyocyte apoptosis. Each of these pathophysiologic responses to ALDOST were prevented by quercetin or cyclosporine A cotreatment.[91] Thus, mitochondria proved to play a central role in initiating the cellular-molecular pathway that leads to necrotic cell death and myocardial scarring. This destructive pathogenetic cycle can be interrupted, and myocardium protected, by mitochondria-targeted strategies, including the flavonoid quercetin. Quercetin has likewise been shown to be cardioprotective in rats: with the metabolic syndrome;[92,93] with abdominal aorta banding;[94] receiving doxorubicin[95,96] or isoproterenol;[97,98] having genetic hypertension;[99,100] or acute myocarditis.[101–103]

This MSTE pathway to necrosis was further investigated in cardiomyocytes and subsarcolemmal mitochondria (SSM) harvested from rats receiving 4 weeks ALDOST. We used mitochondria-targeted pharmaceutical interventions as cardioprotective strategies by designing cotreatment with either carvedilol or nebivolol, third-generation β blockers with additional properties as a β_3 receptor agonist (see Fig. 20.3).[104,105] We found 4 weeks ALDOST to be accompanied by: elevated cardiomyocyte free $[Ca^{2+}]_i$ and SSM free $[Ca^{2+}]_m$; increased H_2O_2 production and 8-isoprostane in SSM, cardiac tissue, and plasma; and enhanced opening of mPTP with myocardial scarring. Increments in antioxidant capacity induced by increased cytosolic free $[Zn^{2+}]_i$ during ALDOST were overwhelmed. However, cotreatment with either carvedilol or nebivolol markedly augmented $[Zn^{2+}]_i$ and attenuated $[Ca^{2+}]_i$ and $[Ca^{2+}]_m$ overloading, preventing oxidative stress and reducing mPTP opening potential.[106] Thus, major components of this pathway to cardiomyocyte necrosis seen with ALDOST include intracellular Ca^{2+} overloading coupled to SSM-based oxidative stress and mPTP opening. This subcellular pathway can be favorably regulated by these unique β blockers through their promoting a marked rise in $[Zn^{2+}]_i$, which bolsters antioxidant defenses and thereby salvages cardiomyocytes and prevents fibrosis.

SUMMARY AND CONCLUSIONS

The CHF syndrome has its pathophysiologic origins rooted in inappropriate neurohormonal activation, where homeostatic responses beget dyshomeostasis (see Fig. 20.4). It includes a prooxidant heart failure phenotype, where the rate of reactive oxygen species generation exceeds their rate of degradation by endogenous antioxidant defenses. An ongoing nonischemic necrosis of cardiomyocytes and ensuing reparative fibrosis with stiff fibrillar collagen preserves the structural integrity of myocardium but at the expense of compromised systolic and diastolic function and arrhythmogenicity. Catecholamine and/or parathyroid hormone-mediated Ca^{2+} entry with excessive intracytosolic Ca^{2+} accumulation gives way to mitochondrial Ca^{2+} overloading and consequent oxidative stress by these organelles and enhanced opening potential of their inner membrane permeability transition pore leading to necrotic cardiomyocyte death.

Endogenous antioxidant defenses can be bolstered by dietary and nondietary antioxidants to counteract the prooxidant heart failure phenotype and to provide cardioprotection. Diet-based antioxidants include: cation supplements with Ca^{2+}, Mg^{2+}, and Zn^{2+}, together with vitamin D; a flavonoid; and sodium nitrate. Nondietary antioxidants include: *N*-acetylcysteine; Zn^{2+} ionophore; agents that interfere with RAAS and ANS effector hormones, including spironolactone, nebivolol, or carvedilol; and Ca^{2+} channel blocker.

The management of the prooxidant CHF phenotype can be optimized and cardioprotection achieved through the effective use of antioxidant nutriceuticals and pharmaceuticals directed at the root causes and consequences of oxidative stress.

SUMMARY POINTS

- Congestive heart failure (CHF) is now the leading admitting diagnosis at hospitals in the USA. Its origins are rooted in a salt-avid state induced by an inappropriate and deleterious homeostatic response – neurohormonal activation – evoked by kidneys underperfused by a failing heart and mediated by effector hormones of the activated renin–angiotensin–aldosterone system (RAAS) and adrenergic nervous system (ANS).
- Endocrine properties of RAAS hormones orchestrate a systemic illness whose features are referred to as a prooxidant and proinflammatory phenotype. Its clinical manifestations include: (i) oxidative stress; (ii) activated peripheral blood mononuclear cells releasing proinflammatory cytokines; and (iii) a wasting of soft tissues and bone, termed cardiac cachexia.
- In addition to this systemic illness there is a progressive pathologic remodeling of the myocardium, including the ongoing loss of cardiomyocytes to necrotic and apoptotic forms of cell death. In this arguably postmitotic organ, incapable of myocyte regeneration, myocyte loss accounts for a pathologic remodeling of myocardium. Another prominent feature of this remodeling is the appearance of foci of microscopic scars, footprints of cardiomyocyte necrosis. Conversely, apoptosis leaves behind no such morphologic footprint. Widespread fibrosis is recognized as the major component of pathologic cardiac remodeling.
- Mechanisms accounting for ongoing myocyte necrosis are therefore of fundamental importance. This turns attention to the pathophysiologic impact of oxidative stress and the role of catecholamines, angiotensin II, and aldosterone in promoting mono- and divalent cation dyshomeostasis with compromised antioxidant defenses, and the susceptibility of cardiomyocytes to necrosis.
- Several micronutrients are deficient in patients with heart failure, including vitamin D, which may play a pathophysiologic role in the genesis of the prooxidant cardiomyopathic phenotype and its progressive nature.
- We address pathophysiologic mechanisms leading to this prooxidant heart failure phenotype and the potential role of dietary and nondietary antioxidants in regulating this phenotype, thereby salvaging cardiomyocytes and preventing myocardial scarring, and, hence, serving as cardioprotectants.

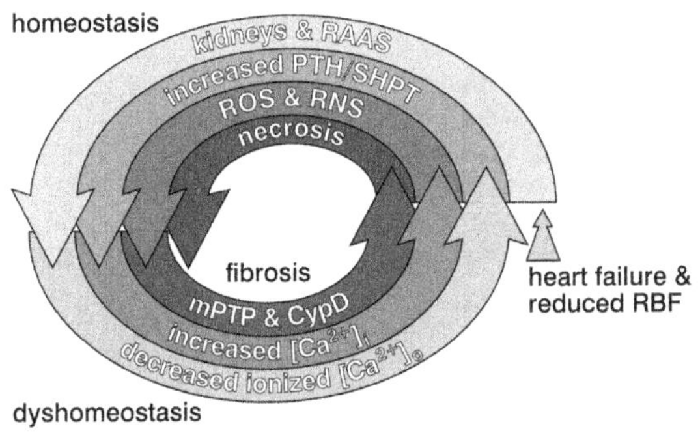

FIGURE 20.4 Homeostatic responses elicited in congestive heart failure that beget dyshomeostasis. There is a progressive inward spiral, where homeostasis begets dyshomeostasis at organ, cellular, and molecular levels, leading to cardiomyocyte necrosis. The cycle begins with heart failure and reduced renal blood flow (RBF), leading to the homeostatic activation of the renin–angiotensin–aldosterone system (RAAS). Ionized hypocalcemia $[Ca^{2+}]_o$ appears due to an accompanying increased excretory loss of Ca^{2+}. In turn, this dyshomeostatic reaction accounts for the subsequent homeostatic response which is initiated by the appearance of secondary hyperparathyroidism (SHPT) with increased circulating PTH. The dyshomeostatic response to SHPT is PTH-mediated intracellular $[Ca^{2+}]_i$ overloading wherein induction of oxidative stress follows with the generation of reactive oxygen and nitrogen species (ROS/RNS). Together, intracellular Ca^{2+} overloading and oxidative stress contribute to the pathologic opening of the mPTP pore and activation of cyclophilin D with the ensuing osmotic injury to mitochondria, and ultimately necrotic cell death. Reprinted with permission from Kamalov G et al J Cardiovasc Pharmacol 2010;56:320–328.

Acknowledgment

This work was supported, in part, by NIH grants R01-HL73043 and R01-HL90867 (KTW). Its contents are solely the responsibility of the authors and do not necessarily represent the official views of the NIH.

References

1. Kamalov G, Bhattacharya SK, Weber KT. Congestive heart failure: where homeostasis begets dyshomeostasis. *J Cardiovasc Pharmacol* 2010;**56**:320–8.
2. Khaper N, Bryan S, Dhingra S, et al. Targeting the vicious inflammation-oxidative stress cycle for the management of heart failure. *Antioxid Redox Signal* 2010;**13**:1033–49.
3. Borkowski BJ, Cheema Y, Shahbaz AU, et al. Cation dyshomeostasis and cardiomyocyte necrosis. The Fleckenstein hypothesis revisited. *Eur Heart J* 2011;**32**:1846–53.

4. Weber KT. The neuroendocrine–immune interface gone awry in aldosteronism. *Cardiovasc Res* 2004;**64**:381–3.
5. Sukhanov S, Semprun-Prieto L, Yoshida T, et al. Angiotensin II, oxidative stress and skeletal muscle wasting. *Am J Med Sci* 2011;**342**:143–7.
6. Yndestad A, Damas JK, Oie E, et al. Systemic inflammation in heart failure – the whys and wherefores. *Heart Fail Rev* 2006;**11**:83–92.
7. Ishii J, Nomura M, Nakamura Y, et al. Risk stratification using a combination of cardiac troponin T and brain natriuretic peptide in patients hospitalized for worsening chronic heart failure. *Am J Cardiol* 2002;**89**:691–5.
8. Peacock 4th WF, De Marco T, Fonarow GC, et al. Cardiac troponin and outcome in acute heart failure. *N Engl J Med* 2008;**358**:2117–26.
9. Taniguchi R, Sato Y, Nishio Y, et al. Measurements of baseline and follow-up concentrations of cardiac troponin-T and brain natriuretic peptide in patients with heart failure from various etiologies. *Heart Vessels* 2006;**21**:344–9.
10. Sukova J, Ostadal P, Widimsky P. Profile of patients with acute heart failure and elevated troponin I levels. *Exp Clin Cardiol* 2007;**12**:153–6.
11. Ilva T, Lassus J, Siirilä-Waris K, et al. Clinical significance of cardiac troponins I and T in acute heart failure. *Eur J Heart Fail* 2008;**10**:772–9.
12. Sato Y, Nishi K, Taniguchi R, et al. In patients with heart failure and non-ischemic heart disease, cardiac troponin T is a reliable predictor of long-term echocardiographic changes and adverse cardiac events. *J Cardiol* 2009;**54**:221–30.
13. Miller WL, Hartman KA, Burritt MF, et al. Profiles of serial changes in cardiac troponin T concentrations and outcome in ambulatory patients with chronic heart failure. *J Am Coll Cardiol* 2009;**54**:1715–21.
14. Beltrami CA, Finato N, Rocco M, et al. Structural basis of end-stage failure in ischemic cardiomyopathy in humans. *Circulation* 1994;**89**:151–63.
15. Cotran RS, Kumar V, Robbins SL. *Robbins pathologic basis of disease*. 4th ed Philadelphia: W B Saunders; 1989.
16. Sun Y, Ramires FJA, Weber KT. Fibrosis of atria and great vessels in response to angiotensin II or aldosterone infusion. *Cardiovasc Res* 1997;**35**:138–47.
17. Matzinger P. The danger model: a renewed sense of self. *Science* 2002;**296**:301–5.
18. Witte KK, Clark AL, Cleland JG. Chronic heart failure and micronutrients. *J Am Coll Cardiol* 2001;**37**:1765–74.
19. Hughes CM, Woodside JV, McGartland C, et al. Nutritional intake and oxidative stress in chronic heart failure. *Nutr Metab Cardiovasc Dis* 2012;**22**:376–82.
20. Soukoulis V, Dihu JB, Sole M, et al. Micronutrient deficiencies an unmet need in heart failure. *J Am Coll Cardiol* 2009;**54**:1660–73.
21. Bhattacharya SK, Ahokas RA, Carbone LD, et al. Macro- and micronutrients in African-Americans with heart failure. *Heart Fail Rev* 2006;**11**:45–55.
22. Shane E, Mancini D, Aaronson K, et al. Bone mass, vitamin D deficiency, and hyperparathyroidism in congestive heart failure. *Am J Med* 1997;**103**:197–207.
23. Alsafwah S, LaGuardia SP, Nelson MD, et al. Hypovitaminosis D in African Americans residing in Memphis, Tennessee with and without heart failure. *Am J Med Sci* 2008;**335**:292–7.
24. Lee AH, Mull RL, Keenan GF, et al. Osteoporosis and bone morbidity in cardiac transplant recipients. *Am J Med* 1994;**96**:35–41.
25. Schmid C, Kiowski W. Hyperparathyroidism in congestive heart failure. *Am J Med* 1998;**104**:508–9.
26. Ogino K, Ogura K, Kinugasa Y, et al. Parathyroid hormone-related protein is produced in the myocardium and increased in patients with congestive heart failure. *J Clin Endocrinol Metab* 2002;**87**:4722–7.
27. Zittermann A, Schleithoff SS, Tenderich G, et al. Low vitamin D status: a contributing factor in the pathogenesis of congestive heart failure? *J Am Coll Cardiol* 2003;**41**:105–12.
28. Sugimoto T, Tanigawa T, Onishi K, et al. Serum intact parathyroid hormone levels predict hospitalisation for heart failure. *Heart* 2009;**95**:395–8.
29. Schierbeck LL, Jensen TS, Bang U, et al. Parathyroid hormone and vitamin D – markers for cardiovascular and all cause mortality in heart failure. *Eur J Heart Fail* 2011;**13**:626–32.
30. Pilz S, Tomaschitz A, Drechsler C, et al. Parathyroid hormone level is associated with mortality and cardiovascular events in patients undergoing coronary angiography. *Eur Heart J* 2010;**31**:1591–8.
31. Hagström E, Hellman P, Larsson TE, et al. Plasma parathyroid hormone and the risk of cardiovascular mortality in the community. *Circulation* 2009;**119**:2765–71.
32. Hagström E, Ingelsson E, Sundström J, et al. Plasma parathyroid hormone and risk of congestive heart failure in the community. *Eur J Heart Fail* 2010;**12**:1186–92.
33. LaGuardia SP, Dockery BK, Bhattacharya SK, et al. Secondary hyperparathyroidism and hypovitaminosis D in African-Americans with decompensated heart failure. *Am J Med Sci* 2006;**332**:112–8.
34. Chhokar VS, Sun Y, Bhattacharya SK, et al. Loss of bone minerals and strength in rats with aldosteronism. *Am J Physiol Heart Circ Physiol* 2004;**287**:H2023–6.
35. Chhokar VS, Sun Y, Bhattacharya SK, et al. Hyperparathyroidism and the calcium paradox of aldosteronism. *Circulation* 2005;**111**:871–8.
36. Rossi E, Perazzoli F, Negro A, et al. Acute effects of intravenous sodium chloride load on calcium metabolism and on parathyroid function in patients with primary aldosteronism compared with subjects with essential hypertension. *Am J Hypertens* 1998;**11**:8–13.
37. Fujita T, Palmieri GM. Calcium paradox disease: calcium deficiency prompting secondary hyperparathyroidism and cellular calcium overload. *J Bone Miner Metab* 2000;**18**:109–25.
38. Weber KT, Sun Y, Bhattacharya SK, et al. Myofibroblast-mediated mechanisms of pathological remodelling of the heart. *Nat Rev Cardiol* 2013;**10**:15–26.
39. Gunnewiek JM, Van Der Hoeven JG. Cardiac troponin elevations among critically ill patients. *Curr Opin Crit Care* 2004;**10**:342–6.
40. Bell NH, Greene A, Epstein S, et al. Evidence for alteration of the vitamin D-endocrine system in blacks. *J Clin Invest* 1985;**76**:470–3.
41. Sawaya BP, Monier-Faugere MC, Ratanapanichkich P, et al. Racial differences in parathyroid hormone levels in patients with secondary hyperparathyroidism. *Clin Nephrol* 2002;**57**:51–5.
42. Zittermann A, Fischer J, Schleithoff SS, et al. Patients with congestive heart failure and healthy controls differ in vitamin D-associated lifestyle factors. *Int J Vitam Nutr Res* 2007;**77**:280–8.
43. Zittermann A, Schleithoff SS, Gotting C, et al. Poor outcome in end-stage heart failure patients with low circulating calcitriol levels. *Eur J Heart Fail* 2008;**10**:321–7.
44. Singal PK, Kirshenbaum LA. A relative deficit in antioxidant reserve may contribute in cardiac failure. *Can J Cardiol* 1990;**6**:47–9.
45. Kirshenbaum LA, Singal PK. Antioxidant changes in heart hypertrophy: significance during hypoxia-reoxygenation injury. *Can J Physiol Pharmacol* 1992;**70**:1330–5.
46. Sharir H, Zinger A, Nevo A, et al. Zinc released from injured cells is acting via the Zn^{2+}-sensing receptor, ZnR, to trigger signaling leading to epithelial repair. *J Biol Chem* 2010;**285**:26097–106.
47. Iwata M, Takebayashi T, Ohta H, et al. Zinc accumulation and metallothionein gene expression in the proliferating epidermis during wound healing in mouse skin. *Histochem Cell Biol* 1999;**112**:283–90.

48. Thomas M, Vidal A, Bhattacharya SK, et al. Zinc dyshomeostasis in rats with aldosteronism. Response to spironolactone. *Am J Physiol Heart Circ Physiol* 2007;**293**:H2361–6.
49. Selektor Y, Parker RB, Sun Y, et al. Tissue ^{65}zinc translocation in a rat model of chronic aldosteronism. *J Cardiovasc Pharmacol* 2008;**51**:359–64.
50. Tubek S. Role of zinc in regulation of arterial blood pressure and in the etiopathogenesis of arterial hypertension. *Biol Trace Elem Res* 2007;**117**:39–51.
51. Eide DJ. The oxidative stress of zinc deficiency. *Metallomics* 2011;**3**:1124–9.
52. Maret W. The function of zinc metallothionein: a link between cellular zinc and redox state. *J Nutr* 2000;**130**:1455S–8S.
53. Oteiza PI. Zinc and the modulation of redox homeostasis. *Free Radic Biol Med* 2012;**53**:1748–59.
54. Golik A, Modai D, Averbukh Z, et al. Zinc metabolism in patients treated with captopril versus enalapril. *Metabolism* 1990;**39**:665–7.
55. Golik A, Zaidenstein R, Dishi V, et al. Effects of captopril and enalapril on zinc metabolism in hypertensive patients. *J Am Coll Nutr* 1998;**17**:75–78.
56. Arroyo M, LaGuardia SP, Bhattacharya SK, et al. Micronutrients in African-Americans with decompensated and compensated heart failure. *Transl Res* 2006;**148**:301–8.
57. Oster O. Trace element concentrations (Cu, Zn, Fe) in sera from patients with dilated cardiomyopathy. *Clin Chim Acta* 1993;**214**:209–18.
58. Topuzoglu G, Erbay AR, Karul AB, Yensel N. Concentrations of copper, zinc, and magnesium in sera from patients with idiopathic dilated cardiomyopathy. *Biol Trace Elem Res* 2003;**95**:11–7.
59. Salehifar E, Shokrzadeh M, Ghaemian A, et al. The study of Cu and Zn serum levels in idiopathic dilated cardiomyopathy (IDCMP) patients and its comparison with healthy volunteers. *Biol Trace Elem Res* 2008;**125**:97–108.
60. Kosar F, Sahin I, Taskapan C, et al. Trace element status (Se, Zn, Cu) in heart failure. *Anadolu Kardiyol Derg* 2006;**6**:216–20.
61. Frustaci A, Sabbioni E, Fortaner S, et al. Selenium- and zinc-deficient cardiomyopathy in human intestinal malabsorption: preliminary results of selenium/zinc infusion. *Eur J Heart Fail* 2012;**14**:202–10.
62. Kamalov G, Deshmukh PA, Baburyan NY, et al. Coupled calcium and zinc dyshomeostasis and oxidative stress in cardiac myocytes and mitochondria of rats with chronic aldosteronism. *J Cardiovasc Pharmacol* 2009;**53**:414–23.
63. Tuncay E, Bilginoglu A, Sozmen NN, et al. Intracellular free zinc during cardiac excitation-contraction cycle: calcium and redox dependencies. *Cardiovasc Res* 2011;**89**:634–42.
64. Turan B. Zinc-induced changes in ionic currents of cardiomyocytes. *Biol Trace Elem Res* 2003;**94**:49–60.
65. Gandhi MS, Deshmukh PA, Kamalov G, et al. Causes and consequences of zinc dyshomeostasis in rats with chronic aldosteronism. *J Cardiovasc Pharmacol* 2008;**52**:245–52.
66. Kamalov G, Ahokas RA, Zhao W, et al. Temporal responses to intrinsically coupled calcium and zinc dyshomeostasis in cardiac myocytes and mitochondria during aldosteronism. *Am J Physiol Heart Circ Physiol* 2010;**298**:H385–94.
67. Kamalov G, Ahokas RA, Zhao W, et al. Uncoupling the coupled calcium and zinc dyshomeostasis in cardiac myocytes and mitochondria seen in aldosteronism. *J Cardiovasc Pharmacol* 2010;**55**:248–54.
68. Schulz R, Rassaf T, Massion PB, et al. Recent advances in the understanding of the role of nitric oxide in cardiovascular homeostasis. *Pharmacol Ther* 2005;**108**:225–56.
69. Carlström M, Larsen FJ, Nyström T, et al. Dietary inorganic nitrate reverses features of metabolic syndrome in endothelial nitric oxide synthase-deficient mice. *Proc Natl Acad Sci USA* 2010;**107**:17716–20.
70. McKnight GM, Smith LM, Drummond RS, et al. Chemical synthesis of nitric oxide in the stomach from dietary nitrate in humans. *Gut* 1997;**40**:211–4.
71. Xi L, Zhu SG, Hobbs DC, Kukreja RC. Identification of protein targets underlying dietary nitrate-induced protection against doxorubicin cardiotoxicity. *J Cell Mol Med* 2011;**15**:2512–24.
72. Xi L, Zhu SG, Das A, et al. Dietary inorganic nitrate alleviates doxorubicin cardiotoxicity: mechanisms and implications. *Nitric Oxide* 2012;**26**:274–84.
73. Goodwin KD, Ahokas RA, Bhattacharya SK, et al. Preventing oxidative stress in rats with aldosteronism by calcitriol and dietary calcium and magnesium supplements. *Am J Med Sci* 2006;**332**:73–78.
74. Zia AA, Komolafe BO, Moten M, et al. Supplemental vitamin D and calcium in the management of African-Americans with heart failure having hypovitaminosis D. *Am J Med Sci* 2011;**341**:113–8.
75. Wang J, Song Y, Elsherif L, et al. Cardiac metallothionein induction plays the major role in the prevention of diabetic cardiomyopathy by zinc supplementation. *Circulation* 2006;**113**:544–54.
76. Chung MJ, Hogstrand C, Lee SJ. Cytotoxicity of nitric oxide is alleviated by zinc-mediated expression of antioxidant genes. *Exp Biol Med (Maywood)* 2006;**231**:1555–63.
77. Chung MJ, Walker PA, Brown RW, Hogstrand C. ZINC-mediated gene expression offers protection against H2O2-induced cytotoxicity. *Toxicol Appl Pharmacol* 2005;**205**:225–36.
78. Singal PK, Dhillon KS, Beamish RE, Dhalla NS. Protective effect of zinc against catecholamine-induced myocardial changes electrocardiographic and ultrastructural studies. *Lab Invest* 1981;**44**:426–33.
79. Singal PK, Kapur N, Dhillon KS, et al. Role of free radicals in catecholamine-induced cardiomyopathy. *Can J Physiol Pharmacol* 1982;**60**:1390–7.
80. Chvapil M, Owen JA. Effect of zinc on acute and chronic isoproterenol induced heart injury. *J Mol Cell Cardiol* 1977;**9**:151–9.
81. Karagulova G, Yue Y, Moreyra A, et al. Protective role of intracellular zinc in myocardial ischemia/reperfusion is associated with preservation of protein kinase C isoforms. *J Pharmacol Exp Ther* 2007;**321**:517–25.
82. Chanoit G, Lee S, Xi J, et al. Exogenous zinc protects cardiac cells from reperfusion injury by targeting mitochondrial permeability transition pore through inactivation of glycogen synthase kinase-3β. *Am J Physiol Heart Circ Physiol* 2008;**295**:H1227–33.
83. Lee S, Chanoit G, McIntosh RV, et al. Molecular mechanism underlying Akt activation in zinc-induced cardioprotection. *Am J Physiol Heart Circ Physiol* 2009;**297**:H569–75.
84. Sun Y, Zhang J, Lu L, et al. Aldosterone-induced inflammation in the rat heart. Role of oxidative stress. *Am J Pathol* 2002;**161**:1773–81.
85. Noda K, Kobara M, Hamada J, et al. Additive amelioration of oxidative stress and cardiac function by combined mineralocorticoid and angiotensin receptor blockers in postinfarct failing hearts. *J Cardiovasc Pharmacol* 2012;**60**:140–9.
86. Lei S, Liu Y, Liu H, et al. Effects of *N*-acetylcysteine on nicotinamide dinucleotide phosphate oxidase activation and antioxidant status in heart, lung, liver and kidney in streptozotocin-induced diabetic rats. *Yonsei Med J* 2012;**53**:294–303.
87. Nagoor Meeran MF. Stanely Mainzen Prince P, Hidhayath Basha R. Preventive effects of *N*-acetyl cysteine on lipids, lipoproteins and myocardial infarct size in isoproterenol induced myocardial infarcted rats: an in vivo and in vitro study. *Eur J Pharmacol* 2012;**677**:116–22.
88. Foltz WU, Wagner M, Rudakova E, Volk T. *N*-acetylcysteine prevents electrical remodeling and attenuates cellular hypertrophy in epicardial myocytes of rats with ascending aortic stenosis. *Basic Res Cardiol* 2012;**107**:290.

89. Mahmoud KM, Ammar AS. Effect of *N*-acetylcysteine on cardiac injury and oxidative stress after abdominal aortic aneurysm repair: a randomized controlled trial. *Acta Anaesthesiol Scand* 2011;**55**:1015–21.
90. Lagoa R, Graziani I, Lopez-Sanchez C, et al. Complex I and cytochrome *c* are molecular targets of flavonoids that inhibit hydrogen peroxide production by mitochondria. *Biochim Biophys Acta* 2011;**1807**:1562–72.
91. Shahbaz AU, Kamalov G, Zhao W, et al. Mitochondria-targeted cardioprotection in aldosteronism. *J Cardiovasc Pharmacol* 2011;**57**:37–43.
92. Rivera L, Morón R, Sánchez M, et al. Quercetin ameliorates metabolic syndrome and improves the inflammatory status in obese Zucker rats. *Obesity (Silver Spring)* 2008;**16**:2081–7.
93. Panchal SK, Poudyal H, Brown L. Quercetin ameliorates cardiovascular, hepatic, and metabolic changes in diet-induced metabolic syndrome in rats. *J Nutr* 2012;**142**:1026–32.
94. He T, Chen L, Chen Y, et al. *In vivo* and *in vitro* protective effects of pentamethylquercetin on cardiac hypertrophy. *Cardiovasc Drugs Ther* 2012;**26**:109–20.
95. Kaiserová H, Simunek T, van der Vijgh WJ, et al. Flavonoids as protectors against doxorubicin cardiotoxicity: role of iron chelation, antioxidant activity and inhibition of carbonyl reductase. *Biochim Biophys Acta* 2007;**1772**:1065–74.
96. Bast A, Haenen GR, Bruynzeel AM, Van der Vijgh WJ. Protection by flavonoids against anthracycline cardiotoxicity: from chemistry to clinical trials. *Cardiovasc Toxicol* 2007;**7**:154–9.
97. Liu H, Zhang L, Lu S. Evaluation of antioxidant and immunity activities of quercetin in isoproterenol-treated rats. *Molecules* 2012;**17**:4281–91.
98. Punithavathi VR. Stanely Mainzen Prince P. The cardioprotective effects of a combination of quercetin and α-tocopherol on isoproterenol-induced myocardial infarcted rats. *J Biochem Mol Toxicol* 2011;**25**:28–40.
99. Galindo P, González-Manzano S, Zarzuelo MJ, et al. Different cardiovascular protective effects of quercetin administered orally or intraperitoneally in spontaneously hypertensive rats. *Food Funct* 2012;**3**:643–50.
100. Sánchez M, Galísteo M, Vera R, et al. Quercetin downregulates NADPH oxidase, increases eNOS activity and prevents endothelial dysfunction in spontaneously hypertensive rats. *J Hypertens* 2006;**24**:75–84.
101. Niwano S, Niwano H, Sasaki S, et al. *N*-acetylcysteine suppresses the progression of ventricular remodeling in acute myocarditis: studies in an experimental autoimmune myocarditis (EAM) model. *Circ J* 2011;**75**:662–71.
102. Arumugam S, Thandavarayan RA, Arozal W, et al. Quercetin offers cardioprotection against progression of experimental autoimmune myocarditis by suppression of oxidative and endoplasmic reticulum stress via endothelin-1/MAPK signalling. *Free Radic Res* 2012;**46**:154–63.
103. Milenković M, Arsenović-Ranin N, Stojić-Vukanić Z, et al. Quercetin ameliorates experimental autoimmune myocarditis in rats. *J Pharm Pharm Sci* 2010;**13**:311–9.
104. Rozec B, Gauthier C. β_3-Adrenoceptors in the cardiovascular system: putative roles in human pathologies. *Pharmacol Ther* 2006;**111**:652–73.
105. Moniotte S, Balligand JL. Potential use of β_3-adrenoceptor antagonists in heart failure therapy. *Cardiovasc Drug Rev* 2002;**20**:19–26.
106. Cheema Y, Sherrod JN, Shahbaz AU, et al. Mitochondriocentric pathway to cardiomyocyte necrosis in aldosteronism: cardioprotective responses to carvedilol and nebivolol. *J Cardiovasc Pharmacol* 2011;**58**:80–6.

CHAPTER

21

Overview of the Role of Antioxidant Vitamins as Protection Against Cardiovascular Disease: Implications for Aging

Koutatsu Maruyama

Department of Basic Medical Research and Education, Ehime University Graduate School of Medicine, Shitsukawa, Toon, Ehime, Japan

Hiroyasu Iso

Public Health, Department of Social and Environmental Medicine, Graduate School of Medicine, Osaka University Suita, Osaka, Japan

List of Abbreviations

CAD Coronary Artery Disease
CHD coronary heart disease
95% CI 95% confidence interval
CVD cardiovascular disease
HDL high-density lipoprotein
HR hazard ratio
LDL low-density lipoprotein
MI myocardial infarction
NO nitric oxide
OR odds ratio
RCT randomized controlled trial
RR relative risk

INTRODUCTION

Cardiovascular diseases (CVDs), including coronary heart disease and stroke, are major causes of mortality, and the mortality rate from CVDs has increased throughout the world.

Oxidative stress, one of the steps in the pathogenesis of CVD, involves an imbalance between (1) the biologic reactions of reactive oxygen species (ROS) and reactive nitrogen species (RNS) and (2) the factors promoting recovery from such damage; it plays a key role in initiating and promoting CVD. The major effects of oxidative stress on the cardiovascular system are as follows: (1) apoptosis of endothelial cells; (2) induction of inflammation by oxidative modification and cell adhesion; (3) reduction of intracellular bioavailability of vasodilator nitric oxide (NO); and (4) oxidative modification of LDL, which stimulates smooth muscle proliferation and platelet aggregation, and then plaque build-up in the arterial wall.[1–3] All of these events contribute to clinical manifestations of CVD.

Water-soluble vitamin C and fat-soluble vitamins A and E, which are well known as antioxidant nutrients, have potential preventive effects on CVD. Evidence for their positive effects has been observed from every field of research, including *in vivo, in vitro,* and epidemiologic studies. According to experimental studies, the mechanisms by which these vitamins prevent the progression of CVD are reduction of platelet aggregation and modulation of the smooth muscle phenotype.[3]

This chapter focuses on the preventive and therapeutic effects of antioxidant vitamins – mainly vitamins A, C, and E – on CVD.

EFFECTS OF AGING ON CARDIOVASCULAR DISEASE

In recent decades, the proportion of elderly people has increased in most developed countries.[4] With aging, arterial thickening and stiffness are likely to develop,

Aging
http://dx.doi.org/10.1016/B978-0-12-405933-7.00021-4

leading to the occurrence of cardiovascular disease (CVD) (Fig. 21.1). Aging enhances exposure to oxidative processes, with reduced endothelial NO synthetase activity and NO from vascular endothelial cells, increased vascular cell adhesion protein-1 (VCAM-1), superoxide, and oxidized LDL – promoting endothelial cell senescence with telomere shortening and DNA damage;[5,6] this leads to apoptosis of endothelial cells, arterial thickening, stiffness, and atherosclerosis. Lifestyles, such as smoking and physical inactivity, may promote these processes, while antioxidant vitamins may inhibit them.[7] Although aging is not avoidable, we can reduce the exposure to oxidative stress by the use of antioxidant vitamins.

PRIMARY PREVENTIVE EFFECTS OF ANTIOXIDANT VITAMINS FOR CVD

Vitamin A

Vitamin A is one of the fat-soluble vitamins – which include pro-vitamin A and carotenoids (mainly carotenes and xanthophyll); vitamin A has shown anti-oxidative effects in the prevention of CVD based on experimental and epidemiologic studies. β-Carotene is located, together with vitamin E, in the lipid core of LDL particles.[8] Previous studies, however, have shown an inconsistent role for carotenoids in oxidative protection, i.e. they have both antioxidant and pro-oxidant properties.[3] The antioxidant property of vitamin A protects against LDL oxidation. The observed pro-oxidant effects have been proposed to be due to the tendency of β-carotene radicals, reacting with oxygen, to give rise to peroxyl radicals that mediate lipid peroxidation.[9] Until now, several meta-analyses of randomized controlled trials (RCTs) and observational studies have examined the associations between (i) the intake, or blood concentration, of vitamin A, and (ii) CVD; the findings, with regard to risk factors, have caused controversy. Table 21.1 summarizes epidemiologic studies that have examined the association between vitamin A and CVD.

Evidence from Observational Studies

A cohort study of a Japanese middle-aged population showed that high serum concentrations of α-carotene and β-carotene were associated with lower CVD mortality (multivariable HR and 95% CI logarithmically transformed values of 0.55 (0.38–0.79) and 0.64 (0.47–0.89), respectively);[10] another cohort study of Japanese showed a non-significant association between vitamin A intake and CVD mortality.[11] One systematic review also showed that high serum concentrations of β-carotene or carotenoids were associated with lower CVD events (Peto's OR = 0.46, 95% CI: 0.37–0.58), while a high intake of β-carotene tended to be inversely associated with

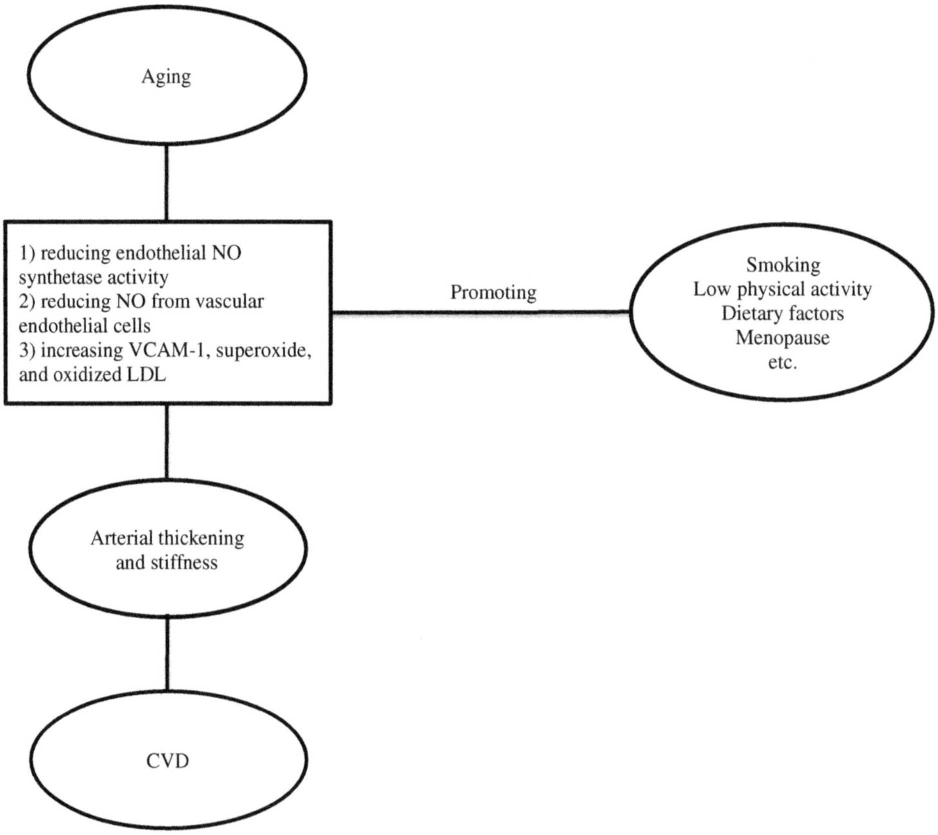

FIGURE 21.1 Mechanisms for aging and CVD. Aging reduces endothelial NO synthetase activity and NO from vascular endothelial cells; it increases (VCAM-1), superoxide, and oxidized LDL, and leads to arterial thickening, stiffness, and atherosclerosis. Lifestyle factors such as smoking and physical inactivity may promote this process. CVD: cardiovascular disease, VCAM-1: vascular cell adhesion protein-1, NO: nitric oxide, LDL: low-density lipoprotein.

TABLE 21.1 Observational Studies, Intervention Studies, and Meta-Analyses for the Association Between Vitamin A and CVD

Study (Authors, Year) (Reference)	Study Design	Population	Exposures (Intake, Concentration of Blood, Treatment)	Outcome/Endpoint	Result
Ito et al 2006 (10)	Cohort study	3061 Japanese men and women	Serum α- and β-carotene and lycopene one logarithmically transformed value increment	CVD mortality	HR (95% CI) of CVD, heart diseases, and stroke α-carotene: 0.55 (0.38–0.79), 0.51 (0.30–0.86), 0.52 (0.31–0.88) β-carotene: 0.64 (0.47–0.89), 0.84 (0.53–1.34), 0.48 (0.31–0.76) Lycopene: 0.77 (0.57–1.04), 0.74 (0.48–1.23), 0.78 (0.51–1.19)
Kubota et al 2012 (11)	Cohort study	58,730 Japanese men and women	Highest (median: men 1439 μg/d, women 1391 μg/d) vs lowest (270 μg/d, 290 μg/d) quintiles of vitamin A intakes	CVD mortality	HR (95% CI) of CVD, CHD, and stroke Men: 1.09 (0.90–1.33), 0.99 (0.65–1.49), 1.15 (0.86–1.53) Women: 1.04 (0.85–1.26), 1.06 (0.68–1.65), 0.89 (0.66–1.21)
Osganian et al 2003 (13)	Cohort study	73,286 US female nurses	Highest vs lowest quintiles of α- and β-carotenes, lutein/zeaxanthin, lycopene, and β-cryptoxanthin intakes	CAD (non-fatal myocardial infarction and fatal CAD) incidence	RR (95% CI) α-carotene: 0.80 (0.65–0.99) β-carotene: 0.74 (0.59–0.93) Lutein/zeaxanthin: 0.90 (0.72–1.12) Lycopene: 0.93 (0.77–1.14) β-cryptoxanthin: 1.17 (0.94–1.44)
Kabagambe et al 2005 (16)	Case–control study	2912 Costa Rican	High vs low quintiles of intake and adipose tissue of α- and β-carotenes, β-cryptoxanthin, lycopene, lutein+zeaxanthin	MI incidence	OR (95% CI) of intake and adipose tissue α-carotene: 1.05 (0.78–1.40), 0.84 (0.62–1.14) β-carotene: 0.96 (0.70–1.31), 0.70 (0.51–0.96) β-cryptoxanthin: 1.33 (0.98–1.79), 0.89 (0.66–1.20) Lutein+zeaxanthin: 1.18 (0.88–1.57), 0.91 (0.67–1.24) Lycopene: 1.05 (0.78–1.42), 1.46 (1.05–2.05)
Hak et al 2003 (14)	Nested case–control study	1063 US male physicians	High vs low quintiles of plasma α- and β-carotenes, β-cryptoxanthin, lutein, lycopene, and retinol concentrations	MI incidence	RR (95% CI) α-carotene: 1.11 (0.64–1.93) β-carotene: 0.74 (0.44–1.26) Among past smokers: 0.55 (0.27–1.12) Among current smokers: 0.46 (0.10–2.06) Among never-smokers: 0.96 (0.46–1.99) β-cryptoxanthin: 1.60 (0.79–3.24) Lutein: 1.25 (0.64–2.46) Lycopene: 1.43 (0.87–2.35) Retinol : 0.94 (0.57–1.53)
Koh et al 2011 (17)	Nested case–control study	840 Chinese	Highest vs lowest quintiles of plasma concentrations of α- and β-carotenes, lycopene, β-cryptoxanthin, lutein, zeaxanthin, and retinol	Acute MI incidence	OR (95% CI) α-carotene: 0.65 (0.35–1.20) β-carotene: 0.03 (0.56–1.89) β-cryptoxanthin: 0.67 (0.37–1.21) Lycopene: 0.97 (0.53–1.77) Lutein: 0.62 (0.35–1.10) Zeaxanthin: 0.96 (0.50–1.83) Retinol: 0.99 (0.55–1.77)
Asplund 2002 (12)	Systematic review of cohort and intervention studies	4–8 cohort studies and 6 RCTs	High vs low intake, blood levels of β-carotene in cohort studies and food supplements in RCTs	CVD events	Peto's OR (95% CI) β-carotene intake (8 cohort studies): 0.88 (0.71–1.01) Blood β-carotene concentration (4 cohort studies): 0.46 (0.37–0.58) Food supplementation (6 RCTs): 1.02 (0.96–1.08)
Ye et al 2008 (15)	Meta-analysis	3 cohort studies	High vs low intake of β-carotene	CHD events	Pooled RR (95% CI): 0.78 (0.53–1.04)

(Continued)

TABLE 21.1 Observational Studies, Intervention Studies, and Meta-Analyses for the Association Between Vitamin A and CVD—cont'd

Study (Authors, Year) (Reference)	Study Design	Population	Exposures (Intake, Concentration of Blood, Treatment)	Outcome/Endpoint	Result
Vivekananthan et al 2003 (19)	Meta-analysis	8 RCTs	Treatment vs control	CVD mortality	Pooled OR (95% CI): 1.1 (1.03–1.17)
Knekt et al 2004 (18)	Pooled meta-analysis	293,172 subjects from 9 cohorts	High vs low quintiles of intake of α- and β-carotenes, lutein, lycopene, β-cryptoxanthin	CHD events	RR (95% CI) for subjects who did not take vitamin supplements α-carotene: 0.90 (0.77–1.04) β-carotene: 0.92 (0.79–1.06) Lutein: 0.89 (0.75–1.04) Lycopene: 0.99 (0.85–1.14) β-cryptoxanthin: 0.94 (0.79–1.12)

HR: hazard ratio, RR: relative risk, 95% CI: 95% confidence intervals, OR: odds ratio, CVD: cardiovascular disease, CAD: coronary artery disease, CHD: coronary heart disease, MI: myocardial infarction, RCT: randomized controlled trial.

CVD events (the OR of the highest β-carotene intake compared with the lowest = 0.88, 95% CI: 0.77–1.01).[12]

Several cohort studies have shown associations between intake or serum concentrations of vitamin A and the risk of CHD and stroke. Women in the highest intake quintiles of α-carotene (1518 μg) and β-carotene (7639 μg) had a lower risk of coronary artery disease (CAD) (non-fatal myocardial infarction and fatal CAD) compared with those in the lowest quintile (α-carotene = 209 μg, β-carotene = 1720 μg). These data were obtained by studying approximately 70,000 female nurses followed-up over a 12-year period. The respective HRs (95% CIs) were 0.80 (0.65–0.99) and 0.74 (0.59–0.93).[13] However, a nested case–control study using the large cohort data showed no association between 5 major carotenoids (α- and β-carotene, β-cryptoxanthin, lutein, and lycopene) and the risk of incident MI. In addition, higher plasma β-carotene tended to be associated with a lower risk of CHD among former and current smokers. In the study, smoking modified the association between plasma β-carotene concentration and MI (interaction with smoking: $P = 0.02$).[14] A meta-analysis of three prospective cohort studies showed that the relative risk (RR) of CHD after a high intake (top third) of β-carotene compared with the bottom third did not reach statistical significance for CHD (RR = 0.78, 95% CI, 0.53–1.04), and that an increase in β-carotene intake of 1 mg/day did not reach statistical significance (RR = 1.0, 95% CI: 0.88–1.14).[15]

A large number of observational studies have also examined the potential preventive effects of other carotenoids: lycopene, lutein, and zeaxanthin. In a case–control study in Italy, intakes of lycopene, total carotenoids, and lutein+zeaxanthin were not associated with a risk of acute MI, while a case–control study in Costa Rica showed that dietary lutein+zeaxanthin tended to be associated with a higher incidence of non-fatal acute MI (OR of highest quintile vs. lowest = 1.18, 95% CI: 0.88–1.57; trend: $P = 0.02$).[16] However, one prospective cohort study among Chinese individuals showed that plasma β-cryptoxanthin and lutein were associated with a lower risk of acute MI. The respective HRs (95% CIs) of the highest vs. lowest quintiles were 0.67 (0.37–1.21; trend: $P = 0.03$) and 0.58 (0.35–0.94; trend: $P = 0.03$) respectively.[17] Additionally, a cohort study that has pooled nine prospective studies followed for 10 years, containing 4647 major incident CHD events registered among 293,172 subjects, showed that energy-adjusted lutein intake was associated with a lower risk of CHD events (pooled RR of highest vs. lowest quintiles = 0.89, 95% CI: 0.75–1.04; trend: $P = 0.03$).[18]

Several cohorts examined vitamin A intake and its supplementation in relation to risk of stroke, but the results from these studies did not find significant associations. A cohort study among middle-aged Japanese individuals showed no significant association between vitamin A intake and mortality from stroke (multivariable HR of highest vs. lowest quintiles = 0.89, 95% CI: 0.66–1.21).[11]

Evidence from RCTs

The evidence gathered from cohort studies supports the notion of protective effects of vitamin A against CVD events, while several meta-analyses of RCTs have shown slightly increased risks for CVD events arising from vitamin A, in particular β-carotene. A meta-analysis of eight RCTs, which administered a β-carotene supplement of 15–50 mg, showed significant increases in CVD deaths (OR = 1.1, 95% CI: 1.03–1.17, $P = 0.003$), and a meta-analysis of three RCTs of β-carotene supplementation also showed no effect on the risk of cerebrovascular accidents (OR compared to control = 1.0, 95% CI: 0.91–1.09).[19]

Vitamin C

Vitamin C is one of the water-soluble vitamins, and it has a strong antioxidant effect. Vitamin C has a potential role in protecting against CVD by modifying LDL oxidation and the NO synthetic pathway, as well as reacting with vitamin E radicals.[1] Vitamin C may be effective in attenuating LDL oxidation by ROS-mediated endothelial dysfunction by quenching aqueous ROS and lowering the affinity of the LDL-bound apolipoprotein B protein. It may also protect normal NO synthesis by modulating the redox states of its components.[1] Further, vitamin C may be most important in the maintenance of vitamin E. It has a role in preventing the consumption of hydrophobic antioxidant vitamins, such as vitamin E and β-carotene, and, importantly, maintaining antioxidative protection. In fact, the combination of vitamins C and E exhibited a stronger protective effect than either vitamin C or vitamin E alone. However, several experimental studies in animal models have indicated that an excess intake of vitamin C might promote atherosclerosis;[20] this corroborates some studies which have shown negative or null associations between vitamin C intake and the risk of CVD.[21,22] Table 21.2 summarizes several epidemiologic studies that have examined the association between vitamin C and CVD.

Evidence from Observational Studies

According to a previous systematic review, dietary intake and blood concentrations of vitamin C are associated with a lower risk of CVD events. In that review, the Peto ORs of high intake and high concentration versus low intake or low concentration of vitamin C using the data from 11 cohort studies were calculated, and they showed a significant negative association between vitamin C and the risk of CVD events; the respective ORs were 0.89 (95% CI: 0.79–0.99) and 0.58 (0.47–0.72).[13] However, vitamin C can also be a pro-oxidant and act to glycate protein under certain circumstances, and thus a high intake of vitamin C may act to promote atherosclerosis.[20] According to a cohort study of approximately 2000 women with diabetes, the risk of CVD of the highest versus lowest vitamin C intakes was significantly higher (HR = 1.84, 95% CI: 1.12–3.01; trend: $P < 0.01$). In particular, the supplementation of high-dose vitamin C (>=300 mg/day from supplements) was associated with an increased risk of CVD, while the high consumption of dietary vitamin C was not.[21]

Several cohort studies have showed independent associations of vitamin C with risks of CHD and stroke.

A meta-analysis of 14 prospective cohort studies showed a significant protective effect for CHD; the RR of the top third versus the bottom third of combined dietary and supplementary vitamin C intake was 0.84 (95% CI: 0.73–0.95), that of dietary vitamin C intake was 0.86 (0.73–0.99), and that of supplementary vitamin C intake was 0.87 (0.63–1.12), respectively.[15] Additionally, a pooled cohort study of 9 cohorts showed that a 700 mg/day or higher vitamin C intake from supplement had the lowest incidence of CHD compared with no intake (HR = 0.75, 95% CI: 0.60–0.93).[18] However, another cohort study of women with diabetes showed that the HR of CHD vitamin C ≥300 mg/day supplement, compared with no intake, was 2.07 (95% CI: 1.27–3.38; trend; $P < 0.01$).[21]

Several observational studies have shown a protective effect of vitamin C on the risk of stroke. A population-based prospective study of 20,649 British men and women aged 40 to 79 years found an HR of the highest versus lowest quartiles of plasma vitamin C concentration of 0.58 (95% CI: 0.43–0.78).[23] Likewise, a large prospective cohort study of approximately 60,000 Japanese men and women aged 40 to 79 years yielded a multivariable HR for the highest versus lowest quintiles of vitamin C intake of 0.70 (95% CI: 0.54–0.92) for total stroke.[11]

Evidence from RCTs

Few clinical trials have examined whether vitamin C supplementation, without other antioxidants, was associated with the risk of CVD. Several RCTs have demonstrated that multivitamin supplementation, including vitamin C, reduces the risk of CVD. The Physicians' Health Study II, a randomized, double-blind, placebo-controlled factorial trial with 14,641 US male physicians enrolled showed that a combined 400 IU/day vitamin E and 500 mg/day vitamin C supplementation for a mean of 8 years had no effect on major cardiovascular events, total MI, or CVD mortality.[22]

Vitamin E

Vitamin E is one of the fat-soluble vitamins and has four isomers (α, β, γ, and δ-tocopherols). Humans cannot synthesize this vitamin and must obtain it from their diet. Vitamin E is widely known as an antioxidant vitamin, and vitamin E in plasma lipoproteins plays an essential role in protecting against oxidative damage of LDL. Oxidized LDL has several potential pro-atherogenesis roles: stimulation of endothelial cells and monocytes to increase production of inflammatory cytokines; stimulation of monocytes and macrophages to increase production of tissue factors; and upregulation of scavenger receptors, eventually leading to foam cell formation of macrophages and the progression of atherosclerotic lesions.[24] Vitamin E is an inhibitor of the oxidation of LDL, and this inhibition contributes protection against the development of atherosclerosis.

Many observational and intervention studies have been conducted to clarify the association between vitamin E and CVD and its risk factors. The many observational studies supported a protective role for dietary and supplementary vitamin E intake on the risk of CVD. However, the results of RCTs are more controversial. Table 21.3 summarizes previous studies that examined the association between vitamin E and CVD.

TABLE 21.2 Observational Studies, Intervention Studies, and Meta-Analyses for the Association Between Vitamin C and CVD

Study (Authors, Year) (Reference)	Study Design	Population	Exposures (Intake, Concentration of Blood, Treatment)	Outcome/Endpoint	Result
Myint et al 2008 (23)	Cohort study	206,49 UK men and women	Highest (>=66 mol/L) vs lowest (<41mol/L) quartiles of plasma vitamin C concentration	Stroke incidence	RR (95% CI): 0.58 (0.43–0.78)
Kubota et al 2012 (11)	Cohort study	58,730 Japanese men and women	Highest (median: men 145 mg/d, women 150 mg/d) vs lowest (52 mg/d, 65 mg/d) quintiles of vitamin C intakes	CVD mortality	HR (95% CI) of CVD, CHD, and stroke Men: 0.88 (0.72–1.07), 0.86 (0.57–1.31), 0.84 (0.62–1.13) Women: 0.79 (0.66–0.94), 0.63 (0.41–0.97), 0.70 (0.54–0.92)
Lee et al 2004 (21)	Cohort study	1923 US postmenopausal women with diabetes	Highest (median: 667 mg/d) vs lowest (85 mg/d) quintiles of vitamin C intakes	CVD mortality	HR (95% CI) of CVD, CHD, and stroke From food and supplements: 1.84 (1.12–3.01), 1.91 (1.05–3.48), 2.57 (0.86–7.66) From food: 1.11 (0.66–1.87), 1.08 (0.57–2.06), 1.89 (0.60–6.03) From supplements (>=300 mg vs 0 mg): 1.69 (1.09–2.44), 2.07 (1.27–3.38), 2.37 (1.01–5.57)
Sesso et al 2008 (22)	RCT	14,641 US male physicians	400 IU of vitamin E every other day and 500 mg of vitamin C daily for mean 8 years	CVD events	RR (95% CI) CVD incidence: 0.99 (0.89–1.11) MI: 1.04 (0.87–1.24) Stroke: 0.89 (0.74–1.07) CVD mortality: 1.02 (0.85–1.21)
Asplund 2002 (12)	Systematic review of cohort and intervention studies	5–11 cohort studies and 2 RCTs	High vs low intake, blood levels of vitamin C in cohort studies and food supplements in RCTs	CVD events	Peto's OR (95% CI) Vitamin C intake (11 cohort studies): 0.89 (0.79–0.99) Blood vitamin C concentration (5 cohort studies): 0.58 (0.47–0.72) Food supplementation (2 RCTs): 0.98 (0.75–1.26)
Ye et al 2008 (15)	Meta-analysis	14 cohort studies	High vs low intake of vitamin C	CHD events	Pooled RR (95% CI): Dietary+supplement: 0.84 (0.73–0.95) Dietary: 0.86 (0.73–0.99) Supplement: 0.87 (0.63–1.12)
Knekt et al 2004 (18)	Pooled meta-analysis	293,172 subjects from 9 cohorts	High vs low quintiles of intake of vitamin C	CHD events	RR (95% CI) Did not take vitamin supplements: 1.23 (1.04–1.45) Incidence for vitamin supplements: 0.75 (0.60–0.93) Mortality for vitamin supplements: 0.76 (0.58–0.99)

HR: hazard ratio, RR: relative risk, 95% CI: 95% confidence intervals, OR: odds ratio, CVD: cardiovascular disease, CHD: coronary heart disease, MI: myocardial infarction, RCT: randomized controlled trial.

Evidence from Observational Studies

A literature review of four cohort studies showed that a high intake of tocopherol was associated with a lower risk of CVD events compared with a low intake. The OR was 0.74 (95% CI: 0.66–0.83).[12]

Several prospective studies have shown an inverse associations between dietary and supplementary vitamin E intake and the risk of CHD. A large cohort study of 39,910 male health professionals aged 40 to 75 years showed that >60 IU/day of vitamin E intake was

TABLE 21.3 Observational Studies, Intervention Studies, and Meta-Analyses for the Association Between Vitamin E and CVD

Study (Authors, Year) (Reference)	Study Design	Population	Exposures (Intake, Concentration of Blood, Treatment)	Outcome/ Endpoint	Result
Rimm et al 1993 (25)	Cohort study	39,910 US male health professionals	Highest (median: 419IU) vs lowest (6.4IU) quintiles of vitamin E intakes	CHD incidence	RR (95% CI): 0.60 (0.44–0.81)
Knekt et al 1994 (26)	Cohort study	5133 Finnish men and women	Highest (men >8.9 mg, women >7.1mg) vs lowest (men =<6.8 mg, women =<5.3 mg) tertiles of vitamin E intakes	CHD mortality	RR (95% CI): Men: 0.68 (0.41–1.11) Women: 0.35 (0.14–0.88)
Kushi et al 1996 (27)	Cohort study	34,486 US postmenopausal women	Highest (>=35.59 IU) vs lowest (<5.68 IU) quintiles of vitamin E intakes	CHD mortality	RR (95% CI): Whole intakes: 0.96 (0.62–1.51) From foods: 0.38 (0.18–0.80) From supplememt: 1.09 (0.67–1.77)
Asplund 2002 (12)	Systematic review of cohort and intervention studies	1–9 cohort studies and 4 RCTs	High vs low intake, blood levels of α-tocopherol in cohort studies and food supplements in RCTs	CVD events	Peto's OR (95% CI) α-tocopherol intake (9 cohort studies): 0.74 (0.66–0.83) Blood concentration (1 cohort study): 1.61 (0.78–3.33) Food supplementation (4 RCTs): 0.96 (0.88–1.04)
Ye et al 2008 (15)	Meta-analysis	9 cohort studies	High vs low intake of vitamin E	CHD events	Pooled RR (95% CI): Diet+supplement: 0.78 (0.66–0.91) Diet: 0.78 (0.60–0.97) Supplement: 0.78 (0.66–0.91)
Vivekananthan et al 2003 (19)	Meta-analysis	6 RCTs	Treatment vs control	CHD mortality	Pooled OR (95% CI): 1.0 (0.94–1.06)
Eidelman et al 2004 (28)	Meta-analysis	7 RCTs	Treatment vs control	CVD events	Pooled OR (95% CI): CVD event: 0.98 (0.94–1.03) Nonfatal MI: 1.00 (0.92–1.09) Nonfatal stroke: 1.03 (0.93–1.14) Ischemic stroke: 1.01 (0.90–1.14) Hemorrhagic stroke: 1.24 (0.96–1.59) CVD death: 1.00 (0.94–1.05)
Bin et al 2011 (29)	Meta-analysis	13 RCTs	Treatment vs control	Stroke	Pooled RR (95% CI): Total stroke: 1.01 (0.96–1.07) Ischemic stroke: 1.01 (0.94–1.08) Hemorrhagic stroke: 1.12 (0.94–1.33) Fatal stroke: 0.94 (0.77–1.14) Nonfatal stroke: 0.99 (0.91–1.08)
Schürks et al 2010 (30)	Meta-analysis	9 RCTs	Treatment vs control	Stroke	Pooled RR (95% CI): Total stroke: 0.98 (0.91–1.05) Ischemic stroke: 0.90 (0.82–0.99) Hemorrhagic stroke: 1.22 (1.00–1.48)

HR: hazard ratio, RR: relative risk, 95% CI: 95% confidence intervals, OR: odds ratio, CVD: cardiovascular disease, CHD: coronary heart disease, MI: myocardial infarction, RCT: randomized controlled trial.

associated with a lower incidence of CHD compared with less than 7.5 IU/day; the multivariate RR was 0.64 (95% CI: 0.49–0.83). That study also showed an inverse association between vitamin E supplementation and the incidence of CHD. The RR of at least 100 IU/day for at least 2 years was 0.63 (95% CI: 0.47–0.84).[25]

A European cohort study also found an inverse association between higher (men: >= 8.9 mg/day, women: >=7.1 mg/day) vitamin E intake and a lower risk of CHD in men (RR of highest compared with lowest tertiles = 0.68, 95% CI: 0.41–0.11, trend: P = 0.01) and in women (RR = 0.35, 95% CI: 0.14–0.88; trend: P < 0.01).[26] In addition, the Iowa Women's Health Study showed an inverse association between dietary vitamin E intake and CHD mortality among 34,486 postmenopausal women (RR = 0.38, 95% CI: 0.18–0.8; trend: P = 0.014).[27] Furthermore, a meta-analysis of 9 prospective cohort studies showed that higher dietary, supplementary, and combined vitamin E intakes were associated with a lower risk of CHD. The respective HRs (95% CI) of top third versus bottom third were 0.76 (0.63–0.89), 0.78 (0.60–0.97), and 0.78 (0.66–0.91). In that meta-analysis, the overall relative risk of CHD for a 30 IU/day (20 mg/day) increment of vitamin E intake was 0.96 (95% CI, 0.94–0.99).[15]

Evidence from RCTs

A meta-analysis of 6 RCTs showed no significant association between vitamin E supplementation and CVD mortality; the pooled OR (95% CI) was 1.0 (0.94–1.06).[19] Another meta-analysis of 7 RCTs also showed similar results; the pooled ORs (95% CI) of cardiovascular events, non-fatal MI, non-fatal stroke, and CVD deaths were 0.98 (0.94–1.03), 1.00 (0.92–1.09), 1.03 (0.93–1.14), and 1.00 (0.94–1.05), respectively.[28]

A meta-analysis of 13 RCTs assessing 166,282 participants showed no significant protective effect in the vitamin E group with respect to total stroke (RR = 1.01, 95% CI: 0.96–1.07, P = 0.663). That meta-analysis conducted a subgroup analysis according to stroke subtypes, and then found no significant protective effect of vitamin E against ischemic stroke (RR = 1.01, 95% CI: 0.94–1.08, P = 0.809), hemorrhagic stroke (RR = 1.12, 95% CI: 0.94–1.33, P = 0.221), fatal stroke (RR = 0.94, 95% CI: 0.77–1.14, P = 0.519), or non-fatal stroke (RR = 0.99, 95% CI: 0.91–1.08, P = 0.846).[29]

On the other hand, another meta-analysis of 9 double-blind RCTs reported that vitamin E supplementation increased the risk of hemorrhagic stroke (pooled RR = 1.22, 95% CI: 1.00–1.48, P = 0.045), while it reduced the risk of ischemic stroke (pooled RR = 0.90, 95% CI: 0.82–0.99, P = 0.02).[30]

Vitamin D

Vitamin D is one of the fat-soluble vitamins, and its sources for humans are diet and endogenous synthesis. This vitamin is one of the major nutrients, having potential protective effects against CVD. Many observational studies have indicated that higher vitamin D intake or blood concentrations associate with lower risks for CVD.[31–37] However, little evidence from experimental studies has supported the notion of an antioxidative property for vitamin D, and its antioxidant potency is much lower than that of vitamin E in plasma (approximately 1/250).[38] Therefore, recent studies have tended to consider that the antioxidant property of vitamin D has little or no effect against CVD.

Folic Acid

Folic acid is one of the B vitamins. The major role of this vitamin is to synthesize DNA by one-carbon transfers in the body. Recent experimental studies have shown that folic acid may have properties of free radical scavenging and antioxidant activity.[39] Additionally, folic acid reduces serum homocysteine concentrations, which promote LDL oxidation and are toxic to the arterial endothelium. Indeed, folic acid supplementation in a dosage greater than 200 μg per day reduces plasma homocysteine levels.[40] A large prospective study of Japanese individuals showed that a high folic acid intake was associated with lower mortality from heart failure in men (multivariable HR = 0.50, 95% CI: 0.27–0.94) and with lower mortality from CHD in women (HR = 0.57, 95% CI: 0.34–0.96).[41] In a large RCT of 5442 women with either a history of CVD or three or more coronary risk factors, individuals were given a combination pill of 2.5 mg folic acid, 50 mg vitamin B6, and 1 mg vitamin B12 for 7.3 years. This treatment showed no significant preventive effects against CVD events (RR = 1.03, 95% CI: 0.90–1.19).[42]

SECONDARY PREVENTIVE EFFECTS AND THERAPEUTIC EFFECTS OF ANTIOXIDANT VITAMINS FOR CVD

Several RCTs of CVD patients have studied whether antioxidant vitamin supplementation prevented their relapse of CVD or improved risk factors or CVD status. The Cholesterol Lowering Atherosclerosis Study of 156 men aged 40–59 years with a history of coronary artery bypass graft surgery reported that subjects with a supplementary vitamin E intake of 100 IU/day or greater showed lower levels of progression of coronary artery lesions compared with those with >100 IU/day, for all lesions (P = 0.04), and for mild/moderate lesions (P = 0.01), while supplementary vitamin C intake had no effect.[43] A double-blind placebo-controlled clinical trial of 45 CVD patients conducted the following three types of intervention: (1) placebo control; (2) 400 IU vitamin

E, 500 mg vitamin C, 12 mg β-carotene; and (3) 800 IU vitamin E, 1000 mg vitamin C, 24 mg β-carotene daily for 12 weeks, and assessed the effect on the susceptibility of LDL oxidation. Reduced susceptibility of LDL to oxidation was estimated by an increase in lag phase (minutes). That trial demonstrated a significantly increased lag phase: from 190.1 minutes at the baseline to 391.1 minutes at 12 weeks among subjects treated with 800 IU vitamin E, 1000 mg vitamin C, and 24 mg β-carotene daily compared to control subjects (from 219.5 minutes to 256.8 minutes) ($P = 0.04$).[44]

In the Cambridge Heart Antioxidant Study (CHAOS) – a double-blind placebo-controlled clinical trial of 2002 patients with angiographically proven coronary atherosclerosis followed up for a median of 510 days – 1035 patients took 800 IU of α-tocopherol and 967 patients received a placebo. The treatment group had lower numbers of major CVD events (RR = 0.53, 95% CI: 0.34–0.83) and non-fatal MI (RR = 0.23, 95% CI: 0.11–0.47), but not of CVD deaths (RR = 1.18, 95% CI: 0.62–2.27).[45] Another large randomized placebo-controlled trial of approximately 20,000 adults with a prior stroke, CHD, other occlusive arterial disease, or diabetes examined the effects of various antioxidant vitamin supplementations (600 mg vitamin E, 250 mg vitamin C, and 20 mg β-carotene daily) on the incidence of vascular disease, cancer, and other serious adverse events. After a 5-year period of supplementation, while plasma concentrations of vitamin C for the vitamins-allocated group were higher than those for the placebo-allocated group (respective mean concentrations were 58.9 and 43.2 μmol/L), they were not significantly different in relation to the risk of stroke between the intervention and placebo-allocated group (proportion of stroke during 5 years were 5.0% in the vitamins-allocated group versus 5.0% in the placebo-allocated group (RR = 0.99, 95% CI: 0.87–1.12).[46] The Women's Antioxidant Cardiovascular Study tested the effects of multivitamin supplementation (500 mg/day vitamin C, 600 IU/day vitamin E, and 50 mg/day β-carotene: 2 (vitamin C or placebo) ×2 (vitamin E or placebo) ×2 (β-carotene or placebo) factorial design) on the combined outcome of MI, stroke, coronary revascularization, and CVD deaths among 8171 female health professionals. After a mean follow-up of 9.4 years, the combined vitamin C and E supplementation reduced the risk of stroke compared with the control group (RR = 0.69, 95% CI: 0.49–0.98); however, there were no associations between the other types of supplementation with CVDs.[47]

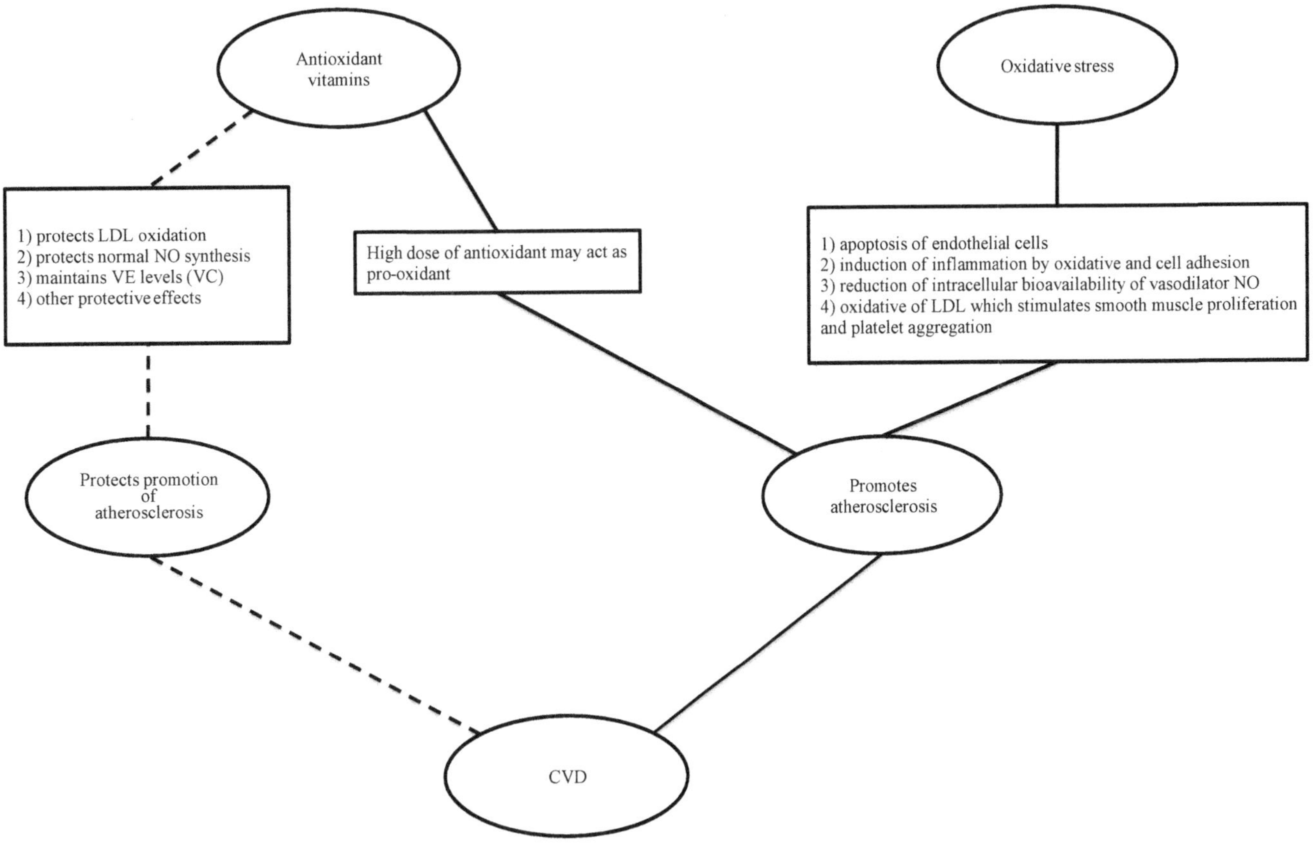

FIGURE 21.2 Mechanisms for oxidative stress, antioxidant vitamins, and CVD. Solid lines mean the pathways of promoting atherosclerosis to CVD. Dotted lines mean the pathways of protection against promotion of atherosclerosis to CVD. CVD: cardiovascular disease, VC: vitamin C, VE: vitamin E, NO: nitric oxide, LDL: low-density lipoprotein.

EFFECT OF MODIFICATION FACTORS

Several previous studies have suggested that modification factors have an effect on the association between antioxidant vitamins and CVD. Smoking is one of the major effect modifiers. Smoking has been shown to increase the levels of oxidative stress and the progression of atherosclerosis.[48] Smokers have lower plasma concentrations of antioxidant vitamins compared with non-smokers. Smoking reduces the levels of catalase and glutathione peroxidase, which protect from oxidative damage. Antioxidant vitamin supplementation may improve the antioxidant status of smokers and reduce their oxidative stress.[3] A nested case–control study examined the association of plasma β-carotene and α-tocopherol with the risk of CHD stratified by smoking status. Among former and current smokers, a high β-carotene concentration tended to be associated with a lower risk of MI; the multivariate-adjusted OR (95% CI) of the highest versus the lowest quintiles of β-carotene concentration was 0.55 (0.27–1.12) for former smokers and 0.46 (0.10–2.06) for current smokers, respectively. However, there was no significant association between β-carotene concentration and risk of MI among non-smokers (OR = 0.96, 95% CI: 0.46–1.99). That study also showed that an association between α-tocopherol concentration and the risk of CHD was not modified by smoking status (interaction: $P = 0.85$).[14] Another large cohort study showed that smoking did not significantly modify this effect. The multivariable HRs (95% CIs) of CAD for the highest versus the lowest quintiles of β-carotene among never, former, and current smokers were 0.84 (0.54–1.29), 0.88 (0.58–1.34), and 0.90 (0.64–1.27), respectively.[13]

There are several potential modification factors: obesity and hypertension, hyperlipidemia, and diabetes.[3] A population-based RCT of 28,519 middle-aged male smokers examined the effects of three types of vitamin supplementations (50 mg α-tocopherol, 20 mg β-carotene, both, and placebo) on the risk of stroke during 6 years of follow-up. That trial demonstrated that α-tocopherol supplementation increased the risk of subarachnoid hemorrhage (RR = 2.45, 95% CI: 1.08–5.55), but decreased the risk of cerebral infarction (RR = 0.70, 95% CI: 0.55–0.89) in hypertensive men. This had no effect in normotensive men. In that trial β-carotene supplementation also increased the risk of subarachnoid hemorrhage in heavy drinkers (RR = 2.66, 95% CI:1.11–6.37).[49] However, few studies have shown that the association between antioxidant vitamin and CVD was modified by alcohol consumption.

Further studies are needed in order to clarify the modification factors related to the associations of antioxidant vitamins with CVD.

CONCLUSION

In conclusion, antioxidant vitamins, mainly vitamins A, C and E, may have potential preventive effects against CVD to balance the CVD-promoting effects of oxidative stress, which cause apoptosis of endothelial cells, induction of inflammation, reduction of intracellular bioavailability of vasodilator NO and LDL oxidation (Fig. 21.2). While most cohort studies have suggested that there are indeed preventive effects of antioxidant vitamins on CVD, related findings from clinical trials has been controversial. Some studies have suggested that antioxidant vitamins may act as pro-oxidants, and thus promote atherosclerosis. There are several effect modification factors that have masked the antioxidant vitamin–CVD association. Therefore, more studies are needed to clarify the association between antioxidant vitamins and CVD.

SUMMARY POINTS

- Aging is a major determinant of cardiovascular disease because it promotes oxidative stress and atherosclerosis.
- Antioxidant vitamins have antioxidant properties as modifiers of LDL oxidation and the nitric oxide synthetic pathway. These properties play a key role in preventing cardiovascular disease.
- The major antioxidant vitamins – vitamins A, C, and E – have been suggested as preventive factors against cardiovascular disease by observational studies, whereas findings from randomized controlled trials have been controversial.
- Several experimental studies have suggested that high doses of antioxidant vitamins may act as pro-oxidants, thus promoting atherosclerosis.
- Folic acid has antioxidant properties, and several observational studies have shown that a high intake of folic acid is associated with a decreased risk of cardiovascular disease.
- Several major randomized controlled trials that examined the secondary prevention of cardiovascular disease by antioxidant vitamins among cardiovascular disease patients have not shown conclusive evidence.
- There are several effect modification factors – i.e. smoking, obesity, medications, and alcohol consumption – which mask the association between antioxidant vitamins and risk of cardiovascular disease.

References

1. Li Y, Schellhorn HE. New developments and novel therapeutic perspectives for vitamin C. *J Nutr* 2007;**137**:2171–84.

2. Madamanchi NR, Vendrov A, Runge MS. Oxidative stress and vascular disease. *Arterioscler Thromb Vasc Biol* 2005;**25**:29–38.
3. Honarbakhsh S, Schachter M. Vitamins and cardiovascular disease. *Br J Nutr* 2009;**101**:1113–31.
4. United Nations. World Population Prospects, the 2012 Revision.
5. North BJ, Sinclair DA. The intersection between aging and cardiovascular disease. *Circ Res* 2012;**110**:1097–108.
6. Wang JC, Bennett M. Aging and atherosclerosis: mechanisms, functional consequences, and potential therapeutics for cellular senescence. *Circ Res* 2012;**111**(2):245–59.
7. Seals DR, Jablonski KL, Donato AJ. Aging and vascular endothelial function in humans. *Clin Sci (Lond)* 2011;**120**:357–75.
8. Esterbauer H, Rotheneder G, Striegl G, et al. Vitamin E and other lipophilic antioxidants protect LDL against oxidation. *Fat Sci Technol* 1989;**91**:316–24.
9. Tsuchihashi H, Kigoshi M, Iwatsuki M, Niki E. Action of beta-carotene as an antioxidant against lipid peroxidation. *Arch Biochem Biophys* 1995;**323**:137–47.
10. Ito Y, Kurata M, Suzuki K, et al. Cardiovascular disease mortality and serum carotenoid levels: a Japanese population-based follow-up study. *J Epidemiol* 2006;**16**:154–60.
11. Kubota Y, Iso H, Date C, JACC Study Group, et al. Dietary intakes of antioxidant vitamins and mortality from cardiovascular disease: the Japan Collaborative Cohort Study (JACC) study. *Stroke* 2011;**42**:1665–72.
12. Asplund K. Antioxidant vitamins in the prevention of cardiovascular disease: a systematic review. *J Intern Med* 2002;**251**:372–92.
13. Osganian SK, Stampfer MJ, Rimm E, et al. Dietary carotenoids and risk of coronary artery disease in women. *Am J Clin Nutr* 2003;**77**:1390–9.
14. Hak AE, Stampfer MJ, Campos H, et al. Plasma carotenoids and tocopherols and risk of myocardial infarction in a low-risk population of US male physicians. *Circulation* 2003;**108**:802–7.
15. Ye Z, Song H. Antioxidant vitamins intake and the risk of coronary heart disease: meta-analysis of cohort studies. *Eur J Cardiovasc Prev Rehabil* 2008;**15**:26–34.
16. Kabagambe EK, Furtado J, Baylin A, Campos H. Some dietary and adipose tissue carotenoids are associated with the risk of nonfatal acute myocardial infarction in Costa Rica. *J Nutr* 2005;**135**:1763–9.
17. Koh WP, Yuan JM, Wang R, et al. Plasma carotenoids and risk of acute myocardial infarction in the Singapore Chinese Health Study. *Nutr Metab Cardiovasc Dis* 2011;**21**:685–90.
18. Knekt P, Ritz J, Pereira MA, et al. Antioxidant vitamins and coronary heart disease risk: a pooled analysis of 9 cohorts. *Am J Clin Nutr* 2004;**80**:1508–20.
19. Vivekananthan DP, Penn MS, Sapp SK, et al. Use of antioxidant vitamins for the prevention of cardiovascular disease: meta-analysis of randomised trials. *Lancet* 2003;**361**:2017–23.
20. Du J, Cullen JJ, Buettner GR. Ascorbic acid: chemistry, biology and the treatment of cancer. *Biochim Biophys Acta* 2012;**1826**:443–57.
21. Lee DH, Folsom AR, Harnack L, et al. Does supplemental vitamin C increase cardiovascular disease risk in women with diabetes? *Am J Clin Nutr* 2004;**80**:1194–200.
22. Sesso HD, Buring JE, Christen WG, et al. Vitamins E and C in the prevention of cardiovascular disease in men: the Physicians' Health Study II randomized controlled trial. *JAMA* 2008;**300**:2123–33.
23. Myint PK, Luben RN, Welch AA, et al. Plasma vitamin C concentrations predict risk of incident stroke over 10 y in 20649 participants of the European Prospective Investigation into Cancer Norfolk prospective population study. *Am J Clin Nutr* 2008;**87**:64–9.
24. Yoshida H, Kisugi R. Mechanisms of LDL oxidation. *Clin Chim Acta* 2010;**411**:1875–82.
25. Rimm EB, Stampfer MJ, Ascherio A, et al. Vitamin E consumption and the risk of coronary heart disease in men. *N Engl J Med* 1993;**328**:1450–6.
26. Knekt P, Reunanen A, Järvinen R, et al. Antioxidant vitamin intake and coronary mortality in a longitudinal population study. *Am J Epidemiol* 1994;**139**:1180–9.
27. Kushi LH, Folsom AR, Prineas RJ, et al. Dietary antioxidant vitamins and death from coronary heart disease in postmenopausal women. *N Engl J Med* 1996;**334**:1156–62.
28. Eidelman RS, Hollar D, Hebert PR, et al. Randomized trials of vitamin E in the treatment and prevention of cardiovascular disease. *Arch Intern Med* 2004;**164**:1552–6.
29. Bin Q, Hu X, Cao Y, Gao F. The role of vitamin E (tocopherol) supplementation in the prevention of stroke. A meta-analysis of 13 randomised controlled trials. *Thromb Haemost* 2011;**105**:579–85.
30. Schürks M, Glynn RJ, Rist PM, et al. Effects of vitamin E on stroke subtypes: meta-analysis of randomised controlled trials. *BMJ* 2010;**341**:c5702.
31. Wang TJ, Pencina MJ, Booth SL, et al. Vitamin D deficiency and risk of cardiovascular disease. *Circulation* 2008;**117**:503–11.
32. Ginde AA, Scragg R, Schwartz RS, Camargo Jr CA. Prospective study of serum 25-hydroxyvitamin D level, cardiovascular disease mortality, and all-cause mortality in older U.S. adults. *J Am Geriatr Soc* 2009;**57**:1595–603.
33. Semba RD, Houston DK, Bandinelli S, et al. Relationship of 25-hydroxyvitamin D with all-cause and cardiovascular disease mortality in older community-dwelling adults. *Eur J Clin Nutr* 2010;**64**:203–9.
34. Parker J, Hashmi O, Dutton D, et al. Levels of vitamin D and cardiometabolic disorders: systematic review and meta-analysis. *Maturitas* 2010;**65**:225–36.
35. Kilkkinen A, Knekt P, Aro A, et al. Vitamin D status and the risk of cardiovascular disease death. *Am J Epidemiol* 2009;**170**:1032–9.
36. Giovannucci E, Liu Y, Hollis BW, Rimm EB. 25-hydroxyvitamin D and risk of myocardial infarction in men: a prospective study. *Arch Intern Med* 2008;**168**:1174–80.
37. Sun Q, Pan A, Hu FB, et al. 25-Hydroxyvitamin D levels and the risk of stroke: a prospective study and meta-analysis. *Stroke* 2012;**43**:1470–7.
38. Stone WL, Krishnaswamy G. The role of fat-soluble nutrients and antioxidants in preventing heart disease. In: Arnoldi A, editor. *Functional foods, cardiovascular disease and diabetes*. Cambridge: Woodhead Publishing; 2004. p 67.
39. Joshi R, Adhikari S, Patro BS, et al. Free radical scavenging behavior of folic acid: evidence for possible antioxidant activity. *Free Radic Biol Med* 2001;**30**:1390–9.
40. Malinow MR, Duell PB, Hess DL, et al. Reduction of plasma homocyst(e)ine levels by breakfast cereal fortified with folic acid in patients with coronary heart disease. *N Engl J Med* 1998;**338**: 1009–15.
41. Cui R, Iso H, Date C, Kikuchi S, Tamakoshi A, Japan Collaborative Cohort Study Group. Dietary folate and vitamin b6 and B12 intake in relation to mortality from cardiovascular diseases: Japan collaborative cohort study. *Stroke* 2010;**41**:1285–9.
42. Albert CM, Cook NR, Gaziano JM, et al. Effect of folic acid and B vitamins on risk of cardiovascular events and total mortality among women at high risk for cardiovascular disease: a randomized trial. *JAMA* 2008;**299**:2027–36.
43. Hodis HN, Mack WJ, LaBree L, et al. Serial coronary angiographic evidence that antioxidant vitamin intake reduces progression of coronary artery atherosclerosis. *JAMA* 1995;**273**:1849–54.
44. Mosca L, Rubenfire M, Mandel C, et al. Antioxidant nutrient supplementation reduces the susceptibility of low density lipoprotein to oxidation in patients with coronary artery disease. *J Am Coll Cardiol* 1997;**30**:392–9.
45. Stephens NG, Parsons A, Schofield PM, et al. Randomised controlled trial of vitamin E in patients with coronary disease: Cambridge Heart Antioxidant Study (CHAOS). *Lancet* 1996;**347**:781–6.

46. Heart Protection Study Collaborative Group. MRC/BHF Heart Protection Study of antioxidant vitamin supplementation in 20536 high-risk individuals: a randomised placebo-controlled trial. *Lancet* 2002;**360**:23–33.
47. Cook NR, Albert CM, Gaziano JM, et al. A randomized factorial trial of vitamins C and E and beta carotene in the secondary prevention of cardiovascular events in women: results from the Women's Antioxidant Cardiovascular Study. *Arch Intern Med* 2007;**167**:1610–8.
48. Perlstein TS, Lee RT. Smoking, metalloproteinases, and vascular disease. *Arterioscler Thromb Vasc Biol* 2006;**26**:250–6.
49. Leppälä JM, Virtamo J, Fogelholm R, et al. Vitamin E and beta carotene supplementation in high risk for stroke: a subgroup analysis of the Alpha-Tocopherol, Beta-Carotene Cancer Prevention Study. *Arch Neurol* 2000;**57**:1503–9.

CHAPTER

22

Hypertension, Menopause and Natural Antioxidants in Foods and the Diet

Maria Grazia Modena

University of Modena and Reggio Emilia, Modena, Italy

List of Abbreviations

ACE angiotensin-converting enzyme
ALCAR acetyl L-carnitine
Ang angiotensin
Arg arginine
BP blood pressure
CHACS Cambridge Heart Antioxidants Study
CoQ coenzyme Q_{10}
ED endothelial dysfunction
Fl flavonoids
HOPE Heart Outcomes Prevention Evaluation Study
HTN hypertension
LA alpha-lipoic acid
LDL low-density lipoprotein (cholesterol)
NADP nicotinamide adenine dinucleotide phosphate
NO nitric oxide
Nox all types of NO and compounds
O_2^- superoxide
OS oxidative stress
PPH polyphenols
Redox reduction–oxidation reactions
ROS reactive oxygen species
RS resveratrol

INTRODUCTION

Worldwide, hypertension (HTN) is the most important cardiovascular risk factor for stroke, myocardial infarction and heart failure.[1]

Endothelial dysfunction (ED), defined as functional and reversible alteration of endothelial cells, is the common early target of multifactor mechanisms implicated in the development of HTN occurring prior to progressive organ damage and its atherosclerotic and thrombotic complications.[2]

Several lines of evidence suggest that ED results from impairment in bioavailability of reactive oxygen species (ROS) and the neutralizing action of the antioxidant system. It is caused by oxidative stress (OS) in the vessels, heart, brain and kidneys, and promotes the local inflammation that underlies vascular changes and the structural remodeling also observed in HTN.[3]

In humans, the relationship between the excessive production of ROS and the development of HTN has been demonstrated in the setting of essential, renovascular, malignant, diet-induced, drug-induced or pre-eclampsia disease.[3–6] Moreover, it has been proposed that increased levels of ROS can give rise to hypertensive organ damage.[7,8]

In conjunction with the evidence of 'female sex fragility' in the development of HTN, higher levels of OS biomarkers were detected in postmenopausal women compared to levels in both males and premenopausal women – suggesting that the main molecular mechanism of HTN in women is the loss of estrogen-dependent antioxidant effects.[9,10]

While the antioxidant effects of classic antihypertensive drugs (beta-adrenergic blockers, angiotensin-converting enzyme (ACE) inhibitors, AT1, receptor blocker and Ca^{2+} channel blockers) are known in both sexes, the benefit of natural antioxidants and their supplementation to prevent/treat HTN has not yet been fully investigated.

This chapter briefly explores the anti-hypertensive mechanisms of common natural antioxidants reported in preclinical and clinical studies, and focuses on new strategies to improve women's health status.

OXIDATIVE STRESS AND HYPERTENSION: HIDDEN MECHANISMS

The normal cellular metabolism involves oxygen reduction processes, generating ROS. The dominant initial ROS produced is superoxide (O_2^-), a short-lived molecule that can oxidize proteins and lipids – or

Aging
http://dx.doi.org/10.1016/B978-0-12-405933-7.00022-6

react with the endothelium-derived nitric oxide (NO), creating reactive nitrogen species (RNS), and react with critical enzymatic cofactors implicated in injurious cellular effects and DNA damage. Usually, the homeostatic levels of ROS are controlled by enzymatic and non-enzymatic systems that prevent or inhibit oxidation of oxidizable biomolecules such as DNA, lipids and proteins.

The increase in OS – due to excess ROS, decreased NO levels and low antioxidant bioavailability – has been implicated in the pathophysiology of many cardiovascular diseases, including hypertension. Changes in redox-sensitive signaling in the vascular system trigger a chronic low-grade inflammation that supports vasoconstriction and the procoagulant status of ED which, in turn, characterizes HTN patients. The development of HTN consequently promotes a modest increase in O_2^- and causes the activation of redox-sensitive pathways at the cerebral and renal level, establishing secondary pathologic mechanisms implicated in both the hypertensive profile and organ damage.[3,11]

Of several enzymatic sources of ROS, the Nox family of NAD(P)H oxidases is the most important system implicated in HTN. The activities of these proteins – which are expressed in virtually all vessel walls as well as in key organs implicated in the regulation of blood pressure – explain several functions, such as differentiation, proliferation, apoptosis, senescence, inflammatory responses and oxygen sensing by signaling molecules. In wild-type mouse models, NOX expression and NOX-dependent ROS production are particularly sensitive to angiotensin II-induced HTN, while in NOX knockout mice, ROS production is not affected by HTN.[12] The link between NOX expression and HTN has also been demonstrated in human studies; these studies revealed that polymorphisms in the genes encoding NAD(P)H oxidase subunits are associated with increased atherosclerosis and hypertension.[13]

Emerging data indicate a complex regulatory crosstalk between NAD(P)H oxidase and other sources of ROS, particularly mitochondrial respiratory enzymes; the effects of these enzymes have been extensively proven in the setting of idiopathic and salt-induced HTN, and xanthine oxidase is upregulated by NAD(P)H oxidase under shear conditions.[14]

MENOPAUSE, OXIDATIVE STRESS AND HYPERTENSION: PINK NETWORKING

Menopause, a consequence of the normal aging process in women, is characterized by intense physiologic and biochemical changes which lead to an increased risk of cardiovascular disease – but also to osteoporosis, cancer and other degenerative diseases.

Recent studies – designed to understand the pathogenesis of these disorders – support the hypothesis that estrogen deficiency (mainly of estrogen E_2) negatively influences the activity of cellular antioxidant enzymes and upregulates the expression of antioxidant and longevity-related genes concurring at the increased of OS resulting from aging process. These data are supported by the detection of lower levels of OS in premenopausal females compared to both men and postmenopausal women, and are confirmed by higher levels of biomarkers of inflammation and OS in elderly women compared to elderly men.[10,15,16]

The mechanisms underlying the development of postmenopausal hypertension remain unrecognized. It is known that the estrogens modulate the activity and expression of NO and Ang II. At the renal level, the hypoestrogenism promotes an imbalance between NO and Ang II, resulting in disturbed handling of renal sodium, oxidative stress, and increased salt sensitivity – with the logical consequence of hypertension, particularly in genetically prone women. In addition, the postmenopausal hormonal status contributes to endothelial dysfunction and to a reduction in arterial elasticity and compliance.[17,18]

NATURAL ANTIOXIDANT AGENTS

The term 'antioxidant' refers to enzymatic and non-enzymatic complexes that prevent or inhibit the oxidation of biomolecules.

The main non-enzymatic antioxidants are biologic compounds found mainly in vegetables and fruits.

Several observational studies support the high intake of fruit and vegetables in the setting of cardiovascular disease, citing evidence that they have a modulating effect on OS and inflammation. However, randomized studies have shown disappointing results for antioxidant dietary supplements, particularly in samples from healthy populations.

With regard to hypertension, the principal antioxidants examined are:

Flavonoids

The flavonoids (FLs) are a class of plant secondary metabolites included in a large group of polyphenols (PPH). They have a C6–C3–C6 structure and are classified by the nature of the C3 element into anthocyanins, chalcones, dihydrochalcones, flavanols, flavanones, flavones, flavonols, isoflavones and proanthocyanidins.

Over 4000 flavonoids have been identified in fruits (apples, bananas, oranges, grapefruit, pears, melons, grapes, strawberries, pineapples, blueberries, blackcurrants, blackberries), vegetables (broccoli, carrots, onions, potatoes, mushrooms, lettuce, tomatoes, cucumber, sweet pepper, spinach, oregano, basil, thyme, mint, coriander, rosemary) and beverages (tea, coffee, beer, wine and fruit drinks).

Epidemiologic studies have shown that the intake of FLs is inversely related to mortality from coronary heart disease and to the incidence of heart attacks.

The cardioprotective effects of FLs are related to several mechanisms involved in ischemic–reperfusion injury, suppression of platelet aggregation, enhanced antioxidant status and increased NO bioavailability.

PPHs act as free-radical scavengers, reducing the oxidation of LDL and the capacity of macrophages to generate free radicals.[19–21] In atherosclerosis plaques, PPHs also inhibit the proliferation of vascular smooth muscle cells (VSMC) and down-regulate the expression of the cyclin A gene.[22]

Moreover, resveratrol (RS) (3,5,40-trihydroxystilbene) and quercetin, the most studied PPHs, have shown an anti-aggregant effect on human platelets – explained by the reduction in platelet cytosolic calcium and the consequent increase in cyclic GMP phosphodiesterase activity.[23,24]

There is also some evidence to suggest that RS is implicated in ischemic preconditioning-mediated cardioprotection by its activation of the transcription factors of antioxidant-encoding genes[25] and its modulating effect on signal-regulated intracellular pathways;[26,27] RS has also been implicated in the improvement of endothelial function by inhibiting the expression of ICAM and VCAM and upregulating the expression of eNOS.[28–30]

In the setting of hypertensive disorders, the chronic and short-term administration of PPHs is associated with a reduction in blood pressure and the attenuation of organ damage in all mouse models (transgenic rats, spontaneously hypertensive rats (SHR) and in deoxycorticosterone acetate–salt hypertensive rats).[31–33]

In humans, oral RS supplementation elicited an acute dose-related improvement in endothelium-dependent vasodilatation – tested by measuring flow-mediated dilatation (FMD) – suggesting an effect on endothelium-derived NO bioavailability.[34]

However, compared to other dietary compounds, the antioxidant effects of these molecules are explained by a higher daily intake (50–800 mg/day).

It is apparent that more studies are needed to understand the real clinical impact of flavonoids.

L-Arginine (Arg)

Arginine is a nitrogenous precursor in the synthesis of nitric oxide by NO synthase. Its content is relatively high in seafood, watermelon juice, nuts, seeds, algae, meats, rice protein concentrate, and soy protein isolate, but only 40% of dietary Arg is degraded by the small intestine in first-pass metabolism.[35]

Low cellular levels of Arg have been demonstrated in human hypertension,[36] and its supplementation has proven efficacy in the reduction of blood pressure when it is combined with *N*-acetylcysteine in healthy human subjects, diabetics and chronic kidney patients.[37–39]

It is suggested that the antihypertensive response to Arg may be mediated in part by its suppressive effects on angiotensin II and endothelin-1, and its potentiating effects on insulin. In fact, experimental data suggest that insulin mediates vascular dilatation and modulates vascular tone through an NO-dependent mechanism and that, in hypertensive patients, the insulin-mediated vasodilatation is impaired.[40] In addition, human clinical studies have provided evidence that Arg may have endocrine actions that improve insulin sensitivity and modulate insulin release.

However, recent concerns about Arg supplementation refer to potential deleterious effects. In fact, the conversion of Arg into creatine may lead to increased homocysteine levels and, thus, to OS .[41,42]

Mitochondrion-Related Antioxidants

Under normal physiologic conditions, mitochondria are a major source of ROS – which are produced by the 1-electron reduction of O_2 to superoxide (O_2^-) in the respiratory chain. Mitochondrial DNA is particularly susceptible to modification by ROS but it is normally protected from oxidative damage by a multilayer network of antioxidant systems:

Coenzyme Q_{10} (CoQ)

CoQ (2,3-dimethoxy-5-meth-6-decaprenyl benzoquinone) is an essential cofactor implicated as a mobile carrier in mitochondrial oxidative phosphorylation and ATP production. Endogenous production is derived from mevalonic acid and phenylalanine, but it could also be derived from animal sources in the normal diet. CoQ is present in the body in both the reduced form (ubiquinol, $CoQ_{10}H_2$) and the oxidized form (ubiquinone, CoQ_{10}).[43] The reduced form is the only endogenously synthesized lipophilic agent that has an antioxidant effect on biologic membranes, and it inhibits lipoprotein peroxidation.[44] The antihypertensive mechanism of CoQ is attributed to a direct effect on the endothelium (NO-mediated) and vascular smooth muscle – resulting in a decrease in peripheral resistance in hypertensive patients; this has not been confirmed in normotensive subjects.[45] A recent meta-analysis of 12 studies (362 patients) reported that oral supplementation of CoQ is associated with a reduction of up to 17 mmHg in systolic blood pressure and 10 mmHg in diastolic blood pressure without significant side effects; this favors its addition to conventional anti-hypertensive therapy.[46]

Alpha-Lipoic Acid (LA)

Alpha-lipoic acid (1,2-dithiolane-3-pentanoic acid or thioctic acid) (LA) is a small molecule with one chiral center containing two oxidized or reduced thiol groups. The reduced form, known as dihydrolipoic acid (DHLA), interacts with ROS, while the oxidized form inactivates free radicals.

Normally, humans synthesize LA from fatty acids and cysteine, but this does not provide an adequate source; thus, it is necessary to obtain LA from animal and plant sources (red meat, liver, heart and kidney, spinach, broccoli, tomatoes, Brussels sprouts, potatoes, garden peas and rice bran). Several factors influencing LA absorption (e.g. substrate competition, transcriptional and post-translational regulatory mechanisms) explain the quite variable pharmacokinetics of LA. The therapeutic level of LA can be achieved by oral supplementation – which usually involves a mixture of both R and S enantiomers. However, recent studies suggest that only the R enantiomer is the appropriate form for oral supplementation.[47] The rationale for using LA therapeutically in hypertension derives from its capacity to increase the levels of tissue-reduced glutathione and to prevent deleterious sulfhydryl group modification in Ca^{2+} channels. In hypertensive rats, LA normalizes the systolic blood pressure and inhibits the renal and vascular overproduction of endothelin-1i, improving endothelial NO synthesis.[48–50] In humans, the administration of LA produces an antihypertensive effect only when it is combined with acetyl L-carnitine (ALCAR) supposing the role of bioavailability to keep effect.[51] Although no side effects were shown until high doses were used, caution has been expressed in relation to oral supplementation. In rats, intraperitoneal administration of racemic LA (100 mg/kg b.w./day for 2 weeks – equivalent to 5–10 grams per day in humans) has been associated with an increase in the levels of plasma lipid hydroperoxide and oxidative damage to proteins, especially in heart and brain.[52,53]

Acetyl L-Carnitine (ALCAR)

Acetyl L-carnitine is an ester of L-carnitine synthesized in the human brain, liver and kidney by the enzyme ALC transferase. It is implicated in the uptake of acetyl CoA during fatty acid oxidation, in acetylcholine production and in the synthesis of membrane protein phospholipids. In *in vitro* experiments, the potential pro-oxidative effect observed when the compound is used alone is eliminated by association with LA.[54] The few data concerning combination therapy with LA show antihypertensive effects in patients with coronary artery disease and/or metabolic syndrome.[51]

Vitamin A

The generic term 'vitamin A' defines both provitamin A (β-carotene and other carotenoids) and physiologically active forms of the preformed vitamin (all-*trans* retinol and its ester). Common sources of vitamin A include liver, dark green leafy vegetables, corn, tomatoes, oranges and oily fish.

The interest in vitamin A derives from epidemiologic data, focused on β-carotene, that show its cardioprotective effects and an inverse correlation between its plasma level and systolic and diastolic blood pressure.[55] However, the initial enthusiasm has been undermined by *in vitro* research suggesting that β-carotene and other carotenoids have adverse effects on mitochondria related to cleavage products with pro-oxidant properties;[56] this has not been confirmed in human studies using physiologic doses of β-carotene provided by food.[57]

Concerning the limited interest in this compound, long-term supplementation (12 years) in a healthy population did not show any benefit in terms of a reduction in blood pressure, and it is not associated with a reduction in the incidence of cardiovascular diseases.[58]

Recently, the interest in carotenoids has turned to lycopene. In a pilot randomized study testing the efficacy of supplementation with tomato extracts, such as lycopene, Engelhard et al reported a reduction in blood pressure in patients with grade 1 HT who were naive to drug therapy.[59] However, these results were not confirmed in a subsequent similar study.[60] Neither study reported important side effects, but the efficacy and the long-term beneficial effect of lycopene on cardiovascular risk factors remain unclear.

Vitamin E

Alpha tocopherol is the predominant and the most biologically active form of vitamin E. It is found in avocados, asparagus, vegetable oils, nuts and leafy green vegetables.

In the CHAOS study, a randomized controlled trial, alpha tocopherol supplementation was associated with a reduction in cardiovascular events in patients with coronary disease.

Later, Yusuf et al showed a dose-dependent relationship between vitamin E supplementation and all causes of mortality, identifying a progressive increase for dosages greater than 150 IU/d and suggesting that it has a role as an independent predictor for heart failure.[61,62] These results were not confirmed in a prospective cohort study of 29,092 male smokers, in which alpha tocopherol (derived mostly from dietary intake) was associated with significantly lower total and cause-specific death rates.[63]

Conflicting results were also obtained in the setting of hypertension, and actually the clinical relevance of vitamin E is not clear. In the HOPE trial, vitamin E supplementation in patients at high risk of cardiovascular disease failed to show any benefit in terms of lower blood pressure and mortality.[61] Several hypotheses have been proposed to explain the variability of these results; it was suggested that the antioxidant effect was related to long-term assumption of this compound that could be required the association to other antioxidants to lead the maximum benefit.

Vitamin C

Vitamin C (ascorbic acid) is an essential micronutrient acquired primarily through the diet. It is implicated in several biologic functions (such as the synthesis of hormones and neurotransmitters, the detoxification of exogenous compounds) and also in the immune system. Its antioxidant activity is performed mainly by quenching of single-electron free radicals, while its potential cardioprotective effects involve improvement in endothelial function through an increase in NO bioavailability.[64]

Observational studies suggest an inverse association between plasma vitamin C levels and blood pressure.[65] However, interventional trials investigating the role of this nutrient in a hypertensive setting report inconclusive and contradictory results among hypertensive,[66] normotensive,[67] and diabetic[68,69] cohorts. Furthermore, a high intake of vitamin C from supplements is associated with an increased risk of mortality from cardiovascular disease in postmenopausal women with diabetes[70] due to pro-oxidant effects.

Recently, Juraschek et al published the results of a meta-analysis of 29 small trials reporting the effects on systolic BP (SBP) or diastolic BP (DBP), or both, of oral vitamin C (a median dose of 500 mg/day). The pooled changes in SBP and DBP were −3.84 mmHg (95% CI: −5.29, −2.38 mmHg; $P < 0.01$) and −1.48 mmHg (95% CI: −2.86, −0.10 mmHg; $P = 0.04$), respectively. In trials in hypertensive participants, corresponding reductions in SBP and DBP were −4.85 mmHg ($P < 0.01$) and −1.67 mmHg ($P = 0.17$). After the inclusion of nine trials with imputed BP effects, BP effects were attenuated but remained significant. The authors concluded that, in short-term trials, vitamin C supplementation reduced SBP and DBP.[71]

Another study showed that vitamin C infusion significantly lowers blood pressure in hypertensive patients but not in normotensive subjects, suggesting that sympathovagal balance and spontaneous baroreflex sensitivity are restored during vitamin C infusion in hypertensive subjects in comparison to healthy normotensive patients.[72]

CONCLUSIONS

The topic of hypertension, menopause and antioxidants in foods and diet is going to open a new scenario in the treatment of the most prevalent risk factor in postmenopausal women. In fact, hypertension in this particular phase of life is due to an increase in body mass index and a deficiency of estrogen. Pharmacologic therapy is very important, but it should be the second or an alternative choice after modification of lifestyle. In this context, antioxidants and diet may help in weight control and in the natural substitution of the effects of estrogens.

SUMMARY POINTS

- Worldwide, hypertension is the most important cardiovascular risk factor for all cardiovascular events.
- Postmenopausal women have an increased risk of endothelial dysfunction and oxidative stress due to a deficiency of estrogen and an increase in body mass index.
- The flavonoids are a class of plant secondary metabolites which include polyphenols; 4000 of them have been identified in fruits, vegetables and beverages.
- 'Antioxidant' refers to enzymatic and non-enzymatic complexes that prevent or inhibit the oxidation of biomolecules.
- Natural antioxidant agents are biologic compounds found mainly in vegetables and fruits.
- L-Arginine is a nitrogenous precursor of NO; there is a high content in seafood, watermelon, nuts, seeds, algae, meats, rice and soy proteins.
- The mitochondrion-related antioxidants – CoQ, LA and AlCAR – have different mechanisms of action.
- Vitamins A, E and C are all antioxidants present in vegetables and fruits; they have different effects in hypertension.
- Pharmacologic therapy is very important in postmenopausal hypopotassiemic women, but antioxidants and diet may also be very useful because ED is the main cause of hypertension.

References

1. World Health Report. *Reducing risks, promoting healthy life.* Geneva, Switzerland: World Health Organization; 2002. http://www.who.int/whr/2002.
2. Cai H, Harrison DG. Endothelial dysfunction in cardiovascular diseases: the role of oxidant stress. *Circ Res* 2000;**87**(10):840–4.
3. Montezano AC, Touyz RM. Molecular mechanisms of hypertension – reactive oxygen species and antioxidants: a basic science update for the clinician. *Can J Cardiol* 2012;**28**(3):288–95.
4. Rodrigo R, Prat H, Passalacqua W, et al. Relationship between oxidative stress and essential hypertension. *Hypertens Res* 2007;**30**:1159–67.
5. Ward NC, Hodgson JM, Puddey IB, et al. Oxidative stress in human hypertension: association with antihypertensive treatment, gender, nutrition, and lifestyle. *Free Radic Biol Med* 2004;**36**:226–32.
6. Raijmakers MT, Dechend R, Poston L. Oxidative stress and preeclampsia: rationale for antioxidant clinical trials. *Hypertension* 2004;**44**:374–80.
7. Chen K, Xie F, Liu S, et al. Plasma reactive carbonyl species: potential risk factor for hypertension. *Free Radic Res* 2011;**45**:568–74.
8. Murphey LJ, Morrow JD, Sawathiparnich P, et al. Acute angiotensin II increases plasma F2-isoprostanes in salt-replete human hypertensives. *Free Radic Biol Med* 2003;**35**:711–8.

9. Arora KS, Gupta N, Singh RA, et al. Role of free radicals in menopausal distress. *J Clin Diagn Res* 2009;**13**:1900–902.
10. Sánchez-Rodríguez MA, Zacarías-Flores M, Arronte-Rosales A, et al. Menopause as risk factor for oxidative stress. *Menopause* 2012; **19**(3):361–7.
11. Ferroni P, Basili S, Paoletti V, Davì G. Endothelial dysfunction and oxidative stress in arterial hypertension. *Nutr Metab Cardiovasc Dis* 2006;**16**(3):222–33.
12. Haque MZ, Majid DS. High salt intake delayed angiotensin II-induced hypertension in mice with a genetic variant of NADPH oxidase. *Am J Hypertens* 2011;**24**(1):114–8. Epub 2010 Aug 12.
13. Sirker A, Zhang M, Shah AM. NADPH oxidases in cardiovascular disease: insights from *in vivo* models and clinical studies. *Basic Res Cardiol* 2011;**106**(5):735–47.
14. Paravicini TM, Touyz RM. NADPH oxidases, reactive oxygen species, and hypertension: clinical implications and therapeutic possibilities. *Diabetes Care* 2008;**31**(Suppl 2):S170–80.
15. Gierach GL, Johnson BD, Bairey Merz CN, et al. WISE Study Group. Hypertension, menopause, and coronary artery disease risk in the Women's Ischemia Syndrome Evaluation (WISE) Study. *J Am Coll Cardiol* 2006;**47**(3 Suppl):S50–58.
16. Zitňanová I, Rakovan M, Paduchová Z, et al. Oxidative stress in women with perimenopausal symptoms. *Menopause* 2011;**18**(11):1249–55.
17. Grossman E. Does increased oxidative stress cause hypertension? *Diabetes Care* 2008;**31**(Suppl 2):S185–9.
18. Lopez-Ruiz Arnaldo, Sartori-Valinotti Julio, Yanes Licy L, et al. Sex differences in control of blood pressure: role of oxidative stress in hypertension in females. *Am J Physiol Heart Circ Physiol* 2008;**295**(2):H466–74.
19. Furman B, Aviram M. Flavonoids protect LDL from oxidation and attenuate atherosclerosis. *Curr Opin Lipidol* 2001;**12**:41–8.
20. Abu-Amsha R, Croft KD, Puddey IB, et al. Phenolic content of various beverages determines the extent of inhibition of human serum and low-density lipoprotein oxidation in vitro: identification and mechanism of action of some cinnamic acid derivatives from red wine. *Clin Sci (Lond)* 1996;**91**:449–58.
21. Goerlik S, Lapidot T, Shaman I, et al. Lipid peroxidant and coupled vitamin oxidation in simulated and humen gastric fluid inhibited by dietary polyphenols: health implications. *J Agric Food Chem* 2005;**53**:3397–402.
22. Iijima K, Yoshizumi M, Hashimoto M, et al. Red wine polyphenols inhibit vascular smooth muscle cell migration through two distinct signaling pathways. *Circulation* 2002;**105**(20):2404–10.
23. Pace-Asciak CR, Rounova O, Hahn SE, et al. Wines and grape juices as modulators of platelet aggregation in healthy human subjects. *Clin Chim Acta* 1996;**246**:163–82.
24. Pace-Asciak CR, Hahn S, Diamandis EP, et al. The red wine phenolics trans-resveratrol and quercetin block human platelet aggregation and eicosanoid synthesis: implications for protection against coronary heart disease. *Clin Chim Acta* 1995;**235**:207–19.
25. Das S, Cordis GA, Maulik N, Das DK. Pharmacological preconditioning with resveratrol: role of CREB-dependent Bcl-2 signaling via adenosine A3 receptor activation. *Am J Physiol Heart Circ Physiol* 2005;**288**:H328–35.
26. Das S, Tosaki A, Bagchi D, et al. Resveratrol mediated activation of cAMP response element-binding protein through adenosine A3 receptor by Akt-dependent and -independent pathways. *J Pharmacol Exp Ther* 2005;**314**:762–9.
27. Juan SH, Cheng TH, Lin HC, et al. Mechanism of concentration-dependent induction of heme oxygenase-1 by resveratrol in human aortic smooth muscle cells. *Biochem Pharmacol* 2005;**69**:41–48.
28. Sumpio BE, Oluwole B, Wang X, Awolesi M. Regulation of nitric oxide synthase expression and activity by hemodynamic forces. In: *The pathophysiology and clinical applications of nitric oxide*. United Kingdom: Harwood Academic Publishers; 1998. pp. 171–93.
29. Wallerath T, Deckert G, Ternes T, et al. Resveratrol, a polyphenolic phytoalexin present in red wine, enhances expression and activity of endothelial nitric oxide synthase. *Circulation* 2002;**106**:1652–8.
30. Iijima K, Yoshizumi M, Hashimoto M, et al. Red wine polyphenols inhibit vascular smooth muscle cell migration through two distinct signaling pathways. *Circulation* 2002;**105**(20):2404–10.
31. Biala A, Tauriainen E, Siltanen A, et al. Resveratrol induces mitochondrial biogenesis and ameliorates Ang II induced cardiac remodeling in transgenic rats harboring human renin and angiotensinogen genes. *Blood Press* 2010;**19**:196–205.
32. Perez-Vizcaino F, Duarte J, Jimenez R, et al. Antihypertensive effects of the flavonoid quercetin. *Pharmacol Rep* 2009;**61**:67–75.
33. Behbahani J, Thandapilly SJ, Louis XL, et al. Resveratrol and small artery compliance and remodeling in the spontaneously hypertensive rat. *Am J Hypertens* 2010;**23**:1273–8.
34. Grassi G, Seravalle G, Scopelliti F, et al. Structural and functional alterations of subcutaneous small resistance arteries in severe human obesity. *Obesity* 2009;**18**:92e8.
35. Wu G, Bazer FW, Davis TA, et al. Arginine metabolism and nutrition in growth, health and disease. *Amino Acids* 2009;**37**(1): 153–68.
36. Wang D, Strandgaard S, Iversen J, Wilcox CS. Asymmetric dimethylarginine, oxidative stress, and vascular nitric oxide synthase in essential hypertension. *Am J Physiol Regul Integr Comp Physiol* 2009;**296**(2):R195–200.
37. Martina V, Masha A, Gigliardi VR, et al. Long-term *N*-acetylcysteine and L-arginine administration reduces endothelial activation and systolic blood pressure in hypertensive patients with type 2 diabetes. *Diabetes Care* 2008;**31**(5):940–4.
38. Siani A, Pagano E, Iacone R, et al. Blood pressure and metabolic changes during dietary L-arginine supplementation in humans. *Am J Hypertens* 2000;**13**(5 Pt 1):547–51.
39. Kelly BS, Alexander JW, Dreyer D, et al. Oral arginine improves blood pressure in renal transplant and hemodialysis patients. *J Parenter Enteral Nutr* 2001;**25**(4):194–202.
40. Gokce N, et al. L-Arginine amd hypertension. *J Nutr* 2004;**134** (10 Suppl):2807S–11S.
41. Persky AM, Brazeau GA. Clinical pharmacology of the dietary supplement creatine monohydrate. *Pharmacol Rev* 2001;**53**(2): 161–76.
42. Tyagi N, Sedoris KC, Steed M, et al. Mechanisms of homocysteine-induced oxidative stress. *Am J Physiol Heart Circ Physiol* 2005;**289**(6):H2649–56.
43. James AM, Cochemé HM, Smith RA, Murphy MP. Interactions of mitochondria-targeted and untargeted ubiquinones with the mitochondrial respiratory chain and reactive oxygen species. Implications for the use of exogenous ubiquinones as therapies and experimental tools. *J Biol Chem* 2005;**280**(22):21295–312.
44. Molyneux SL, Florkowski CM, Lever M, George PM. Biological variation of coenzyme Q10. *Clin Chem* 2005;**51**:455–7.
45. Digiesi V, Cantini F, Oradei A, et al. Coenzyme Q10 in essential hypertension. *Mol Aspects Med* 1994;**15**(Suppl.):S257–63.
46. Rosenfeldt FL, Haas SJ, Krum H, et al. Coenzyme Q10 in the treatment of hypertension: a meta-analysis of the clinical trials. *J Hum Hypertens* 2007;**21**(4):297–306.
47. Breithaupt-Grogler K, Niebch G, Schneider E, et al. Dose-proportionality of oral thioctic acid – coincidence of assessments via pooled plasma and individual data. *Eur J Pharm Sci* 1999;**8**:57–65.
48. Louhelainen M, Merasto S, Finckenberg P, et al. Lipoic acid supplementation prevents cyclosporine-induced hypertension and nephrotoxicity in spontaneously hypertensive rats. *J Hypertens* 2006;**24**:947–56.
49. Takaoka M, Kobayashi Y, Yuba M, et al. Effects of alpha-lipoic acid on deoxycorticosterone acetate-salt-induced hypertension in rats. *Eur J Pharmacol* 2001;**424**:121–9.

50. Petersen Shay K, Moreau RF, Smith EJ, Hagen TM. Is alpha-lipoic acid a scavenger of reactive oxygen species in vivo? Evidence for its initiation of stress signaling pathways that promote endogenous antioxidant capacity. *IUBMB Life* 2008;**60**: 362–7.
51. McMackin CJ, Widlansky ME, Hamburg NM, et al. Effect of combined treatment with alpha-lipoic acid and acetyl-Lcarnitine on vascular function and blood pressure in patients with coronary artery disease. *J Clin Hypertens (Greenwich)* 2007;**9**: 249–55.
52. Cakatay U, Kayali R, Sivas A, Tekeli F. Prooxidant activities of alpha-lipoic acid on oxidative protein damage in the aging rat heart muscle. *Arch Gerontol Geriatr* 2005;**40**:231–40.
53. Kayali R, Cakatay U, Akcay T, Altug T. Effect of alpha-lipoic acid supplementation on markers of protein oxidation in post-mitotic tissues of ageing rat. *Cell Biochem Funct* 2006;**24**:79–85.
54. Kizhakekuttu Tinoy J, Widlansky Michael E. Natural antioxidants and hypertension: promise and challenges. *Cardiovasc Ther* 2010;**28**(4):e20–32.
55. Stamler J, Liu K, Ruth KJ, et al. Eight-year blood pressure change in middle-aged men: relationship to multiple nutrients. *Hypertension* 2002;**39**(5):1000–1006.
56. Siems W, Sommerburg O, Schild L, et al. Beta-carotene cleavage products induce oxidative stress in vitro by impairing mitochondrial respiration. *FASEB J* 2002;**16**(10):1289–91.
57. Haskell MJ. The challenge to reach nutritional adequacy for vitamin A: β-carotene bioavailability and conversion-evidence in humans. *Am J Clin Nutr* 2012;**96**(5):1193S–203S.
58. Hennekens CH, Buring JE, Manson JE, et al. Lack of effect of long-term supplementation with beta carotene on the incidence of malignant neoplasms and cardiovascular disease. *N Engl J Med* 1996;**334**:1145.
59. Engelhard YN, Gazer B, Paran E. Natural antioxidants from tomato extract reduce blood pressure in patients with grade-1 hypertension: a double-blind, placebo-controlled pilot study. *Am Heart J* 2006;**151**(1):100.
60. Ried K, Frank OR, Stocks NP. Dark chocolate or tomato extract for prehypertension: a randomized controlled trial. *BMC Complement Altern Med* 2009;**9**:22.
61. Yusuf S, Dagenais G, Pogue J, et al. Vitamin E supplementation and cardiovascular events in high-risk patients. The Heart Outcomes Prevention Evaluation Study Investigators. *N Engl J Med* 2000;**342**:154–60.
62. Lonn E, Bosch J, Yusuf S, et al. Effects of long-term vitamin E supplementation on cardiovascular events and cancer: a randomized controlled trial. *JAMA* 2005;**293**(11):1338–47.
63. Wright ME, Lawson KA, Weinstein SJ, et al. Higher baseline serum concentrations of vitamin E are associated with lower total and cause-specific mortality in the Alpha-Tocopherol, Beta-Carotene Cancer Prevention Study. *Am J Clin Nutr* 2006;**84**(5):1200–207.
64. Juraschek Stephen P, Guallar Eliseo, Appel Lawrence J, Miller Edgar R III. Effects of vitamin C supplementation on blood pressure: a meta-analysis of randomized controlled trials. *Am J Clin Nutr* 2012;**95**:1079–88.
65. McCarron DA, Morris CD, Henry HJ, Stanton JL. Blood pressure and nutrient intake in the United States. *Science* 1984;**224**:1392–8.
66. Duffy SJ, Gokce N, Holbrook M, et al. Treatment of hypertension with ascorbic acid. *Lancet* 1999;**354**:2048–9.
67. Fotherby MD, Williams JC, Forster LA, et al. Effect of vitamin C on ambulatory blood pressure and plasma lipids in older persons. *J Hypertens* 2000;**18**:411–5.
68. Mullan BA, Young IS, Fee H, McCance DR. Ascorbic acid reduces blood pressure and arterial stiffness in type 2 diabetes. *Hypertension* 2002;**40**(6):804–9.
69. Darko D, Dornhorst A, Kelly FJ, et al. Lack of effect of oral vitamin C on blood pressure, oxidative stress and endothelial function in type II diabetes. *Clin Sci* 2002;**103**:339–44.
70. Lee Duk-Hee, Folsom Aaron R, Harnack Lisa, et al. Does supplemental vitamin C increase cardiovascular disease risk in women with diabetes? *Am J Clin Nutr* 2004;**80**:1194–200.
71. Juraschek Stephen P, Guallar Eliseo, Appel Lawrence J, Miller Edgar III R. Effects of vitamin C supplementation on blood pressure: a meta-analysis of randomized controlled trials 1–3. *Am J Clin Nutr* 2012;**95**:1079–88.
72. Bruno Rosa M, Elena Daghini, Lorenzo Ghiadoni, et al. Effect of acute administration of vitamin C on muscle sympathetic activity, cardiac sympathovagal balance, and baroreflex sensitivity in hypertensive patients. *Am J Clin Nutr* 2012;**96**:302–8.

CHAPTER

23

Aging and Arthritis: Oxidative Stress and Antioxidant Effects of Herbs and Spices

M.S. Mekha, R.I. Shobha, C.U. Rajeshwari, B. Andallu

Sri Sathya Sai Institute of Higher Learning, Anantapur, A.P., India

List of Abbreviations

ARDs autoimmune rheumatic diseases
ATP adenosine triphosphate
CVD coronary vascular disease
DNA deoxyribonucleic acid
ESR erythrocyte sedimentation rate
GSH reduced glutathione
GSSG oxidized glutathione
H_2O_2 hydrogen peroxide
NADPH reduced nicotinamide adenine dinucleotide phosphate
NIDDM non-insulin-dependent diabetes mellitus
O_2 oxygen
$^1O_2$ singlet oxygen
O_2^- superoxide anion
OA osteoarthritis
$OH^\bullet$ hydroxyl radical
OS oxidative stress
PD Parkinson's disease
RA rheumatoid arthritis
ROS reactive oxygen species.

INTRODUCTION

Free radicals are hazardous substances produced in the body along with toxins and wastes which are formed during the usual metabolic processes of the body. Oxidation of carbohydrates, fats, and proteins produces energy through both aerobic and anaerobic processes – which also lead to the production of free radicals.[1] The progression of aerobic metabolic processes – such as respiration and photosynthesis – inevitably leads to the production of reactive oxygen species (ROS) in mitochondria, chloroplasts, and peroxisomes. A feature common to the different types of ROS is their ability to cause oxidative damage to proteins, deoxyribonucleic acid (DNA), and lipids. These cytotoxic properties of ROS explain the evolution of complex arrays of non-enzymatic and enzymatic detoxification mechanisms in plants. Increasing evidence indicates that ROS also function as signaling molecules in plants and are implicated in regulating development and also defense responses against pathogens.[2]

TYPES OF FREE RADICAL AND THEIR GENERATION

Free radicals that can be formed within the body include: the superoxide anion (O_2^-), the hydroxyl radical ($OH^\bullet$), singlet oxygen ($^1O_2$), and hydrogen peroxide (H_2O_2). Superoxide anions are formed when oxygen (O_2) acquires an additional electron, leaving the molecule with only one unpaired electron. Within the mitochondria, O_2^- is continually formed, and the rate of formation depends on the amount of oxygen flowing through the mitochondria at any given time. Hydroxyl radicals are short-lived, but are the most damaging radicals within the body (Fig. 23.1);[3] they can be formed from O_2^- and H_2O_2 via the Haber–Weiss reaction. The interaction of copper or iron and H_2O_2 also produces $OH^\cdot$ as first observed by Fenton. Hydrogen peroxide, produced *in vivo* by many reactions, is unique as it can be converted to the highly damaging hydroxyl radical or can be catalyzed and excreted harmlessly as water. Glutathione peroxidase is indispensable for the conversion of reduced glutathione (GSH) to oxidized glutathione (GSSG), during which H_2O_2 is reduced to water.[4] If H_2O_2 is not converted into water, singlet oxygen is formed; it is not a free radical but it can be formed during radical reactions and also causes further reactions. When oxygen is energetically excited, one of the electrons can jump to an empty orbital, creating unpaired electrons. Singlet oxygen can then transfer the energy to a new molecule and can act as a catalyst for free radical formation. The molecule can also interact with other molecules, leading to the formation of a new free radical.[5]

Aging
http://dx.doi.org/10.1016/B978-0-12-405933-7.00023-8

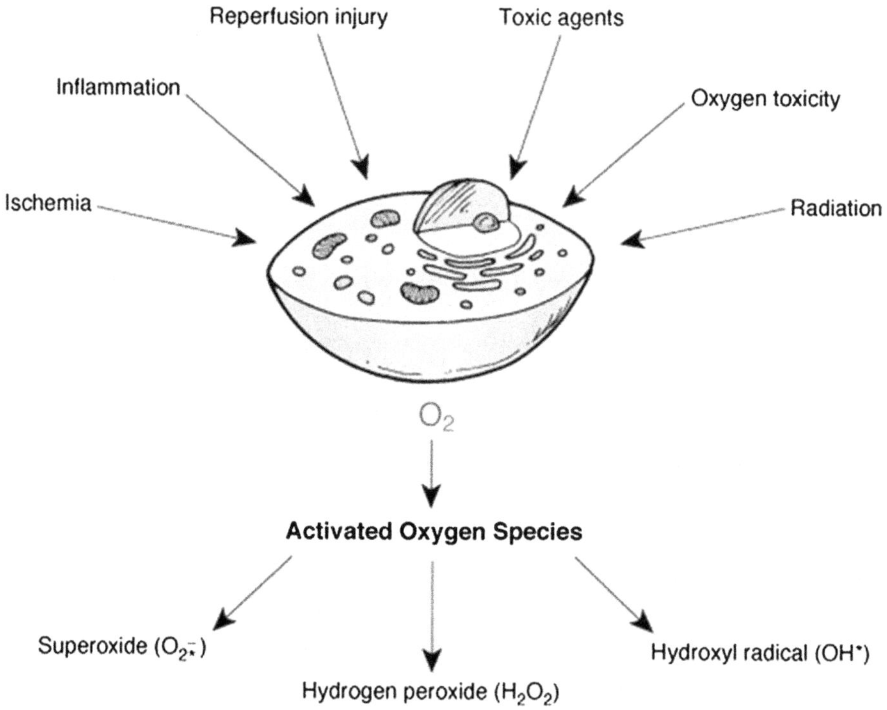

FIGURE 23.1 Formation of free radicals (see text). *From Halliwell & Gutteridge (1985).*[3]

THE PHENOMENON OF OXIDATIVE STRESS

Chemically, oxidative stress is coupled with the increased production of oxidizing species or a significant decrease in the potential of antioxidant defense, such as glutathione. The effects of oxidative stress depend upon the size of these changes, with a cell being able to overcome small perturbations and regain its original state. However, more severe oxidative stress can cause cell death, and even moderate oxidation can trigger apoptosis, while more intense stress may cause necrosis.[6] A particularly destructive aspect of oxidative stress is the production of reactive species, which include free radicals and peroxides. Some of the less reactive species (e.g. superoxide) can be renewed by oxido-reduction reactions with transition metals or other redox cycling compounds (quinones) into more destructive radical species that can cause intense cellular damage. The major portion of long-term effects is inflicted by damage to DNA; under the severe levels of oxidative stress that cause necrosis, the damage causes ATP depletion, preventing restricted apoptotic death and causing the cell to cleanly fall apart.[7]

OXIDATIVE STRESS IN DISEASES

In all disease states, the chance of ROS formation is much elevated, compared with the typical physiologic state. A relationship between oxidative stress and the common neurodegenerative disorders and diabetes is observed. Diabetic complications are found to be induced by the formation of advanced glycation end products (AGEs) which interact with specialized receptors and promote ROS production. It is known that alloxan, a compound largely used for the induction of experimental diabetes, is a redox-cycling compound which can damage insulin-producing pancreatic β-cells. Alzheimer's and Parkinson's diseases are allied with the mutilation of mitochondrial function, resulting in enhanced ROS generation. The key proteins composing protein aggregates in Parkinson's and Alzheimer's diseases, α-synuclein and β-amyloid, respectively, were found to be capable of producing ROS themselves.[8]

AGING AND DISEASES

The free radical theory of aging, conceived in 1956, is rapidly attracting the interest of biological research. It includes phenomenologic measurements of age linked to oxidative stress, interspecies comparisons, dietary restriction, the manipulation of metabolic activity and oxygen tension, treatment with dietary and pharmacologic antioxidants, *in vitro* senescence, classic and population genetics, molecular genetics, transgenic organisms, the study of human diseases of aging, epidemiologic studies, and ongoing elucidation of the role of active oxygen in biology.[9] The chief diseases associated with aging are arthrits, Parkinson's disease, Alzhemier's disease, cognitive dysfunction etc.

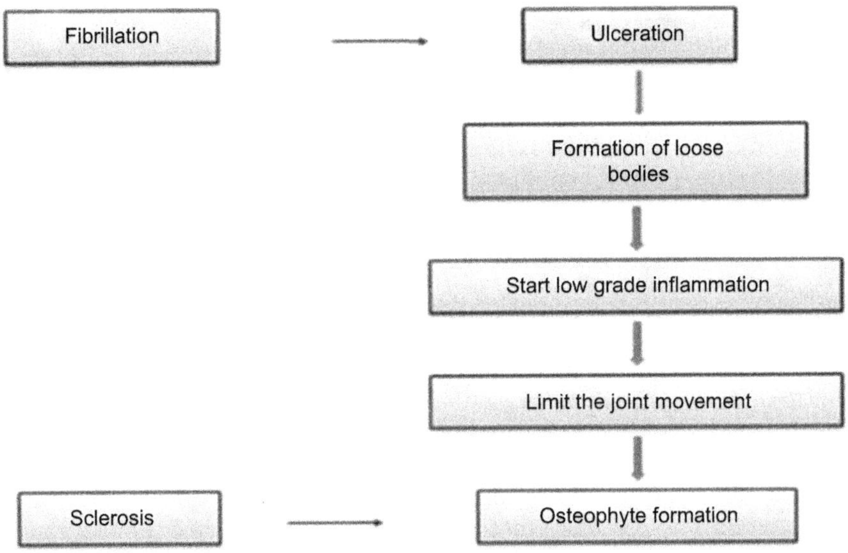

FIGURE 23.2 Pathophysiology of osteoarthritis. The pathology of OA starts from fibrillation and ends with sclerosis. *From Yudoh & Karasawa (2004).*[14]

Arthritis

Arthritis, a chronic, inflammatory multisystem disease, is one of the oldest known diseases. It occurs in all the age groups and consists of such a varied array of chronic conditions, affecting the joints and surrounding tissues, that it has been difficult for population-based surveys to precisely measure its prevalence, severity, and associated effects on health. The joints most commonly affected in arthritis are weight-bearing joints, such as those in feet, knees, hips and spine, as well as others, such as those in the fingers and thumbs.[10]

Oxidative Stress and Osteoarthritis (OA)

Osteoarthritis (OA) is a progressive rheumatic disease characterized by the deterioration of articular cartilage. OA is the most common chronic medical condition in people aged 65 years and older, affecting approximately 85% of adults aged 75–79 years. The current concept holds that osteoarthritis involves the entire joint, including the subchondral bone, menisci, ligaments, periarticular muscle, capsule, and synovium. Its most prominent feature is the progressive destruction of articular cartilage, which results in impaired joint motion, severe pain, and the structural and functional failure of synovial joints.[11] Oxidative stress can produce major unified derangements of cell metabolism, including DNA strand breakage (often an early event), rises in intracellular 'free' Ca^{2+}, damage to membrane ion transporters and/or other specific proteins, and peroxidation of lipids. Propagative lipid peroxidation is a degenerative process that affects cell membranes and other lipid-containing structures under conditions of oxidative stress. The reactive oxygen species (ROS) which are increasingly formed in arthritis are implicated in both cartilage aging and the pathogenesis of osteoarthritis.[12]

Onset and progression of disease depends on various factors such as age, obesity, joint injury, metabolic diseases, bone and joint malformations, genetic factors, and nutritional factors. Weight reduction, through dietary or other means, may reduce the risk or progression of OA. Obese osteoarthritis patients deficient in nutrients such as vitamin D, folacin, vitamin B_6, zinc and pantothenic acid were benefitted by supplementation with these nutrients.[13]

The articular cartilage matrix undergoes substantial structural, molecular, and mechanical changes with aging, including surface fibrillation, alteration in proteoglycan structure and composition, increased collagen cross-linking, and decreased tensile strength and stiffness. Recently, attention has been given to the suggestion that cartilage aging and chrondocyte senescence play an important role in the pathogenesis and development of osteoarthritis. The pathology of OA starts from fibrillation and ends at sclerosis, as shown in Fig. 23.2.[14]

Oxidative Stress in Cartilage Senescence and the Development of Osteoarthritis

Several reports revealed that chondrocyte senescence contributes to the risk for cartilage degeneration by decreasing the ability of chondrocytes to sustain and repair the articular cartilage tissue. The mitotic and synthetic activity of chondrocytes declines with advancing age. In addition, human chondrocytes become less receptive to anabolic mechanical stimuli with aging and exhibit an age-related decline in response to growth factors such as the anabolic cytokine insulin-like growth factor-I. These findings provide evidence supporting the concept that chondrocyte senescence may be involved in the progression of cartilage degeneration.[15]

Mechanical and chemical stresses are thought to stimulate free radical production, consequently leading to oxidative damage to the tissue. Oxidative damage can not only initiate apoptosis through caspase activation but may also lead to irrevocable growth arrest, similar to replicative senescence. Furthermore, it has been reported that oxygen free radicals (O_2^- and peroxynitrite) directly injure the guanine repeats in the telomere DNA, indicating that oxidative stress directly leads to telomere erosion, regardless of active division by the cell. Generally, it is now thought that oxidative stress/antioxidative capacity may be prominent among factors that control telomere length. These findings strongly suggest that oxidative stress could induce chondrocyte telomere instability, with no requirement for cell division, in articular cartilage, leading to chondrocyte senescence.[16]

Oxidative Stress and Rheumatoid Arthritis (RA)

Rheumatoid arthritis is a progressive, elapsing, and chronic inflammatory disease. It causes annihilation of cartilage and periarticular bone due to the release of neutral proteases and matrix metalloproteinases as part of the inflammatory process within the synovium.[17] Although RA originates in the synovial tissues of the joints, it is a systemic disease that, in its more severe form, may include rapidly progressive multisystem inflammation. RA is characterized by progressive, erosive, and chronic polyarthritis. Cellular proliferation of the synoviocytes and neoangiogenesis leads to the formation of a pannus which destroys the articular cartilage and the bone.[18]

The blood serum and joint fluid often contain an antibody (rheumatoid factor) that binds to immunoglobulin G (IgG). The inception of rheumatoid arthritis is usually slow. The synovium of the joints becomes swollen and damaged, and joint cartilage is eroded. Production of synovial fluid, the natural joint lubricant, is increased but its viscosity is much below normal and its lubrication capacity is greatly diminished. The decrease in viscosity is due to breakdown of the polymer hyaluronic acid. Exposure of synovial fluid to oxygen (O_2) produced by chemical systems or by activated phagocytes *in vitro* produces a similar breakdown, which can be accredited to the O_2-dependent formation of hydroxyl radical by an iron-catalyzed Haber–Weiss reaction in the synovial fluid.[19]

Nonetheless, ROS have been reported to play an important role in RA pathogenesis. ROS, which can be produced as a result of normal aerobic metabolism, and whose production is increased by active neutrophils during inflammation, have recently attracted increasing attention.[9]

Oxidative stress plays an extensive role not only in the pathogenesis of autoimmune rheumatic diseases (ARDs) and their complications but also in specific disease activity. Thus, discrepancy in the pro- and anti-inflammatory molecules due to the dysregulation of redox homeostasis oxidants/antioxidants could play a role in the pathophysiology of ARDs.[20] Oxygen metabolism has a significant role in the pathogenesis of rheumatoid arthritis. Reactive oxygen species (ROS) have been implicated as mediators of tissue damage in patients with rheumatoid arthritis. As ROS (created in the course of cellular oxidative phosphorylation and by activated phagocytic cells during oxidative bursts) exceed the physiologic buffering capacity, the result is oxidative stress. ROS are formed in oxidative processes that usually occur at moderately low levels in all cells and tissues. Under normal conditions, a diversity of antioxidant mechanisms serves to control this ROS production. In contrast, high doses and/or insufficient removal of ROS result in oxidative stress, which may cause severe metabolic malfunctions and damage to biologic macromolecules and matrix components.[21]

The synovial fluid of the inflamed rheumatoid joint swarms with activated neutrophils, which produce large amounts of ROS. Further, decomposition of peroxidized lipids yields a wide variety of end products, including malondialdehyde. Elevated levels of malondialdehyde have been observed in the serum (or plasma) and synovial fluid of rheumatoid arthritis patients. In recent years, increasing attention has been given to the role of reactive oxygen metabolites in the pathogenesis of inflammatory diseases such as rheumatoid arthritis. Increased activity of free radicals, the unstable molecules associated with cell damage, is theorized to underlie the mucosal injury commonly seen in various inflammatory diseases (Fig 23.3).[22]

Gouty Arthritis

Gout is a common type of arthritis caused by an increased concentration of uric acid in biologic fluids, the final product of purine metabolism. In gout, uric acid crystals are deposited in joints, tendons, kidneys, and other tissues, where they cause considerable inflammation and damage. Gout is classified into two major categories: primary and secondary. Primary gout accounts for about 90% of all cases; secondary gout accounts for only 10%. Cytokines, chemokines, proteases, and oxidants involved in acute inflammation contribute to chronic inflammation leading to chronic synovitis, cartilage loss, and erosion. Even during remissions of acute flares, low-grade synovitis exists and joints may persist with ongoing intra-articular phagocytosis of crystals by leukocytes.[23]

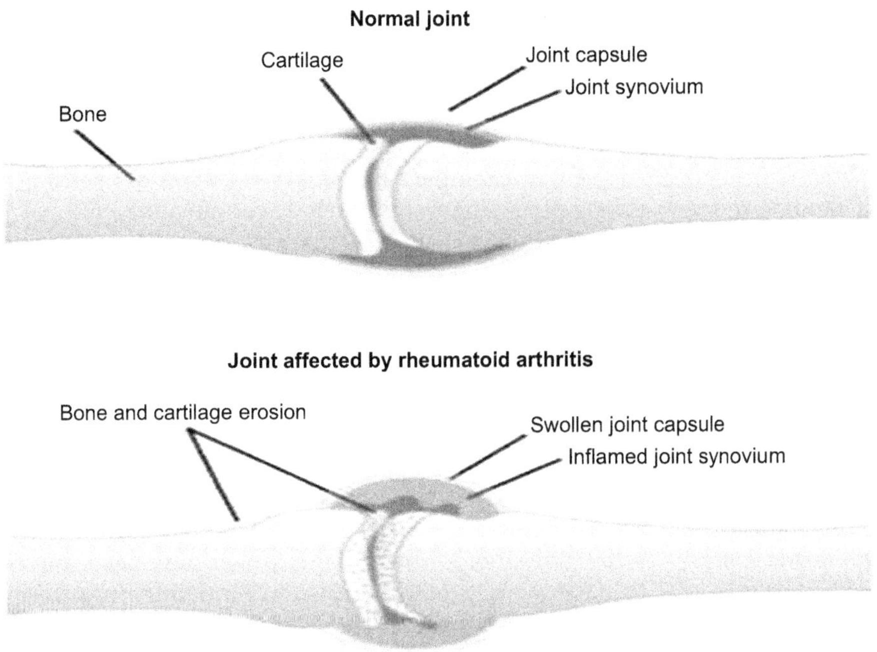

FIGURE 23.3 Normal joint and joint affected by rheumatoid arthritis. Increased activity of free radicals, the unstable molecules associated with cell damage, is theorized to underlie the mucosal injury commonly seen in various inflammatory diseases. *From Kuloglu et al (2002).*[22]

ANTIOXIDANT SYSTEMS

Several antioxidant systems have been reported which have different activities. Circulating human erythrocytes have the ability to scavenge $O_2^{\cdot -}$ and H_2O_2 generated extracellularly by activated neutrophils, superoxide dismutase, catalase, and glutathione peroxidase-dependent mechanisms. Superoxide dismutase, the first line of defense against ROS, catalyzes the dismutation of the superoxide anion into hydrogen peroxide. Catalase can then transform hydrogen peroxide into H_2O and O_2. Glutathione peroxidase is a selenoprotein, which reduces lipidic or non-lipidic hydroperoxides as well as H_2O_2 while oxidizing glutathione. An understanding of the complex interactions involved in these pathways might allow the development of novel therapeutic strategies for rheumatoid arthritis.[14]

Antioxidants and Arthritis

High doses of vitamin E administered to RA patients were effective in reducing pain symptomatology. Furthermore, a theoretical field of intervention takes into account foods rich in antioxidants; in fact, in the *in vitro* studies, anti-inflammatory effects are shown to be linked to the down-regulation of NF-κB.[20]

Many animal studies strongly imply an anti-inflammatory role for antioxidants such as superoxide dismutase and vitamin E in experimentally induced arthritis. Antioxidant therapy strategies have been proposed for the prevention and treatment of RA. Vitamin E seems to uncouple joint inflammation and joint destruction in the transgenic KRN/NOD mouse model of RA, with a beneficial effect on joint destruction.[24] The therapeutic value of adding a high dose of vitamin E, or an antioxidant combination, to the treatment regimen for rheumatoid disease suggested that the symptoms of arthritis were better controlled from the first month, and that by the end of the second month better control of disease was achieved.[25] Another study by Cerhan et al[26] suggested that the intake of certain antioxidant micronutrients, particularly β-cryptoxanthine and supplemental zinc – and possibly a diet rich in fruits and cruciferous vegetables – has a protective role against the development of RA.

TRADITIONAL MEDICINE AND HERBS

Traditional medicine is the combination of knowledge, skills, and practices based on the theories, beliefs, and experiences indigenous to different cultures used in the maintenance of health, prevention of diseases, and improvement of physical and mental illness. In practice, traditional medicine refers to: acupuncture (China), Ayurveda (India), Unani (Arabic countries), traditional birth attendant's medicine, mental healer's medicine, herbal medicine, and various other forms of indigenous medicine. Complementary or alternative medicine refers to a broad set of healthcare practices that are not part of a country's own tradition and are not integrated into the dominant healthcare system. Traditional medicine has maintained its popularity in all regions of the developing world, and its use is rapidly spreading in industrialized countries.[27]

The therapeutic use of herbs is as old as human civilization and has evolved along with it. The majority of people on this planet still rely on their indigenous system of medicine and use herbal drugs. The Indian and Chinese systems of medicine are well established, with written records going back around 3000 years. Medicinal plant drug discovery continues to provide new and significant leads against various pharmacologic targets, including cancer, malaria, cardiovascular diseases, and neurologic disorders. Interest in herbal drugs and natural medicine is undergoing resurgence at the present time. The medicinal properties of plants are due to the presence of active principles, which are the bioactive secondary metabolites.[28]

Knowledge of plants and of healing has been closely linked from the time of human beings' earliest social and cultural groupings. Knowledge of the medicinal plants used in the drugs of traditional systems of medicine (TSM) has been of great significance, especially as a lead for the discovery of new single-molecule medicines for the modern system of medicine. To determine the chemical nature of such compounds, isolation of a substance in pure form using various separation techniques, chemical properties, and spectral characteristics is a prerequisite for establishing its correct structure. Thus, medicinal plants are used in crude or purified form in the preparation of drugs in different systems. In countries such as India, China, and others with well-founded traditional systems of medicine, plant-based formulations occupy an important place in health management.[29]

Plant cells produce two types of metabolite, i.e., primary and secondary. Primary metabolites are involved directly in growth and metabolism, viz. carbohydrates, lipids, and proteins, which are produced as a result of photosynthesis and are additionally involved in cell component synthesis. Most natural products are compounds derived from primary metabolites and are generally categorized as secondary metabolites. Secondary metabolites are considered products of primary metabolism and are generally not involved in metabolic activities, viz. alkaloids, phenolics, essential oils and terpenes, sterols, flavonoids, lignins, tannins, etc. These secondary metabolites are the major source of pharmaceuticals, food additives, fragrances, and pesticides.[30] Secondary metabolites are synthesized by two principal pathways: shikimic acid or aromatic amino acid, and mevalonic acid.[28]

Some plants are rich in secondary plant products, and it is because of these compounds that these plants are termed 'medicinal' or 'officinal' plants. These secondary metabolites exert a profound physiologic effect on mammalian systems; thus, they are known as the active principle of plants. With the discovery of the physiologic effect of a particular plant, efforts are being made to know the exact chemical nature of these drugs (called the active principle) and, subsequently, to obtain these compounds by chemical synthesis. Besides secondary plant products, several primary metabolites exert strong physiologic effects. In this category, proteins are the principal compounds – having diverse functions such as blood agglutinants from Fabaceae, hormones (e.g. insulin), various snake-venom poisons, ricin from *Ricinus communis*, and abrine and precatorine from *Abrus precatorius*. Other examples of primary metabolites exerting a strong physiologic effect include certain antibiotics, vaccines, and several polysaccharides acting as hormones or elicitors.[31]

The priority of developed countries is different from that of developing countries in relation to medicinal plants. Developed countries are looking for leads to develop drugs from medicinal plants, while developing countries would like to have cheap herbal formulations as these countries cannot afford the long path of drug discovery using pure compounds. The indigenous system of medicine is officially renowned in India.[31]

In laboratory settings, plant extracts have been shown to have, among other effects, anti-inflammatory, vasodilatory, antimicrobial, anticonvulsant, sedative, and antipyretic effects. In a typical study, an infusion of lemongrass leaves produced a dose-dependent reduction of experimentally induced hyperalgesia in rats. Studies in human subjects also confirm specific therapeutic effects of particular herbs. Randomized controlled trials support the use of ginger for treating nausea and vomiting, feverfew (*Chrysanthemum parthenium*) for migraine prophylaxis, and ginkgo (*Ginkgo biloba*) for intermittent claudication and dementia. The best-known evidence for a herbal product concerns St John's wort (*Hypericum perforatum*) for treating mild-to-moderate depression. A systematic review of 23 randomized controlled trials found this herb to be significantly superior to placebo and therapeutically equivalent to, but with fewer side effects than, antidepressants such as amitriptyline. There is still, however, little evidence on the efficacy of herbalism as practiced – that is, using principles such as combining herbs and unconventional diagnosis.[32] There are a number of herbs that work synergistically to reduce chronic joint inflammation in cases such as osteoarthritis, rheumatoid arthritis, and other types of arthritis. Basic scientific research has uncovered the mechanisms by which some plants afford their therapeutic effects. Many studies have revealed that herbs show potential for the treatment of arthritis.[33]

In animal studies, ashwagandha was found more valuable than the prescription drug phenylbutazone in controlling inflammation. The ashwagandha-treated group decreased inflammatory proteins, whereas animals treated with phenylbutazone, as well as the control groups, had increased inflammatory proteins. Similar results were achieved in carrageenan-induced inflammation. Ashwagandha extract showed far superior (almost double) results when compared to the drug hydrocortisone to decrease inflammation. One of the studies showed a noteworthy decrease in the swelling of an arthritic paw due to ashwagandha, which may have been due to cox-2 inhibition. The reduction in swelling and degeneration was better than in hydrocortisone-treated animals. In a different study, *Withania* root extract (1000 mg/kg, orally, daily for 15 days) caused significant reduction in both paw swelling and bony degenerative changes in Freund's adjuvant-induced arthritis in rats as observed by radiologic examination. The reductions were better than those produced by the reference drug, hydrocortisone 40. In one of the studies it was found that *Withania* inhibited the granuloma formation in cotton-pellet implantation in rats, and the effect was comparable to that produced by treatment with hydrocortisone sodium succinate (5 mg/kg).[34]

In another study, ashwagandha root powder given to 46 patients with rheumatoid arthritis, in doses of 4, 6, or 9 grams for 3–4 weeks, relieved the pain and swelling completely in 14 patients, and showed considerable improvement in 10 patients and mild improvement in 11 patients. In one double-blind, placebo-controlled study, the combination of ashwagandha with turmeric and zinc showed positive effects in osteoarthritis cases. Patients showed significant improvement in pain severity and disability score. The free radical mechanism is one of the mechanisms considered to contribute to many inflammatory diseases. Oral administration of ashwagandha has demonstrated powerful antioxidant action by preventing lipid peroxidation.[35]

The bark of Boswellia (*Boswellia serrata*, Burseraccea) is sweet, cooling, and tonic, containing boswellic acid. It is good in vitiated conditions of pitta, cough, and asthma. It is useful in fevers, urethrorrhea, diaphoresis, convulsions, chronic laryngitis, and jaundice, and is analgesic, antihyperlipidemic, and antiatherosclerotic. Adults and children with rheumatoid arthritis both experienced effective relief from the symptoms when treated with *Boswellia*, despite having responded poorly in the past to standard therapies such as non-steroidal anti-inflammatory drugs (NSAIDs).[36] The anti-inflammatory and anti-arthritic activities of *Boswellia* have been mainly attributed to a component in the resin containing [beta]-boswellic acid. In an animal-model study, boswellic acids significantly reduced the infiltration of leukocytes in the knee joint and, in turn, significantly reduced inflammation.[37]

A randomized double-blind placebo-controlled crossover study was conducted to assess the efficacy, safety, and tolerability of *Boswellia serrata* extract in 30 patients with osteoarthritis of the knee, 15 each receiving active drug or placebo for 8 weeks. All receiving *Boswellia* reported a decrease in knee pain, increased knee flexion, and a decrease in the frequency of swelling in the knee joint.[38] A study on the mechanisms underlying the anti-inflammatory actions of boswellic acid derivatives in experimental colitis demonstrated that ϱ-selectin-mediated recruitment of inflammatory cells is a major site of action for this novel anti-inflammatory agent.[39]

Aujaie, one of the herbal products prepared by Hamdard Laboratories (Waqf) Pakistan, is believed to have the potential for providing relief from joint pain (arthritis with or without swelling of joints), gout, lumbago, sciatica, and stiffening of joints, and it also helps in the excretion of uric acid. This product is prepared from nine different plants; most of these plants have medicinal value in traditional medicine for treating rheumatism and gout – such as *Balsamodendron mukul*, which is very potent for various types of joint problem, such as rheumatoid arthritis, osteoarthritis, and gout and is used for reducing pain, swelling, and tenderness in inflammatory joints.[40] Khan et al[41] reported that *Colchicum luteum*, *Curculigo orchioides*, and *Zingiber officinalis* have antirheumatic, anti-gout, and anti-inflammatory actions.

Ptychotis ajowan relieves rheumatic and neuralgic pain, *Pistacia lentiscus* resolves inflammation, and *Withania somnifera* is useful in rheumatoid arthritis and rheumatic fever; it reduces the discomfort associated with arthritis and is also used to prevent tumors and inflammation.[42] The properties of these plants cannot determine the pharmacology of the product, Aujaie, as such. In order to understand the pharmacologic basis in the treatment of inflammatory diseases, a study was designed to investigate the anti-inflammatory as well as the antinociceptive activity of the product, Aujaie, in mice. Aujaie has been confirmed to possess significant anti-inflammatory and antinociceptive activity in a dose-dependent manner. An inhibitory effect was seen at 1 hour after carrageenan injection, which is attributed to the release of histamine and serotonin. A marked inhibition of edema formation was also observed at the third and fifth hour, suggesting an inhibition of the release of kinins or cyclooxygenase, one of the enzymes involved in the formation of the prostaglandins that induce inflammation.[43]

The antinociceptive activity of Aujaie was also evaluated using the writhing test in mice. Acetic acid, which is used in the writhing syndrome, causes algesia by liberation of endogenous substances, which then excite the pain nerve endings. Aujaie exerted a significant inhibitory activity on the writhing response in a dose range of 60–240 mg/kg body weight. It is therefore concluded that Aujaie plays a vital role in the treatment of inflammation. Aujaie, up to the dose of 300 mg/kg/p.o. in mice and about 6000 mg/kg/p.o. in rats, did not produce any toxic symptoms.[44]

Rumalaya tablet, a combination of several herbs, contains *Mahayograj guggul*, *Hibiscus abelmoschus*, *Rubia cordifolia*, *Moringa pterygosperma*, and *Tinospora cordifolia* as its main ingredients. All of these ingredients claim to have analgesic and anti-inflammatory actions, besides antispasmodic activity and antiseptic actions. Some constituents stimulate uric acid excretion, and prevent gouty arthritis, osteoporosis and degenerative changes in the joints. Thus, Rumalaya tablet was subjected to clinical evaluation in patients with symptoms of chronic arthritis and related pain. The study aimed at analyzing the effect of Rumalaya tablet, an Ayurvedic preparation, with one of the commonly used analgesics – Diclofenac sodium. The analgesic effect of Rumalaya tablet was comparable to that of Diclofenac tablet. Rumalaya tablet contains many herbal ingredients; together, these play a vital role in reducing inflammation and pain of arthritic disorders. The advantage of Rumalaya tablet is its good gastric tolerance. Earlier studies confirmed the efficacy of Rumalaya in long-term use, showing it to be safe for prolonged use in arthritis patients. Rumalaya tablet is thus suitable for patients with arthritis and non-articular rheumatic conditions associated with pain and restricted movements, especially for elderly patients and patients with a history of hypersensitivity to Diclofenac.[45]

Conventional (allopathic) anti-inflammatory drugs are the main stay of treatment for a variety of immune disorders, including rheumatoid arthritis (RA).[46] The non-steroidal anti-inflammatory drugs (NSAIDs) and biologics (e.g. antitumor necrosis factor (TNF)-α antibody and the decoy TNF-α receptor) represent a prominent group of such drugs. However, the usage of these drugs is associated with severe adverse effects, including gastrointestinal bleeding and cardiovascular complications. Owing to the side effects and the high cost of conventionally used anti-inflammatory drugs, patients with arthritis are increasingly using complementary and alternative medicine (CAM) modalities of treatment. Traditional Chinese medicine, Ayurvedic medicine, Kampo, and homeopathy are among the major contributors to the natural products consumed by patient populations. However, despite the increasing usage and popularity of CAM products in the western world, one of the main limitations of their use is the meagre information about their mechanisms of action and objectivity in evaluating efficacy. This also is one of the main reasons for skepticism about CAM in the minds of both the lay public and professionals. Thus, there is a need for continued studies on the mechanistic aspects of action of CAM products.[47]

The use of complementary and alternative medical (CAM) therapies is widespread among patients, including those with rheumatic diseases. Herbal medications are often utilized with little to no physician guidance or knowledge. An appreciation of this information will help physicians to counsel patients concerning the utility of CAM therapies. An understanding and elucidation of the mechanisms by which CAM therapies may be efficacious can be instrumental in discovering new molecular targets in the treatment of diseases. Various studies suggest that the extracts of various herbs or compounds derived from them may provide a safe and effective adjunctive therapeutic approach for the treatment of arthritis.[33]

A diverse group of diseases is characterized by inflammation that can be triggered not only by foreign microbial antigens but also by self-antigens. The response to self-antigens results in autoimmune inflammation. Therefore, like the infectious diseases, the autoimmune diseases (such as multiple sclerosis (MS), type-1 diabetes mellitus (T1D), RA, and atherosclerosis) are also associated with inflammation. Considering that autoimmune diseases result from a dysregulated immune system, it is imperative to examine and unravel the immunologic basis of the therapeutic activity of CAM products against autoimmune disorders as well as other conditions involving inflammation.[48]

One study focused on the immunomodulation of autoimmune arthritis by herbal CAM products. Adjuvant arthritis (AA) was described as a prototypic experimental model of RA. Conceptually, the main immune effector pathways in AA are broadly representative of various other animal models of arthritis; for example, collagen-induced arthritis (CIA), streptococcal cell wall-induced arthritis (SCWIA), and proteoglycan-induced arthritis. Specific immune pathways in arthritis are modulated by a variety of herbal preparations originating from plants native to different regions of the world. These immune mechanisms include the cellular and humoral responses, the cytokine response/balance, and the cellular migration into the target organ. The abovementioned immunologic events in the pathogenesis of arthritis also offer many promising targets for therapeutic intervention (Figs 23.4 and 23.5).[49]

Spices

Spice is a dried seed/fruit/root/bark of a plant or a herb used in small quantities for flavor, color, or as a preservative. Many of these substances are also used in traditional medicines. The term 'spice' can be broadly defined as a compound that has a pungent flavor or coloring activity or one that increases appetite or enhances digestion.[50]

Since ancient times, there has been an awareness of the fact that there is much more to herbs and spices than merely culinary function as seasonings, used to improve the sensory properties of food. In cooking, the term 'herb' refers to the leaves of a plant, while spices refer to any other part of plants.[51]

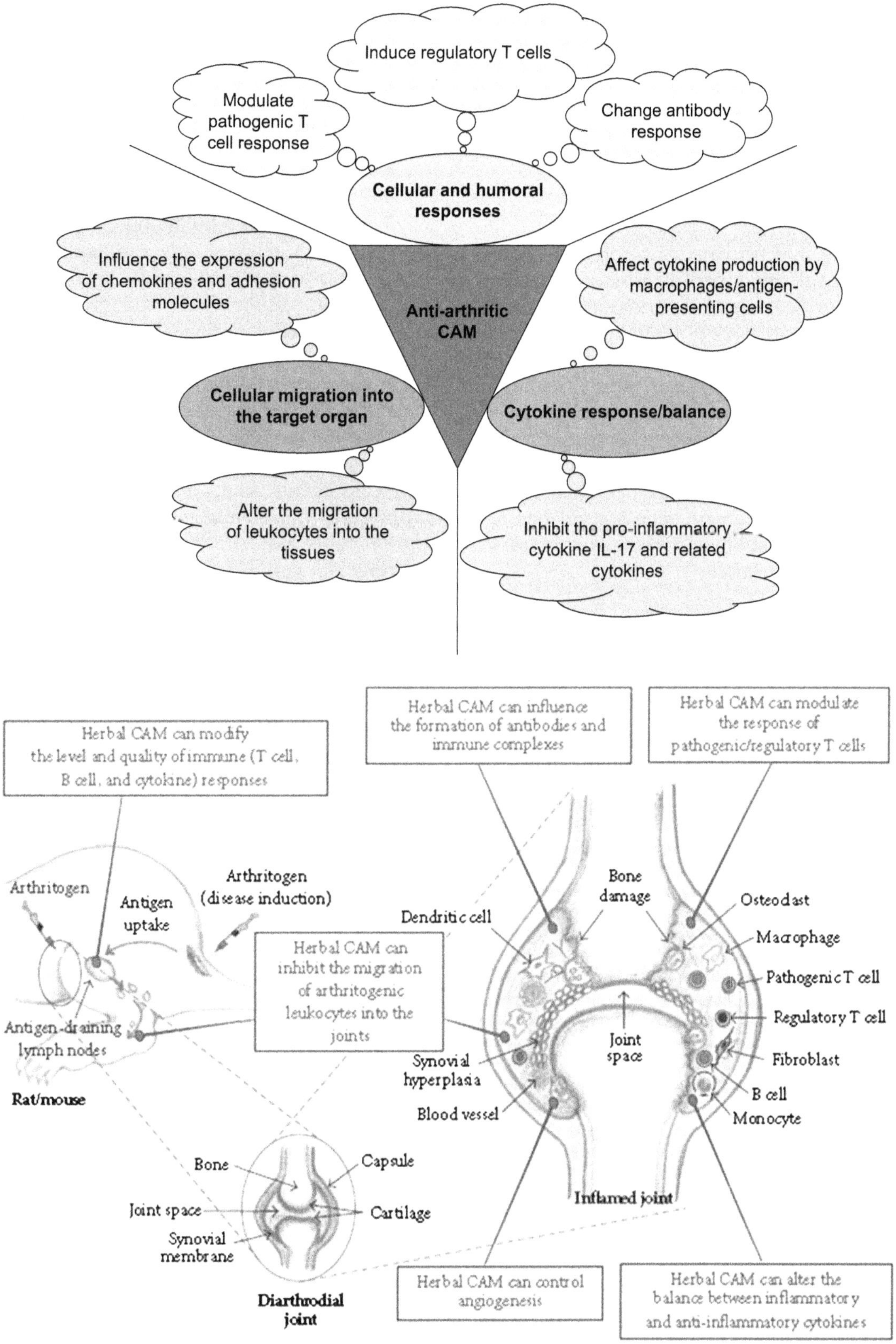

FIGURE 23.4 & 23.5 Figure 23.4: Immunologic mechanisms that mediate the anti-arthritic activities of different complementary and alternative medicine (CAM) modalities (see text); Figure 23.5: Herbal CAM can intervene at multiple steps in the pathogenesis of autoimmune arthritis (see text). *Both figures are from Shivaprasad et al (2010).*[49]

Apart from the traditional use, a host of beneficial physiologic effects have been brought to the fore by extensive animal studies during the past three decades. Among these are their beneficial influence on lipid metabolism, their efficacy as antidiabetics, their ability to stimulate digestion, their antioxidant property, and their anti-inflammatory potential.[52]

Anti-Inflammatory Property

Lipid peroxides and activated macrophages play a crucial role in arthritis and other inflammatory diseases. Both *in vitro* and *in vivo* animal experiments have documented the anti-inflammatory potential of spice principles suxh as curcumin, capsaicin, and eugenol. Turmeric happens to be the earliest anti-inflammatory drug known in the indigenous system of medicine in India. Turmeric extracts, curcuminoids, and volatile oil of turmeric have been found to be effective in experiments with mice, rats, rabbits, and pigeons.[53] Recently, capsaicin has received considerable attention as a pain reliever and has also been suggested for the initial management of neuralgia consequent to herpes infection.[54]

Antioxidant Potential

Oxidative damage at the cellular or subcellular level is now considered to be an important event in disease processes such as coronary vascular disease (CVD), inflammatory disease, carcinogenesis, and aging. Reactive oxygen radicals are detrimental to cells at both membrane and genetic levels. They cause lipid peroxidation in cellular membranes, generating lipid peroxides that cause extensive damage to membranes in terms of cross linking of membrane components, leaks and lysis, and membrane-mediated chromosomal damage. Spice principles, capsaicin, curcumin, and eugenol, inhibit lipid peroxidation by quenching oxygen free radicals and by enhancing the activity of endogenous antioxidant enzymes such as superoxide dismutase, catalase, glutathione peroxidase, and glutathione-*S*-transferase.[55]

Curry leaf extract displayed scavenging activity against radicals, viz. 2,2-diphenyl-1-picrylhydrazyl (DPPH), nitric oxide, and superoxide, in a concentration-dependent manner and inhibited $FeSO_4$-induced lipid peroxides and hydroperoxides in an erythrocyte membrane model. This is supported by significantly decreased lipid peroxidation in plasma and erythrocytes, significantly elevated levels of non-enzymatic antioxidants (β-carotene, vitamin A, C, & E) in serum, and significantly decreased activity of catalase, elevated activities of superoxide dismutase and glutathione-*S*-transferase, and raised levels of reduced glutathione (GSH) in erythrocytes in curry leaves-treated aged subjects. Thus, the effect is a result of synergistic action of antioxidant phytochemicals, i.e. carotenoids, flavonoids, oleoresins etc. present in the leaves. Curry leaves (*Murraya koenigi*) can be prescribed as an adjunct to diet for controlling oxidative stress that causes chronic diseases.[56]

Another study by Rajeshwari et al[57] proved that supplementation of coriander leaves could significantly influence almost all parameters in arthritic patients without any detrimental effects. This was evidenced by elevated levels of serum non-enzymatic antioxidants, viz. β-carotene and vitamin C, and erythrocyte non-enzymatic antioxidant-glutathione, improved activity of erythrocyte antioxidant enzyme, glutathione-*S*-transferase; lipid peroxidation was also very effectively decreased in erythrocytes and plasma in arthritis patients. Supplementation also decreased the activity of serum alkaline phosphatase, elevated serum calcium levels, and decreased the erythrocyte sedimentation rate (ESR) – indicating the anti-arthritic efficacy of coriander leaves. These effects are a result of the co-ordinated action of bioactive compounds, viz. apigenin, caffeic acid, chlorogenic acid, protocatechuric acid, ascorbic acid, β-carotene, quercetin, rhamnetin, rutin, terpinen, *trans*-anethole, umbelliferone, borneol etc. reported[58] to be present in coriander leaves. Normalcy in liver and kidney function tests indicated safety of the treatment with coriander leaves.

Spices as Nutraceuticals

Hypolipidemic/Hypocholesterolemic Effect

The importance of serum cholesterol levels and of lipoproteins in relation to atherosclerosis and coronary heart disease is well known. Consumption of a high-fat diet may lead to an increase in serum cholesterol and plasma fibrinogen levels which, in turn, may result in decreased fibrinolytic activity and blood coagulation time. These changes would also increase the risk of atherosclerosis and heart diseases.[52]

The spices fenugreek, red pepper, turmeric, garlic, onion, and ginger were found to be effective as hypocholesterolemic agents under various conditions of experimentally induced hypercholesterolemia/hyperlipemia. Curcumin and capsaicin, the active principles of turmeric and red pepper, respectively, are also efficacious at doses comparable to calculated human daily intake. The anti-platelet aggregation, the anti-platelet adhesion, and the anti-proliferation properties of aged garlic extracts appear to contribute more to cardiovascular protection than do the hypolipidemic properties.[55]

Antilithogenic Effect

Studies on the experimental induction of cholesterol gallstones in mice and hamsters by feeding a lithogenic diet revealed a 40–50% smaller incidence of gallstones in the animals maintained on a diet containing 0.5% curcumin or 5 mg% capsaicin. Animal studies also revealed significant regression of preformed cholesterol gall stones by these spice principles in a 10-week feeding trial. The anti-lithogenicity of curcumin and capsaicin is considered to be due to lowering of the cholesterol concentration and enhancement of the bile acid concentration, both of which contribute to lowering of the cholesterol saturation index and, hence, reduction in crystallization.[55]

Diet has been recognized as a corner stone in the management of diabetes mellitus. Fenugreek, turmeric, or the active principle curcumin, onion or its active principle allylpropyl disulfide, garlic, and cumin were observed to improve glycemic status in diabetic animals and in non-insulin-dependent diabetes mellitus (NIDDM) patients. Studies have unequivocally demonstrated the anti-diabetic potential of fenugreek in both type I and type II diabetes. The spice probably delays gastric emptying by direct interference with glucose absorption, and the gel-forming dietary fiber reduces the release of insulinotropic hormones and gastric inhibitory polypeptides.[53] The hypoglycemic potency of garlic and onion has been attributed to the sulfur compounds, namely, (2-propenyl) disulfide and 2-propenylpropyl disulfide respectively.[59]

Turmeric is another spice with a beneficial hypoglycemic effect which has improved glucose tolerance in a limited number of studies.[60] Nephropathy is a common complication in chronic diabetes. High blood cholesterol is an added risk factor that determines the rate of decline of kidney function in diabetics. Dietary curcumin and onion have been found to have a promising ameliorating influence on the severity of renal lesions in streptozotocin diabetic rats.[61]

Digestive Stimulant Action

Spices are well recognized to stimulate gastric function. They are generally believed to intensify salivary flow and gastric juice secretion and, hence, aid in digestion.[60] Spices such as ginger, mint, ajowan, cumin, fennel, coriander, and garlic are used as ingredients of commercial digestive stimulants as well as of home remedies for digestive disorders and intestinal disorders. Animal studies have revealed that a good number of spices, when consumed through diet, bring about an enhanced secretion of bile with a higher bile acid content, which plays a vital role in fat digestion and absorption. Spices such as curcumin, capsaicin, piperine, ginger, and mint have been shown to stimulate pancreatic digestive enzymes, such as lipase, amylase, trypsin, and chymotrypsin, which play a crucial role in food digestion. A few spices have been shown to have a beneficial effect on the terminal digestive enzymes of small intestinal mucosa. Thus, many of the common spices act as digestive stimulants by enhancing biliary secretion of bile acids, which are vital for fat digestion and absorption, and by stimulating the activities of pancreatic and intestinal enzymes involved in digestion.[55]

Antimutagenic and Anticarcinogenic Properties

Food mutagens are formed under certain cooking and processing conditions. These harmful products can be modified by the presence of antimutagens in food spices that have antioxidant properties – and can function as antimutagens. Because mutation is one of the mechanisms by which cancer is caused, an antimutagenic substance is likely to prevent carcinogenesis. Among phytochemicals, curcumin had been shown to be antimutagenic in several experimental systems. The bioactive compounds of spices exert their anticarcinogenic effect by one or more of the following mechanisms: by virtue of their antioxidant property, by deactivating the carcinogens, or by enhancing the tissue levels of protective enzymes in the body.[54]

Role in Platelet Aggregation

Platelets in blood have an active role in hemostasis during damage to blood vessels and thrombosis by forming compact and adhesive aggregates. Compounds that counter platelet aggregation have a protective role against thrombotic disorders. Water extracts of coriander and cumin were also evidenced to have a significant inhibitory effect on platelet aggregation.[55]

Antimicrobial Activity

The evaluation for antimicrobial agent of plant origin begins with thorough biological evaluation of plant extracts to ensure efficacy and safety followed by identification of active principles, dosage formulations, efficacy and pharmacokinetic profile of the new drug. Many plants have been used because of their antimicrobial traits and antimicrobial properties of plants have been investigated by a number of researchers worldwide. Ethno pharmacologists, botanists, microbiologists and natural product chemists are searching the world for phytochemicals which could be developed for treatment of infectious diseases. The antibacterial property of turmeric is well known. Its active principle, curcumin, is a known bacteriostatic agent, and the essential oil of turmeric is also bacteriostatic and fungistatic.

This paragraph is not for anti-microbial activity. It is just a concluding remark for this review. The components of spices responsible for the quality attributes have been designated as active principles, and in many instances they are also responsible for the health beneficial physiologic effects. Exceptionally, fenugreek seeds are rich in dietary fiber, which makes up 52% of the seeds; hence, their liberal consumption in certain Indian diets contributes health-giving soluble fiber. In view of their several health beneficial attributes, spices and their active principles should be valued as nutraceuticals.[63]

SUMMARY POINTS

- Oxidative stress is associated with the increased production of oxidizing species or a significant decrease in the capability of antioxidant defense.
- Aging is generally defined as a process of deterioration in the functional capacity of a person occurring after maturity and resulting from structural changes related to the process in chronologic order.
- Deleterious actions of oxygen-derived radicals are responsible for the functional deterioration associated with aging.
- Arthritis is a chronic, inflammatory multisystem disease involving the wearing down of cartilage, which cushions the ends of the bones.
- Dietary spices maintain human health by their antioxidative, chemopreventive, antimutagenic, anti-inflammatory, immunomodulatory effects. They also have a wide array of putative beneficial effects on human health via their action on gastrointestinal, cardiovascular, respiratory, metabolic, reproductive, neural, and other systems.
- Spices can be used as natural products to delay the aging process and to prevent age-related diseases as well as to treat autoimmune diseases such as rheumatoid arthritis.

References

1. Sen S, Chakraborthy RC, Sridhari YS, et al. Free radicals, antioxidants, diseases and phytomedicines: current status and future prospect. *Int J Pharm Sci Rev Res* 2010;**3**(1):91–100.
2. Apel K, Hirt H. Reactive oxygen species: metabolism, oxidative stress and signal transduction. *Annu Rev Plant Biol* 2004;**55**:373–99.
3. Halliwell B, Gutteridge JMC. The chemistry of oxygen radicals and other oxygen-derived species. In: Halliwell B, Gutteridge JMC, editors. *Free radicals in biology and medicine*. New York: Oxford University Press; 1985. pp. 20–64.
4. Alessio HM, Blasi ER. Physical activity as a natural antioxidant booster and its effects on a healthy lifestyle. *Res Q Exerc Sport* 1997;**68**(4):292–302.
5. Karlsson J. Exercise, muscle metabolism and the antioxidant defense. *World Rev Nutr Diet* 1997;**82**:81–100.
6. Lennon SV, Martin SJ, Cotter TG. Dose-dependent induction of apoptosis in human tumour cell lines by widely diverging stimuli. *Cell Prolif* 1991;**24**(2):203–14.
7. Lee YJ, Shacter E. Oxidative stress inhibits apoptosis in human lymphoma cells. *J Biol Chem* 1999;**274**(28):19792–8.
8. Wang C, Liu L, Zhang L, et al. Redox reactions of the α-synuclein–Cu^{2+} complex and their effects on neuronal cell viability. *Biochemistry* 2010;**49**(37):8134–42.
9. Jasin HE. Mechanisms of tissue damage in rheumatoid arthritis. In: Koopman WJ, Moulund LW, editors. *Arthritis and allied conditions: a textbook of rheumatology*. 15th ed. Philadelphia: Lippincot Williams and Wilkins; 2005. pp. 1141–64.
10. Strange CJ. *Coping with arthritis in its many forms*; 1996. FDA consumer, USA: Food and Drug Administration Publication No. (FDA) 97–1237.
11. Yelin E. The economics of osteoarthritis. In: Brandt KD, Doherty M, Lohmander LS, editors. *Osteoarthritis*. Oxford: Oxford University Press; 2003. pp. 17–21.
12. Tiku ML, Shah R, Allison GT. Evidence linking chondrocyte lipid peroxidation to cartilage matrix protein degradation. Possible role in cartilage aging and the pathogenesis of osteoarthritis. *J Biol Chem* 2000;**275**(26):20069–76.
13. White O'Connar B, Sobal J. Nutrient intake and obesity in a multidisciplinary assessment of osteoarthritis. *Clin Ther* 1986;**9**(Suppl B): 30–42.
14. Yudoh K, Karasawa R. Statin prevents chondrocyte aging and degeneration of articular cartilage in osteoarthritis (OA). *Aging* 2004;**2**(12):990–8.
15. Martin JA, Ellerbroek SM, Buckwalter JA. Age-related decline in chondrocyte response to insulin-like growth factor-I: the role of growth factor binding proteins. *J Orthopaedic Res* 1997;**15**:491–8.
16. von Zglinicki T. Oxidative stress shortens telomeres. *Trends Biochem Sci* 2002;**27**:339–44.
17. Fox DA. Etiology and pathogenesis of rheumatoid arthritis. In: Koopman WJ, Howland LW, editors. *Arthritis and allied conditions: a textbook of rheumatology*. 15th ed. Philadelphia: Lippincot Williams and Wilkins; 2005. pp. 1089–115.
18. Edwards CRW, Bouchier IAD, editors. *Davidsons' principles and practice of medicine*. 16th ed. Edinburgh: Churchill Livingstone; 1994.
19. Greenwald RA, Moy WW. Effect of oxygen-derived free radicals on hyaluronic acid. *Arhritis Rheum* 1980;**23**:455–63.
20. Sukkar SG, Rossi E. Oxidative stress and nutritional prevention in auto immune rheumatic diseases. *Autoimmun Rev* 2004;**3**:199–206.
21. Lledias F, Rangel P, Hansberg W. Oxidation of catalase by singlet oxygen. *J Biol Chem* 1998;**273**:10630–7.
22. Kuloglu M, Ustundag B, Atmaca M, et al. Lipid peroxidation and antioxidant enzyme levels in patients with schizophrenia and bipolar disorder. *Cell Biochem Funct* 2002;**20**:171–5.
23. Pascual E, Battle-Gualda E, Martinez A, et al. Synovial fluid analysis for diagnosis of intercritical gout. *Ann Intern Med* 1999;**131**:756–9.
24. Bandt MD, Grossin M, Driss F. Vitamin E uncouples joint destruction and clinical inflammation in a transgenic mouse model of rheumatoid arthritis. *Arthritis Rheum* 2002;**46**:522–32.
25. Helmy M, Shohayeb M, Helmy MH. Antioxidants as adjuvant therapy in rheumatoid disease. A preliminary study. *Arzneimittelforschung* 2001;**51**:293–8.
26. Cerhan JR, Sagg KG, Merlino LA. Antioxidant micronutrients and risk of rheumatoid arthritis in a cohort of older women. *Am J Epidemiol* 2003;**157**:345–54.
27. Ramawat KG, Goyal S. The Indian herbal drugs scenario in global perspectives. In: Ramawat KG, Merillon JM, editors. *Bioactive molecules and medicinal plants*. Berlin: Springer; 2008. p. 323.
28. Ramawat KG, Dass S, Meeta M. The chemical diversity of bioactive molecules and therapeutic potential of medicinal plants. In: Ramawat KG, editor. *Ethnomedicine to modern medicine*. Berlin: Springer; 2009. pp. 79115–7.
29. Jia W, Zhang L. Challenges and opportunities in the Chinese herbal drug industry. In: Demain AL, Zhang L, editors. *Natural products: drug discovery and therapeutic medicine*. Totowa: Humana; 2005. p. 229.
30. Patwardhan B, Vaidya ADB, Chorghade M. Ayurveda and natural products drug discovery. *Curr Sci* 2004;**86**:789–99.
31. Audi J, Belson M, Patel M, et al. Ricin poisoning: a comprehensive review. *J Am. Med. Assoc* 2005;**294**:2342–51.
32. Andrew V. Herbal medicine. *BMJ* 1999;**319**:1050–3.
33. Patwardhan KK, Kaumudee SB, Sameer SG. Coping with arthritis using safer herbal options. *Int J Pharm Sci* 2010;**2**(1):1–8.
34. Gupta GL, Rana AC. *Withania somnifera* (Ashwagandha): a review. *PHCOG MAG Plant Rev* 2007;**1**(1):129–36.
35. Begum VH, Sadique J. Long term effect of herbal drug *Withania somnifera* on adjuvant induced arthritis in rats. *Indian J Exp Biol* 1988;**26**(11):877–82.

36. Sharma PC, Yelne MB, Dennis TJ. *Database on medicinal plants used in Ayurved*. vol. 1. p. 120. New Delhi: Central Council for Research in Ayurveda and Siddha; 2001.
37. James J, Gormley HP, Ammon. *Boswellia serrata*: an ancient herb for arthritis, cholesterol, and more better nutrition. *Planta Med* 1991;**57**(3):203–7.
38. Kimmatkar N, Thawani V, Hingorani L, Khiyani R. Efficacy and tolerability of *Boswellia serrata* extract in treatment of osteoarthritis of knee – a randomized double blind placebo controlled trial. *Phytomed* 2003;**10**(1):3–7.
39. Anthoni C, Laukoetter MG, Rijcken E. Mechanisms underlying the anti-inflammatory actions of boswellic acid derivatives in experimental colitis. *Indian J Exp Biol* 2003;**41**(12):1460–2.
40. Vatsyayan R. *Ayurvedacharya 'Health Bulletin' Guggul: Nature's mighty weapon against arthritis*. Chandigarh, India: *The Tribune* online edition; 2001.
41. Khan U., Saeed A, Alam MT *Indusyunic medicine, traditional medicine of herbal, animal and mineral origin in Pakistan*, 1st ed. Department of Pharmacognosy, Faculty of Pharmacy, University of Karachi; 1997. pp. 169–171, 196–197, 451–452.
42. Said M, Saeed A, D'Silva LA, et al. *Medicinal herbal*. 1st ed. Pakistan: Hamdard Foundation; 1996. pp. 1, 188–190, 264–266.
43. Vinegar R, Schreiber W, Hugo R. Biphasic development of carrageenan edema in rats. *J Pharmacol Expt Ther* 1969;**166**:96–103.
44. Di Rosa M, Giroud JP, Willoughby DA. Studies of the acute inflammatory response induced in rats in different sites by carrageenan and turpentine. *J Pathol* 1971;**104**:15–29.
45. Rastogi S, Bansal R. Efficacy of rumalaya tablets in arthritis – a double blind placebo controlled trial. *The Antiseptic* 2001;**5**(98):172–3.
46. Lipsky PE. Rheumatoid arthritis. In: Kasper DL, et al., editor. *Harrison's principles of internal medicine*. 16th ed. New York: McGraw–Hill; 2005. pp. 1968–77.
47. Ben-Arye E, Frenkel M, Klein A, Scharf M. Attitudes toward integration of complementary and alternative medicine in primary care: perspectives of patients, physicians and complementary practitioners. *Patient Education Counseling* 2008;**70**(3):395–402.
48. Watanabe T, Yamamoto T, Yoshida M. The traditional herbal medicine saireito exerts its inhibitory effect on murine oxazolone-induced colitis via the induction of Th1-polarized immune responses in the mucosal immune system of the colon. *Int Arch Allergy Immunol* 2009;**151**(2):98–106.
49. Venkatesha SH, Rajaiah R, Berman BM, Kamal DM. Immunomodulation of autoimmune arthritis by herbal CAM. *Evidence-Based Complement Alternat Med* 2010:1–13.
50. Kunnumakkara AB, Koca C, Dey S, et al. *Traditional uses of spices: an overview, molecular targets and therapeutic uses of spices*. Singapore: World Scientific Publishing; 2009. pp. 1–24.
51. Tapsell L, Hemphill I, Cobiac L, et al. Health benefits of herbs and spices: the past, the present, the future. *Med J Australia* 2006;**185** (4 Suppl):S4–24.
52. Srinivasan K, Sambaiah K, Chandrasekhara N. Spices as beneficial hypolipidemic food adjuncts. *Food Rev Intr* 2004;**20**:187–220.
53. Hussain MS, Chandrasekhara N. Effect of curcumin and capsaicin on the regression of pre-established cholesterol gallstones in mice. *Nutr Res* 1994;**14**:1561–74.
54. Surh YJ. Anti-tumor promoting potential of selected spice ingredients with antioxidative and anti-inflammatory activities: a short review. *Food Chem Toxicol* 2002;**40**:1091–7.
55. Srinivasan K. Role of spices beyond food flavoring: nutraceuticals with multiple health effects. *Food Rev Intr* 2005;**21**:167–88.
56. Andallu B, Mahalakshmi S, Rajeshwari CU, Vinaykumar AV. Efficacy of curry (*Murraya koenigii*) leaves on scavenging free radicals *in vitro* and controlling oxidative stress *in vivo*. *Biomed Prev Nutr* 2011;**1**:263–7.
57. Rajeshwari CU, Siri S, Andallu B. Antioxidant and antiarthritic potential of coriander (*Coriandrum sativum* L.) leaves. *e-SPEN J* 2012;**7**:e223–8.
58. http://www.ars-grin.gov/duke// (accessed on January 2013)
59. Augusti KT, Sheela CG. Antiperoxide effect of S-allyl cysteine sulfoxide, an insulin secretagogue in diabetic rats. *Experientia* 1996;**52**:115–9.
60. Tank R, Sharma R, Sharma T, Dixit VP. Antidiabetic activity of *Curcuma longa* in alloxan-induced diabetic rats. *Indian Drugs* 1990;**27**:587–9.
61. Babu PS, Srinivasan K. Renal lesions in streptozotocin-induced diabetic rats maintained on onion or capsaicin diet. *J Nutr Biochem* 1999;**10**:477–83.
62. Tanaka Y, Chen C, Maher JM, Klaassen CD. Kupffer cell-mediated downregulation of hepatic transporter expression in rat hepatic ischemia-reperfusion. *Transplantation* 2006;**82**:258–66.
63. Govindarajan VS. Turmeric: chemistry, technology and quality. *CRC Crit Rev Food Sci Nutr* 1980;**18**:199–301.

CHAPTER

24

Lycopene and Other Antioxidants in the Prevention and Treatment of Osteoporosis in Postmenopausal Women

L.G. Rao

Department of Medicine, St Michael's Hospital and University of Toronto, Toronto, Canada

N.N. Kang, A.V. Rao

Department of Nutritional Sciences, University of Toronto, Toronto, Canada

List of Abbreviations

CTX crosslinked C-telopeptides of type I collagen
NTx crosslinked N-telopeptides of type I collagen
ROS reactive oxygen species
TBARS thiobarbituric acid reactive substances
TEAC Trolox equivalent antioxidant capacity

INTRODUCTION

Oxidative stress caused by reactive oxygen species (ROS) results in the development of chronic diseases, one of which is osteoporosis.[1] There is ample evidence to show that oxidative stress is responsible for bone loss, as revealed through epidemiologic studies,[2] and it is therefore a risk factor for osteoporosis. This chapter includes an overview of osteoporosis, the role of oxidative stress in the bone cells – osteoclasts and osteoblasts – oxidative stress as a risk factor in the development of osteoporosis, and a review of studies on the use of antioxidants in counteracting oxidative stress in the prevention of osteoporosis.

OSTEOPOROSIS

Osteoporosis is a metabolic bone disease accompanied by the gradual loss of bone occuring over many years. Because there are usually no noticeable symptoms until the bones are so fragile that a fracture occurs, it is known as 'the silent thief'.[3] Bone, as a dynamic tissue, continually renews itself throughout life by the process of bone remodeling; this involves the two major bone cells: the bone-forming cells (osteoblasts) and the bone-resorbing cells (osteoclasts). The coupled processes of bone formation and bone resorption in mature, healthy bone are tightly regulated. Disturbances in the remodeling process can lead to metabolic bone diseases. One such disturbance, caused partly by oxidative stress, and shown to control the functions of both osteoclasts and osteoblasts, may contribute to the pathogenesis of the skeletal system, including osteoporosis, the most prevalent metabolic bone disease.[4]

Some of the risk factors for osteoporosis[5,6] are presented in Table 24.1. The risk factors that are of interest in this review are oxidative stress-generating factors, including smoking, alcohol intake, low antioxidant status, nutritive deficiency, excessive sports activity and excessive caffeine intake; all of these factors have been shown to increase the rate of bone loss that eventually leads to osteoporosis.

OXIDATIVE STRESS

Oxidative stress is caused by reactive oxygen species (ROS); these are the main by-products formed in the cells of aerobic organisms. ROS can initiate autocatalytic reactions in such a way that they cause a chain of damage to target molecules after converting them into free

Aging
http://dx.doi.org/10.1016/B978-0-12-405933-7.00024-X

TABLE 24.1 Risk Factors for Osteoporosis

Cannot Be Changed	Can Be Changed
Race	Chronic inactivity
Sex	Low body weight
Age	Low lifetime calcium intake
Frame size	Medication used
Family history (e.g. osteoporosis, susceptibility to fracture)	Oxidative-stress-related low antioxidant status excessive sports activity excessive alcohol consumption smoking excessive caffeine consumption

radicals.[7] Oxidative stress results from over-production of ROS and/or the weakening of antioxidant defenses in the body. ROS are highly reactive because they contain one or more unpaired electrons, a state that makes them seek out another electron to fill their orbital, which then stabilizes their electron balance.[8] ROS are a family of highly reactive free radicals and oxygen-containing molecules which include singlet oxygen, hydroxyl radical ($OH\cdot^-$), superoxide radical ($O_2\cdot^-$), hydrogen peroxide (H_2O_2) and lipid peroxides.[9] Because ROS have an extremely short half-life, they are difficult to measure in humans; the only way to measure them is through the damage they cause to proteins, lipids and DNA – which is manifested as chronic diseases, including osteoporosis.[10] This damage can occur by inhibition of cellular proliferation, cell-cycle arrest, modulation of cellular differentiation, and finally, by the induction of apoptosis.[11]

The endoplasmic reticulum, electron transport chains in the mitochondria, and nuclear membranes are the major intracellular sites where ROS are generated.[12] Oxidative stress may result from: normal metabolic activity; low antioxidant status; lifestyle factors such as cigarette smoking, high alcohol intake, excessive sports activity, excessive caffeine; nutritive deficiency; acute or chronic immune responses; and environmental factors such as ultraviolet radiation, chemicals, pollution and toxins [reviewed in reference 5]. It is now evident that ROS production increases with age[13] and is associated with several chronic diseases, including osteoporosis.

ANTIOXIDANTS

Cells can fight free radical attack or oxidative stress under normal physiologic conditions by promoting antioxidant defenses. Several mechanisms, present in the body for endogenous defense, include the antioxidant enzymes glutathione peroxidase (GPx), catalase (CAT), and superoxide dismutase (SOD), as well as the metal-chelating proteins.[14] Exogenous antioxidants are those from dietary sources present in fruits and vegetables containing several phytonutrient antioxidants; they include the water-soluble antioxidant polyphenols; the carotenoids potent antioxidant lipid-soluble lycopene; and vitamins such as E and C.[15] The repair enzymes come into play in cases where the endogenous antioxidants or antioxidants from the diet fail to prevent oxidative damage; they include DNA repair enzymes, proteases, lipases and transferases.[12] Diseases associated with oxidative stress develop when oxidative stress prevails over the activity of antioxidants; such diseases include cancer, diabetes, cardiovascular disease, neurologic diseases and osteoporosis.[15]

The phytochemical antioxidants are naturally present in animal- and plant-derived foods; they include the water-soluble antioxidants such as polyphenols and the lipid-soluble carotenoids, a group to which the potent antioxidant lycopene belongs.[15] Figure 24.1 is a cartoon illustrating how oxidative stress from ROS, lifestyle factors and environmental pollution can cause damaging effects on DNA, lipids and proteins – which subsequently result in chronic diseases; the figure also shows the protection afforded by antioxidants.

Polyphenols: Water-Soluble Antioxidants

Polyphenols are water-soluble molecules found naturally in plants. They are defined as compounds having molecular masses ranging from 500 to 3000–4000 Da with 12 to 16 phenolic hydroxy groups on five to seven aromatic rings per 1000 Da of relative molecular mass.[16] It is estimated that there are 10,000 different phytonutrients (phyto, meaning from plants). To date, over 8000 polyphenols have been identified.[17] Polyphenols can be divided into two main groups: flavonoids and non-flavonoids.[18] The health benefits associated with fruits, vegetables, red wine, tea and Mediterranean diets are probably linked to the polyphenol antioxidants that they contain.[19,20]

Although previous data have shown that it is difficult to quantify the exact dietary intake of polyphenols, it is known that the average human consumes approximately 1.5 g a day. This is more than that of all other known dietary antioxidants; it is 10 times greater than the amount of vitamin C and 100 times greater than the amounts of vitamin E and carotenoids.[21] The polyphenol content claimed to be present in plant foods and beverages varies in different scientific publications. There are countless factors that influence the value of the polyphenol content and profiles in plants, including: plant variety or cultivar, growth conditions (climate and soil), crop management (irrigation and fertilization), state of maturity at harvest, postharvest handling, storage and processing.[22] Due to this variability, it is understandable that it has been challenging for researchers to determine

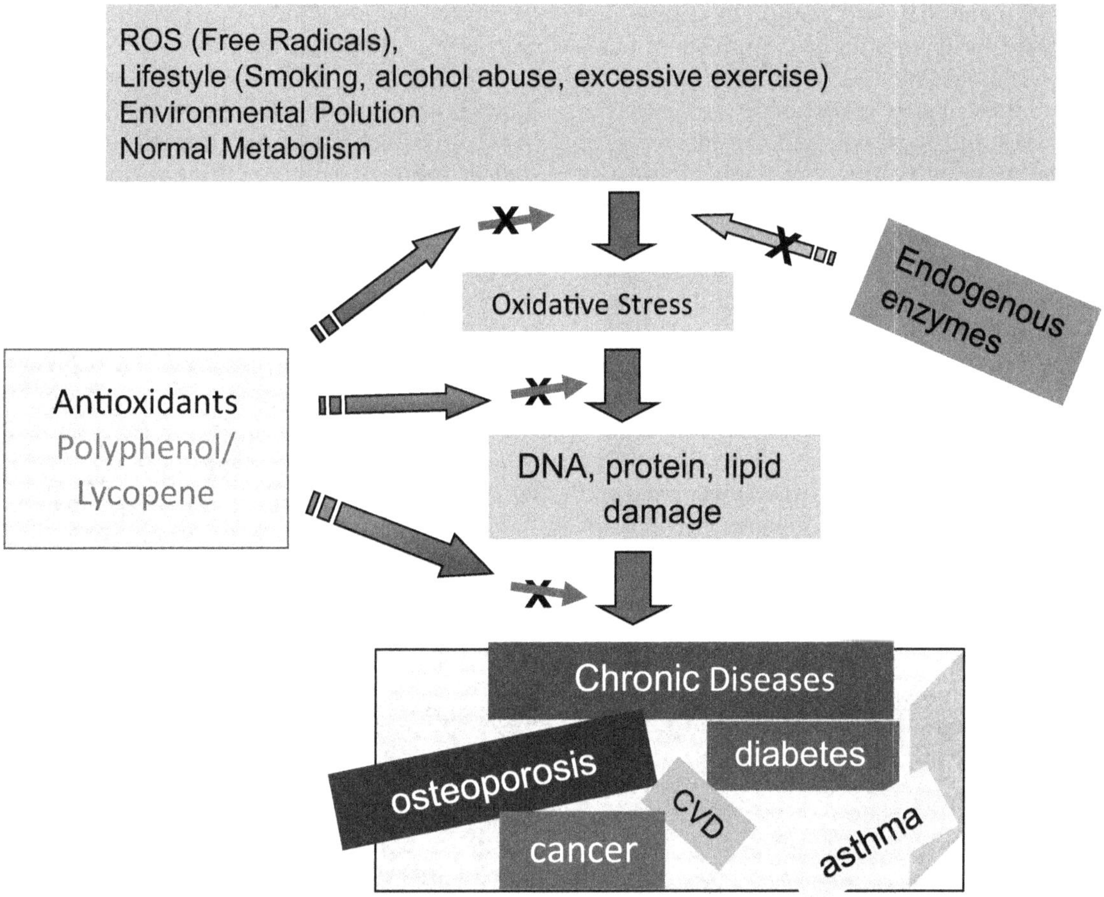

FIGURE 24.1 Cartoon demonstrating the damaging effects of oxidative stress and the beneficial effects of antioxidants to prevent the development of chronic diseases.

the exact value of the content of a given phenolic compound in a given food.

The polyphenols of interest in our study are a mixture of flavonoids, such as quercetin, apigenin, kaempferol and luteolin present in the supplement greens+™;[23] greens+™ in combination with another supplement, bone builder™, was used by Rao et al[23] on osteoblast cells and also in clinical intervention studies on the prevention of risk of osteoporosis in postmenopausal women. Studies on polyphenols and bone are reviewed in later sections of this chapter.

Lycopene: A Carotenoid Lipid-Soluble Antioxidant

Lycopene is an acyclic isomer of β-carotene, with no vitamin A activity, that is not synthesized in the body.[24] A highly unsaturated, straight-chain hydrocarbon, it has a total of 13 double bonds, of which 11 are conjugated, thus making it one of the most potent antioxidants.[15] The chemistry and antioxidant properties of lycopene have been comprehensively reviewed.[25] The singlet oxygen-quenching ability of lycopene is twice that of β-carotene and 10 times that of α-tocopherol.[26] The dietary source of lycopene for 85% of North Americans is tomatoes and processed tomato. Lycopene can also be obtained from watermelon, pink grapefruit and pink guavas.[15] The all-*trans* configuration of lycopene found in raw tomatoes is not readily absorbed. Heat processing converts lycopene from the all-*trans* to the *cis*-isomeric configuration, thus making it more efficiently absorbed from processed tomato products than from raw tomatoes.[27] The presence of small amounts of lipids further enhances its absorption because lycopene is a lipid-soluble compound that is absorbed via a chylomicron-mediated mechanism.[15] Although the health benefits of lycopene may be due to its potent antioxidant property, there is evidence for other mechanisms, such as its effects on cell cycling[28] and gap-junction communication.[29] There is no official recommended daily intake of lycopene, but a daily intake of 7 mg is suggested based on published research.[15] The reported average daily intake levels of lycopene vary considerably from country to country, from 25 mg per day in Canada to as low as 0.7 mg per day in Finland; a generally accepted universal level of daily intake is 2.5 mg.

The role of lycopene in the prevention of human diseases is supported by the results of a number of clinical studies.[5] The initial epidemiologic observations by Giovannicci on the inverse relationship between the intake of tomatoes and lycopene and the incidence of prostate cancer[30] triggered a flurry of epidemiologic as well as clinical intervention studies on the role of lycopene in the prevention of cancers at other sites, as well as hypertension, diabetes, coronary heart disease, male infertility, macular degenerative disease, and neurodegenerative disease.[5] The role of lycopene in bone health has so far been based on the well-known role of oxidative stress in bone health, the effects of lycopene on bone cells in culture (see below) and, more recently, the results of epidemiologic studies.[31,32] To date, the studies of Rao et al at St Michael's Hospital on the role of lycopene and the elucidation of its mechanism in lowering the risk for osteoporosis in postmenopausal women (aged 50 to 60 years) are so far the only clinical intervention studies reported in the literature.

STUDIES ON THE ANTIOXIDANTS POLYPHENOLS AND LYCOPENE

Studies on Polyphenols

It is now evident that polyphenols have a role in the prevention of chronic diseases such as cancers, diabetes, cardiovascular diseases, neurodegenerative diseases and osteoporosis. There has been increased interest in polyphenols and bone health in the last 10 years.[34] Trzeciakiewicz et al[33] reviewed the mechanisms of action of polyphenols in osteoclasts and osteoblast function, while Horcajada & Offord[34] have recently reviewed the anabolic role of phytonutrients, especially polyphenols. Green-tea polyphenols are now widely known to have beneficial effects on bone.[35]

Knowledge about polyphenols and their effects on bone has emerged from *in vitro* and *in vivo* animal studies, but very few clinical studies have been reported. As the protective effects of polyphenol consumption against osteoporosis have been comprehensively reviewed,[36] only a few studies will be reviewed here.

In Vitro *Studies on Polyphenols in Bone Cells*

Numerous lines of evidence have shown that oxidative stress increases the differentiation and function of osteoclasts[37–39] as well as inhibiting differentiation of osteoblasts.[40,41]

The most commonly studied polyphenol, abundant in green tea, is epigallocatechin-3-gallate (EGCG).[42] The following mechanisms have been proposed for the action of polyphenols: induction of apoptosis via caspase activation in osteoclast differentiation;[43] involvement of complex networks of anabolic signaling pathways such as BMPs or estrogen-receptor-mediated pathways;[34] modulation of the expression of transcription factors such as runt-related transcription factor-2 (Runx2) and Osterix, NFkappaB and activator protein-1 (AP-1) in osteoblasts;[33] and action on cellular signaling, such as mitogen-activated protein kinase (MAPK), bone morphogenetic protein (BMP), estrogen receptor and osteoprotegerin/receptor activator of NF-kappaB ligand (OPG/RANKL) in osteoblast functions.[34]

Other polyphenols/sources of polyphenols which were found to have beneficial effects on bone cells include the dried plum black tea polyphenol, phenolic leaf extract of *Heimia myrtifolia* (Lythraceae), oleuropein and the polyphenol component of red wine. A number of animal studies have been reported and these were reviewed by Rao et al.[36]

The effects of combinations of polyphenols have also been frequently studied. One such combination studied by Rao et al is the nutritional supplement greens+™, a blend of several herbal and botanical products containing a substantial amount of polyphenols, including quercetin, apigenin and luteolin,[23] which act as antioxidants and therefore should be able to counteract oxidative stress. Rao et al[44] have shown that the polyphenolic extracts from greens+™ have a stimulatory effect on mineralized bone nodule formation in human osteoblast cells in a time- and dose-dependent manner and are more effective than epicatechin (EC). It was further shown that this stimulatory effect is accompanied by decreases in the reactive oxygen species H_2O_2.[45] These results prove that greens+™ is able to counteract oxidative stress in human osteoblastic cells and may therefore be a good candidate as a nutritional supplement to reduce the risk of osteoporosis.

Two additional nutritional supplements have since been formulated and may prove to be good for bone health. These are the bone builder™ and the greens+bone builder™; the latter is the original greens+™ product that has been supplemented with the bone builder™ formula containing several compounds, including vitamins, minerals and antioxidants. These various components have been separately shown to have some beneficial effect on bone.[46] Using the human osteoblast SaOS-2 cells, it was shown that, similarly to greens+™, the water-soluble bone builder™ extract had a significant dose-dependent stimulatory effect on bone nodule formation.[44] Additionally, it was shown that when the two supplements, greens+™ and bone builder™, were tested in combination, the effects were six times more effective than either one alone in stimulating bone formation in osteoblast culture.[47]

Clinical Intervention Studies on Polyphenol

Clinical studies on polyphenols and osteoporosis have recently been reviewed.[36] Only the more recent

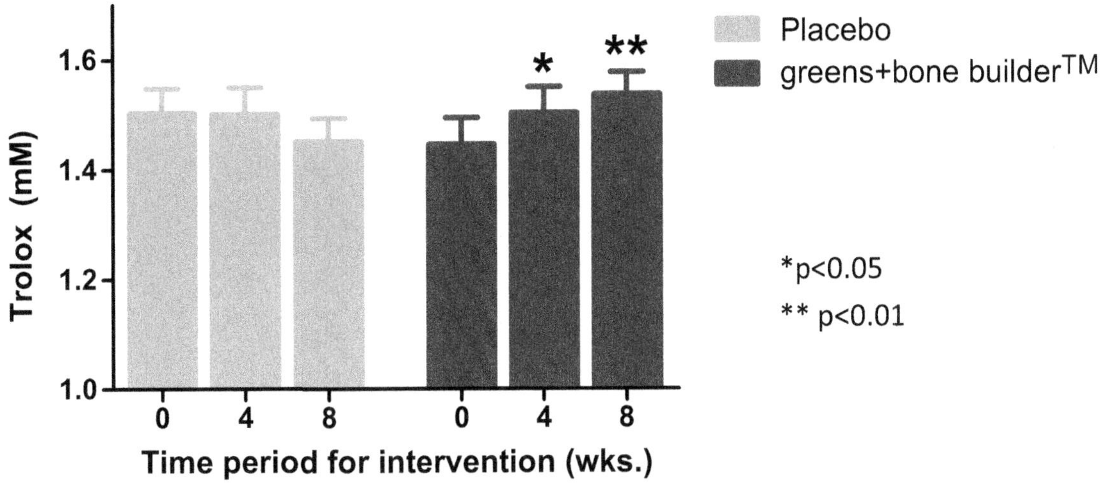

FIGURE 24.2 Graph showing the increase in total antioxidant capacity in the serum after 4 and 8 weeks of greens+bone builder™ supplementation without an increase in the placebo group.

reports will be reviewed here. Shen et al[48] showed that a dietary supplement in the form of green tea combined with tai-chi, a mind–body exercise, can alleviate bone loss in osteopenic women. It was found that catechin was negatively associated with bone-resorption markers, that total flavonoid intake was associated with BMD at the femoral neck and lumbar spine and that procyanidins and catechins are associated with annual percent change in BMD of perimenopausal Scottish women.[49]

Other than these clinical studies in the last 2 years, there have not been any reported clinical studies on polyphenols in human subjects except those of Rao et al.[36,50,51]

The *in vitro* effects of greens+™, bone builder™, and greens+bone builder™ on bone formation in osteoblasts encouraged Rao et al to form the rationale for their clinical studies to test whether these products can reduce the risk of osteoporosis in postmenopausal women. greens+bone builder™ was chosen because it was six times more effective than the other two and that of the three products, it gave the greatest stimulatory effect on bone formation. Rao et al[5] therefore carried out a randomized cross-sectional clinical intervention study to test whether a daily supplementation with greens+bone builder™ may be important in reducing the oxidative stress parameters in postmenopausal women at risk for osteoporosis. A sample of postmenopausal women, 50–60 years of age, was randomized to either the Treatment group, consuming 1 scoop (equivalent to ¼ cup) daily of greens+bone builder™ (n = 23), or the Placebo group (n = 24) for a period of 8 weeks. Participants were asked to give blood samples at 0, 4 and 8 weeks of supplementation for analyses. Results revealed that total antioxidant capacity, as measured by the Trolox value, increased significantly after 4 weeks ($p<0.05$) and 8 weeks ($p<0.01$) of supplementation as shown in Figure 24.2. Also noted were significant decreases in protein oxidation (measured by thiol increase) at both 4 weeks ($p<0.01$) and 8 weeks ($p<0.001$) as shown in Figure 24.3. Lipid peroxidation, as measured by TBARS (as shown in Figure 24.4), was found to be decreased over a 4- and 8-week period ($p<0.0001$) of intervention with greens+bone builder™ while the placebo control showed no changes in any of the parameters measured. This suggests that the nutritional supplement may have a beneficial effect on bone health by counteracting the effects of oxidative stress.[50]

Rao et al carried out a second study in order to test whether the antioxidant properties of greens+bone builder™ can prevent the risk of osteoporosis in postmenopausal women by also measuring the serum bone turnover markers – C-terminal telopeptide of type I collagen (CTX) as an indicator of bone resorption, and procollagen type I N-terminal propeptide (PINP) as an indicator of bone formation – and correlated the results with data on serum antioxidant capacity, and the oxidative stress parameters lipid peroxidation and protein oxidation. Results revealed that, at 8 weeks of treatment, the greens+bone builder™ supplemented group significantly decreased the bone resorption marker CTX ($p <0.05$), while the Placebo group showed no significant changes; the supplemented group was also significantly different from that of the Placebo group (Figure 24.5). Thus, the findings that the decrease in CTX correlated with the increase in their serum total antioxidant capacity and decreases in oxidative stress parameters protein oxidation and lipid peroxidation may suggest that a daily supplementation with polyphenols and micronutrients may be important in reducing oxidative damage by reducing bone resorption, thereby reducing the risk of osteoporosis in postmenopausal women.[51]

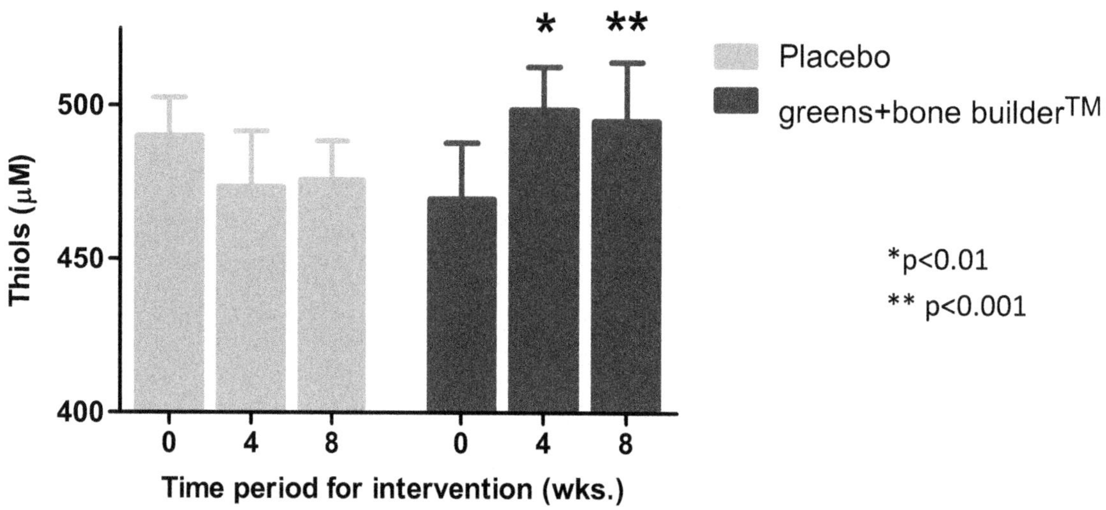

FIGURE 24.3 Graph showing the decrease in protein oxidation as demonstrated by increased thiol over time in greens+bone builder™-treated postmenopausal women without any change in the placebo-treated controls; treated values were also higher than in the placebo group as determined by the Mann–Whitney test (*$p<0.01$, **$p<0.001$).

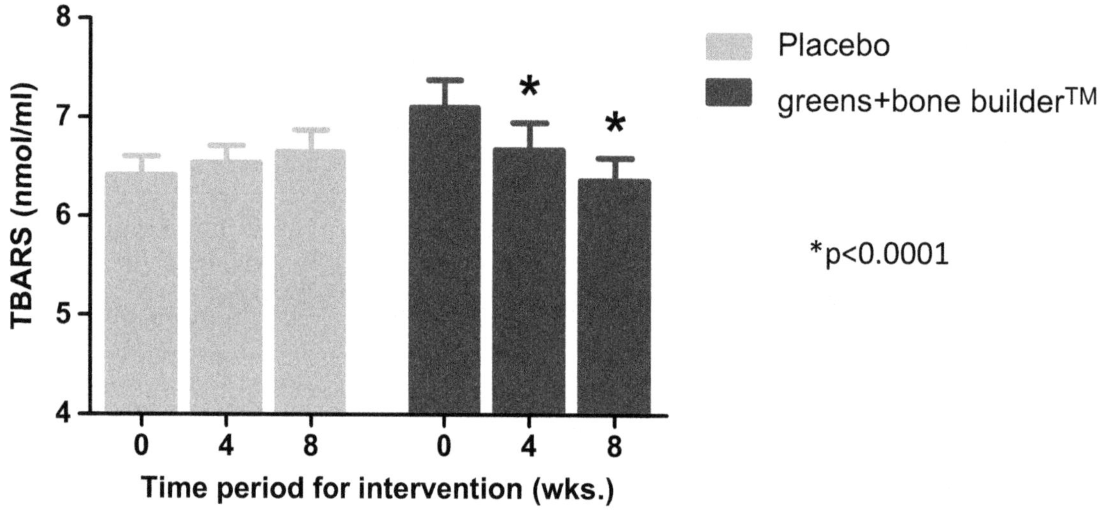

FIGURE 24.4 Graph showing the significant decrease in lipid peroxidation expressed as TBARS after 4 and 8 weeks of greens+bone builder™ supplementation in postmenopausal women without any change in the placebo-treated controls; treated values were also lower than in the placebo group. Mann–Whitney test (*$p<0.0001$). Values are mean ± SEM.

Concluding Remarks

Although, in the last 10 years, there has been an explosion of reports in the literature on the role of polyphenols in bone, most of the reports involved *in vitro* studies on osteoclasts and osteoblasts, animal studies, and epidemiologic studies. Evidence from the excellent studies reported revealed that oxidative stress is one of the primary culprits responsible for the pathogenesis of osteoporosis via its role in osteoclastic resorption and the detrimental effects on the bone-forming osteoblasts. To date, only a few clinical intervention studies have been reported. It is easy to see why it is very difficult to evaluate the role of polyphenols because, as outlined in this chapter, at least 8000 different polyphenols have been identified to date, each one probably having different effects on humans. Additionally, the data on the beneficial effects on bone health of nutrients in food, other than polyphenols, make it difficult to narrow down the effects of polyphenols alone. Rao's laboratory combined the effects of a number of polyphenols present in the nutritional supplement greens+™ with the nutritional components present in bone builder™ – such as minerals, vitamins and other nutrients. Their conclusion is that it is possible that the effects of greens+bone builder™ in increasing total antioxidant capacity, decreasing the oxidative stress markers, protein oxidation and lipid peroxidation – which may have led to the decrease in bone turnover marker for bone resorption – are a result of the

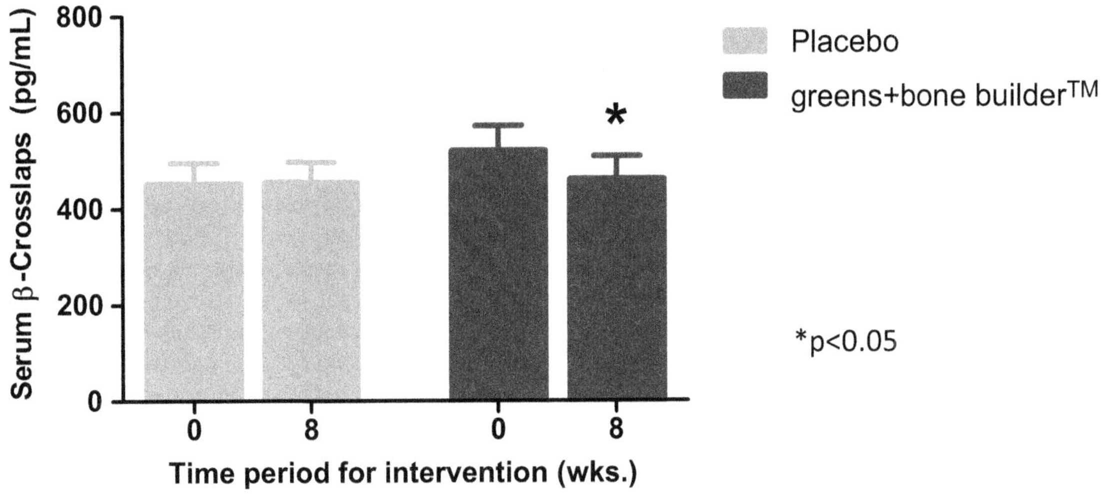

FIGURE 24.5 Graph showing the decrease in serum concentrations of CTX in greens+bone builder™-treated postmenopausal women after 8 weeks (meaning decreased bone resorption marker) compared to that of the placebo-treated controls (a paired *t*-test (*p<0.05).

combined effects of the different polyphenols contained in greens+™ with those of the other nutritional components present in the bone builder™.[50,51] It remains for future studies to zero in on specific components that are responsible for the beneficial effects of greens+bone builder™. The results, however, point to the possible use of greens+bone builder™ in the reduction of risk of osteoporosis in postmenopausal women, either as an alternative to, or in addition to the presently approved drugs.

Studies on Lycopene

The reports on the direct role of lycopene in osteoblasts and osteoclasts no doubt have opened up research into human studies on the role of lycopene in the prevention and treatment of osteoporosis. This role is now further supported by both epidemiologic and clinical intervention with lycopene in postmenopausal women who are at risk of osteoporosis.

In Vitro *Studies on Lycopene in Osteoblasts and Osteoclasts*

Only a few studies on the effects of lycopene in osteoblasts have been reported. The most likely explanation could be that lycopene is not soluble in the culture medium and needs to be solubilized in an organic solvent before it can be added to the cell culture. Kim et al[52] used the lyc-o-mato preparation that is partially dispersed in micelle form in water and showed that lycopene had a stimulatory effect on cell proliferation as well as a stimulatory effect on alkaline phosphatase activity when added to mature human osteoblast-like SaOS-2 cells. These findings comprised the first report on the effect of lycopene on human osteoblasts.[52] Other effects of lycopene on mouse cells have also been reported.[53] Subsequently, Rao et al showed that the *cis* isomers, but not the all-*trans* isomers, of lycopene are capable of preventing and repairing the damaging effects of H_2O_2-induced oxidative stress on the formation of mineralized bone nodules.[5,54]

To date, there have been only two studies on the effects of lycopene in osteoclasts.[55,56] Rao et al cultured cells from bone marrow prepared from rat femur in 16-well, calcium phosphate-coated Osteologic™ multitest slides. Lycopene was shown to inhibit the formation of multinucleated osteoclast cells as well as the formation of ROS-secreting osteoclasts.[56] The effect of lycopene on osteoclast formation and bone resorption was also reported by Ishimi et al in murine osteoclasts formed in co-culture with calvarial osteoblasts.[55]

Epidemiologic Studies on Lycopene

The Mediterranean diet has been a model diet for disease prevention [195]. The major components of this diet are plant foods, including fruits and vegetables.[57] It is therefore possible that the active components in its ability to prevent diseases are lycopene[58] and polypenols.[59] Epidemiologic evidence supports the beneficial role of tomatoes and tomato products present in the Mediterranean diet in the prevention of osteoporosis in the Mediterranean population.[19]

Epidemiologic and intervention studies on the role of lycopene in the prevention of risk for osteoporosis have been reported and recently reviewed.[5,54] In summary, several epidemiologic studies link the role of lycopene, as an antioxidant in the prevention of oxidative stress, with corresponding increase in bone mineral density in postmenopausal women.[7,31,60]

A cross-sectional study was next carried out by Rao et al[32] in which 33 postmenopausal women, aged 50–60 years, were recruited. The participants provided 7-day dietary records and blood samples for analyses of: (i) total antioxidant capacity, (ii) oxidative stress parameters, including protein oxidation and lipid peroxidation, and (iii) bone turnover markers, including bone resorption marker NTx and bone formation marker; values were higher than in those who took less lycopene. The results also showed that the estimated dietary lycopene had a significant and direct correlation with serum lycopene, suggesting that lycopene from the diet is bioavailable. The conclusion that the higher serum lycopene was associated with a low NTx ($p<0.005$), and lower protein oxidation ($p<0.05$), supports the view that lycopene's antioxidative properties are involved in its mechanism of action in bone.[32]

The overall conclusions that can be derived from the reported epidemiologic studies support the beneficial role of lycopene in the prevention of risk for osteoporosis. Further clinical studies, described below, support this conclusion.

Clinical Intervention Studies on Lycopene

Rao et al carried out four different clinical studies to establish the role of lycopene in the prevention of risk for osteoporosis in postmenopausal women. First, a study was carried out to determine the effects of a lycopene-restricted diet on oxidative stress parameters and bone turnover markers in postmenopausal women.[61] To avoid the effects of compounding factors with antioxidants, participants who smoke, use medications which may affect bone metabolism or have antioxidant properties, and those who could not avoid consuming tomatoes and tomato products were excluded. Results revealed that lycopene restriction resulted in significant decreases in serum lycopene, α-/β-carotene and lutein/zeaxanthin, but the overall change in the serum carotenoids was not as high as that seen for lycopene. After lycopene restriction, all configurations of lycopene (all-*trans*, 5-*cis*- and other *cis* lycopene) were found to be decreased and the antioxidant enzymes SOD and CAT were also significantly depressed. These changes were accompanied by a significant increase in the bone resorption marker NTx.

The above is the first reported study on the dramatic effects of dietary lycopene restriction on increasing the risk for osteoporosis in postmenopausal women, proving that lycopene may be beneficial in reducing this risk. It is possible that the significant increase in the bone resorption marker NTx could lead to a long-term decrease in BMD and increased fracture risk, as was observed by Brown et al;[62] a longer restriction period may be detrimental to a group of postmenopausal women already at high risk for osteoporosis. It was further speculated that a shorter wash-out period of no lycopene consumption is all that is needed in clinical trials examining the effects of lycopene on bone health.[61]

A clinical, fully randomized controlled intervention study was next carried out by Mackinnon et al[63] to investigate directly the effects of lycopene supplementation on decreasing the risk for osteoporosis. Participants included 60 postmenopausal women, 50–60 years of age. The exclusion criteria were similar to those in the first study. After a wash-out period of 1 month without lycopene consumption, participants were assigned to one of the following four groups (n = 15 per group): group consuming[1] regular tomato juice,[2] lycopene-rich tomato juice,[3] tomato lycopene capsules or[4] placebo capsules, twice daily for total lycopene intakes of 30, 70, 30 and 0 mg/day, respectively, for 4 months. Serum was collected and assayed for total antioxidant capacity, oxidative stress parameters and bone turnover markers. Results showed that lycopene supplementation for 4 months significantly increased serum lycopene compared to placebo ($p<0.001$). Because the increase in serum lycopene was similar for all three supplements, the results were pooled into a 'LYCOPENE-supplemented' and PLACEBO-supplemented group for further statistical analyses. After the 4-month period, the LYCOPENE-supplemented group had a significant increase in total antioxidant capacity (Fig. 24.6), a decrease in oxidative stress parameters protein oxidation, as shown by the increase in thiol values (Fig. 24.7A), and a decrease in lipid peroxidation, as shown by TBARS (Fig. 24.7B). These changes correlated with a decrease in NTx (Fig. 24.8), and the changes were significantly different from those in the PLACEBO group. These findings suggest that lycopene obtained in the form of tomato juice, or in a capsule, exerted equivalent antioxidant potency in reducing the risk of osteoporosis in postmenopausal women.[63]

A third study was undertaken by Mackinnon et al[54] to compare the data on serum lycopene, bone turnover markers and oxidative stress parameters between postmenopausal women who were supplemented with lycopene and those who obtained either a low or high intake of lycopene from their daily food. The purpose of the study was to determine whether the elevated dose obtained through supplementation was more beneficial in reducing bone turnover markers than intakes typically obtained from the usual daily diet. Results showed that women who consumed a lycopene supplement had significantly lower TBARS values than participants who obtained a low intake or high intake of lycopene through their usual daily diets. These differences in TBARS values may be attributed to a significantly higher concentration of serum 5-*cis* in lycopene-supplemented participants compared to participants who obtained their lycopene from their usual (low or high) daily diet. This suggests that it is the 5-*cis* isomer, with the most potent antioxidant capacity, which, at higher concentrations, decreases bone turnover markers due to its ability to provide

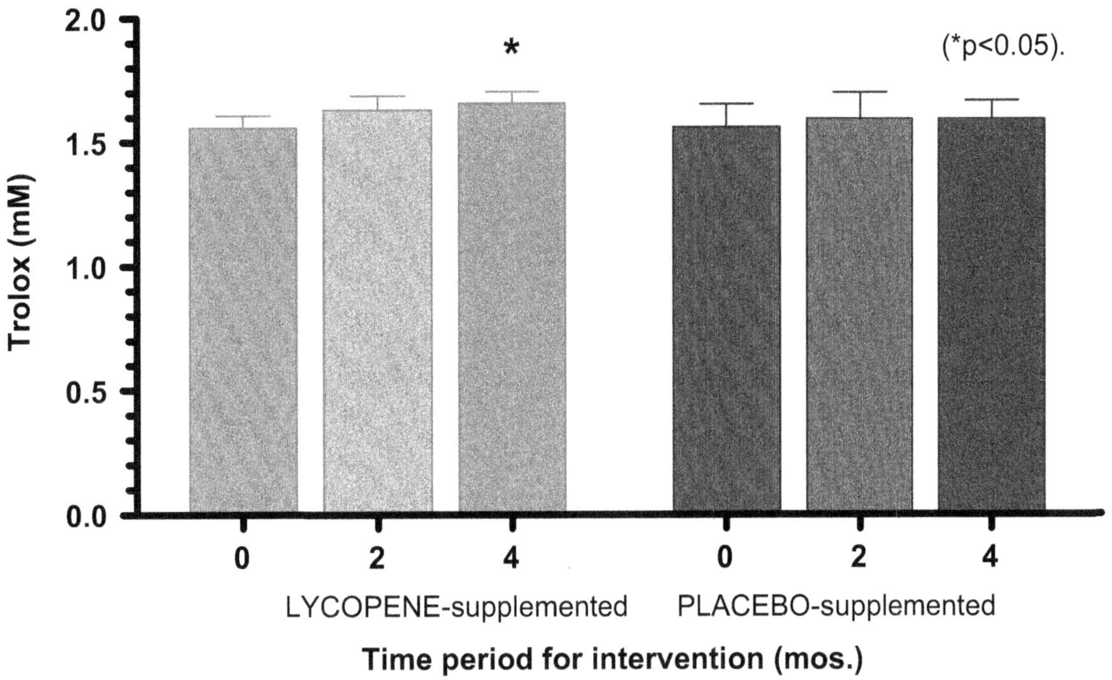

FIGURE 24.6 Graph showing the increase in total antioxidant capacity determined using the TEAC assay in the serum of 60 postmenopausal female participants supplemented with LYCOPENE for a period of 4 months. No change was noted in the PLACEBO group. Values are mean ± SEM and were compared within the supplemented group using repeated-measures ANOVA (*$p<0.05$).

the greatest protection against oxidative stress. It also appears to show that supplementation with lycopene may be necessary in spite of the daily intake of lycopene.

The role of 172T→A or 584A→G polymorphisms in the paraoxonase 1 gene (PON 1) in modulating the effects of serum lycopene on antioxidant capacity, oxidative stress parameters and bone turnover markers, and in women between the ages of 25 and 70 years, was studied by Mackinnon et al.[64] Their results showed that the PON1 polymorphism modified the association between lycopene and NTx and between lycopene and BAP, an interaction that may also moderate the risk of osteoporosis.[64]

In another study, these authors showed that there is a significant interaction between the PON1 genotype and the change in TBARS ($p<0.05$), suggesting that supplementation with lycopene resulted in decreased lipid peroxidation, which interacted with the PON1 genotype to decrease bone resorption markers in postmenopausal women. These findings provide a mechanistic evidence of how intervention with lycopene may act to decrease lipid peroxidation and thus the risk of osteoporosis in postmenopausal women.[54,65]

Concluding Remarks

It is now evident that antioxidants such as lycopene can counteract the damaging effects of oxidative stress, brought about by ROS, that lead to the development of several chronic diseases, including osteoporosis. The evidence includes the results of studies on their role in osteoclastic resorption and disruption of osteoblastic bone formation, epidemiologic studies and, more recently, clinical intervention studies. Due to the increasing numbers of reports on the adverse side effects of conventional therapy (e.g. HRT and bisphosphonates), there is a demand for the use of other natural food components in the management of postmenopausal osteoporosis. The results of studies reviewed here indicate that lycopene may be useful either as a dietary alternative to drug therapy or as a complement to the drugs presently approved for use by women at risk of osteoporosis.

GENERAL SUMMARY AND CONCLUSION

This chapter clearly demonstrates that[1] oxidative stress is caused by reactive oxygen species (ROS);[2] oxidative stress is shown to be associated with the development of osteoporosis;[3] the damaging effect caused by oxidative stress may be preventable by supplementation with the antioxidants polyphenols and lycopene;[4] these conclusions are supported by results of *in vitro* studies in osteoblasts and osteoclasts, and epidemiologic and clinical intervention studies on the antioxidants polyphenols and lycopene;[5] and polyphenol and lycopene may have the potential for use as an alternative or complementary agent with other established drugs approved for the prevention or treatment of osteoporosis in postmenopausal women.

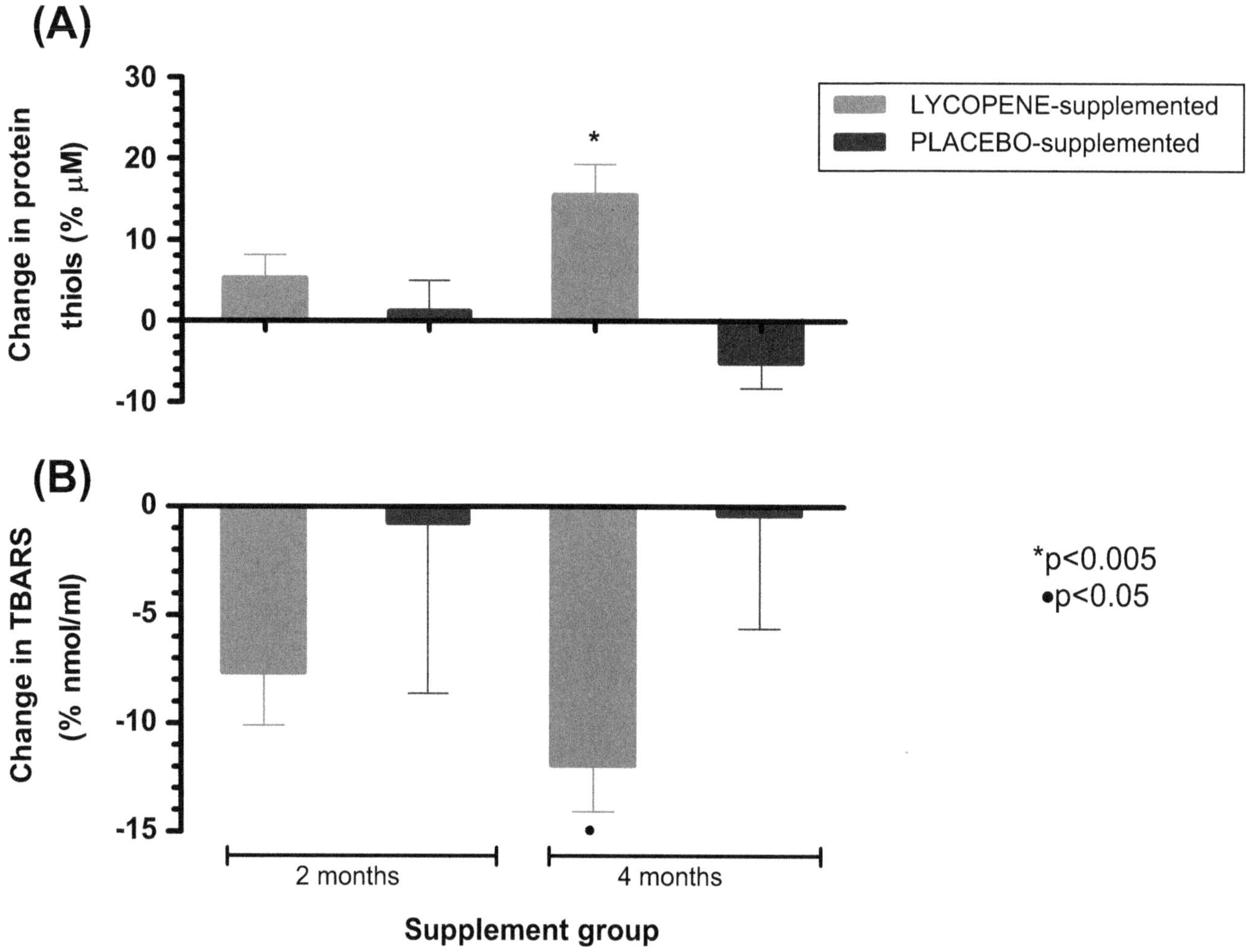

FIGURE 24.7 Graph showing the significant decreases in protein oxidation (7A) (i.e. increase in thiol concentration) and lipid peroxidation (7B) from baseline values over the 4-month supplement period in the LYCOPENE group without any change in the PLACEBO group. Values are mean % change ± SEM for each supplement group and were compared using an unpaired *t*-test (*p<0.005 and •p<0.05, respectively).

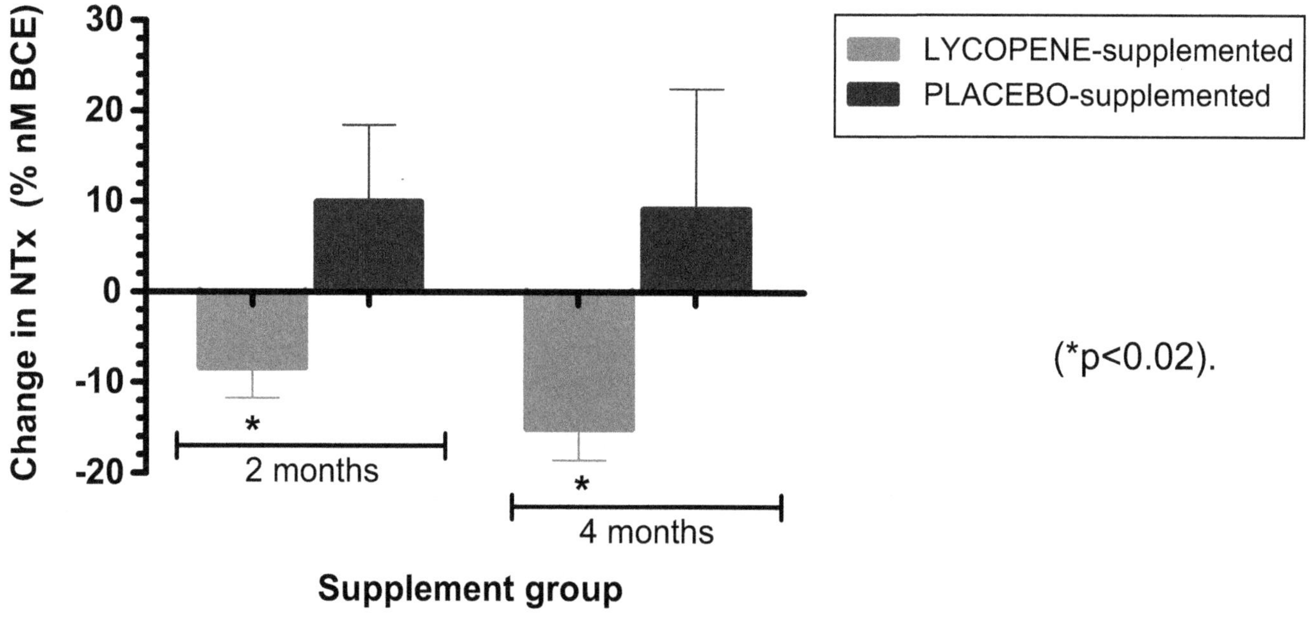

FIGURE 24.8 Graph showing the significant decrease in bone resorption marker NTx over the 4-month supplement period in the LYCOPENE group without any change in the PLACEBO group. Values are mean % change ± SEM for each supplement group and were compared using an unpaired *t*-test (*p<0.02).

Acknowledgments

Funding for this research into Oxidative Stress, Antioxidants and Bone Health is shared by Genuine Health Ltd (Canada), the HJ Heinz Co (USA), Millenium Biologix Inc. (Canada), Kagome Co. (Japan) and LycoRed Natural Product Industries, Ltd. (Israel) and matched by the Canadian Institutes of Health Research (CIHR). We are sincerely grateful for the valuable contributions to this research by the following undergraduate summer students, graduate students, postdoctoral fellow and staff at the Calcium Research Laboratory, Department of Medicine at St Michael's Hospital and the University of Toronto and Department of Nutritional Sciences, University of Toronto: Dr Bala Balachandran, Jaclyn Beca, Dawn Snyder, Loren Chan, Honglei Shen, Salva Sadeghi, Ayesha Quireshi, Dr Erin Mackinnon and Nancy Kang. Their contributions were based on their experimental data, written reports and published/in press manuscripts. We would also like to thank Dr Alan Logan and Dr RG Josse for providing us with medical expertise as well as allowing us access to the list of patients we were able to recruit. Special thanks to Dr H Vandenberghe for carrying out the CTX assay and for her valuable suggestions.

References

1. Raman K. Studies on free radicals, antioxidants, and co-factors. *Clin Interventions Aging* 2007;**2**(2):219–36.
2. Manolagas S, Parfitt A. What old means to bone. *Trends Endocrinol Metab* 2010;**21**(6):369–74. Epub 2010 Mar 11.
3. Ahmed S, Elmantaser M. Secondary osteoporosis. *Endocr Dev* 2009;**16**:170–90.
4. Riggs B, Melton LI. The worldwide problem of osteoporosis: insights afforded by epidemiology. *Bone* 1995;**17**(Suppl):505S–11S.
5. Rao L, Rao AV. Oxidative stress and antioxidants in the risk of osteoporosis – role of the antioxidants lycopene and polyphenols. In: Valdes Flores M, editor. *New findings in osteoporosis*. 2013. in press.
6. Stetzer E. Identifying risk factors for osteoporosis in young women. *Int J Allied Health Sci Practice* 2011;**9**(4):1–8.
7. Yang Z, Zhang Z, Penniston K, et al. Serum carotenoid concentrations in postmenopausal women from the United States with and without osteoporosis. *Int J Vitamin Nutr Res* 2008;**78**(3):1050–111.
8. Sahnoun Z, Jamoussi K, Zeghal K. Free radicals and antioxidants: human physiology, pathology and therapeutic aspects. *Therapie* 1997;**52**. 251–70.
9. Juránek I, Š Bezek S. Controversy of free radical hypothesis: reactive oxygen species – cause or consequence of tissue injury? *Gen Physiol Biophys* 2005;**24**:2263–78.
10. Semba RD, Ferrucci L, Sun K, et al. Oxidative stress is associated with greater mortality in older women living in the community. *J Am Geriatr Soc* 2007;**55**(9):1421–5.
11. Lee D, Lim B, Lee YK, Yang HC. Effects of hydrogen peroxide (H_2O_2) on alkaline phosphatase activity and matrix mineralization of odontoblast and osteoblast cell lines. *Cell Biol Toxicol* 2006;**22**:39–46.
12. Willcox J, Ash S, Catignani G. Antioxidants and prevention of chronic disease. *Crit Rev Food Sci Nutr* 2004;**44**:275–95.
13. Zhang Y, Zhong Z, Hou G, et al. Involvement of oxidative stress in age-related bone loss. *J Surg Res* 2011;**169**(1):e37–42. Epub 2011 Mar 21.
14. Mate JM, Perez-Gomez C, Nunez de Castro I. Antioxidant enzymes and human diseases. *Clin Biochem* 1999;**32**(8):595–603.
15. Rao A, Ray M, Rao L. Lycopene. In: Taylor SL, editor. *Advances in food and nutrition research*. New York: Academic Press; 2006. pp. 99–164.
16. Quideau S, Deffieux D, Douat-Casassus C, Pouysegu L. Plant polyphenols: chemical properties, biological activities, and synthesis. *Angew Chem Int (English edition)* 2011;**50**(3):586–521. Epub 2011/01/13.
17. Hendrich A. Flavonoid-membrane interactions: possible consequences for biological effects of some polyphenolic compounds. *Acta Pharmacologica Sinica* 2006;**27**(1):27–40.
18. Manach C, Scalbert A, Morand C, et al. Polyphenols: food sources and bioavailability. *Am J Clin Nutr* 2004;**79**:727–47.
19. Puel C, Coxam V, Davicco M. Mediterranean diet and osteoporosis prevention. *Med Sci (Paris)* 2007;**23**:756–60.
20. Urquiaga I, Strobel P, Perez D, et al. Mediterranean diet and red wine protect against oxidative damage in young volunteers. *Atherosclerosis* 2010;**211**(2):694–9. Epub 2010 Apr 21.
21. Scalbert A, Manach C, Morand C, et al. Dietary polyphenols and the prevention of diseases. *Crit Rev Food Sci Nutr* 2005;**45**(4):287–306.
22. Hollman PC, Cassidy A, Comte B, et al. The biological relevance of direct antioxidant effects of polyphenols for cardiovascular health in humans is not established. *J Nutr* 2011;**141**(5):989S–1009S. Epub 2011/04/01.
23. Rao A, Balachandran B, Shen H, et al. In vitro and in vivo antioxidant properties of the plant-based supplement greens+. *Int J Mol Sci* 2011;**12**:4896–908.
24. Agarwal S, Rao AV. Tomato lycopene and its role in human health and chronic diseases. *CMAJ* 2000;**163**(6):739–44.
25. Khachik F, Carvallo L, Bernstein P, et al. Chemistry, distribution and metabolism of tomato carotenoids and their impact on human health. *Exp Biol Med* 2002;**227**(10):845–51.
26. Di Mascio P, Kaiser S, Sies H. Lycopene as the most effcient biological carotenoid singlet oxygen quencher. *Arch Biochem Biophys* 1989;**274**:532–8.
27. Stahl W, Sies H. Uptake of lycopene and its geometrical isomers is greater from heat-processed than from unprocessed tomato juice in humans. *J Nutr* 1992;**122**:2161–6.
28. Amir HKM, Giat J, et al. Lycopene and 1,25(OH)2D3 cooperate in the inhibition of cell cycle progression and induction of HL-60 leukemic cells. *Nutr Cancer* 1999;**33**:105–12.
29. Zhang L, Cooney R, Bertram J. Carotenoids enhance gap junctional communication and inhibit lipid peroxidation in C3H/10T1/2 cells: relationship to their cancer chemopreventive action. *Carcinogenesis* 1991;**12**:2109–14.
30. Giovannucci E. Tomatoes, tomato-based products, lycopene, and cancer: review of the epidemiologic literature. *J Natl Cancer Inst* 1999;**91**:317–31.
31. Sahni S, Hannan M, Blumberg J, et al. Protective effect of total carotenoid and lycopene intake on the risk of hip fracture: a 17-year follow-up from the Framingham Osteoporosis Study. *J Bone Mineral Res* 2009;**24**(6):1086–94.
32. Rao L, Mackinnon E, Josse R, et al. Lycopene consumption decreases oxidative stress and bone resorption markers in postmenopausal women. *Osteoporosis Int* 2007;**18**(1):109–15.
33. Trzeciakiewicz A, Habauzit V, Horcajada M. When nutrition interacts with osteoblast function: molecular mechanisms of polyphenols. *Nutr Res Rev* 2009;**22**(1):68–81. Epub 2009 Feb 26.
34. Horcajada M, Offord E. Naturally plant-derived compounds: role in bone anabolism. *Curr Mol Pharmacol* 2012;**5**(2):205–18.
35. Cabrera CRA, Giménez R. Beneficial effects of green tea – a review. *J Am Coll Nutr* 2006;**25**(2):79–99.
36. Rao L, Kang N, Rao A. Polyphenols and bone health: a review. In: Rao A, editor. *Phytochemicals*. Rijeka (Croatia): InTech Open Access Publisher; 2012. pp. 958–73.
37. Garrett IR, Boyce BF, Oreffo RO, et al. Oxygen-derived free radicals stimulate osteoclastic bone resorption in rodent bone in vitro and in vivo. *J Clin Invest* 1990;**85**(3):632–9. Epub 1990/03/01.
38. Wauquier F, Leotoing L, Coxam V, et al. Oxidative stress in bone remodelling and disease. *Trends Mol Med* 2009;**15**(10):468–77. Epub 2009/10/09.
39. Baek KH, Oh KW, Lee WY, et al. Association of oxidative stress with postmenopausal osteoporosis and the effects of hydrogen peroxide on osteoclast formation in human bone marrow cell cultures. *Calcif Tissue Int* 2010;**87**(3):226–35. Epub 2010/07/09.
40. X-c Bai, Lu D, Bai J, et al. Oxidative stress inhibits osteoblastic differentiation of bone cells by ERK and NF-κB. *Biochem Biophys Res Commun* 2004;**314**(1):197–207.

41. Mody N, Parhami F, Sarafian TA, Demer LL. Oxidative stress modulates osteoblastic differentiation of vascular and bone cells. *Free Radic Biol Med* 2001;**31**(4):509–19.
42. Vali B, Rao L, El-Sohemy A. Epigallocathechin-3-gallate (EGCG) increases the formation of mineralized bone nodules by human osteoblast-like cells. *J Nutr Biochem* 2007;**18**(5):341–7. Epub 2006 ep 8.
43. Yun J, Kim C, Cho K, et al. (-)-Epigallocatechin gallate induces apoptosis, via caspase activation, in osteoclasts differentiated from RAW 264.7 cells. *J Periodontal Res* 2007;**42**(3):212–8.
44. Rao L, Balachandran B, Rao A. Polyphenol extract of Greens+™ nutritional supplement stimulates bone formation in cultures of human osteoblast-like SaOS-2 cells. *J Diet Suppl* 2008;**5**(3):264–82.
45. Rao L, Balachandran B, Rao A. The stimulatory effect of the polyphenols in the extract of Greens+™ herbal preparation on the mineralized bone nodule formation (MBNF) of SaOS-2 cells is mediated via its inhibitory effect on the intracellular reactive oxygen species (iROS). 27th Annual Meeting of the American Society of Bone and Mineral Research; September 23-27, 2005; Nashville, Tennesse, 2005.
46. Graci S. *The bone building solution*. Mississauga, Ontario: John Wiley; 2006:392.
47. Snyder DM, Rao AV, Bashyam B, et al. Extracts of the nutritional supplement, bone builder™, and the herbal supplement, greens+™, synergistically stimulate bone formation by human osteoblast cells in vitro. *J Bone Mineral Res* 2010;**25**(Suppl 1).
48. Shen C, Chyu M, Yeh J, et al. Effect of green tea and Tai Chi on bone health in postmenopausal osteopenic women: a 6-month randomized placebo-controlled trial. *Osteoporosis Int* 2012;**23**(5):1541–52. Epub 2011 Jul 16.
49. Hardcastle A, Aucott L, Reid D, Macdonald H. Associations between dietary flavonoid intakes and bone health in a Scottish population. *J Bone Mineral Res* 2011;**26**(5):941–7.
50. Kang N, Rao A, De Asis K, et al. Antioxidant effects of a nutritional supplement containing polyphenols and micronutrients in postmenopausal women: a randomized controlled study. *J Aging: Res Clin Practice* 2012;**1**(3):183–7.
51. Kang N, Rao A, Josse R et al. Dietary polyphenols and nutritional supplements significantly decreased oxidative stress parameters and the bone resorption marker collagen type 1 cross-linked C-telopeptide in postmenopausal women. Annual Meeting of the Canadian Nutrition Society May 25, 2012; Vancouver, British Columbia, 2012.
52. Kim L, Rao A, Rao L. Lycopene II. Effect on osteoblasts: the carotenoid lycopene stimulates cell proliferation and alkaline phosphatase activity of SaOS-2 cells. *J Medicinal Food* 2003;**6**:79–88.
53. Park CK, Ishimi Y, Ohmura M, et al. Vitamin A and carotenoids stimulate differentiation of mouse osteoblastic cells. *J Nutr Sci Vitaminol* 1997;**43**:281–96.
54. MacKinnon E. *The role of the carotenoid lycopene as an antioxidant to decrease osteoporosis risk in women: Clinical and in vitro studies [dissertation for PhD in Institute of Medical Science]*. Toronto, Ontario: University of Toronto; 2010.
55. Ishimi Y, Ohmura M, Wang X, et al. Inhibition by carotenoids and retinoic acid of osteoclast-like cell formation induced by bone resorbing agents in vitro. *J Clin Biochem Nutr* 1999;**27**:113–22.
56. Rao L, Krishnadev N, Banasikowska K, et al. Lycopene I. Effect on osteoclasts; lycopene inhibits basal and parathyroid hormone (PTH)-stimulated osteoclast formation and mineral resorption mediated by reactive oxygen species (ROS) in rat bone marrow cultures. *J Medicinal Food* 2003;**6**:69–78.
57. Serra-Majem L, Roman B, Estruch R. Scientific evidence of interventions using the Mediterranean diet: a systematic review. *Nutr Rev* 2006;**64**(2):S27–47.
58. Hagfors L, Leanderson P, Skoldstam L, et al. Antioxidant intake, plasma antioxidants and oxidative stress in a randomized, controlled, parallel, Mediterranean dietary intervention study on patients with rheumatoid arthritis. *Nutr J* 2003;**2**:5.
59. Leighton F, Cuevas A, Guasch V, et al. Plasma polyphenols and antioxidants, oxidative DNA damage and endothelial function in a diet and wine intervention study in humans. *Drugs Exp Clin Res* 1999;**25**:133–41.
60. Sugiura M, Nakamura M, Ogawa K, et al. Bone mineral density in post-menopausal female subjects is associated with serum antioxidant carotenoids. *Osteoporos Int* 2008;**19**:211–9. Epub 2011 Jul 16.
61. Mackinnon E, Rao A, Rao L. Dietary restriction of lycopene for a period of one month resulted in significantly increased biomarkers of oxidative stress and bone resorption in postmenopausal women. *J Nutr Health Aging* 2011;**15**(2):133–8.
62. Brown J, Albert C, Nassar B, et al. Bone turnover markers in the management of postmenopausal osteoporosis. *Clin Biochem* 2009;**42**(10-11):929–42. Epub 2009 Apr 10. Review.
63. Mackinnon E, Rao A, Josse R, Rao L. Supplementation with the antioxidant lycopene significantly decreases oxidative stress parameters and the bone resorption marker N-telopeptide of type I collagen in postmenopausal women. *Osteoporosis Int* 2011;**22**(4):1091–101.
64. Mackinnon E, El-Sohemy A, Rao A, Rao L. Paraoxonase 1 polymorphisms 172T→A and 584A→G modify the association between serum concentrations of the antioxidant lycopene and bone turnover markers and oxidative stress parameters in women 25–70 years of age. *J Nutrigenet Nutrigenom* 2010;**3**(1):1–8.
65. Rao L, Mackinnon E, El-Sohemy A, Rao V. Postmenopausal women with PON1 172TT genotype respond to lycopene intervention with a decrease in oxidative stress parameters and bone resorption marker NTx. Annual Meeting of the American Society for Bone and Mineral Research (ASBMR); October 13–16; Toronto, Ontario, 2010.

CHAPTER

25

Zinc, Oxidative Stress in the Elderly and Implications for Inflammation

Ananda S. Prasad

Department of Oncology, Wayne State University School of Medicine and Barbara Ann Karmanos Cancer Institute, Detroit, MI, USA

List of Abbreviations

AREDS Age-related Eye Disease Study
C/EBP CCAAT/enhancer binding protein
CMI skin-test reactivity to common antigens
HAE 4-hydroxyalkenals
IFN-γ interferon-γ
LDL low-density lipoprotein
LPS lipopolysaccharide
NF-κB nuclear factor κ-B
PMNC peripheral blood mononuclear cell
SCD sickle-cell disease
SP-1 specificity protein 1
TNF-α tumor necrosis factor-α

INTRODUCTION

The importance of zinc in humans was recognized only 50 years ago.[1,2] We have now witnessed tremendous advances in the field of zinc metabolism in humans. We estimate that nearly 2 billion people in the developing world may be suffering from zinc deficiency. The major manifestations of zinc deficiency include growth retardation, testicular hypofunction, cell-mediated immune disorders, cognitive impairment, increased oxidative stress and increased generation of inflammatory cytokines.[3-8] The consequences of zinc deficiency, if severe and unrecognized, may be fatal.

Zinc has now been used successfully to treat and prevent diarrhea in infants and children throughout the world, and this has resulted in the saving of millions of lives.[9,10] Zinc is an effective therapeutic agent for the treatment of two genetic disorders: Wilson's disease and acrodermatitis enteropathica (AE).[11,12] Zinc acetate lozenges are effective in decreasing the duration and severity of the common cold.[13,14] Zinc has also been used successfully to prevent the progression of age-related macular degeneration (AMD) and blindness in the elderly population.[15–17] The most recent publication of the AREDS group shows that during 10 years of follow-up, the mortality due to cardiovascular events in the elderly was significantly decreased in the group taking zinc.[18] All of these clinical effects of zinc are highly impressive and have a large impact on human health.

Remarkable advances have also taken place in the basic science area. We now know that there are over 300 enzymes that require zinc for their activation or structural stability, and there are over 2000 transcription factors involved in gene expression of proteins that are zinc dependent.[7] We now know that zinc is a molecular signal for immune cells, and that homeostasis of intracellular Zn^{2+} levels is maintained by 14 ZIP (SLC 39A) and 10 ZNT (SLC 30A) transporters.

In this chapter, I briefly describe the history of the discovery of zinc as an essential element for humans, present the clinical effects of zinc deficiency and discuss in detail the role of zinc in cell-mediated immunity and its roles as an antioxidant and as an anti-inflammatory agent as they relate to the health of the elderly.

DISCOVERY OF ZINC DEFICIENCY IN HUMANS

Although the essentiality of zinc for the growth of microorganisms, plants and animals has been known for many decades, its essentiality for humans was recognized only in 1963.[1,2] In the fall of 1958, following my training under Dr CJ Watson at the University of Minnesota Medical School as a clinical investigator, I went to Shiraz, Iran.

Aging
http://dx.doi.org/10.1016/B978-0-12-405933-7.00025-1

The story of zinc in human health began when an Iranian physician presented to me a 21-year-old male patient who looked like a 10-year-old boy who was severely anemic. He had infantile genitalia, rough and dry skin, mental lethargy, hepatosplenomegaly, and indulged in clay eating (geophagia). He ate only bread made from whole-wheat flour, and there was no intake of animal protein. He ate one pound of clay daily. The habit of geophagia was very common in the villages around Shiraz.

This patient was severely iron deficient but had no blood loss. His hemoglobin was approximately 5 g per cent. The anemia was hypochromic-microcytic. Later, I discovered that this syndrome was common in the villages near Shiraz.[1]

Figure 25.1 shows a picture of some of these dwarfs in Iran. On the left is the photograph of our radiologist colleague who volunteered for this picture.

I had two clinical dilemmas. How did this patient become so iron deficient without blood loss? And how could growth retardation and hypogonadism be explained on the basis of iron deficiency, as growth retardation and testicular atrophy are not observed in iron-deficient experimental animals? An examination of the Periodic table suggested to me that deficiency of another transitional element, perhaps zinc, may also have been present, which could account for the growth retardation and hypogonadism. We hypothesized that a high phosphate content of the diet and geophagia may have decreased the gastrointestinal absorption of both iron and zinc, resulting in a deficiency of both elements.[1]

I subsequently moved to Cairo, Egypt, and carried out extensive studies of zinc metabolism in these dwarfs at the US Naval Medical Research Unit No. 3.

The zinc concentrations in plasma, red blood cells and hair, and the total zinc content in 24-hour urine, were decreased in the Egyptian dwarfs.[2] With the use of ^{65}Zn, we showed that the plasma zinc turnover rate was increased, and the 24-hour exchangeable zinc pool decreased, in the dwarfs, as compared to controls.[2] We also showed that zinc supplementation to these dwarfs resulted in 12.7–15.2 cm of growth in 1 year, and that the genitalia reached normal size within 3–6 months of supplementation.[3] Thus, our studies showed for the first time that zinc was essential for humans and that zinc deficiency occurred in some Middle Eastern villagers.[2]

The details of circumstances leading to the discovery of human zinc deficiency in the Middle East have recently been published.[19] For nearly one decade, the idea that zinc deficiency occurred in humans remained very controversial. In 1974, however, the National Research council of the National Academy of Sciences declared zinc to be an essential element for humans, and established a recommended dietary allowance (RDA).[20] In 1978, the FDA made it mandatory to include zinc in total parenteral nutritional fluids.[21]

The current WHO estimate is that nearly two billion people in the developing world may have a nutritional deficiency of zinc. This is because most of the population consumes mainly bread made of whole wheat flour which contains high quantities of phytate, an organic phosphate compound which binds both iron and zinc and makes them unavailable for absorption. A phytate-to-zinc molar ratio greater than 20 in a diet is very unfavorable for zinc absorption and may lead to its deficiency.

In developed countries, such as the USA, zinc deficiency is prevalent in some populations. In the USA, nearly 30% of elderly subjects, 25–30% of African-Americans

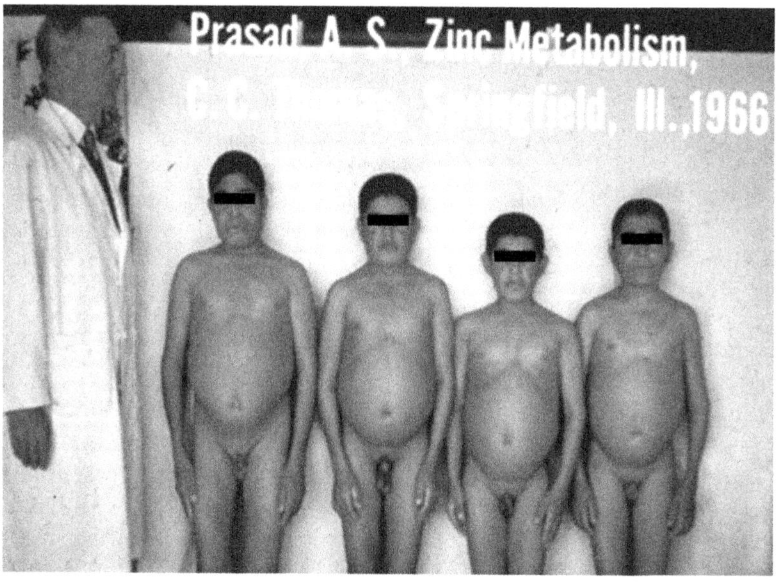

FIGURE 25.1 Zinc-deficient Iranian dwarfs aged 16 to 21 years.

and Mexican-Americans, and approximately 25% of premenopausal women of child-bearing age are vulnerable to zinc deficiency.[22]

Besides nutritional deficiency of zinc, there are many cases of conditioned deficiency of zinc in the USA and other developed countries. Some examples of conditional deficiency of zinc in humans include patients with alcoholic cirrhosis of the liver, chronic alcoholism with hyperzincura, chronic renal disease, malabsorption syndrome and those with sickle-cell disease (SCD) who exhibit hyperzincuria due to loss of zinc caused by chronic hemolysis.[7] Many subjects with chronic diseases such as malignancies are also vulnerable to zinc deficiency, inasmuch as their appetite is poor and intake of food is inadequate. Thus it is obvious that deficiency of zinc in clinical practice is not rare. Unfortunately, physicians do not recognize this problem and as such many patients are not properly managed.

CLINICAL MANIFESTATIONS OF ZINC DEFICIENCY

Severe Deficiency of Zinc

We now recognize a spectrum of clinical deficiencies of zinc in humans. On the one hand the deficiency may be very severe, as is often seen in patients with acrodermatitis enteropathica (AE), a genetic disorder which affects infants of Italian, Armenian or Iranian lineage.[7,11] The dermatologic manifestations include bullous pustular dermatitis of the extremities and around orifices, and alopecia. Ophthalmic signs include blepharitis, conjunctivitis, photophobia and corneal opacities. Neuropsychiatric signs include irritability, emotional instability, tremors and occasional cerebellar ataxia. Weight loss, growth retardation and male hypogonadism are also prominent features. Congenital malformation of fetuses and infants born of pregnant women with AE has frequently been observed.[7,23]

AE patients have an increased susceptibility to infections. Thymic hypoplasia, absence of germinal centers in lymph nodes and plasmacytosis in the spleen are seen commonly. All T cell-mediated functional abnormalities are corrected with zinc supplementation. Without zinc supplementation, the clinical course is downhill, with failure to thrive and complications due to inter-current bacterial, viral, fungal and other opportunistic infections. Gastrointestinal manifestations include diarrhea, malabsorption, steatorrhea and lactose intolerance. Zinc supplementation results in complete recovery.

The AE gene has been localized to a ~3.5 cm region on the 8 q 24 chromosome. The gene encodes ZIP-4, a family of transmembrane proteins, known as zinc transporters. In AE, mutations of this gene have been documented.[24]

Severe deficiency of zinc has been reported in patients receiving total parenteral nutrition for prolonged periods without zinc.[25] In the USA, zinc is now being routinely included in parenteral fluids for subjects who receive such therapy for prolonged periods.

Severe deficiency of zinc has been reported in a patient with Wilson's disease who received therapy with pencillimine as a de-coppering agent.[26]

Moderate Deficiency of Zinc

In this condition, the clinical manifestations include growth retardation, male hypogonadism in adolescents, rough skin, poor appetite, mental lethargy, delayed wound healing, cell-mediated immune dysfunctions and neurosensory disorders.[7] These manifestations have been reported in subjects with nutritional deficiency of zinc observed in many developing countries globally. Females are equally susceptible to zinc deficiency, and their ovarian functions are also adversely affected. These manifestations are also seen in many cases of conditioned deficiency of zinc.

Mild Deficiency of Zinc

Recognition of mild deficiency of zinc is difficult. We, therefore, developed an experimental model of human zinc deficiency in order to define mild deficiency.[27]

We induced a state of mild human zinc deficiency in a group of adult volunteers by dietary means.[27–29] A semi-purified diet which supplied approximately 3.0 to 5.0 mg of zinc daily was used to induce deficiency of zinc.[27–29]

In this model, as a result of the zinc deficiency, we observed decreased serum testosterone levels, oligospermia, decreased natural killer cell (NK cell) lytic activity, decreased IL-2 activity of T helper cells, decreased serum thymulin activity (thymulin is a thymic hormone essential for the development and maturation of T cells), hyperammonemia, hypogeusia, decreased dark adaptation and decreased lean body mass.[27–29] This study clearly established that even a mild deficiency of zinc in humans adversely affects clinical, biochemical and immunologic functions.

ZINC DEFICIENCY IN ELDERLY SUBJECTS

It is well known that, in the USA and other developing countries, the dietary intake of zinc declines with advancing age.[30] It has been estimated that daily zinc intake in the elderly population is approximately 8 to 9 mg whereas the RDA is 15 mg in males and 12 mg in females.[30] This is due to an increased preference of cereal proteins by the elderly, decreased caloric intake and decreased intake of animal protein due to low income. Cereal proteins contain increased amounts of phytate, an organic phosphate compound which decreases the absorption of both iron and zinc.[7]

In the past, many investigators used only plasma zinc for the assessment of zinc status, and several contradictory reports were published prior to 1993.[30] Two studies, however, suggested that zinc supplementation may have some beneficial effects in the elderly. Duchateau et al[31] observed a significant improvement in the numbers of circulating T lymphocytes. Improved delayed cutaneous hypersensitivity reactions to purified protein derivatives (candida, streptokinase and streptodornase) and improved immunoglobulin G antibody response to tetanus vaccine in the elderly after supplementation with zinc sulfate 220 mg twice daily for 1 month were reported. In another limited study, Newsome et al[15] reported significant beneficial effects of zinc supplementation (zinc sulfate 100 mg twice daily for up to 24 months) on age-related macular degeneration. These two studies, however, did not define the baseline zinc status of their subjects, and inasmuch as the dosage of zinc supplementation was much higher than the RDA, it was not clear whether the effects of zinc supplementation were pharmacologic responses or simply a correction of dietary zinc deficiency.

We studied elderly subjects in the Detroit area in 1987–1993 with respect to their zinc status in extensive detail.[30] The dietary mean zinc intake of 180 healthy elderly subjects (who were ambulatory and free of chronic disease) was 9.06 mg/day, and the mean copper intake was 1.45 mg/day.

We assessed IL-1 activity by testing the ability of supernatants of cultured mononuclear cells to augment phytohemagglutinin (PHA)-induced proliferation of murine thymocytes.[30] Production of IL-1 by mononuclear cells by our technique was significantly decreased in the elderly subjects ($p = 0.01$).

Figure 25.2 shows the data for zinc concentration in lymphocytes and granulocytes in 118 elderly subjects aged 60 to 80 years when both determinations were available. Thirty six subjects had zinc concentrations in the deficient range, in both cell lines, in comparison to normal controls. Thus, according to our cellular zinc assay criteria, 30% of the elderly subjects were zinc deficient in our study. We also observed decreased taste acuity and a reduced response to the CMI skin antigen panel compared with the responses of younger adults.

We carried out a controlled zinc supplementation trial in the elderly subjects.[5,6] The results are shown in Table 25.1. The subjects first received placebo for 3 months, at the end of which blood samples were drawn and all baseline parameters were repeated. After this, they received zinc 30 mg as zinc gluconate daily for 6 months. The dietary history was checked in this group during baseline, pre-supplementation and post-supplementation periods. No changes in dietary intake of nutrients were observed during the study, except for the

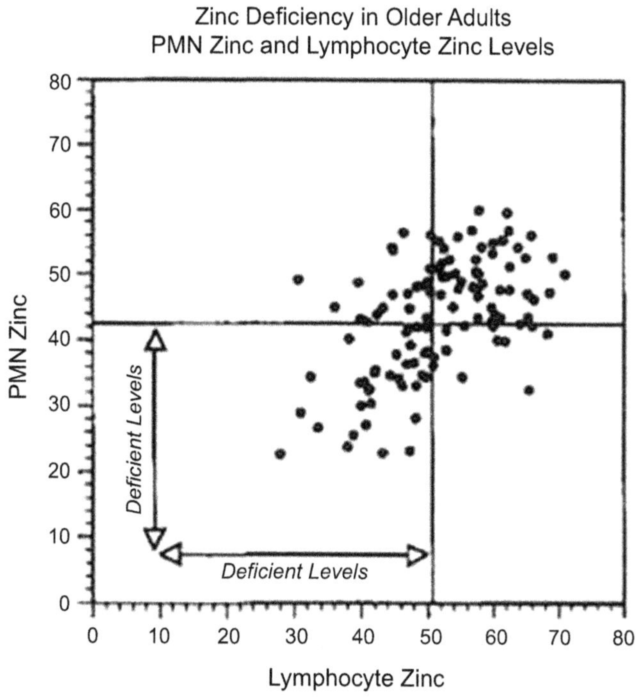

FIGURE 25.2 Zinc deficiency in older adults: PMN zinc and lymphocyte zinc levels. Granulocyte and lymphocyte zinc concentrations in elderly people. Based on decreased levels of zinc in both cell lines, 36 of 118 subjects were zinc deficient. The correlation coefficient between granulocyte and lymphocyte zinc was $r = 0.52$, $p = 0.001$.

TABLE 25.1 Zinc in Granulocytes, Lymphocytes, Platelets, Erythrocytes and Plasma, and Plasma Copper and IL-1 Production by Mononuclear Cells[a]

Measurement	Elderly	Younger Healthy Control	p
Zinc			
Granulocytes (μg/10^{10} cells)	44.20 ± 9.50 (149)	47.54 ± 5.20 (62)	<0.01
Lymphocytes (μg/10^{10} cells)	52.33 ± 9.60 (135)	56.57 ± 6.42 (65)	<0.01
Platelets (μg/10^{10} cells)	2.75 ± 0.56 (161)	2.71 ± 0.48 (69)	NS
Erythrocytes (μg/g Hb)	35.26 ± 6.35 (174)	36.32 ± 4.38 (61)	NS
Plasma (μg/dL)	110.57 ± 13.35 (180)	108.28 ± 12.14 (72)	NS
Plasma copper (μg/dL)	156.18 ± 39.71 (179)	116.09 ± 16.20 (63)	<0.01
IL-1 production (U/mL)	7.96 ± 8.18 (135)	15.55 ± 8.08 (22)	<0.01

[a]*Values are means ± SD with in parentheses.*
Data from Ref. 30.

prescribed supplementation. No significant changes in cellular zinc levels occurred by the end of the 3-month placebo supplementation. We analyzed the data for differences between pre- and post-zinc supplementation periods. We also compared the data of elderly subjects with those of a younger age group (20 to 59 years).

Following zinc supplementation, the zinc concentration in plasma, granulocytes and lymphocytes increased significantly (Table 25.1). Plasma copper significantly decreased after zinc supplementation. A significant increase in plasma thymulin activity and lymphocytes Ecto 5′ NT (nucleotidase) was observed in the elderly subjects following zinc supplementation. Ecto 5′ NT is zinc-dependent and is a biochemical marker of lymphocyte maturity.[32]

Thus, our studies, for the first time, showed conclusively that nearly 30% of the well-to-do ambulatory elderly subjects living in Michigan were zinc deficient, suggesting that zinc deficiency in the elderly may be a common problem even in the well developed countries. It may be even more prevalent in lower-income and/or chronically ill elderly subjects than in the subjects we studied. In view of the important roles of zinc in growth and development, cell-mediated immunity, neurosensory and cognitive functions, it is important to maintain optimal zinc status in the elderly.

Effect of Zinc Supplementation in the Elderly: Cell-Mediated Immunity and Incidence of Infection

Zinc deficiency and susceptibility to infections due to cell-mediated immune dysfunction have been observed in the elderly.[6] Oxidative stress is an important contributing factor for several chronic diseases attributed to aging, such as atherosclerosis and related cardiac disorders, cancer, neurodegenerative and immunologic disorders, and even the aging process.[6] Together, $O_2^{\cdot-}$, H_2O_2 and $\cdot OH$ are known as reactive oxygen species (ROS) and they are continually produced *in vivo* under aerobic conditions. In eukaryotic cells, the mitochondrial respiratory chain, microsomal cytochrome P_{450} enzymes, flavoprotein oxidases and peroxisomal fatty acid metabolism are the most significant intracellular sources of ROS.[6] The nicotinamide adenine dinucleotide phosphate oxidases are a group of plasma membrane-associated enzymes which catalyze the production of $O_2^{\cdot-}$ from oxygen by using NADPH as the electron donor.[6] Zinc is an inhibitor of NADPH oxidases which results in decreased generation of ROS. Zinc is also a co-factor for superoxide dismutase (SOD), an enzyme that catalyzes the dismutation of $O_2^{\cdot-}$ to H_2O_2. Zinc induces the generation of metallothionein, which is very rich in cysteine and is an excellent scavenger of $\cdot OH$.

Although the role of zinc as an antioxidant in cell cultures and animal models has been reported earlier, our studies were the first to show the role of zinc as an antioxidant in the elderly.[5,6] Inflammatory cytokines such as tumor necrosis factor (TNF-α) and interleukin-1β (IL-1β), generated by activated monocytes and macrophages, are also known to generate significant amounts of ROS.[5,6]

Zinc supplementation to healthy human subjects aged 20–50 years decreased the concentration of oxidative stress markers such as malondialdehyde (MDA), 4-hydroxyalkenals (HAE) and 8-hydroxydeoxyguanine in the plasma, inhibited the *ex vivo* induction of TNF-α and IL-1β mRNA in mononuclear cells[33] and provided protection against TNF-α-induced nuclear factor-κB activation in isolated mononuclear cells (MNCs).[33] Zinc increases the expression of A20 and the binding of A20 transactivating factor to DNA, which results in the inhibition of induced nuclear factor κB activation.[33] NF-κB is involved in the expression of genes encoding TNF-α and IL-1β in monocytes and macrophages in humans and HL-60 cells (a human promyelocytic leukemia cell line which differentiates to the monocyte and macrophage phenotype in response to Phorbol Myristate Acetate (PMA). The effect of zinc inhibiting the expression of genes encoding TNF-α and IL-1β is cell-specific.[34]

To understand the mechanism by which zinc may affect cell-mediated immune functions, we utilized the reverse transcriptase (RT)-polymerase chain reaction (PCR) analysis to assess phytohemagglutinin-induced expression of IL-2 mRNA in isolated MNCs obtained from elderly subjects before and after zinc supplementation. Because zinc supplementation in younger subjects decreased the generation of inflammatory cytokines and decreased oxidative stress markers,[33] we hypothesized that zinc supplementation in the elderly would not only increase the generation of IL-2 mRNA in MNCs but also decrease the generation of inflammatory cytokines, such as TNF-α and IL-1β, and decrease oxidative stress markers.

We recruited 50 healthy elderly subjects of both sexes (aged 55–87 years) and from all ethnic groups from a senior citizen center in Detroit, MI, to participate in a randomized, placebo-controlled trial of the efficacy of zinc with respect to the incidence of infections and the effect on *ex vivo* generated inflammatory cytokines and plasma concentration of oxidative stress markers. One participant in the zinc group dropped out on the second day, so we collected data from only 49 subjects.

Exclusion criteria included: life expectancy of <8 months, progressive neoplastic diseases, severe cardiac dysfunction, significant kidney disease and significant liver disease. We excluded those who were self-supplementing with zinc, who were not mentally competent or who could not provide informed consent.

Elderly subjects were randomly assigned in pairs to the zinc-supplemented or the placebo group. The zinc-supplemented group received 45 mg elemental zinc, as the gluconate, daily. The placebo group received an identical placebo capsule.

TABLE 25.2 A Comparison of Selected Variables in Young Adults (18–54 Years Old) vs Older Adult Subjects (>55 years)

Variable[a]	Young Adults	Older Subjects	p Value[c]
Plasma zinc (μg/dL)	101.4 ± 10.0[1] (31)[b]	94.3 ± 11.4 (49)	0.046
Plasma ICAM-1 (ng/mL)	538 ± 112.7 (25)	652.6 ± 169.8 (47)	0.001
Plasma VCAM-1 (ng/mL)	1766 ± 480.4 (25)	2209 ± 890.5 (46)	0.008
Plasma E-Selectin (ng/mL)	32.2 ± 13.1 (19)	84.6 ± 47.6 (69)	<0.001
Plasma NO (μM)	42.7 ± 10.9 (24)	55.6 ± 14.7 (36)	<0.001
Plasma MDA (μM)	0.36 ± 0.10 (16)	0.49 ± 0.15 (34)	<0.001
IL-1β (% cells)	8.5 ± 9.2 (28)	17.4 ± 23.5 (48)	0.023
IL-1β generated (pg/mL)	679.5 ± 110.9 (31)	938.3 ± 423.3 (28)	0.004
TNF-α (% cells)	10.18 ± 10.86 (22)	18.25 ± 20.5 (48)	0.035
TNF-α generated (pg/mL)	1522 ± 390 (26)	1882 ± 722.6 (24)	0.036

[a]*Values represent mean ± SD.*
[b]*Number of subjects.*
[c]*t-test.*
Data from Ref. 5.

A comparison of baseline data between younger subjects and the elderly subjects is shown in Table 25.2. Plasma zinc was lower and the percentages of cells producing IL-1β and TNF-α and the generated levels of these cytokines were significantly higher in the elderly subjects. Intercellular adhesion molecules (ICAM), vascular endothelial cell adhesion molecules (VCAM) and E-selectin in the plasma also were significantly higher in the elderly. IL-10 generated by Th2 cells, which is known to produce a negative effect on IL-2 generation by Th1 cells, was significantly higher in the elderly. The oxidative stress markers were also significantly higher in the elderly than in the younger adults.

Table 25.3 shows the effect of zinc supplementation on clinical variables. The mean incidence of infections for subjects, in 12 months, was significantly lower ($p < 0.01$) in the zinc-supplemented group (0.29 ± 0.46) than in the placebo group (1.4 ± 0.95, effect size 1.46). In the zinc-supplemented group, the total incidence of infections was 7 whereas in the placebo group, the total incidence of infection was 35.

With time (12 months of supplementation), the plasma zinc in the zinc group increased significantly, whereas in the placebo group it tended to be lower (Table 25.4). Also with time (12 months of zinc supplementation), the *ex vivo* generation of TNF-α decreased significantly in the zinc-supplemented group and increased significantly in the placebo group (Table 25.4). The reduction in TNF-α concentration was maximal at the end of 6 months. The *ex vivo* generation of IL-10 decreased non-significantly with time in the zinc group. The percentage of cells positive for TNF-α, IL-1β or IL-10 showed no change with time in either the placebo or zinc-supplemented group (Table 25.5).

The changes in plasma molecular markers of oxidative stress (MDA + HAE and 8-OHdG) between baseline

TABLE 25.3 Effect of Zinc and Placebo Supplementation on Clinical Variables[a]

Variable	Percentage of Subjects Affected in One Year		
	Zinc Group (n = 24)	Placebo Group (n = 25)	Chi Square Fishers Exact Test p
Infection	29	88	<0.001
Upper respiratory tract infection	12	24	0.136
Tonsillitis	0	8	0.255
Common cold	16	40	0.067
Cold sores	0	12	0.124
Flu	0	12	0.124
Fever	0	20	0.027
One infection/year	29%	52%	
Two infections/year	0%	24%	
Three infections/year	0%	8%	
Four infections/year	0%	4%	
Received antibiotics	8%	48%	

[a]*Each person could appear in more than one sub-category of infection.*
Data from Ref. 5.

TABLE 25.4 Effect of Zinc and Placebo Supplementation on Plasma Zinc and Copper Levels

	Baseline	12 Months	p = (Time × Group[a])
Plasma zinc (μg/dL)			
Zinc suppl.[b]	92.9 ± 9.45[c]	104 ± 16.69	
			0.0002
Placebo suppl.[b]	95.7 ± 13.09	88.5 ± 9.66	
Plasma copper (μg/dL)			
Zinc suppl.	182.2 ± 50.5	210.7 ± 60.7	
			0.750
Placebo suppl.	193.4 ± 61.6	215.4 ± 58.7	

[a]*p value for change in groups over time. Multivariate repeated measures analyses were used to examine measures over time.*
[b]*Zinc (n = 24) or placebo (n = 25) supplemented subjects.*
[c]*Values represent mean ± SD. There were no significant differences in plasma zinc or plasma copper between the two groups at baseline.*
Data from Ref. 5.

TABLE 25.5 Effect of Zinc and Placebo Supplementation on the Generation of Cytokines

	Baseline	6 Months	12 Months	p = (Time × Group[a])
A. PERCENTAGE OF CELLS POSITIVE FOR SELECTED CYTOKINE				
TNF-α				
Zinc suppl.[b]	18 ± 18[c]	19 ± 19	18 ± 18	
				0.060
Placebo suppl.[c]	17 ± 21	24 ± 24	38 ± 34	
IL-1β				
Zinc suppl.	20 ± 24	23 ± 24	23 ±26	
				0.240
Placebo suppl.	14 ± 23	10 ± 11	24 ± 28	
IL-10				
Zinc suppl.	10 ± 11	7 ± 7	8 ± 8	
				0.170
Placebo suppl.	6 ± 6	11 ± 15	12 ± 18	
B. CYTOKINES GENERATED EX VIVO *(PG/ML)*[d]				
TNF-α				
Zinc suppl.	1897 ± 1004	1344 ± 544	1411 ± 786	
				0.018
Placebo suppl.	1728 ± 498	1923 ± 782	2698 ± 785	
IL-1β				
Zinc suppl.	892 ± 372	984 ± 373	766 ± 295	
				0.137
Placebo suppl.	878 ± 188	881 ± 206	955 ± 223	
IL-10				
Zinc suppl.	1916 ± 1277	952 ± 785	934 ± 873	
				0.056
Placebo suppl.	917 ± 608	858 ± 483	1018 ± 834	

[e]n =12 and 14 in the zinc and placebo groups, respectively. Means within group are repeated measures. The repeated measures by group (zinc and placebo) for TNF-α showed a significant change (p = 0.012 and 0.042, respectively), which means that, over time, TNF-α changed from baseline in both groups.

[a]*TNF-α, tumor necrosis factor-α; IL, interleukin. No significant differences* (t-test) *were found in the percentage of cells positive for TNF-α, IL-1β or IL-10 or in generated TNF-α and IL-1β between the two groups at baseline. The generated IL-10 concentration was, however, higher in the zinc group than in the placebo group* (p = 0.015) *at baseline, the explanation for which is unclear.*

[b]*p value for change in groups over time (time × group interaction) (multivariate repeated-measures analyses).*

[c]*n = 24 in both the zinc and placebo groups.*

[d]*x ± SD (all such values).*

Data from Ref. 5.

and at the end of 6 months of zinc supplementation showed a more significant change compared to the placebo group (Table 25.6).

TABLE 25.6 Effect of Zinc and Placebo Supplementation on Plasma Oxidative Stress Markers

	Baseline	6 Months	p = (Time × Group[a])
MDA+HAE (μmol/L)			
Zinc suppl.[b]	1.66 ± 0.34[c]	1.35 ± 0.18	
			0.0002
Placebo suppl.[c]	1.70 ± 0.30	1.71 ± 0.35	
8-OHdG (ng/mL)			
Zinc suppl.	0.63 ± 0.16	0.50 ± 0.14	
			0.030
Placebo suppl.	0.66 ± 0.13	0.68 ± 0.13	
Nitric oxide (μmol/L)			
Zinc suppl.	87.34 ± 8.08	79.02 ± 10.96	
			0.180
Placebo suppl.	89.43 ± 11.72	86.74 ± 9.28	

[a]*p value for change in groups over time. Multivariate repeated measures analyses were used to examine measures over time.*

[b]*Zinc (n = 13) or placebo (n = 11) supplemented subjects.*

[c]*Values represent mean ± SD. There were no significant differences* (t-test) *in oxidative stress markers between the two groups at baseline.*

Data from Ref. 5.

In MNCs isolated from zinc-deficient elderly people, zinc supplementation increased the *ex vivo* PHA-induced IL-2 mRNA expression and plasma zinc concentration above the values found in zinc-deficient subjects who were given placebo (p <0.05; Table 25.7). At baseline, the plasma zinc and IL-2 mRNA levels did not differ significantly between the zinc and placebo group (Table 25.7). In the zinc-supplemented group, both of these variables increased during the 6-month period, whereas no change was observed in the placebo group. During the 6-month observation period, two subjects in the zinc-treated group had one episode of bronchitis each. In the placebo group, four subjects had infection – one had laryngitis, one had flu, one had upper respiratory tract infection and one had a common cold.

Our study showed that zinc supplementation to the elderly resulted in a significant decrease in the incidence of infections. The enhancing effect on the immune response of zinc should translate into improved host defense and increased resistance to pathogens in zinc-deficient subjects. Zinc deficiency not only adversely affected the thymulin activity (a thymic hormone) and decreased the generation of IL-2 and IFN-γ from Th1 cells, it also decreased IL-12 production from macrophages.[35] IFN-γ, along with IL-12, is required for optimal phagocytic activity of macrophages against parasites, viruses and bacteria. We previously reported that in marginally zinc-deficient subjects, *ex vivo* generation of IL-1β

TABLE 25.7 Effect of Zinc and Placebo Supplementation on IL-2 mRNA and Plasma Zinc Levels in Zinc-Deficient Elderly Subjects

	Baseline	6 Months	p = (Time × Group[a])
IL-2 mRNA[b]			
Zinc suppl.[c]	0.38 ± 0.07[d]	0.63 ± 0.03	<0.001
Placebo suppl.[d]	0.40 ± 0.05	0.39 ± 0.04	
Plasma Zn[e]			
Zinc suppl.	84.0 ± 3.03	97.6 ± 5.98	<0.0088
Placebo suppl.	86.8 ± 2.04	89.2 ± 3.06	

[a]*p value for change in groups over time. Multivariate repeated measures analyses were used to examine measures over time.*
[b]*Relative expression of IL-2 mRNA/18S.*
[c]*Zinc-supplemented (n = 6) or placebo-supplemented (n = 6) zinc deficient elderly subjects.*
[d]*Values represent mean ± SD. There was no significant difference (t-test) in IL-2 mRNA between the two groups at baseline. In spite of the random assignment of zinc deficient subjects into zinc and placebo groups, the plasma zinc level was lower in the zinc group in comparison to the placebo group (p = 0.016).*
[e]*Plasma zinc levels (µg/dL).*
Data from Ref. 5.

is increased, suggesting that zinc deficiency, *per se*, may activate monocytes and macrophages to generate inflammatory cytokines and increase oxidative stress.[27] In our study, we observed that zinc supplementation improved cytokine production by Th1 cells, decreased the generation of inflammatory cytokines and decreased oxidative stress plasma biomarkers. Figure 25.3 summarizes our concept of the role of zinc in cell-mediated immunity.

Other studies have also shown benefits of zinc supplementation with respect to infectious diseases in humans. Controlled trials of zinc supplementation in infants and children showed a 25–30% reduction in the incidence and duration of acute and chronic diarrhea and a reduction of up to 50% in the incidence of pneumonia.[9,10] AE patients have fewer infections when they are treated with therapeutic levels of zinc supplementation.[7,11] In patients with SCD, IL-2 production by mononuclear cells is decreased, and this is corrected by orally administering 75 mg elemental zinc, as acetate, daily. Also, zinc supplementation decreased the incidence of infections, resulting in fewer days in hospital and fewer pain crises in SCD patients.[36,37] Our studies on SCD patients revealed that 50–75 mg of elemental zinc daily decreased the relative levels of TNF-α and IL-1β mRNAs as well as TNF-α-induced NF-κB activation *ex vivo* in PMNCs and plasma oxidative stress markers, in comparison to the placebo control.[37] Our studies on patients with common cold showed that the duration and severity of the common cold was decreased by 50% when they were treated with zinc acetate lozenges (13–14 mg elemental zinc) every 3–4 hours while awake.[13,14]

Oral administration of zinc decreased the incidence of infection and decreased the severity of radiation-induced oropharyngeal mucositis in patients with head and neck cancer.[38]

The administration of 45 mg elemental zinc daily for up to 1 year in our study did not decrease plasma copper in the elderly subjects. We recommended that if the dose of elemental zinc is higher than 50 mg daily, and if it is administered for more than 12 weeks, one must monitor copper status. Zinc is relatively non-toxic and non-mutagenic, except that, in higher doses, if administered for a prolonged period, copper status must be monitored and 1 mg of elemental copper, as sulfate, must be administered to meet the daily requirement of copper.

Effect of Zinc Supplementation on Oxidative Stress and Inflammatory Cytokines in the Elderly

Oxidative stress and chronic inflammation have been implicated in many chronic diseases in the elderly, including atherosclerosis. Atherosclerosis is a slowly progressive chronic inflammatory condition characterized by focal arterial lesions that ultimately occlude the entire blood vessel and may lead to angina, myocardial infarction, stroke and sudden death. Classic risk factors for atherosclerosis include advanced age, smoking, hypertension, diabetes and obesity. Recently, chronic inflammation has been implicated in the development of atherosclerosis.

Nutritional deficiency of zinc is common not only in developing countries but also in developed countries and it has been reported in many chronic diseases such as rheumatoid arthritis, diabetes and cancers which are associated with chronic inflammation and oxidative stress.[6]

In healthy, elderly, zinc-deficient subjects we showed increased concentrations of plasma lipid peroxidation by-products and endothelial cell adhesion molecules compared with those in younger adults.[5,6] Zinc was proposed to have an atheroprotective function because of its anti-inflammatory and antioxidant properties.[5,6]

In another study we examined the effect of zinc supplementation on (1) concentrations of plasma C-reactive protein (CRP), interleukin-6 (IL-6), macrophage chemoattractant protein-1 (MCP-1), vascular endothelial cell adhesion molecules and oxidative stress markers in elderly subjects, and (2) A20, peroxisome proliferator-activated receptor-α (PPAR-α) and nuclear transcription factor (NF-κB) in human vascular endothelial cell lines and monocytes.

We supplemented 20 elderly subjects aged 65 ± 7 years with 45 mg elemental zinc, as zinc gluconate, daily for 6 months. Twenty elderly subjects aged 65 ± 7 years received placebo. This was a double-blind placebo-controlled trial.[6] Plasma zinc increased significantly in the zinc group but there was no change in plasma zinc in the

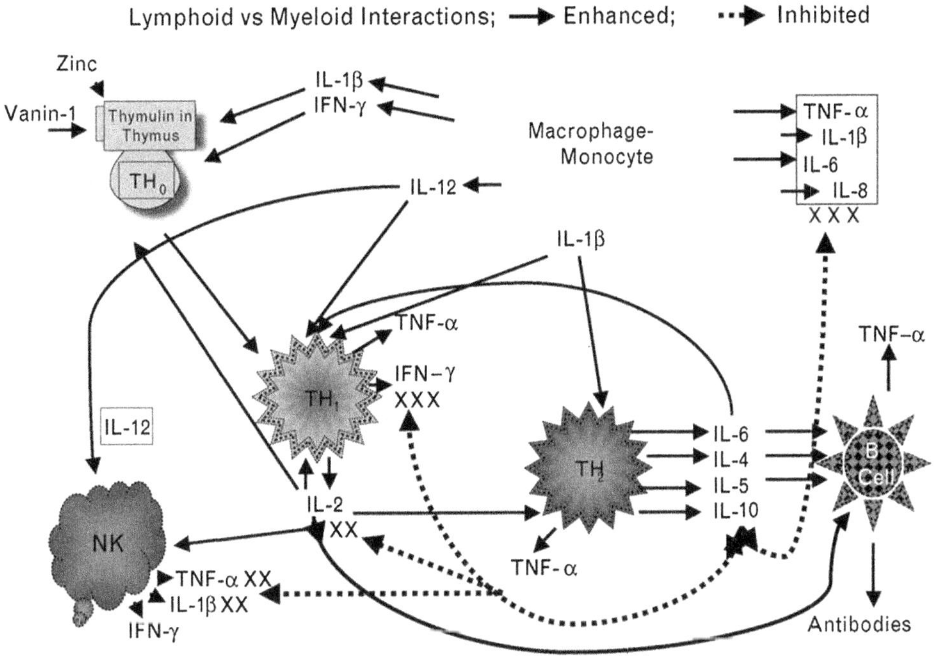

FIGURE 25.3 Landscape of zinc effects on immune cells. Zinc deficiency decreased thymulin activity and decreased Th1 production of IL-2 and INF-γ. Decreased IL-2 leads to decreased NK cell and T cytolytic cell activities. Macrophages-monocytes produce IL-12 (a zinc-dependent cytokine), which, along with INF-γ, kills parasites, viruses and bacteria. Th2 cytokines were not affected by zinc deficiency – except IL-10, which may be increased in zinc-deficient elderly individuals. Increased IL-10 from Th2 cells may affect Th1 functions adversely. Thus, in zinc deficiency there is a shift from Th1 to Th2 functions and cell-mediated immune functions are impaired. Zinc deficiency also leads to stress and activation of macrophages-monocytes, resulting in increased generation of inflammatory cytokines, IL-1β, IL-6, IL-8 and TNF-α. Solid lines indicate pathways leading to generation of selected cytokines, and dotted lines represent pathways which lead to inhibition of cytokine generation. NK represents natural-killer cells; Th1 represents activated Th1 type T cells and secreted cytokines (small triangles); Th2 represents activated Th2 type T cells and secreted cytokines (small circles); B-cell represents B-cell lineages and associated immunoglobulins (triangles).

placebo group. Plasma antioxidant power, represented by ascorbate equivalent units, in μ/mL, increased significantly in the zinc group, and MDA + HAE, plasma oxidative stress markers decreased in the zinc group following zinc supplementation ($p = 0.0011$) (Tables 25.8 and 25.9).

Zinc supplementation resulted in significant decreases in hsCRP, plasma IL-6 concentration, MCP-1, SPLA (secretory phospholipase A), SE-selectin (soluble E-selectin) and sVCAM-1 after 6 months of supplementation compared with the placebo group.[6] Our analysis also showed that the changes in plasma zinc concentrations in the elderly subjects were inversely correlated with the changes in concentrations of plasma hsCRP, MCP-1, MDA+HAE and VCAM-1 after 6 months of zinc supplementation (Tables 25.8 and 25.9).

A high CRP concentration is a risk factor that is independent of total cholesterol, LDL cholesterol, age, smoking, body mass index and hypertension.[6] CRP is widely used as a biomarker of the clinical course and prognosis of atherosclerosis. In our study, we demonstrated that the supplementation of 45 mg elemental zinc/day as the gluconate decreased plasma hsCRP concentration in elderly subjects compared with those who took placebo. This is the first demonstration of down-regulation of CRP by *in vivo* zinc administration in humans.

The increased generation of ROS and the activation of redox-dependent signaling cascades are involved in atherosclerosis.[6] ROS, *per se*, can initiate NF-κB-mediated transcriptional activation of genes involved in inflammation,[6] thereby potentially acting as independent triggers of atherosclerosis. Zinc supplementation decreased oxidative stress in cell culture models, animal models, and humans.[6] Thus, decreased oxidative stress by zinc in the elderly may decrease LDL oxidation and exhibit an atheroprotective effect. In our study we also showed that the concentration of IL-6, sVCAM and SE-selectin increased in the placebo group of the elderly but decreased in the zinc-supplemented group following supplementation. CRP and MDA+HAE concentrations were significantly decreased in the zinc group but not in the placebo group. Antioxidant power was also increased significantly in the zinc group only.

Zinc Trials in Age-Related Macular Degeneration (AMD)

A multi-centered, randomized controlled clinical trial has been conducted by the National Eye Institute, NIH, to assess the long-term effects (10 years) of the formulation

TABLE 25.8 Changes in Plasma Zinc, Oxidative Stress Markers and Inflammatory Cytokines/Molecules in Zinc-Supplemented (zn supp) Elderly Subjects[a]

Group	n	Pre	Post	p^b	Change[c]	Δp^d
Zinc (μM)						
Placebo	20	92.0 ± 3.8[e,f]	90.8 ± 5.0	0.134	−1.17 ± 4.59	<0.0001
Zn supp	20	91.9 ± 7.4[e]	101.5 ± 9.2	0.00006	9.52 ± 8.88	
MDA+HAE (μM)						
Placebo	14	1.66 ± 0.37[e]	1.68 ± 0.35	0.357	0.019 ± 0.186	0.002
Zn supp	14	1.59 ± 0.40[e]	1.29 ± 0.26	0.0011	−0.30 ± 0.30	
Antioxidant power (U/mL)						
Placebo	20	6.6 ± 2.2[e]	6.5 ± 1.7	0.417	−0.11 ± 2.32	0.0258
Zn supp	20	6.0 ± 1.6[e]	7.6 ± 1.9	0.0001	1.56 ± 2.23	
hsCRP (μg/L)						
Placebo	20	2.14 ± 1.71[e]	2.49 ± 1.94	0.149	0.36 ± 1.45	0.0298
Zn supp	20	2.46 ± 1.91[e]	1.90 ± 1.51	0.015	−0.55 ± 1.05	
IL-6 (pg/mL)						
Placebo	20	5.42 ± 3.47[e]	7.15 ± 4.56	0.026	1.74 ± 3.76	0.0031
Zn supp	20	8.34 ± 7.13[e]	5.44 ± 4.85	0.013	−2.94 ± 5.46	
MCP-1 (pg/mL)						
Placebo	20	496.5 ± 154.0[e]	570.4 ± 205.4	0.011	74.1 ± 133.3	0.0113
Zn supp	20	531.5 ± 142.7[e]	506.8 ±131.0	0.136	−24.25 ± 97.2	
sPLA (U/mL)						
Placebo	20	76.0 ± 25.8[e]	100.6 ± 28.8	0.001	24.6 ± 30.9	0.006
Zn supp	20	73.3 ± 34.6[e]	70.0 ± 32.2	0.314	−3.23 ± 29.5	
sVCAM-1 (ng/mL)						
Placebo	20	2102.9 ± 415.1[e]	2297.6 ± 358.2	0.0024	194.7 ± 273.1	<0.0001
Zn supp	20	2208.0 ± 345.6[e]	2035.0 ± 267.8	0.001	−171.5 ± 218.5	
sICAM-1 (ng/mL)						
Placebo	20	321.1 ± 89.1[e]	302.7 ± 105.6	0.307	−18.40 ± 160.4	0.830
Zn supp	20	301.3 ± 68.9[e]	292.4 ± 75.6	0.365	−8.90 ± 113.6	
sE-selectin (ng/mL)						
Placebo	20	54.7 ± 8.3[e]	57.5 ± 7.3	0.023	2.73 ± 5.70	0.0068
Zn supp	20	57.2 ± 9.7[e]	51.6 ± 8.6	0.023	−5.65 ± 11.7	

[a]*Pre, baseline; Post, after 6 months; MDA+HAE, malondialdehyde and hydroxyalkenals; hsCRP, high-sensitivity C-reactive protein; IL-6, interleukin-6; MCP-1, macrophage chemoattractant protein 1; sPLA, secretory phospholipase A2; sVCAM-1, soluble vascular cell adhesion molecule 1; sICAM-1, soluble intercellular adhesion molecule1; sE-selectin, soluble E-selectin.*

[b]*Pre compared with Post values (paired t-test).*

[c]*Post minus Pre values.*

[d]*p for differences (Pre compared with Post) between placebo and zinc groups (t-test).*

[e]*There was no significance of plasma zinc and all biomarkers between placebo and Zn supp groups before supplementation (Pre).*

[f]*Mean ± SD (all such values).*

Data from Ref. 6.

of high-dose antioxidants and zinc supplement on the progression of age-related macular degeneration. Eleven retinal specialty clinics enrolled 4757 participants in AREDS from 1992 to 1998. Participants were 55 to 80 years of age at enrollment and had best-corrected visual acuity of 20/33 or better in at least one eye.[18]

TABLE 25.9 Effect of Zinc on Tumor Necrosis Factor-α (TNF-α), Interleukin-1β, Vascular Cell Adhesion Molecule 1 (VCAM-1) and Malondialdehyde and Hydroxyalkenals (MDA+HAE) in HL-60 and THP-1 Cells and Human Aortic Endothelial Cells (HAECs)[a]

	No Stimulation			oxLDL Stimulation		
	Zn–	Zn+	p^b	Zn-	Zn+	p^b
HL-60 cells						
TNF-α (pg/mL)	26.5 ± 25.4	12.1 ± 12.1	0.21	317.2 ± 119.7	152.7 ± 96.4	0.007
IL-1β (pg/mL)	1.4 ± 1.2	0.8 ± 0.7	0.52	3.9 ± 1.4	1.3 ± 0.5	0.042
VCAM-I (pg/mL)	18.3 ± 4.7	14.2 ± 1.7	0.073	69.9 ± 2.8	32.1 ± 4.3	0.006
MDA+HAE (μM)	2.6 ± 0.7	1.5 ± 0.6	0.001	5.6 ± 1.4	23 ± 0.5	0.046
THP-1 cells						
TNF-α (pg/mL)	32.2 ± 15.1	23.8 ± 12.0	0.022	181.2 ± 13.9	121.0 ± 17.9	0.027
IL-1β (pg/mL)	1.5 ± 0.1	0.9 ± 0.4	0.027	4.4 ± 0.7	1.7 ± 0.9	0.004
MDA+HAE (μM)	1.4 ± 0.6	1.0 ± 0.5	0.013	4.5 ± 0.6	2.0 ± 0.7	0.004
HAECs						
TNF-α (pg/mL)	8.0 ± 6.6	4.2 ± 5.0	0.06	22.6 ± 2.3	13.6 ± 2.1	0.034
IL-β (pg/mL)	5.8 ± 5.7	2.8 ± 2.7	0.11	13.1 ± 4.8	6.2 ± 1.8	0.028
VCAM-1 (ng/mL)	3.5 ± 1.5	4.5 ± 2.1	0.247	23.8 ± 4.7	13.8 ± 2.1	0.016
MDA+HAE (μM)	1.16 ± 0.36	1.02 ± 0.20	0.18	3.33 ± 1.02	1.45 ± 0.77	0.028

[a]*All values are means ± SDs. Zn–, zinc deficient; Zn+, zinc sufficient.*
[b]*For differences between Zn– (1 μM Zn) and Zn+ (15 μM Zn) cell groups (Student's* t-test; *n = 3).*
Data from Ref. 6.

Participants were recruited based upon the severity of AMD and were placed into four AREDS AMD categories according to the size and extent of drusen in each eye, the presence of advanced AMD, and visual acuity, as previously described.[18]

Participants in the clinical trial were randomly assigned to one of four treatment groups: placebo, zinc, antioxidants (vitamins C and E, β-carotene) and vitamins plus zinc. The doses were as follows: vitamin C 500 mg, vitamin E 400 IU and β-carotene 15 mg. Zinc was given as zinc oxide 80 mg along with cupric oxide 2 mg daily to prevent copper deficiency. Median follow-up in the randomized trial was 6.5 years. Following the termination of the clinical trial, participants were invited to continue in a follow-up observational study when the AREDS formulation became available for distribution. Participants with at least intermediate AMD (AREDS category 3) were offered the antioxidant plus zinc formulation, and the effect of treatment was monitored.

Comparison of the participants originally assigned to placebo in AREDS categories 3 and 4 at baseline with those originally assigned to AREDS formulation at 10 years demonstrated a statistically significant ($p < 0.01$) odds reduction in the risk of developing advanced AMD or the development of neovascular (NV) AMD (odds ratios and 99% confidence intervals: OR 0.66, CI: (0.53–0.83) and OR 0.60, CI (0.47–0.78) respectively). No statistically significant reduction was seen for the CGA (central geographic atrophy). A statistically significant reduction for the development of moderate vision loss was seen in the AREDS formulation treated group. No adverse effects were associated with the AREDS formulation.

Most importantly, mortality was reduced in participants who received only zinc, and this decrease was due to a reduction in death due to circulatory diseases, suggesting a beneficial effect of zinc on atherosclerosis.[18]

Zinc and NF-κB Activation

Cell Culture Study

The effect of zinc on inflammatory cytokines and oxidative stress markers in HL-60 (human promyelocytic leukemia cell line), THP-1 (human monocytic leukemic cell line) and HAEC (human aortic endothelial cells) was studied.[6] We observed that zinc significantly decreased the generation of tumor necrosis factor-α (TNF-α), IL-1β, VCAM-1 and MDA+HAE in HL-60 and THP-1 cells and HAECs after incubation with oxLDL for 24 hours compared with zinc-deficient cells ($p < 0.05$). Zinc increased A20 and PPAR-α generation in oxLDL stimulated THP-1 cells and HAECs compared with zinc-deficient cells.

There was no significant difference in NF-κB activation by either NF-κB-driven luciferase reporter gene assay or

EMSA assay between the non-stimulated THP-1 cells or HAECs incubated in zinc-deficient media or zinc-sufficient media. However, after 24 hours of oxLDL stimulation, zinc-sufficient THP-1 cells and HAECs showed a significant decrease in NF-κB activation compared with zinc-deficient cells.[6]

NF-κB is one of the major immune response transcription factors involved in atherosclerosis.[6] Zinc plays an important role in the activation of NF-κB. The regulation of NF-κB activation by zinc is, however, cell specific.[33] It has been reported earlier that zinc is required for NF-κB DNA binding in purified or recombinant NF-κB p50 protein or T helper cell lines. Other studies have shown that zinc decreases LPS-, ROS- or TNF-α-induced NF-κB activation in endothelial cells and cancer cells. We also reported that normal healthy volunteers, who were supplemented with 45 mg zinc/day compared with placebo, had a significant decrease in TNF-α and IL-1β messenger RNAs and TNF-α-induced NF-κB DNA binding in isolated peripheral blood mononuclear cells.[6] Additionally, zinc upregulated the expression of A20 in HL-60 cells. In our study we showed that zinc decreased oxLDL-induced generation of TNF-α, IL-1β and VCAM-1, oxidative stress markers in the plasma and activation of NF-κB, and increased A20 and PPAR-α protein in human monocytic and vascular endothelial cells.[6] It was proposed that zinc inhibited NF-κB activation via A20, a zinc-finger transactivating factor that plays an important role in down-regulating IL-1β and TNF-α-induced NF-κB activation.[39] A20 was originally thought to protect cells from TNF-α-induced cytotoxicity by inhibiting the activation of NF-κB, resulting in decreased IL-1β and TNF-α signaling in endothelial cells.[6,39] It was reported that A20 inhibits NF-κB signaling by TNF-α and IL-1β via TNF receptor associated factor pathways in endothelial cells.[6,39]

The peroxisome proliferator-activated receptor-α (PPAR-α) and PPAR-γ of nuclear receptors – the mediators of lipoprotein metabolism, inflammation and glucose homeostasis – were shown to play an important protective role in the development and progression of atherosclerosis.[6] The mechanism by which zinc may exhibit an atheroprotective role is most likely due to its anti-inflammatory effect. We showed that zinc-sufficient HAEC cells had an increase in PPAR-α concentration compared with zinc-deficient HAEC cells, suggesting that zinc increases the expression of PPAR-α protein which may contribute to down-regulation of inflammatory cytokines and adhesion molecules. We conclude that down-regulation of NF-κB activation by zinc via A20 and PPAR-α signaling pathways results in decreased generation of inflammatory cytokines which protects the endothelial cells from atherosclerosis (see Figs 25.3 and 25.4).

PROPOSED CONCEPT OF MECHANISM OF ZINC ACTION AS AN ANTIOXIDANT AND ANTI-INFLAMMATORY AGENT

Our concept of the mechanism of zinc action as an antioxidant and anti-inflammatory agent is shown in Figure 25.5. Inflammation generates oxidative stress by increasing ROS, resulting in oxidation of LDL. Oxidized LDL activates the NF-κB-inducible kinase/Ik-β kinase/NF-κB signaling pathway and upregulates its downstream target genes, such as those encoding inflammatory cytokines, CRP, adhesion molecules, inducible nitric oxide synthase, cyclo-oxygenase 2, fibrinogen and tissue factor. These cytokines and molecules attract neutrophils, monocytes-macrophages and platelets, induce coagulation and initiate the development of atherosclerosis. Our study showed that zinc supplementation increased plasma antioxidant power, and decreased plasma inflammatory cytokines and oxidative stress biomarkers in the elderly subjects.

Zinc decreased NF-κB activation and its target genes, such as those encoding TNF-α, IL-1β and VCAM, and increased the expression of genes encoding A20 and PPAR-α, the two zinc-finger proteins with anti-inflammatory properties, in HL-60 and THP-1 cells and HAECs, after oxLDL stimulation. Thus, zinc decreased the expression of these cytokines and molecules by inhibition of NF-κB activation via A20 and PPAR-α pathways.

Mechanism of Zinc Action as an Antioxidant

Zinc also functions as an antioxidant by different mechanisms. First, zinc competes with iron (Fe) and copper (Cu) ions for binding to cell membranes and proteins, displacing these redox-active metals which catalyze the production of ·OH from H_2O_2. Second, zinc binds to sulfhydryl (SH) groups of biomolecules, protecting them from oxidation. Third, zinc increases the activation of antioxidant proteins, molecules and enzymes, such as glutathione (GSH), catalase and superoxide dismutase (SOD), and it also reduces the activities of oxidant-promoting enzymes, such as inducible nitric acid synthase (iNOS) and NADPH oxidase, and inhibits the generation of lipid peroxidation products.[40] Fourth, zinc induces the expression of a metal-binding protein, metallothionein (MT), which is very rich in cysteine and is an excellent scavenger of ·OH ions.[41]

Several *in vitro* experiments have revealed that human lung fibroblast cells cultured under zinc-deficient conditions not only increase oxidative stress and DNA damage but also lose the capacity for DNA repair.[42] Adequate zinc levels are, therefore, important for maintaining DNA integrity and preventing DNA damage. Following exposure to TNF-α, zinc-deficient porcine vascular endothelial cells showed an increase in the production

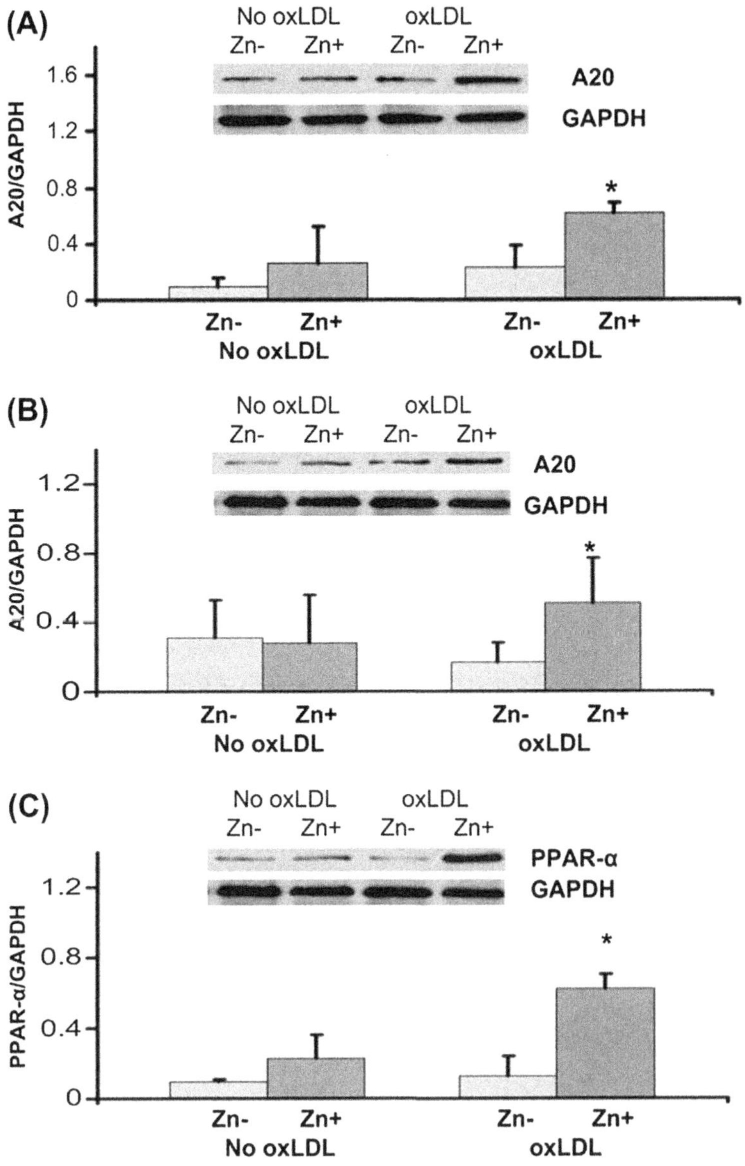

FIGURE 25.4 Effect of zinc on A20 and peroxisome proliferator-activated receptor-α (PPAR-α) in THP-1 cells (A), and human aortic endothelial cells (HAECs) (B and C), after oxidized LDL (oxLDL) stimulation. The cells were incubated either in zinc-deficient (Zn^{2+}, 1 μM) or zinc-sufficient (Zn^{2+}, 15 μM) medium for 8 days (for THP-1) and for 6 days (for HAECs), followed by 24 hours of stimulation with 50 μg oxLDL/mL. A20 and PPAR-α proteins were measured by Western blot analysis. *$p < 0.05$ for Zn− compared with Zn+ (n = 3). GAPDH = glyceraldehyde 3-phosphate dehydrogenase.

of oxidative stress and IL-6 as well as an activation of NF-κB and AP-1 (activator protein 1), compared to zinc-sufficient cells.

Decreased oxidative stress biomarkers have been reported following zinc supplementation in patients with type 2 diabetes mellitus, in hemodialysis patients, and in smokers with colorectal adenoma.[43–45]

Zinc as a Molecular Signal Regulating Oxidative Stress and Chronic Inflammation

NF-κB is a well known major transcription factor regulating the expression of genes encoding various inflammatory cytokines. This transcription factor is activated by many intrinsic and extrinsic stimuli, such as inflammatory cytokines (IL-1β, IL-6 and TNF-α), lipopolysaccharide (LPS), protein kinase C activator (phorbol ester), ROS, ultraviolet light and ionizing radiation, and other cellular stresses.[46,47] NF-κB activation regulates the expression of several genes encoding inflammatory cytokines, chemokines (IL-8, MCP-1), oxidant-promoting enzymes (iNOS) and inducible cyclooxygenases, transforming growth factor 2, adhesion molecules (ICAM-1 and E-selectin), receptors (IL-2 receptor α) and other molecules, thereby controlling several immune responses, the stress response and cell survival and proliferation. TNF-α and IL-1β not only activate NF-κB but are also induced by NF-κB activation, leading to a positive feedback loop of amplification with chronic activation of NF-κB in many cells, including cancer cells. NF-κB can also cooperate with other transcription factors – such as AP-1, SP-1 and c/EBP (NF-κ-bZIP interacting) – to increase the expression of many different genes; this may explain the wide range of genes expressed in a cell-type-dependent pattern.[48]

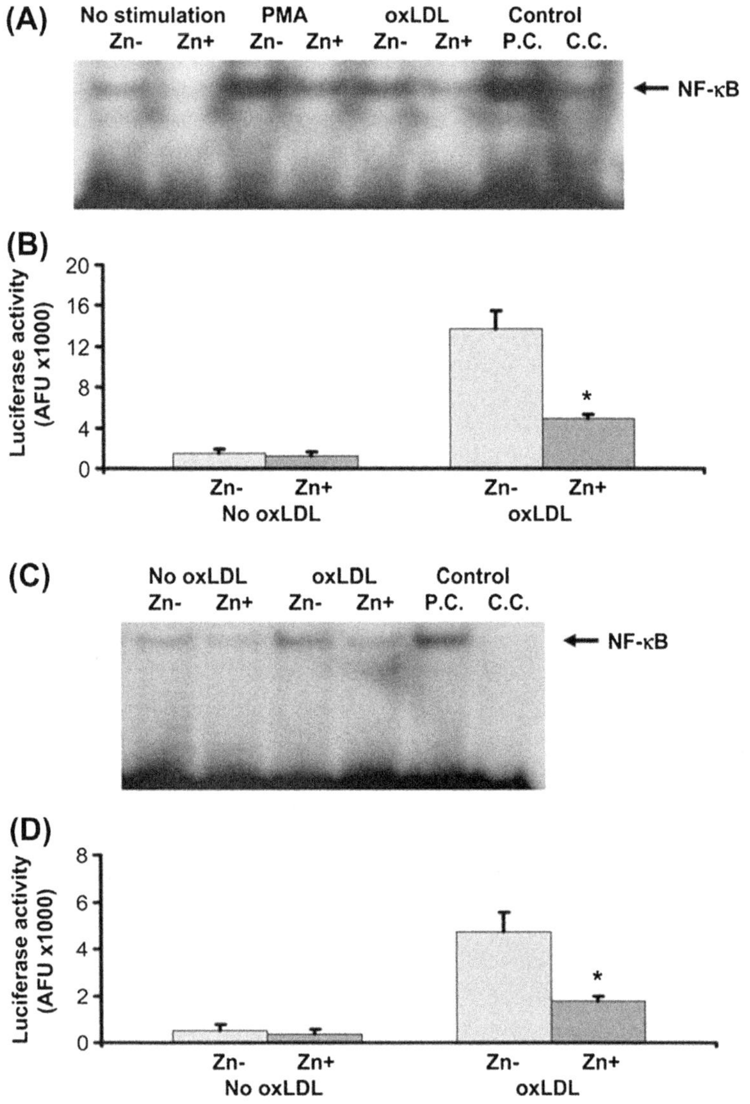

FIGURE 25.5 (A and B) Effect of zinc on nuclear transcription factor κB (NF-κB) activation in THP-1 cells after oxidized LDL (oxLDL) or phorbol myristate acetate (PMA) stimulation. Zinc-deficient (Zn–) THP-1 cells and zinc-sufficient (Zn+) THP-1 cells were used for the measurement of NF-κB activation by electrophoretic mobility shift assay (EMSA; A) and luciferase reporter gene assay (B). (C and D) Effect of zinc on NF-κB activation in human aortic endothelial cells (HAECs) after oxLDL stimulation. Zn– HAECs and Zn+ HAECs were used for the measurement of NF-κB activation by EMSA (C) and luciferase reporter gene assay (D). *$p < 0.05$ for Zn– compared with Zn+ (n = 3). AFU = arbitrary fluorescent unit/β-galactosidase U/100 μg protein; P.C. = positive control; C.C. = competition control.

There are also other inhibitors of NF-κB activation. A20, a zinc-finger protein, inhibits the TNF-α- and IL-1β-induced NF-κB signaling pathway via the TNF-receptor-associated factor (TRAF) in endothelial cells.Thus, A20 is an anti-inflammatory protein which acts by negatively targeting the NF-κB signaling pathway. A20 has also been identified as a tumor suppressor in the development and progression of several B-cell lymphomas, and over-expression of A20 in human salivary adenoid cystic carcinoma cells inhibits tumor cell invasion and cell growth and NF-κB activation.[49] These findings suggest that A20 may function as an anti-tumor molecule in some cells. Our studies have shown that zinc down-regulates the production of inflammatory cytokines via, in part, the A20 signaling pathway, which down-regulates NF-κB activation.[39]

Nuclear factor erythroid 2-related factor 2 (Nrf2), a family member of Cap'n' collar/basic leucine zipper (CNC-bZIP) proteins, is a critical transcription factor that regulates the gene expression of antioxidant proteins and enzymes such as GSH and SOD, as well as detoxifying enzymes such as glutathione S-transferase-1 (GSTA1) and hemeoxygenase-1 (HO-1), by binding to an antioxidant responsive element (ARE) in the promoter region of the target genes.[50–52] These antioxidant molecules and enzymes play an important role in controlling oxidative stress. Several studies have shown that zinc may have a regulatory role in Nrf2. Zinc upregulates Nrf2 activity and inhibits the generation of oxidative stress.[50–52]

SUMMARY POINTS

- Zinc deficiency in humans was first suspected in Iranian villagers in 1961, and the deficiency was confirmed by zinc metabolism studies in Egypt in 1963.

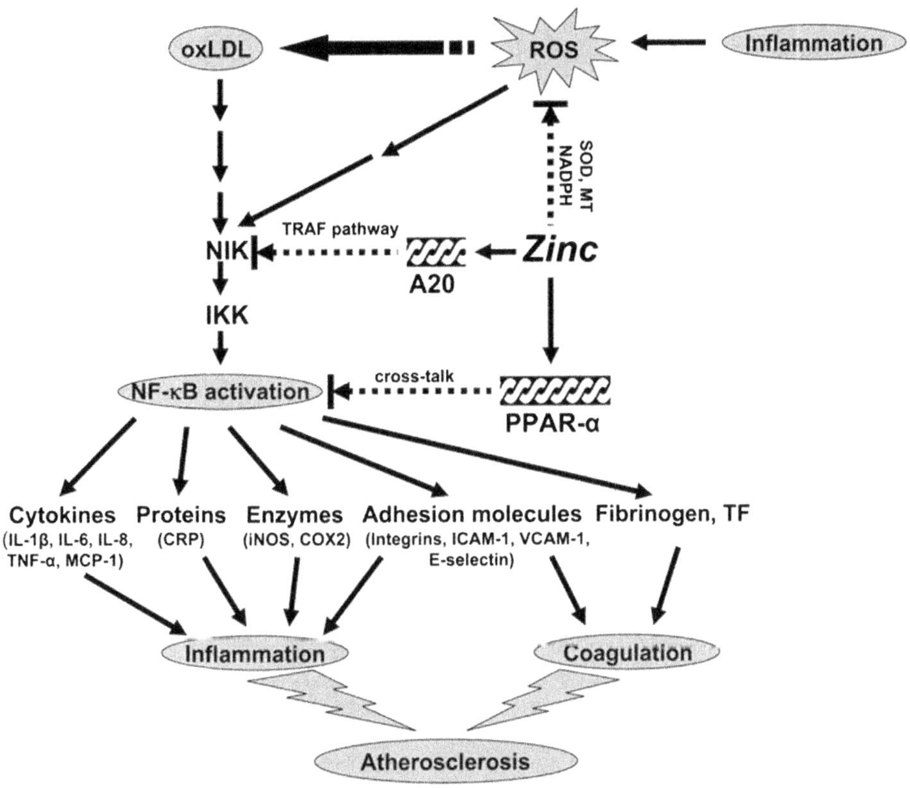

FIGURE 25.6 Zinc reduces atherosclerotic inflammatory markers. Signaling pathway for zinc's prevention of atherosclerosis in monocytes/macrophages and vascular endothelial cells: a proposed hypothesis. Reactive oxygen species (ROS), induced by many stimuli, modify LDL into oxidized LDL (oxLDL) in macrophages and vascular endothelial cells. oxLDL or ROS can activate the nuclear transcription factor κB (NF-κB) pathway via NF-κB-inducible kinase (NIK) activation, which eventually results in inflammation and progression of atherosclerosis. Zinc might have an atheroprotective function by the following mechanisms: (1) inhibition of ROS generation via increased metallothionein (MT), activation of superoxide dismutase (SOD) and inhibition of NADPH oxidase, and (2) down-regulation of atherosclerotic cytokines/molecules, such as inflammatory cytokines, adhesion molecules, inducible nitric oxide synthase (iNOS), cyclooxygenase 2 (COX2), fibrinogen and tissue factor (TF), through inhibition of NF-κB activation by A20-mediating tumor necrosis factor (TNF)-receptor-associated factor (TRAF) signaling and peroxisome proliferator-activated receptor α (PPAR-α) – mediating crosstalk signaling. The black arrows indicate upregulation; arrows with a broken line indicate down-regulation or the inhibitory pathway. IKK = IκB kinase; IL = interleukin; MCP-1= macrophage chemoattractant protein 1; CRP = C-reactive protein; ICAM-1 = intercell adhesion molecule 1; VCAM-1 = vascular cell adhesion molecule 1.

- Severe deficiency of zinc has been recognized in patients with AE, a genetic disorder caused by mutations in the ZIP-4 gene.
- Moderate deficiency of zinc in growing children is characterized by growth retardation, immune dysfunctions and cognitive impairment, as has been reported in cases of nutritional deficiency of zinc in developing countries.
- In an experimental model of mild deficiency of zinc in human volunteers, T cell dysfunction, decreased testosterone level, neurosensory disorders and hyperammonemia were reported.
- Zinc deficiency in well-to-do ambulatory elderly subjects in Detroit, MI, USA, was first reported by us in 1993.
- A zinc supplementation trial in the elderly showed that the incidence of infections was significantly decreased, and the plasma oxidative stress markers and generation of inflammatory cytokines were decreased, in the zinc-supplemented group in comparison to the placebo group.
- In another zinc-supplementation trial in the elderly, we observed a significant decrease in hsCRP, plasma IL-6 concentration, MCP-1, SPLA, SE-selectin and sVCAM-1 after 6 months of zinc supplementation in comparison to the placebo group.
- In the AREDS study, antioxidants (vitamins C and E, β-carotene) plus zinc at 10 years follow-up showed a significant reduction in the risk of developing advanced AMD, and the group receiving zinc alone showed decreased mortality due to cardiovascular events.
- From our cell culture studies we conclude that down-regulation of NF-κB activation by zinc, via A20 and PPAR signaling pathways, results in decreased generation of inflammatory cytokines and oxidative stress which protects the endothelial cells from atherosclerosis.

- Zinc is a molecular signal which regulates oxidative stress and chronic inflammation via NF-κB and A20 signaling pathways.

References

1. Prasad AS, Halsted JA, Nadimi M. Syndrome of iron deficiency anemia, hepatosplenomegaly, hypogonadism, dwarfism, and geophagia. *Am J Med* 1961;**31**:532–46.
2. Prasad AS, Miale A, Farid Z, et al. Zinc metabolism in patients with the syndrome of iron deficiency anemia, hypogonadism and dwarfism. *J Lab Clin Med* 1963;**61**:537–49.
3. Sandstead HH, Prasad AS, Schulert AR, et al. Human zinc deficiency, endocrine manifestations and response to treatment. *Am J Clin Nutr* 1967;**20**:422–42.
4. Prasad AS, Miale A, Farid Z, et al. Biochemical studies on dwarfism, hypogonadism, and anemia. *Arch Intern Med* 1963;**111**: 407–28.
5. Prasad AS, Beck FWJ, Bao B, et al. Zinc supplementation decreases incidence of infections in the elderly: effect of zinc on generation of cytokines and oxidative stress. *Am J Clin* 2007;**85**:837–44.
6. Bao B, Prasad AS, Beck FWJ, et al. Zinc decreases C-reactive protein, lipid peroxidation, and implication of zinc as an atheroprotective agent. *Am J Clin Nutr* 2010;**91**:1634–41.
7. Prasad AS. *Biochemistry of zinc.* New York: Plenum Press; 1993.
8. Sandstead HH, Penland JG, Alcock NW, et al. Effects of repletion with zinc and other micronutrients on neuropsychologic performance and growth of Chinese children. *Am J Clin Nutr* 1998;**68**(Suppl):470S–5S.
9. Sazawal S, Black RE, Bhan MK, et al. Zinc supplementation in young children with acute diarrhea in India. *N Eng J Med* 1995;**333**:839–44.
10. Fisher Walker CL, Lamberti L, Roth D, Black RE. In: Rink L, editor. *Zinc in human health.* Amsterdam: IOS Press; 2011. pp. 234–53.
11. Barnes PM, Moynahan EJ. Zinc deficiency in acrodermatitis enteropathica. *Proc R Soc Med* 1973;**66**:327–9.
12. Brewer GJ, Yuzbasiyan-Gurkan V. Wilson disease. *Medicine* 1992;**71**:139–64.
13. Prasad AS, Fitzgerald JT, Bao B, et al. Duration of symptoms and plasma cytokine levels in patients with the common cold treated with zinc acetate. *Ann Int Med* 2000;**133**:245–52.
14. Prasad AS, Beck FWJ, Bao B, et al. Duration and severity of symptoms and levels of plasma interleukin-1 receptor antagonist, soluble tumor necrosis factor receptor, and adhesion molecule in patients with common cold treated with zinc acetate. *J Infect Dis* 2008;**197**:795–802.
15. Newsome DA, Miceli MV, Tats DJ, et al. Zinc content of human retinal pigment epithelium decreases with age and macular degeneration but superoxide dismutase activity increases. *J Trace Elem Exper Med* 1996;**8**:193–9.
16. Age-Related Eye Disease Study Research group (AREDS Report No. 8). A randomized, placebo controlled, clinical trial of high-dose supplemented with vitamins C and E, beta-carotene, for age-related macular degeneration and vision loss. *Arch Ophthalmol* 2001;**119**:1417–36.
17. AREDS Report No.13. Association of mortality with ocular disorders and an intervention of high dose antioxidants and zinc in the age-related eye disease study. *Arch Ophthalmol* 2004;**122**:716–26.
18. Chew EY, Clemons TE, Elvira-Agron MA, et al. For the Age-related Eye Disease Study (AREDS) group. Long term effects of vitamins C, E, beta-carotene and zinc on age-related macular degeneration. *AREDS report No. 35, Ophthalmol* 2004 (in press).
19. Sandstead HH. Human zinc deficiency: discovery to initial translation. *Adv Nutr* 2013;**4**:76–81.
20. Recommended Dietary Allowance. *National Academy of Sciences Trace Elements, Zinc, 8th revised edition*; 1974. pp. 99–101.
21. Guidelines for essential trace element preparation for parenteral use. (1979). A statement by an expert panel. AMA Department of Foods and Nutrition. Expert panel: Shils, M.E. et al *JAMA* 241(19), 2051–2054.
22. Sandstead HH, Prasad AS, Penland JG, et al. Zinc deficiency in Mexican American children: influence of zinc and other micronutrients on T cells, cytokines, and anti-inflammatory plasma proteins. *Am J Clin Nutr* 2008;**88**:1067–73.
23. Cavdar AO, Babacan E, Arcasoy A, Ertein U. Effect of nutrition on serum zinc concentration during pregnancy in Turkish women. *Am J Clin Nutr* 1980;**33**:542–4.
24. Wang K, Zhou B, Kuo YM, et al. A novel member of a zinc transporter family is defective in acrodermatitis enteropathica. *Am J Hum Genet* 2002;**71**:66–73.
25. Kay RG, Tasman-Jones C. Zinc deficiency and intravenous feeding. *Lancet* 1975;**2**:605–6.
26. Klingberg WG, Prasad AS, Oberleas D. Zinc deficiency following penicillamine therapy. In: Prasad AS, editor. *Trace elements in human health and disease*, **vol. 1**. New York: Academic Press; 1976. pp. 51–65.
27. Prasad AS, Rabbani P, Abbasi A, et al. Experimental zinc deficiency in humans. *Ann Intern Med* 1978;**89**:483–90.
28. Beck FWJ, Prasad AS, Kaplan J, et al. Changes in cytokine production and T cell subpopulations in experimentally induced zinc deficient humans. *Am J Physiol Endocrinol Metab* 1977;**272**:1002–7.
29. Prasad AS, Meftah S, Abdallah J, et al. Serum thymulin in human zinc deficiency. *J Clin Invest* 1988;**82**:1202–10.
30. Prasad AS, Fitzgerald JT, Hess JW, et al. Zinc deficiency in the elderly patients. *Nutrition* 1993;**9**:218–24.
31. Duchateau J, Delepesse G, Vrijens R, Collet H. Beneficial effects of oral zinc supplementation on the immune response of old people. *Am J Med* 1981;**70**:1001–4.
32. Meftah S, Prasad AS, Lee DY. Ecto-5′nucleotidase (5′NT) as a sensitive indicator of human zinc deficiency. *J Lab Clin Med* 1991;**118**:309–16.
33. Prasad AS, Bao B, Beck FWJ, et al. Antioxidant effect of zinc in humans. *Free Rad Biol Med* 2004;**37**:1182–90.
34. Prasad AS, Bao B, Beck FW, Sarkar FH. Zinc activates NF-kappaB in HUT-78 cells. *J Lab Clin Med* 2001;**138**:250–6.
35. Bao B, Prasad AS, Beck FWJ, et al. Intracelluarlr free zinc up-regulates IFN-γ and T-bet essential for Th1 differentiation in con-Astimulated HUT-78 cells. *BBRC* 2011;**407**:703–7.
36. Prasad AS, Beck FWJ, Kaplan J, et al. Effect of zinc supplementation on incidence of infections and hospital admissions in sickle cell disease (SCD). *Am J Hematol* 1999;**61**:194–202.
37. Bao B, Prasad AS, Beck FWJ, et al. Zinc supplementation decreased oxidative stress, incidence of infection and generation of inflammatory cytokines in sickle cell disease patients. *Translational Res* 2008;**152**:67–80.
38. Ertekin MV, Koc M, Karslioglu I, Sezen O. Zinc sulfate in the prevention of radiation-induced oropharyngeal mucositis: a prospective, placebo-controlled, randomized study. *Int J Radiat Oncol Biol Phys* 2004;**58**:167–74.
39. Prasad AS, Bao B, Beck FWJ, Sarkar FH. Zinc-suppressed inflammatory cytokines by induction of A20-medated inhibition of nuclear factor-κB. *Nutrition* 2010;**27**:816–23.
40. Dimitrova AA, Strashimirov DS, Russeva AL, et al. Effect of zinc on the activity of Cu/Zn superoxide dismutase and lipid profile in Wistar rats. *Folia Med (Plovdiv)* 2005;**47**:42–6.
41. Kagi JH, Schaffer A. Biochemistry of metallothionein. *Biochemistry* 1988;**27**:8509–15.
42. Ho E, Ames BN. Low intracellular zinc induces oxidative DNA damage, disrupts p53, NFkappa B, and AP1 DNA binding, and affects DNA repair in a rat glioma cell line. *Proc Natl Acad Sci USA* 2002;**99**:16770–5.
43. Roussel AM, Kerkeni A, Zouari N, et al. Antioxidant effects of zinc supplementation in Tunisians with type 2 diabetes mellitus. *J Am Coll Nutr* 2003;**22**:316–21.

44. Candan F, Gultekin F, Candan F. Effect of vitamin C and zinc on osmotic fragility and lipid peroxidation in zinc-deficient haemodialysis patients. *Cell Biochem Funct* 2002;**20**:95–8.
45. Hopkins MH, Fedirko V, Jones DP, et al. Antioxidant micronutrients and biomarkers of oxidative stress and inflammation in colorectal adenoma patients: results from a randomized, controlled clinical trial. *Cancer Epidemiol Biomarkers Prev* 2010;**19**:850–8.
46. Barnes PJ. Nuclear factor-kappa B. *Int J Biochem Cell Biol* 1997;**29**:867–70.
47. Perkins ND. Achieving transcriptional specificity with NF-kappa B. *Int J Biochem Cell Biol* 1997;**29**:1433–48.
48. Baldwin Jr AS. The NF-kappa B and I kappa B proteins: new discoveries and insights. *Annu Rev Immunol* 1996;**14**:649–83.
49. Malynn BA, Ma A. A20 takes on tumors: tumor suppression by an ubiquitin-editing enzyme. *J Exp Med* 2009;**206**:977–80.
50. Hybertson BM, Gao B, Bose SK, McCord JM. Oxidative stress in health and disease: the therapeutic potential of Nrf2 activation. *Mol Aspects Med* 2011;**32**:234–46.
51. Lee JH, Khor TO, Shu L, et al. Dietary phytochemicals and cancer prevention: Nrf2 Signaling, epigenetics, and cell death mechanisms in blocking cancer initiation and progression. *Pharmacol Ther* 2013;**137**:153–71.
52. Zhou S, Ye W, Zhang M, Liang J. The effects of nrf2 on tumor angiogenesis: a review of the possible mechanisms of action. *Crit Rev Eukaryot Gene Expr* 2012;**22**:149–60.

CHAPTER

26

Antioxidant Supplementation in the Elderly and Leukocytes

David Simar

Inflammation and Infection Research, School of Medical Sciences, Faculty of Medicine, University of New South Wales, Sydney, NSW, Australia

Corinne Caillaud

Exercise Physiology and Nutrition, Faculty of Health Sciences, University of Sydney, Lidcombe, NSW, Australia

List of abbreviations

Akt protein kinase B
CD45 cluster of differentiation 45
DTH delayed hypersensitivity test
ERK1/2 extracellular signal-regulated kinases ½
FLMP receptor *N*-formyl-methionine-leucine-phenylalanine receptor
GM-CSF receptor granulocyte macrophage colony-stimulating factor receptor
GSH glutathione
IFNγ interferon gamma
IKK IkappaB kinase
IL interleukin
JAK/STAT janus kinase/signal transducers and activators of transcription
JNK c-Jun NH_2-terminal kinase
MAPK mitogen-activated protein kinase
NAC *N*-acetylcysteine
NADPH-oxidase nicotinamide adenine dinucleotide phosphate-oxidase
NF-κB nuclear factor κB
NK natural killer
oxLDL oxidized low-density lipoprotein
PBMCs peripheral blood mononuclear cells
PGE_2 prostaglandin E_2
PI3k phosphoinositide 3-kinase
ROS reactive oxygen species
TCR T cell receptor
Th T helper
TLR toll-like receptor
TNF-α tumor necrosis factor alpha

INTRODUCTION

Aging is characterized by a progressive decline in the main biologic functions of the human body. The immune system is similarly affected by this process, as evidenced by the dramatic changes observed in leukocytes from older individuals, culminating in immune dysfunction – also referred to as immune senescence.[1–3] The most common manifestations of these age-induced immune alterations include a progressively impaired response to vaccination, increased susceptibility to infections, and an elevated prevalence of neoplasia and autoimmune diseases in the elderly (Table 26.1).[2,4–9] As aging is associated with a plethora of chronic conditions that may themselves trigger a defect in immune functions, the etiology of those changes leading to immune senescence remains unclear, and the distinction between primary and secondary alterations is complex.[10] Among the different mechanisms potentially involved, oxidative stress – through the accumulation of oxidative damage during the aging process – has received significant attention, notably in regard to the critical role that reactive oxygen species (ROS) play in the regulation of leukocyte signaling, function, and survival.[4,6,11,12]

CHANGES IN THE IMMUNE SYSTEM DURING AGING

Aging is associated with dramatic changes that affect the number and phenotype of leukocytes, signal transduction, and cellular functions as well as metabolic capacity and viability.[1,3] There is general agreement that most of these aspects are negatively affected by aging, altering both the innate and adaptive immune functions. However, some controversies still remain in terms of the specific changes (impairment or improvement)

Aging
http://dx.doi.org/10.1016/B978-0-12-405933-7.00026-3

TABLE 26.1 Characteristics of Immune Senescence[a]

Manifestations of Immune Senescence	Related Mechanisms
Impaired response to vaccination	Oxidative stress
Increased susceptibility to infections	Thymic involution
Increased prevalence of neoplasm	Altered DNA repair
Increased onset of autoimmune diseases	Telomere attrition
Chronic low-grade systemic inflammation	Impaired transcription factors
Impaired immune surveillance	

[a]*Immune senescence is characterized by multiple alterations in the immune system, including impaired immune surveillance and response to vaccination as well as increased prevalence of neoplasms or susceptibility to infections, increased low-grade inflammation, and onset of autoimmune diseases. Although many leading mechanisms have been identified, oxidative stress seems to play a major role in the development of immune senescence.*

TABLE 26.2 Changes in Innate and Adaptive Immunity During Aging[a]

	Function	Effect of Aging
INNATE IMMUNITY		
Neutrophils	Chemotaxis	Decreased or unchanged
	Phagocytosis	Decreased
	Oxidative burst	Decreased
	Response to cytokines	Decreased
	Propensity to apoptosis	Increased
Monocytes/macrophages	Phagocytosis	Decreased
	Cytokine production	Decreased or increased
	TLR4 expression	Decreased
Natural killer cells	Cytotoxicity	Decreased
	Cytokines/chemokines production	Decreased
ADAPTIVE IMMUNITY		
T lymphocytes	Naïve/memory T cells	Increased
	T cell repertoire	Decreased
	Regulatory T cells	Decreased or increased
	Signal transduction	Impaired
	TCR-mediated activation	Impaired
	Cytokine production	
	– IL-2	Decreased
	– IL-4/IL-10	Increased
	– IFNγ	Increased (memory cells)
	Propensity to apoptosis	Increased
B lymphocytes	Naïve/memory B cells	Increased
	B cell repertoire	Decreased
	Signal transduction	Impaired
	Antibody specificity	Decreased
	Affinity maturation	Decreased

[a]*Changes in immunity during aging affect both the innate and the adaptive immune systems. The changes are characterized by cell- and function-specific alterations that contribute to immune senescence. TLR: toll-like receptor, TCR: T cell receptor, IL-2: interleukin 2, IFNγ: interferon gamma.*

that affect leukocytes[1,3] (Table 26.2). In addition, the widely reported state of chronic systemic inflammation observed in older adults, also known as 'inflamm-aging',[4,13] seems to go against the hypothesis that all immune functions are impaired during aging. The increase in low-grade inflammation in older adults could, in fact, suggest an increased production of pro-inflammatory cytokines and provide some support to the existence of cell- and compartment-specific immune alterations during aging.

Innate Immunity During Aging

Polymorphonuclear cells, and in particular neutrophils, are largely involved in the immune response against parasitic or microbial infections. Two critical steps – tightly regulated in this process – are (i) the initiation of the inflammatory response, supporting appropriate clearance of the infectious agents, and (ii) the resolution phase, limiting non-specific tissue damage and preventing the emergence of a state of chronic inflammation.[14] The increased prevalence of infections in the elderly, as well as the progressive development of systemic inflammation in this population, strongly support a functional decline in neutrophils during aging.[4,5,7,8] However, it must be noted that this defect does not seem to be related to a reduction in the absolute number of circulating precursors or mature neutrophils.[15] Investigations focusing on neutrophil recruitment, through adherence to the endothelium and chemotaxis, have reported conflicting results. While some studies have shown a defect, others have observed a maintained recruitment capacity of neutrophils in older but otherwise healthy donors.[15–17] Nonetheless, major impairments in neutrophil phagocytic activity and the production of ROS have been widely reported in the literature, dramatically limiting their antimicrobial activity.[15–17] In addition, neutrophils from healthy older donors seem to show an impaired response to cytokines and an increased propensity to undergo apoptosis, further limiting their capacity to efficiently contribute to the immune response.[15] These functional defects have been linked to impaired signal transduction that develops during the aging process in these cells.[16,17]

The increased prevalence of atherosclerosis, cardiovascular, and neurodegenerative disorders in the elderly is believed to be linked to the impaired function of monocytes and macrophages,[13,14,18] and, as observed in neutrophils, impaired phagocytosis is a common feature of monocytes from older people.[18] Studies on another major component of the monocytic functional machinery, the Toll-like receptor (TLR) signaling, have yielded inconsistent results in terms of age-related changes.[14] Although reduced expression of different TLRs and impaired TLR-mediated signaling have been reported in older adults,[14,18] increased production of pro-inflammatory cytokines by monocytes from old adults, both in unstimulated cells and upon TLR4 stimulation, suggests that some monocytic functions are preserved during aging.[13,18] In addition, the increased frequency of intermediate and non-classic monocytes could provide a potential mechanism for the chronic inflammation observed in this population.[18]

The effect of aging on the function of natural killer (NK) cells has received far less attention when compared to the attention given to neutrophils and monocytes/macrophages, at least in humans. The primary function of NK cells involves cytotoxic activity directed against cancerous or infected cells, together with the production of cytokines. Both aspects have been reported to be impaired during aging, potentially contributing to the increased prevalence of cancer or infections in the elderly.[1,3,14]

Adaptive Immunity During Aging

Among the numerous changes associated with immune senescence, impairment of the adaptive immune system is widely reported, with the earliest studies focusing on the defects specifically affecting T cells.[3,19–21] While the number of circulating T cells is maintained during aging, the progressive involution of the thymus results in a net reduction in the export of naïve T cells.[2] However, this decline is insufficient to explain the reduction in the number of circulating naïve T cells,[22] which is also associated with a decreased proliferative capacity of these lymphocytes. Altogether, this contributes to the decline in the pool of naïve T cells, further leading to a contraction of the T cell repertoire.[1] In addition, the cumulative exposure to foreign pathogens and antigens results in the progressive accumulation of memory T cells, which are characterized, in older donors, by a relative unresponsiveness. Although the reduction in reactive naïve T cells and the increased frequency of memory T cells could both contribute to the impaired cell-mediated immunity, it has been recently suggested that impairment in T cell signaling is the primary mechanism leading to this defect.[23] Indeed, signal transduction in T cells is not only responsible for supporting proliferation and clonal expansion, but it also plays a critical role in memory and effector functions. In line with the altered signaling of T cells during aging, impaired T cell receptor (TCR)-mediated activation of T cells has been reported in older individuals.[16,23] Another major T cell function is the production of cytokines. Although it was initially thought that the overall capacity of T cells to produce cytokines was impaired during aging, cumulative evidence supports a remodeling of the cytokine network rather than a general defect that seems to be T cell-subtype-specific.[24] Finally, impaired signaling in T cells has also be linked to increased apoptosis and could represent a contributing factor to immune senescence.[20]

B cells also show a progressive functional decline and a shift in their profile during aging. One of the most supportive arguments for this age-related dysfunction is the reduced capacity to mount an antibody response after vaccination; this has been widely reported in the elderly.[5] Similarly to what has been observed in T cells, the diversity of the B cell repertoire is dramatically reduced.[1] In addition, the specificity of antibodies progressively shifts from foreign to autologous antigens, contributing to the increased prevalence of autoimmune diseases. Observations focusing on the effect of aging on the number of circulating naïve and memory B cells have been inconclusive, but a progressive decline in clonal expansion capacity has been hypothesized to contribute to the age-related defect observed in the humoral immune response.[1]

As reported in T cells, the effect of aging on the cytokine network still remains to be elucidated. It is nonetheless widely accepted that aging is associated with a progressive systemic low-grade inflammatory process, probably resulting from different mechanisms affecting both the innate and the adaptive immune system.[4]

THE ROLE OF OXIDATIVE STRESS IN IMMUNE SENESCENCE

The free radical theory of aging is now more than 50 years old.[25] This theory links the accumulation of oxidative damage to most age-related changes reported in old animals, organs, tissues, or cells and has gained substantial interest in the past few decades, supported by a significant number of publications.[26] It is based on the observation that oxidative damage progressively accumulates in aging organisms due to the overproduction of ROS, a failure of the antioxidant defence system, or a combination of both. Not surprisingly, leukocytes from older donors are also characterized by increased oxidative damage, and this has been linked to the decline in the immune system or immune senescence.[4,6,11] Thus, higher levels of oxidative stress in peripheral blood lymphocytes,[27,28] decreased levels of glutathione (GSH), the major scavenger of ROS, in both neutrophils and lymphocytes,[29] increased DNA fragmentation in these leukocytes,[30] or increased ROS levels in neutrophils[31] have been reported in older adults. Interestingly, these changes in ROS levels, antioxidant capacity, and the resulting oxidative damage have been associated with impaired function, viability, and intracellular signaling in leukocytes.

In healthy postmenopausal women, both neutrophils and lymphocytes showed increased levels of oxidative damage when compared to those of younger women.[29] This was associated with impaired neutrophil chemotaxis and phagocytosis and increased production of superoxide anion (Fig. 26.1). Lymphocytes were also affected, as evidenced by reduced chemotactic and proliferative capacities, as well as depressed natural killer activity. The reduction in oxidative damage by antioxidant supplementation restored the functions of both neutrophils and lymphocytes, supporting a critical role for oxidative stress in the alteration of leukocyte function during aging.[29] Similarly, increased levels of prostaglandin (PG) E_2 and oxidative damage have been associated with decreased lymphocyte proliferation and impaired cell-mediated immunity in the elderly. Lowering PGE_2 levels by the use of antioxidants restored not only lymphocyte proliferation but also the production of interleukin (IL)-2, supporting an improvement in cell-mediated immunity.[32] In addition, oxidative stress is believed to dramatically influence the cytokine network, as depletion of GSH inhibits the production of Th1 cytokines and promotes the Th2 response.[12] This effect is fully reversible upon restoration of intracellular GSH stores in antigen-presenting cells. This is consistent with the role of

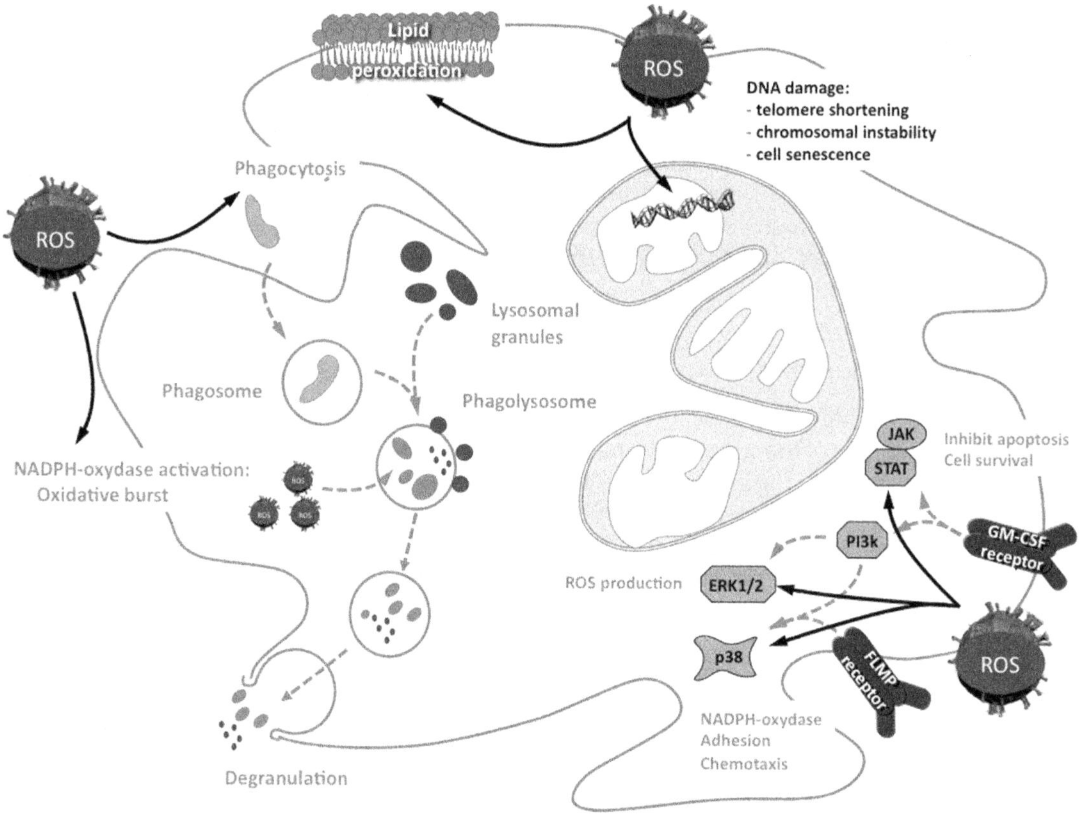

FIGURE 26.1 Role of reactive oxygen species on neutrophil dysfunction. Primary functions of neutrophils include antimicrobial and cytotoxic activities. Neutrophils can ingest antigens by phagocytosis and then degrade them through the generation of ROS by NADPH-oxidase during the oxidative burst, before degranulation. These functions, as well as adhesion, chemotaxis, and survival, are tightly regulated through signal transduction. Overproduction of ROS can lead to impaired signaling, resulting in NAPDH-oxidase dysfunction, an altered oxidative brust, increased adhesion, and impaired chemotaxis. In addition, the accumulation of oxidative damage to the plasma membrane or direct damage to DNA can lead to cellular dysfunction and increased apoptosis. ROS: reactive oxygen species, NADPH: nicotinamide adenine dinucleotide phosphate-oxidase, JAK/STAT: janus kinase/signal transducers and activators of transcription, GM-CSF: granulocyte macrophage colony-stimulating factor, FLMP: *N*-formyl-methionineleucine-phenylalanine, PI3k: phosphoinositide 3-kinase, ERK1/2: extracellular signal-regulated kinases. Dotted gray arrows denote normal function. Plain black arrows denote functions impaired by oxidative stress.

GSH levels in macrophages, which regulate the release of interleukin IL-6, IL-12, and PGs, consequently affecting T helper polarization (the Th1/Th2 ratio).[33]

Although the exact mechanisms linking oxidative damage to leukocyte dysfunction during aging are not fully understood, it is now acknowledged that the preferential oxidation of critical proteins involved in the regulation of leukocyte function is of major importance.[27,28] Oxidative alterations can directly affect proteins involved in the control of leukocyte function or impair intracellular signal transduction by specifically targeting signaling molecules or switching the redox state of the cell. Many critical proteins involved in leukocyte signal transduction are redox sensitive, including protein tyrosine phosphatases (e.g., cluster of differentiation of CD45), mitogen-activated protein kinase (MAPK: p38 and c-Jun NH_2-terminal kinase or JNK), and nuclear factor κB (NF-κB). Although physiologic changes in the intracellular redox state are necessary to modulate their activity, aberrant alterations are associated with impaired intracellular signaling and consequently leukocyte dysfunction (Figs. 26.1 and 26.2).[12] Oxidative inactivation of CD45 is known to be responsible for ineffective signaling in response to TCR stimulation. This defect could be the leading mechanism responsible for the impaired responsiveness of T cells to antigen stimulation during aging (Fig. 26.2).[12] Another major consequence of compromised intracellular signaling is the aberrant production of pro-inflammatory cytokines resulting from the overactivation of NF-κB, a redox-sensitive transcription factor, whose activity is increased in older individuals. In a murine model of aging, inactivation of NF-κB by vitamin E supplementation not only decreased oxidative stress but also reduced the production of pro-inflammatory cytokines or enzymes, highlighting the critical role of oxidative stress in the development of inflamm-aging.[34]

Oxidative stress is also believed to contribute to immune senescence by directly affecting the number of apoptotic leukocytes. The accumulation of oxidative damage in T cells during aging results in an increase in damage-induced cell death.[4] Similarly, cumulative exposure to oxidized low-density lipoproteins (oxLDL) or glycated proteins can result in increased apoptosis in macrophages.[4] As both oxLDL and oxidized protein

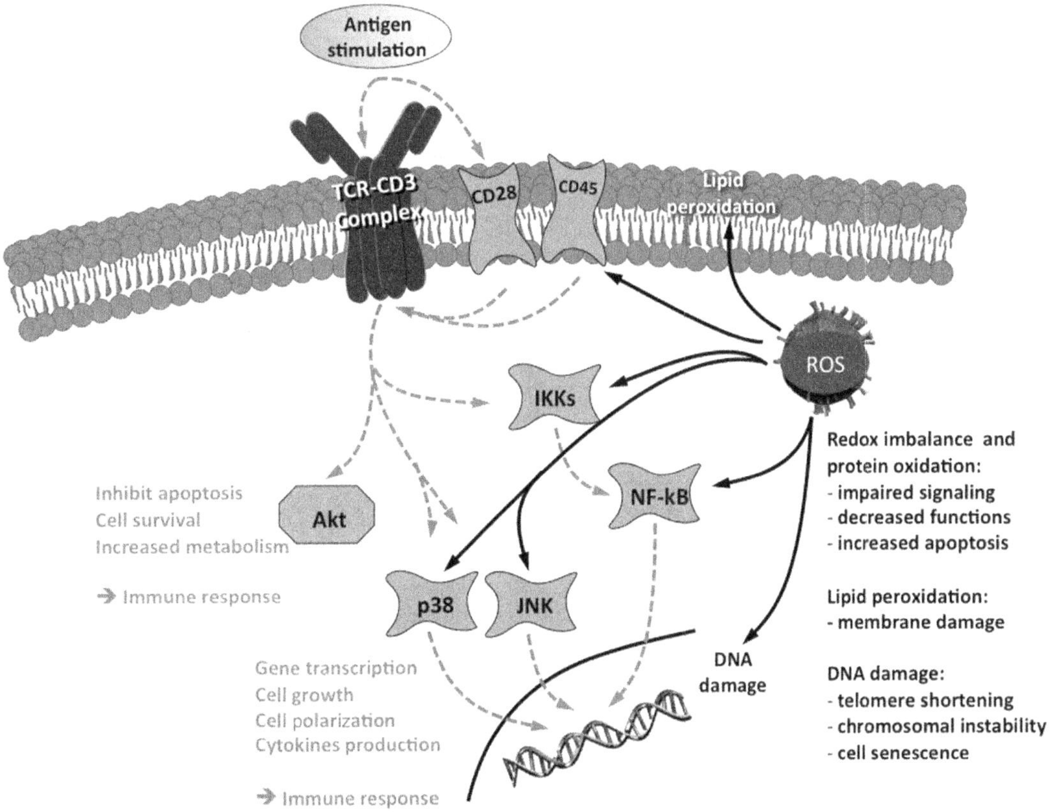

FIGURE 26.2 Role of reactive oxygen species in T cell dysfunction. The immune response in T cells is initiated by co-stimulation of the TCR/CD3 complex and CD28. This leads to the activation of intracellular signaling pathways responsible for gene transcription to support increased metabolism and cell growth, T cell polarization, and effector functions (e.g. cytokine production). Increased levels of ROS can directly inactivate those signaling pathways by modifying the redox state of those key proteins. In addition, the accumulation of oxidative damage in the plasma membrane, as well as DNA damage, can lead to cell dysfunction and apoptosis. ROS: reactive oxygen species, TCR: T cell receptor, CD3: cluster of differentiation 3, Akt: protein kinase B, IKK: IkappaB kinase, JNK: c-Jun NH_2-terminal kinase, NF-κB: nuclear factor κB. Dotted gray arrows denote normal function. Plain black arrows denote functions impaired by oxidative stress.

levels are elevated in plasma from older people, it is hypothesized that this mechanism could play a critical role in the development of immune senescence. Oxidative stress is also known to cause DNA damage, resulting in shortening of telomere length and contributing to apoptosis, chromosomal instability, and cell senescence. In the elderly, an increased level of oxidative damage has been linked to shorter telomere length in leukocytes and could thus contribute to impaired immune function in this population.[35]

EFFECTS OF ANTIOXIDANTS ON THE IMMUNE SYSTEM DURING AGING

Given the critical role oxidative stress plays in the development of immune senescence, the possibility of improving immune function by limiting oxidative damage in leukocytes through an increase in their antioxidant capacity has generated much interest.[36] Although not all interventions using antioxidant supplementations have shown significant improvements in the immune system, there is general agreement that reducing oxidative stress through supplementation using antioxidants can benefit leukocyte functions.

Vitamins

α-tocopherol is a liposoluble antioxidant that represents the most biologically active form of vitamin E. Its main role in the antioxidant system is to act as a scavenger for highly reactive peroxylipids, preventing the propagation of lipid peroxidation in the plasma membrane.[37] Vitamin E supplementation in older adults exerts its beneficial effects on various immune functions through the reduction of oxidative damage. In healthy elderly men and women (70-yr-old), 200 mg of vitamin E daily for 3 months significantly improved the functionality of both neutrophils and lymphocytes.[31] Immediately post-supplementation, the chemotactic and phagocytic capacities of neutrophils were restored to levels similar to those reported in healthy younger adults. Adherence and superoxide anion production were abnormally increased in the healthy elderly before the intervention, but were normalized to young adults' levels after 3 months of vitamin E. The same supplementation also improved lymphocyte proliferative and chemotactic capacities, as well as natural killer function and IL-2 production. Of interest, most of the beneficial adaptations were lost 6 months after the end of the supplementation, highlighting the need for a continuous provision of vitamin E to maintain normal immune function during aging. It should also be noted that this population of healthy and non-institutionalized elderly people consumed a well-balanced Mediterranean diet and were not considered at risk of vitamin E deficiency. A higher concentration (800 mg daily) for 30 days successfully increased α-tocopherol levels in peripheral blood mononuclear cells (PBMCs) from healthy non-institutionalized elderly men and women (70-yr-old).[32] This improvement was associated with increased production of IL-2 by lymphocytes, suggesting enhanced cell-mediated immunity. In a different cohort of healthy and free-living adults (70-yr-old), a significant improvement in T cell-mediated immune response was also reported after 235 days of vitamin E supplementation.[38] Three different concentrations were used in that study (60, 200, and 800 mg per day), and although all resulted in beneficial adaptations, only 200 mg per day led to an improvement in antibody titer to tetanus vaccine. In addition, 200 and 800 mg resulted in a greater increase in serum vitamin E levels. Individuals with serum levels in the upper tertile showed the highest improvement in cell-mediated immunity, highlighting the importance of maintaining an optimal level of vitamin E to preserve cell-mediated immunity.[38] This is consistent with a previous report that failed to show a positive effect of vitamin E when lower concentrations, 50 mg and 100 mg daily, were used[39] or a recent study that showed a positive effect only for a concentration of at least 200 mg daily.[40] This could suggest that doses higher than the recommended daily intake are more beneficial. However, this should be carefully considered, as an excessively high dosage of vitamin E supplementation has been linked to an increase in all causes of mortality.[41] It is noteworthy that in this last intervention,[40] the association of vitamin E supplementation with fish oil limited the beneficial adaptations observed in response to vitamin E alone, potentially due to a lesser increase in serum vitamin E. As previously discussed,[37,42,43] the most likely protective effect of vitamin E might be related to its central role in preventing the propagation of membrane lipid peroxidation or the direct oxidation of proteins involved in signal transduction, thus improving signaling in T and B cells and supporting cell-mediated immunity.[42,43]

Though vitamin C has not received as much attention as vitamin E in the aging population, it is believed to effectively reduce oxidative damage, potentially contributing to the restoration of immune function during aging.[37] In older adults, one of the first reports to show an effect of ascorbic acid on leukocytes provided evidence that incubation of lymphocytes with vitamin C could restore their mitogen-induced proliferative capacity, a function known to be impaired in the elderly.[44] However, in the same study, the ingestion of a high dose of vitamin C (2 g daily) for 3 weeks failed to improve any of the immune markers measured in a small group of 65-yr-old and older persons diagnosed and treated for cardiovascular diseases. The lack of effect in that particular study could have been due to the short duration of the supplementation. Indeed, in a different study, where only 500 mg of ascorbic acid per day was given to healthy elderly over a month, both the *in vitro* proliferation of T cells and the tuberculin skin hypersensitivity *in vivo* were improved.[45] It was previously reported in a study using 1 g of vitamin C daily, in combination with vitamin E, that there was only a marginal additional increase in plasma ascorbic acid compared to results with 500 mg of vitamin C daily.[41] In addition, neither of the two concentrations affected the level of DNA damage measured in leukocytes from healthy elderly (67-yr-old). It seems that although some beneficial effects on immune function have been observed in response to vitamin C supplementation, data from clinical trials is still conflicting, and the optimal modalities to implement a supplementation are yet to be identified.

Due to its provitamin A activity, β-carotene supplementation has been considered as a promising strategy to limit oxidative damage. In addition, it has been well established that vitamin A has a major immunostimulatory role by regulating the macrophage oxidative burst and phagocytic activity, as well as lymphocyte proliferation and polarization.[46] However, only a limited number of studies have investigated its effect on leukocyte function, and the results generated are inconsistent.[47–50] One of the potential explanations involves the fact that β-carotene seems to act in a cell-type-specific manner. Two studies involving supplementations, ranging from 3 weeks to up to 12 years, failed to show any beneficial adaptations in T cell-mediated immunity,[47,49] whereas the same supplementation (50 mg every other day for 10–12 years) significantly improved the activity of NK cells,[50] independently of the production of cytokines by PBMCs.[48] This suggests that although β-carotene supplementation seem to exert some beneficial effects on the immune system during aging, there is a clear need for further work to establish its mode of action and confirm the best modalities of administration.

Supplementations using a single vitamin have generated variable results in terms of their effects on oxidative stress and leukocyte function (Table 26.3). Vitamin E seems to be the most effective supplement to improve leukocyte function, although vitamins C and A also provide some cell-specific adaptations. Some of the data generated suggest the existence of a minimal dose that is required to generate positive adaptations, and individuals with low vitamin levels are more likely to benefit from these supplementations.

Trace Elements and Glutathione

Trace elements, such as zinc and selenium, have a pivotal role in the regulation of the enzymatic antioxidant system through superoxide dismutase and glutathione peroxidase, respectively. They can also have direct immunomodulatory roles (in particular zinc), and depletion in these elements has been associated with increased oxidative stress and impaired immune function.[7,12] Depletion in trace elements has been reported during aging, suggesting that supplementations may be beneficial to restore normal levels. In a cohort of elderly males and females who presented reduced levels of zinc, a supplementation using 45 mg of zinc gluconate daily for 12 months successfully increased zinc plasma levels.[7] It also contributed to lowering systemic levels of oxidative stress as well as the production of pro-inflammatory cytokines and increased IL-2 mRNA levels in mononuclear cells. Those positive adaptations undoubtedly contributed to the reduction of infections that was reported in the supplemented group. Selenium is another trace element that is believed to play a critical role in the maintenance of the antioxidant capacity during aging. Its effect on the immune system in older adults has received limited attention. Nonetheless, when selenium supplementation was carried out in a population of elderly people diagnosed with chronic kidney disease, a condition characterized by high levels of oxidative stress, patients who received the supplementation (200 μg daily for 3 months) not only showed increased selenium levels but also decreased DNA damage in lymphocytes.[51]

Glutathione is the main intracellular non-enzymatic reducing agent, and the balance between its reduced and oxidized forms determines the intracellular redox state in most cells. *N*-acetylcysteine (NAC), as a GSH precursor, has been shown to efficiently increase intracellular GSH levels, providing protection to lymphocytes exposed to high levels of oxidative stress.[52] In postmenopausal women, increased oxidative stress, GSH depletion, and impaired immune function have been reported.[29] NAC supplementation (600 mg daily for 4 months) successfully increased intracellular GSH levels in both lymphocytes and neutrophils and reduced systemic oxidative stress.[29] These improvements were associated with increased chemotactic and phagocytic capacities, as well as decreased adherence and superoxide anion production in neutrophils. In lymphocytes, reduced adherence was also observed, as well as improved chemotaxis, proliferation, and natural killer cell activity. It must be noted that women, whether below or above the age of 70, showed similar improvements, suggesting that these beneficial adaptations could be achieved even late in life. In addition, not only were those positive adaptations reported as early as 2 months into the supplementation, they were also maintained for up to 3 months after the end of the supplementation, suggesting a prolonged positive effect of NAC supplementation on the immune function in the elderly.

TABLE 26.3 Effects of Vitamin or Trace Element Supplementation on Leukocyte Function During Aging[a]

Antioxidant	Duration	Population Characteristics	Leukocyte Function Tested	Effects	Ref
Vit E (200 mg)	3 months	Independently living healthy elderly (70-yr-old)	Neutrophil and lymphocyte adherence Neutrophil chemotaxis, phagocytosis $O_2^{\cdot-}$ production, lymphocyte chemotaxis, proliferation, NK function, IL-2 production	Decreased All improved	31
Vit E (800 mg)	30 days	Independently living healthy elderly (70-yr-old)	DTH, lymphoproliferation, IL-2 production, PGE_2 production	Improved Decreased	32
Vit E (60/200/800 mg)	4 months	Independently living healthy elderly (70-yr-old)	DTH, antibody response to vaccines, autoantibodies to DNA	Improved DTH for all doses, higher Ab titer in 200 (HepB/tetanus) and 800 mg (HepB)	38
Vit E (50/100 mg)	6 months	Independently living healthy elderly (70-yr-old)	DTH, IL-2, IL-4, IFNγ production	No effect	39
Vit C (2 g)	3 weeks	Elderly patients with CVD (>65-yr-old)	DTH, mitogen-stimulated lymphoproliferation	No effect	44
Vit C (500 mg)	1 month	Healthy elderly	DTH, proliferation of T cells	Improved	45
β-carotene (8.2 mg)	12 weeks	Free living healthy elderly (70-yr-old)	Lymphocyte subsets, monocyte adhesion molecules and MHC class II molecules, lymphoproliferation, cytokine production	No effect	47
β-carotene (90/25 mg)	3 weeks/ 12 years	Healthy elderly females (70-yr-old/63-yr-old)	DTH, lymphoproliferation, IL-2/PGE_2 production, lymphocyte subsets/ activation	No effect	49
β-carotene (25 mg)	12 years	Healthy elderly men (57-yr-old and 74-yr-old)	NK cell percentage, NK cell cytotoxicity, IL-2/PGE_2 production	NK cell cytotoxicity improved	50
Zinc (200 mg)	12 months	Independently living healthy elderly (65-yr-old)	% of positive cells for cytokine and cytokine production, Th1/Th2	Decreased TNF-α production and infections, increased IL-2 mRNA	7
Selenium (200 ug)	3 months	Elderly patients with chronic kidney disease (60-yr-old)	DNA damage	Decreased	51
NAC (600 mg)	4 months	Postmenopausal women (70-yr-old)	Neutrophil and lymphocyte adherence Neutrophil chemotaxis, phagocytosis $O_2^{\cdot-}$ production, lymphocyte chemotaxis, proliferation, NK function, IL-2 production	Decreased adherence and all other functions improved	29

All doses are given daily. $O_2^{\cdot-}$: superoxide anion, NK: natural killer, IL-2: interleukin 2, DTH: delayed hypersensitivity test, PGE_2: prostaglandin E2, Ab: antibody, HepB: hepatitis B, IFNγ: interferon gamma, CVD: cardiovascular diseases, Th1: T helper 1, TNF-α: tumor necrosis factor alpha, NAC: *N*-acetylcysteine.
[a]*Supplementations using single vitamin or trace element antioxidants have shown beneficial effects on different components of the immune system.*

It must be acknowledged that evidence for the beneficial effects of additional micronutrients, including phenolic compounds or extracts from fruits showing antioxidant properties, have also been reported.[53] However, the number of studies is still quite limited, and further work is required to definitely establish the role of these compounds on leukocyte function in the elderly.

Combination of Multiple Vitamins and Trace Elements

Due to its strong antioxidant activity, α-tocopherol represents an ideal candidate to be used in supplementations, but it can also act as an oxidant when oxidized. Ascorbic acid is a known reducing agent for oxidized vitamin E, and it can then limit its accumulation in its oxidized form.[37] This observation has led to the hypothesis that supplementations combining multiple antioxidants could provide greater benefits than isolated vitamins or trace elements. This has since been supported by reports that have identified a link between the use of multivitamins and telomere length in leukocytes and in particular a strong association with vitamin C and E intake.[54] Thus, several studies have investigated the effects of supplementations mixing multiple vitamins and/or trace elements in older adults, reporting various beneficial adaptations.[41,55–58]

In independently living, healthy elderly people, combining vitamin C and E, using either a high (1000 mg) or moderate (500 mg) dose of ascorbic acid with 400 IU of α-tocopherol daily for 6 months, leads to an increase in systemic levels of both vitamins, with no difference between the two doses of ascorbic acid.[41] However, both supplementations failed to improve plasma total antioxidant capacity or to reduce systemic oxidative damage or DNA damage in lymphocytes. We have since reported consistent findings in independently living healthy elderly people supplemented with 500 mg of vitamin C and 100 mg of vitamin E daily for 8 weeks.[58] In our cohort, although we observed an increase in systemic levels of both vitamins and reduced levels of oxidative damage, these adaptations did not lead to a reduction in intrinsic or exercise-induced apoptosis in lymphocytes, monocytes, or neutrophils. In a different cohort of institutionalized, older adults, when vitamin A was combined with vitamins C and E, a significant improvement in immune function was observed.[56] In those patients, associating 64 mg of vitamin A with 100 mg of vitamin C and 50 mg of vitamin E daily for 28 days led to an increase in the total number of T cells, the number of T helper cells (CD4 positive), and T cytotoxic cells (CD8 positive) and also the ratio CD4/CD8. In addition, the mitogen-stimulated proliferative capacity of lymphocytes was also improved. It must be noted that this cohort was considered to be malnourished and at high risk of immunosuppression. The same combination, though at lower concentrations (vitamin A 6 mg, vitamin C 120 mg, vitamin E 15 mg daily), given to long-term institutionalized elderly patients for 2 years, resulted in a significant though limited increase in β-carotene, vitamin C, and α-tocopherol.[55] However, this increase in antioxidant vitamins failed to improve any of the immune markers measured in that population. These studies indicate that a multivitamin combination could be beneficial to cell-mediated immunity in older adults who present low levels of those vitamins, while it might have very discrete effects on elderly people meeting the recommended daily intake of β-carotene, vitamin C, and α-tocopherol.

Interestingly, a combination of zinc and selenium sulfate (20 mg and 100 μg, respectively) has been shown to improve vaccination-induced humoral response, and this adaptation was at least in part maintained when the cocktail of trace elements (zinc and selenium) was combined with vitamins A, C, and E.[55] This association of vitamins with trace elements has since been reported to exert positive effects not only on oxidative damage but also on immune functions.[46] Lymphocytes from healthy elderly people have been shown to be more prone to intrinsic apoptosis than those from children.[59] When older adults were supplemented with a cocktail of vitamins and micronutrients (3000 IU retinol, 1.5 mg β-carotene, 200 mg α-tocopherol, 500 mg ascorbic acid, and 400 μg selenium), a significant reduction in both intrinsic and UV-induced apoptosis in lymphocytes was observed. This is in agreement with another study, which showed that a mixture of vitamins and minerals could reduce intrinsic and hydroperoxide-induced DNA damage in lymphocytes from old adults.[60]

Although the ideal combination of vitamins and trace elements is still yet to be determined, the studies presented here strongly support the positive effect of antioxidant supplementations on leukocyte function during aging (Table 26.4). In particular, supplementations combining both vitamins and trace elements seem to generate greater adaptations. In addition, old adults who are at risk or who are experiencing deficiency in those antioxidants are highly likely to benefit from these supplementations.

CONCLUSION

The progressive decline in the immune function during aging, also know as immune senescence, has been linked to the accumulation of oxidative damage in leukocytes, resulting in progressive dysfunction and apoptosis. Therapeutic strategies to preserve or improve the immune system in older adults should aim at reducing oxidative stress. Antioxidant supplementations, including vitamins and/or minerals, result in significant improvements in leukocyte function and viability and should be strongly supported in apparently healthy or clinical populations at risk of immune suppression. Although the optimal combination and dosage of the different antioxidants still needs to be determined, the association of multiple vitamins and minerals seems to provide clear benefits for the immune system, in particular in individuals deficient in some of these elements.

SUMMARY POINTS

- Aging is associated with a continuous decline in the immune system, also called immune senescence.
- Immune senescence is characterized by a progressive loss of functions and a dramatic switch in the phenotype of leukocytes.

TABLE 26.4 Effects of Multiple Vitamins and Trace Elements Supplementations on Leukocyte Functions During Aging[a]

Antioxidants	Duration	Population Characteristics	Leukocyte Function Tested	Effect	Refs
Vit C (1000/500 mg) Vit E (400 IU)	6 months	Independently living healthy elderly (67-yr-old)	DNA damage	No effect	41
Vit C (500 mg) Vit E (100 mg)	8 weeks	Independently living healthy elderly (75-yr-old)	Intrinsic and exercise induced apoptosis in lymphocytes, monocytes and neutrophils	No effect	58
Vit A (64 mg) Vit C (100 mg) Vit E (50 mg)	28 days	Institutionalized due to stroke but not medicated, considered malnourished and deficient in some vitamins (84-yr-old)	Total number of T cells, CD4 and CD8 T cells, CD4/CD8, mitogen-induced proliferation	All improved	56
Vit A (6 mg) Vit C (120 mg) Vit E (15 mg) Zinc (20 mg) Selenium (100 ug) 3 groups: V, T, VT	2 years	Long-term institutionalized patients in nursing homes (84-yr-old)	Delayed-type hypersensitivity skin test response to 7 different antigens Antibody titer against influenza virus Seroprotected patients after vaccination Infectious events	No effect Improved in T Improved in T and VT No effect	55
Retinol (3000 IU) β-carotene (1.5 mg) Vit C (500 mg) Vit E (200 mg) Selenium (400 μg)	2 months	Healthy elderly (65-yr-old)	Intrinsic and UV-induced apoptosis in lymphocytes	Decreased	59
Vit A, B1, B2, B6, B12, C, D3, E, K3 + 12 different minerals	4 weeks	Healthy adults (51-yr-old)	Hydroperoxide-induced DNA damage	Decreased	60

[a]*Combining multiple antioxidants, including vitamins and/or trace elements, can result in improved immune function in the elderly. Benefits vary depending on the type or duration of the supplementation as well as the marker of immune function that is being measured. All doses are given daily. For reference 55, V: supplemented with vitamins only, T: supplemented with trace elements only, VT: supplemented with a combination of vitamins and trace elements.*

- The exact mechanisms responsible for these changes are still unclear, but an impaired redox homeostasis has been linked to altered leukocyte function and increased apoptosis.
- The use of antioxidant supplementation not only limits oxidative stress but also protects leukocytes from oxidative stress-related dysfunction.
- Supplementing even healthy elderly people represents a powerful strategy to prevent the development of leukocyte dysfunction and to improve immune function during aging.
- Supplementations generate better results when combining multiple vitamins and minerals and when specifically targeting individuals with suboptimal diet and deficiencies in vitamins and minerals.

References

1. Desai A, Grolleau-Julius A, Yung R. Leukocyte function in the aging immune system. *J Leukocyte Biol* 2010;**87**:1001–9.
2. Grubeck-Loebenstein B, Wick G. The aging of the immune system. *Adv Immunol* 2002;**80**:243–84.
3. Miller RA. The aging immune system: primer and prospectus. *Science* 1996;**273**:70–4.
4. Cannizzo ES, Clement CC, et al. Oxidative stress, inflamm-aging and immunosenescence. *J Proteom* 2011;**74**:2313–23.
5. Chen WH, Kozlovsky BF, Effros RB, et al. Vaccination in the elderly: an immunological perspective. *Trends Immunol* 2009;**30**:351–9.
6. Ponnappan S, Ponnappan U. Aging and immune function: molecular mechanisms to interventions. *Antioxid Redox Signal* 2011;**14**:1551–85.
7. Prasad AS, Beck FWJ, Bao B, et al. Zinc supplementation decreases incidence of infections in the elderly: effect of zinc on generation of cytokines and oxidative stress. *Am J Clin Nutr* 2007;**85**:837–44.
8. Prelog M. Aging of the immune system: a risk factor for autoimmunity? *Autoimmun Rev* 2006;**5**:136–9.
9. Sansoni P, Vescovini R, Fagnoni F, et al. The immune system in extreme longevity. *Exp Gerontol* 2008;**43**:61–5.
10. Wick G, Grubeck-Loebenstein B. The aging immune system: primary and secondary alterations of immune reactivity in the elderly. *Exp Gerontol* 1997;**32**:401–13.
11. Peters T, Weiss JM, Sindrilaru A, et al. Reactive oxygen intermediate-induced pathomechanisms contribute to immunosenescence, chronic inflammation and autoimmunity. *Mech Ageing Develop* 2009;**130**:564–87.
12. Rider DA, Young SP. A radical view of immunosenescence: does chronic redox depletion interfere with immune cell signalling and function? *Rev Clin Gerontol* 2000;**10**:5–15.
13. Fagiolo U, Cossarizza A, Scala E, et al. Increased cytokine production in mononuclear cells of healthy elderly people. *Eur J Immunol* 1993;**23**:2375–8.
14. Panda A, Arjona A, Sapey E, et al. Human innate immunosenescence: causes and consequences for immunity in old age. *Trends Immunol* 2009;**30**:325–33.
15. Fortin CF, McDonald PP, Lesur O, et al. Aging and neutrophils: there is still much to do. *Rejuven Res* 2008;**11**:873–82.
16. Fülöp T. Signal transduction changes in granulocytes and lymphocytes with ageing. *Immunol Lett* 1994;**40**:259–68.
17. Fulop T, Larbi A, Douziech N, et al. Signal transduction and functional changes in neutrophils with aging. *Aging Cell.* 2004;**3**:217–26.

18. Hearps AC, Martin GE, Angelovich TA, et al. Aging is associated with chronic innate immune activation and dysregulation of monocyte phenotype and function. *Aging Cell* 2012;**11**:867–75.
19. Goronzy JJ, Li G, Yu M, Weyand CM. Signaling pathways in aged T cells – a reflection of T cell differentiation, cell senescence and host environment. *Semin Immunol* 2012;**24**:365–72.
20. Gupta S, Su H, Bi R, et al. Life and death of lymphocytes: a role in immunesenescence. *Immun Aging* 2005;**2**:12.
21. Haynes L, Swain SL. Aged-related shifts in T cell homeostasis lead to intrinsic T cell defects. *Semin Immunol* 2012;**24**:350–5.
22. Murray JM, Kaufmann GR, Hodgkin PD, et al. Naive T cells are maintained by thymic output in early ages but by proliferation without phenotypic change after age twenty. *Immunol Cell Biol* 2003;**81**:487–95.
23. Fulop T, Larbi A, Dupuis G, Pawelec G. Ageing, autoimmunity and arthritis: perturbations of TCR signal transduction pathways with ageing – a biochemical paradigm for the ageing immune system. *Arthritis Res* 2003;**5**:290–302.
24. Alberti S, Cevenini E, Ostan R, et al. Age-dependent modifications of Type 1 and Type 2 cytokines within virgin and memory CD4+ T cells in humans. *Mech Ageing Develop* 2006;**127**:560–6.
25. Harman D. Aging: a theory based on free radical and radiation chemistry. *J Gerontol* 1956;**11**:298–300.
26. Beckman KB, Ames BN. The free radical theory of aging matures. *Physiologic Rev* 1998;**78**:547–81.
27. Poggioli S, Bakala H, Friguet B. Age-related increase of protein glycation in peripheral blood lymphocytes is restricted to preferential target proteins. *Exp Gerontol* 2002;**37**:1207–15.
28. Poggioli S, Mary J, Bakala H, Friguet B. Evidence of preferential protein targets for age-related modifications in peripheral blood lymphocytes. *Ann N Y Acad Sci* 2004;**1019**:211–4.
29. Arranz L, Fernández C, Rodríguez A, et al. The glutathione precursor *N*-acetylcysteine improves immune function in postmenopausal women. *Free Radic Biol Med* 2008;**45**:1252–62.
30. Espino J, Bejarano I, Paredes SD, et al. Melatonin is able to delay endoplasmic reticulum stress-induced apoptosis in leukocytes from elderly humans. *AGE* 2010;**33**:497–507.
31. De la Fuente M, Hernanz A, Guayerbas N, et al. Vitamin E ingestion improves several immune functions in elderly men and women. *Free Radic Res* 2008;**42**:272–80.
32. Meydani SN, Barklund MP, Liu S, et al. Vitamin E supplementation enhances cell-mediated immunity in healthy elderly subjects. *Am J Clin Nutr* 1990;**52**:557–63.
33. Peterson JD, Herzenberg LA, Vasquez K, Waltenbaugh C. Glutathione levels in antigen-presenting cells modulate Th1 versus Th2 response patterns. *Proc Natl Acad Sci USA* 1998;**95**:3071–6.
34. Daynes RA, Enioutina EY, Jones DC. Role of redox imbalance in the molecular mechanisms responsible for immunosenescence. *Antioxid Redox Signal* 2003;**5**:537–48.
35. de Vos-Houben JMJ, Ottenheim NR, Kafatos A, et al. Telomere length, oxidative stress, and antioxidant status in elderly men in Zutphen and Crete. *Mech Ageing Develop* 2012;**133**:373–7.
36. De la Fuente M. Effects of antioxidants on immune system ageing. *Eur J Clin Nutr* 2002;**3**(Suppl. 56):S5–8.
37. Carr AC, Zhu BZ, Frei B. Potential antiatherogenic mechanisms of ascorbate (vitamin C) and alpha-tocopherol (vitamin E). *Circulation Res* 2000;**87**:349–54.
38. Meydani SN, Meydani M, Blumberg JB, et al. Vitamin E supplementation and in vivo immune response in healthy elderly subjects: a randomized controlled trial. *JAMA* 1997;**277**:1380–6.
39. Pallast EG, Schouten EG, de Waart FG, et al. Effect of 50- and 100-mg vitamin E supplements on cellular immune function in noninstitutionalized elderly persons. *Am J Clin Nutr* 1999;**69**:1273–81.
40. Wu D, Han SN, Meydani M, Meydani SN. Effect of concomitant consumption of fish oil and vitamin E on T cell mediated function in the elderly: a randomized double-blind trial. *J Am Coll Nutr* 2006;**25**:300–6.
41. Retana-Ugalde R, Casanueva E, Altamirano-Lozano M, et al. High dosage of ascorbic acid and alpha-tocopherol is not useful for diminishing oxidative stress and DNA damage in healthy elderly adults. *Ann Nutr Metab* 2008;**52**:167–73.
42. Molano A, Meydani SN. Vitamin E, signalosomes and gene expression in T cells. *Mol Aspects Med* 2012;**33**:55–62.
43. Serafini M. Dietary vitamin E and T cell-mediated function in the elderly: effectiveness and mechanism of action. *Int J Develop Neurosci* 2000;**18**:401–10.
44. Delafuente JC, Prendergast JM, Modigh A. Immunologic modulation by vitamin C in the elderly. *Int J Immunopharmacol* 1986;**8**:205–11.
45. Kennes B, Dumont I, Brohee D, et al. Effect of vitamin C supplements on cell-mediated immunity in old people. *Gerontology* 1983;**29**:305–10.
46. Maggini S, Wintergerst ES, Beveridge S, Hornig DH. Selected vitamins and trace elements support immune function by strengthening epithelial barriers and cellular and humoral immune responses. *Br J Nutr* 2007;**98**(Suppl. 1):S29–35.
47. Corridan BM, O'Donoghue M, Hughes DA, Morrissey PA. Low-dose supplementation with lycopene or beta-carotene does not enhance cell-mediated immunity in healthy free-living elderly humans. *Eur J Clin Nutr* 2001;**55**:627–35.
48. Santos MS, Gaziano JM, Leka LS, et al. Beta-carotene-induced enhancement of natural killer cell activity in elderly men: an investigation of the role of cytokines. *Am J Clin Nutr* 1998;**68**:164–70.
49. Santos MS, Leka LS, Ribaya-Mercado JD, et al. Short- and long-term beta-carotene supplementation do not influence T cell-mediated immunity in healthy elderly persons. *Am J Clin Nutr* 1997;**66**:917–24.
50. Santos MS, Meydani SN, Leka L, et al. Natural killer cell activity in elderly men is enhanced by beta-carotene supplementation. *Am J Clin Nutr* 1996;**64**:772–7.
51. Zachara BA, Gromadzinska J, Palus J, et al. The effect of selenium supplementation in the prevention of DNA damage in white blood cells of hemodialyzed patients: a pilot study. *Biol Trace Element Res* 2010;**142**:274–83.
52. Sen CK, Rankinen T, Väisänen S, Rauramaa R. Oxidative stress after human exercise: effect of *N*-acetylcysteine supplementation. *J Appl Physiol* 1994;**76**:2570–7.
53. Ryan-Borchers TA, Park JS, Chew BP, et al. Soy isoflavones modulate immune function in healthy postmenopausal women. *Am J Clin Nutr* 2006;**83**:1118–25.
54. Xu Q, Parks CG, DeRoo LA, et al. Multivitamin use and telomere length in women. *Am J Clin Nutr* 2009;**89**:1857–63.
55. Girodon F, Galan P, Monget AL, et al. Impact of trace elements and vitamin supplementation on immunity and infections in institutionalized elderly patients: a randomized controlled trial. MIN. VIT. AOX. geriatric network. *Arch Intern Med* 1999;**159**:748–54.
56. Penn ND, Purkins L, Kelleher J, et al. The effect of dietary supplementation with vitamins A, C and E on cell-mediated immune function in elderly long-stay patients: a randomized controlled trial. *Age Ageing* 1991;**20**:169–74.
57. Wolvers DA, van Herpen-Broekmans WM, Logman MH, et al. Effect of a mixture of micronutrients, but not of bovine colostrum concentrate, on immune function parameters in healthy volunteers: a randomized placebo-controlled study. *Nutr J* 2006;**5**:28.
58. Simar D, Malatesta D, Mas E, et al. Effect of an 8-weeks aerobic training program in elderly on oxidative stress and HSP72 expression in leukocytes during antioxidant supplementation. *J Nutr Health Aging* 2012;**16**:155–61.
59. Ma AG, Ge S, Zhang M, et al. Antioxidant micronutrients improve intrinsic and UV-induced apoptosis of human lymphocytes particularly in elderly people. *J Nutr* 2011;**15**:912–7.
60. Ribeiro ML, Arçari DP, Squassoni AC, Pedrazzoli Jr J. Effects of multivitamin supplementation on DNA damage in lymphocytes from elderly volunteers. *Mech Ageing Develop* 2007;**128**:577–80.

CHAPTER

27

Metabolic Mobilization Strategies to Enhance the Use of Plant-Based Dietary Antioxidants for the Management of Type 2 Diabetes

Dipayan Sarkar, Kalidas Shetty

Department of Plant Sciences, Loftsgard Hall, NDSU, Fargo, ND, USA

List of Abbreviations

ACE angiotensin-converting enzyme
DPPH 2,2-diphenyl-1-picryl hydrazyl
G6PDH glucose-6-phosphate dehydrogenase
PPP pentose phosphate pathway
ROS reactive oxygen species.

INTRODUCTION

The evolution of higher eukaryotic organisms with mitochondrion-based respiration emerged in the presence of oxygen in the earth's atmosphere. Oxygen is essential as it supports the aerobic life of our planet, but it can be a cellular burden as it can impose oxidative stress and damage cell membranes and cellular organelles through the generation of reactive oxygen species (ROS). ROS, or free radicals, are molecules that have unpaired electrons that react with various other cellular molecules at the site of their formation. The oxidation–reduction cascade from ROS is the foundation of both catabolic and anabolic reactions in higher organisms and as a result can potentially enhance the formation of more ROS in different cellular compartments. In animals, aerobic respiration is the major source of ROS, and any mitochondrial dysfunction can significantly enhance oxidative stress in the cells. Free radicals, such as hydroxyl radical, peroxides, singlet oxygen, and nitrogen monoxide and dioxide, are constantly forming in cells due to cellular oxidation–reduction reactions.[1–4]

ROS are also an integral part of normal cellular functions, and – in both plants and animals – they maintain a complex relationship with different signaling pathways that control growth and the development of cell and tissue systems. They play a significant role in biochemical processes, including cellular differentiation, arresting growth, apoptosis, immunity and defense against microorganisms, and intracellular messaging.[2] But excess ROS production and breakdown of the cellular redox balance – as a result of biotic and abiotic stresses – can lead to apoptosis (cellular death).[3] Aerobic organisms, such as plants and animals, have evolved a very organized and complex antioxidant defense system to minimize oxidative damage. This system comprises various hydrophilic and lipophilic metabolites, enzymes, vitamins, and secondary metabolites such as phenolics (plants only), which scavenge free radicals and reduce the levels of ROS in cells.[3,4]

Oxidative stress plays a significant role in the development of degenerative diseases, including type 2 diabetes, and aging. Worldwide, diabetes reduces life expectancy in the young as well as in the adult population. The role of diabetes as a precursor of aging is complex, and it involves different biochemical and physiologic mechanisms. There is clear evidence that aging is accelerated in diabetics; this is due mainly to the degeneration of different organs as a result of the pathogenic process. Glucose impairment slowly affects different organs and leads to multi-organ failure. Oxidative stress alone, or together with other metabolic diseases, can accelerate the process of aging in diabetics.

THE ROLE OF DIETARY ANTIOXIDANTS AND PLANT PHENOLICS IN THE MANAGEMENT OF TYPE 2 DIABETES

A calorie-restricted diet, with balanced micronutrients and phytochemical-enriched foods, has the potential to mitigate oxidative stress and maintain the

Aging
http://dx.doi.org/10.1016/B978-0-12-405933-7.00027-5

cellular redox balance, resulting in a reduction in the incidence of degenerative diseases. Diet can act as a preventative measure by reducing the cellular ROS load and thereby inhibiting the eventual oxidation of cellular organelles. Dietary antioxidants, particularly those from plants, are the most important components in the early management of type 2 diabetes.[5] Animals, including humans, depend largely on external plant-based antioxidants to counter oxidative stress and its consequences. A healthy diet, with the correct nutritional composition–particularly from plant foods–is important for better regulation of glucose metabolism. Many *in vitro* and *in vivo* studies have shown that dietary antioxidants, taken either as extracts or in food itself, have beneficial effects on glucose metabolism.[5] The effect of some of these dietary antioxidants also mimics certain protective mechanisms induced by low to moderate stress – stimulating the antioxidant defense system by triggering the necessary signals.[6] Different pathways and enzymatic reactions are involved in the generation of cellular ROS, and dietary antioxidants can even reduce ROS formation by inhibiting enzymes responsible for its synthesis[7] (Fig. 27.1). The interaction between dietary antioxidants and beneficial microorganisms can also potentially suppress oxidative reactions in cellular and extracellular compartments. Dietary antioxidants do not act alone; they generally participate in a complex and dynamic system of cellular metabolism.

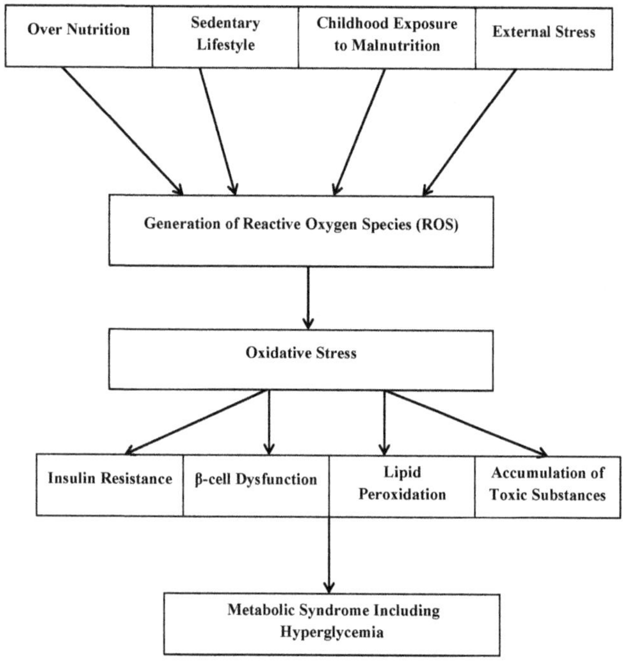

FIGURE 27.1 Pathogenicity of hyperglycemia involving oxidative stress. The figure was created from the overall understanding of oxidative stress-induced pathogenicity in the development of type 2 diabetes.[7]

Phenolic compounds or phenolic phytochemicals are secondary metabolites synthesized by plants to counter biotic or pathogenic attack or diverse biotic or environmental stresses. These diverse compounds contain at least one aromatic ring and usually one or more substituted hydroxyl groups; they have evolved in different plant lineages to address specific needs – mostly as defense and signal compounds.[7] Phenolic compounds help in the overall adaptive strategy for natural selection in plants in diverse niche environments; plants thus accumulate a vast number of these compounds. Owing to their free radical scavenging properties, phenolics can work as powerful antioxidants and have potential in therapeutic measures against oxidative-stress-related diseases. The structural chemistry of polyphenols suggests different antioxidative properties, such as (i) high reactivity as a hydrogen or electron donor, (ii) the ability to stabilize or delocalize an unpaired electron, and (iii) the ability to chelate transition metal ions[5] (Fig. 27.2).

Polyphenols are present in virtually all plant foods, but their levels vary significantly among diets, depending on the type, quantity, and source of the plant-based food. Plant phenolics are a good source of natural antioxidants, and they can be used for dietary management of non-communicable chronic diseases, including type 2 diabetes.[7] Phenolic phytochemicals can scavenge harmful free radicals as well as inhibit oxidative reactions, and they can also protect cells from oxidative damage by stimulating the response of antioxidant enzymes. This diverse group of phytochemicals has a potential preventative effect against specific diseases, particularly in the early stages of disease development when oxidative stress

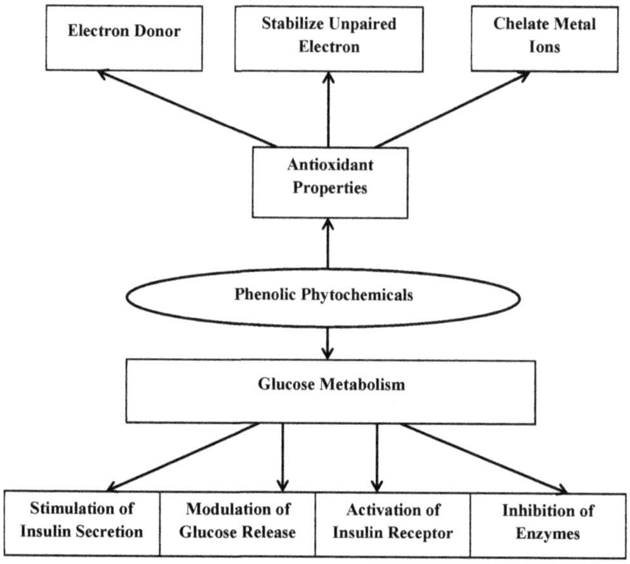

FIGURE 27.2 The role of phenolic phytochemicals in managing oxidative stress and influencing glucose metabolism. This figure was created from the scientific understanding of the role of polyphenols in glucose metabolism described by Hanhineva et al.[5]

is, in part, involved in initiation and progression. Polyphenols have different beneficial properties, including anti-inflammatory, antioxidative, chemopreventive, and neuroprotective activities, and thus they are associated with a lowered risk for major chronic diseases, such as diabetes, cardiovascular diseases, and cancer.[5,7] Dietary polyphenols from different plant-based sources influence glucose metabolism in several ways – for example, inhibition of carbohydrate digestive breakdown and glucose absorption in the intestine, stimulation of insulin secretion from the pancreatic β-cells, modulation of glucose release from the liver, activation of insulin receptors and glucose uptake in the insulin-sensitive tissues, and modulation of hepatic glucose output[5] (Fig. 27.2).

The maintenance of glucose homeostasis is very important for diabetics, and a healthy diet with proper nutritional composition is a key factor in the regulation of glucose metabolism and in mitigating the aging process.[5] The total antioxidant activity of plant-based foods positively correlates with its total soluble phenolic content; an understanding of this relationship can be used to achieve better management of type 2 diabetes-linked oxidative stress (Table 27.1). Carbohydrate digestion (starch breakdown) and glucose absorption are obvious targets for better glycemic control after high-carbohydrate meals; α-amylase and α-glucosidases are key enzymes responsible for the digestion of dietary starch and uptake of glucose.[5] Dietary soluble starches are generally hydrolyzed by pancreatic α-amylase, with absorption via the small intestine aided by α-glucosidase.[8] Important therapeutic approaches available for managing early stages of type 2 diabetes involve controlling the absorption of glucose through the reduction of starch hydrolysis by inhibiting pancreatic α-amylase and limiting the absorption of glucose by inhibiting intestinal α-glucosidase. Many *in vitro* studies report that polyphenols, including flavonoids (anthocyanins, catechins, flavanones, flavonols, flavones, and isoflavones), phenolic acids, and tanins (proanthocyanidinins and ellagitannins), inhibit α-amylase and α-glucosidase.[5]

Phenolics from different plant-based food sources show a positive correlation with total antioxidant activity. Most of the dietary polyphenols are metabolized by the colonic microbiota (microflora) before absorption. Hydrolysis, ring-cleavage, reduction, decarboxylation, and demethylation are different mechanisms through which gut bacteria regulate phenolic mobilization and absorption.[5] The biologic activity of polyphenols depends largely on synergistic action, and it is affected by other constituents present in the diet as well as endogenous factors. Except for α-amylase inhibitory activity in vegetables such as yellow pepper, eggplant, and colored Chilean potatoes, phenolics from different plant-based foods such as berries (strawberries, raspberries, blueberries, blackberries, and black currants), fruits from the *Rosaceae* family (apple, pear, and cherry), vegetables (pumpkin, beans, and sweet potato), and small grains showed significant α-amylase and α-glucosidase inhibition in *in vitro* studies (Table 27.2). Polyphenols also influence glucose transporters and thus mediate intestinal absorption of glucose. Studies with animal models showed that flavonoids – such as chlorogenic,

TABLE 27.1 The Total Soluble Phenolic Content and Total Antioxidant Activity of Selected Fruits and Vegetables[a]

Selected Fruits and Vegetables	Total Phenolic Content	Total Antioxidant Activity (DPPH)
Yellow pepper[19]	200 μg/ml	80%
Eggplant[20]	20–80 μg/ml	10–45%
Colored Chilean potato[21]	6–14 mg GAE/g DW	30–70%
Mung bean (solid-state bioconversion)[39]	20–35 g/100 g DW	70–80%
Grain sprouts[38]	1.5–3 mg/g FW	42–75%
Strawberry[26]	0.3–0.7 mg/g FW	42–90%
Raspberry[27]	0.25–0.9 mg/g FW	50–90%
Cherry (fermented)[25]	0.65 mg GAE/ml	80–90%
Apple (pulp, peel)[31]	37–61, 266–556 μg GAE/g	9–23%, 42–73%
Blackcurrant[29]	13 mg/g FW	50%
Redcurrant[29]	7 mg/g FW	45%
Red gooseberry[29]	4 mg/g FW	50%

GAE: gallic acid equivalent; FW: fresh weight; DW: dry weight; DPPH: 2,2 diphenyl-picryl hydrazyl.
[a]*A reference for each source of data is included in the table.*

TABLE 27.2 Inhibitory Activity (%) of Selected Fruits and Vegetables on α-Amylase and on α-Glucosidase[a]

Selected Fruits and Vegetables	Inhibitory Activity (%) on α-Amylase	Inhibitory Activity (%) on α-Glucosidase
Yellow pepper[19]	0%	30%
Eggplant[20]	0%	40–50%
Colored Chilean potato[21]	0%	10–50%
Grain sprouts[38]	20–55%	15–43%
Strawberry[26]	15–50%	70–90%
Raspberry[27]	30-80%	80–99%
Cherry (fermented)[25]	0%	90–98%
Apple (pulp-peel)[31]	60–80%	40–80%
Blackcurrant[29]	16%	60%
Redcurrant[29]	85%	90–99%
Red gooseberry[29]	30%	99%

[a]*A reference for each source of data is included in the table.*

ferulic, caffeic, and tannic acids, quercetin monoglucosides, tea catechins, and naringenin – could inhibit Na^+-dependent SGLT1-mediated glucose transport and alter the postprandial blood glucose response.[5,9,10] In rats, administration of glucose with quercetin showed a significant reduction in hyperglycemia.[9] Hanamura et al.[10] observed a reduction in the plasma glucose level in mice after administration of maltose with crude *Acerola* polyphenol fraction, suggesting inhibition of α-glucosidase activity and intestinal glucose transport. Berries, a rich source of anthocyanins, significantly decreased the peak glucose increment by reducing the rate of sucrose digestion or absorption from the gastrointestinal tract. Soybean isoflavonoids have shown a positive impact on β-cell function in some recent studies.[11] Choi et al.[12] found that isoflavonoids, such as genistein and daidzein, preserve insulin production by β-cells in mice. Flavonoids, such as quercetin, luteolin, and apigenin, also showed a protective function against β-cells.[5] Dietary polyphenols may also influence glucose metabolism by stimulating peripheral glucose uptake in both insulin-sensitive and insulin-insensitive tissues. Regulation of the expression of genes involved in glucose uptake and insulin signaling pathways was altered by green tea polyphenolic extract in the muscle tissue of metabolic syndrome-induced rats.[13]

In addition to positive modulation of glucose metabolism, flavonoid intake is also able to lower the risk of coronary disease and cancer.[14] A 50% reduction in the coronary heart disease mortality rate was found to be associated with a high intake of flavonoids (approximately 30 mg/day).[14] Several mechanisms of action of flavonoids – including inhibition of LDL oxidation, inhibition of platelet aggregation and adhesion, inhibition of cholesterol esterification, and intestinal lipoprotein secretion – are involved in the reduction in coronary heart disease.[14] Polyphenols with high antioxidant activity can regulate cellular functions through the inhibition of pro-oxidant enzymes, induction of antioxidant enzymes, and inhibition of the redox-sensitive transcription factors.[7] Other epidemiologic studies also suggest that polyphenols can have a therapeutic role, with health-protective benefits, by acting as modifiers of many physiologic functions in the human body.[5,14-18]

INHIBITORY ACTIVITIES OF DIFFERENT PLANT-BASED FOODS ON α-AMYLASE AND α-GLUCOSIDASE

Most of the drugs available for combating type 2 diabetes have negative side effects at high doses. The development of natural and safe antihyperglycemic agents from plant-based food sources is a major focus of current antidiabetic research. Fruits, vegetables, and herbs are rich sources of dietary antioxidants and can have a significant impact on glucose metabolism through the inhibition of enzymes such as α-amylase and α-glucosidase. A high consumption of fruits and vegetables can lower the risk of type 2 diabetes significantly.[15] The consumption of fruits and vegetables is not only a preventive measure: it can also offer better management of blood glucose levels after the disease has started. One study showed that consumption of five portions a day of fruits and vegetables reduced the risk of type 2 diabetes by 40% in women.[15] The risk of diabetes was diminished by 30% after a higher intake of fruits and berries in Finnish men and women.[16] One report described an inverse relationship between the intake of vegetables and the incidence of type 2 diabetes in Chinese women.[17] Green leafy vegetables contain high concentrations of natural antioxidants – such as β-carotene, vitamin C, and polyphenols – and are a good source of magnesium and α-linolenic acid. One study reported that a high intake of green leafy, or dark yellow, vegetables was associated with a reduced risk of type 2 diabetes among overweight women.[18]

Other than epidemiologic or cohort studies, many *in vitro* studies have shown that a given food can have a significant impact on carbohydrate-breaking enzymes. Vegetables from the *Solanaceae* family – including pepper, eggplant, tomato, and colored potato – are a rich source of phenolic phytochemicals with high antioxidant activity. Kwon et al.[19] found high α-glucosidase inhibitory activity and low α-amylase inhibitory activity in several colored peppers (green, red, orange, yellow, Cubanelle, red sweet, yellow sweet, long hot, and jalapeño). They observed that yellow, Cubanelle, and red peppers also had higher ACE-inhibitory activity, which signified relevance for hypertension management. Eggplant is also recommended by the National Institutes of Health, the Mayo clinic, and the American Diabetes Association as an important dietary source for the management of type 2 diabetes. Phenolic-enriched extracts of eggplant with moderate antioxidant activity showed high inhibitory activity against α-glucosidase and moderate to high ACE-inhibitory activity.[20] Saleem et al.[21] found antihyperglycemic potential (moderate inhibition of α-glucosidase) and antihypertensive potential (moderate to high ACE-inhibitory activity) in selected lines of subtropical cultivars of Chilean potatoes. These findings could help to develop vegetable-based whole foods for the dietary management of type 2 diabetes and hypertension.

Like vegetables, grains, and legumes, fruits are also a rich source of vitamins, minerals, dietary fiber, and polyphenols. These bioactive compounds, present in fruits, include flavonoids, isoflavonoids, and phenolic acids; they are well known as free-radical scavengers and are associated with a reduced risk of degenerative diseases, including type 2 diabetes. The antioxidant potential of a fruit-based diet depends largely on an interaction between the various phytochemicals present in different fruits. A combination of different fruits such as orange, apple, grape, and blueberry resulted in enhanced synergistic antioxidant activity compared to each fruit alone.[18] A study on rodents reported increased stimulation of insulin secretion in pancreatic β-cells after administration of different anthocyanin compounds.[22] Cherries are a rich source of bioactive substances, including anthocyanins, and thus have potential for reducing oxidation-linked chronic diseases, including cardiovascular disease, cancer, and hyperlipidemia.[23] Seymour et al.[24] observed that the administration of a cherry-enriched diet for 90 days significantly reduced fasting blood glucose, hyperlipidemia, hyperinsulinemia, and fatty liver in rats. High phenolic content, high total antioxidant activity, and high α-glucosidase-inhibitory activity were observed in different sweet and tart cherry cultivars. Under natural acidic conditions, increased DPPH-linked antioxidant activity and α-glucosidase-inhibitory activity was observed in Northstar sweet cherry after fermentation.[25] Owing to its enriched phenolic profile, tart and sweet cherry could potentially reduce several phenotypic risk factors that are associated with metabolic syndrome and associated type 2 diabetes.[24, 25]

Berries are a rich source of polyphenols – such as anthocyanins, flavonols, phenolic acids, ellagitannins, and procanthocyanidines; they show disease-preventive properties and can be used against diabetes-linked cardiovascular complications. The phenolic content and phenolic profile vary widely among different berries and even within different cultivars of the same species. A polyphenol-rich extract of blueberries, blackcurrants, strawberries, and raspberries has been shown to inhibit α-glucosidase and α-amylase activity in an *in vitro* assay. Cheplick et al.[26] found significant differences in both phenolic-linked antioxidant activity and enzyme-inhibitory activities against enzymes relevant in type 2 diabetes in several strawberry cultivars. In another study, they found high inhibitory activity against α-glucosidase in a yellow raspberry cultivar, suggesting that α-glucosidase may be affected more by specific anthocyanins rather than by the total phenolic content in the extract.[27] Pinto et al.[28] studied the effects on the *in vitro* inhibition of α-amylase and α-glucosidase of different cultivars of Brazilian strawberry; they found that strawberries had a high inhibitory activity against α-glucosidase and a low inhibitory activity against α-amylase. Ellagic acid, quercetin, and chlorogenic acid are major phenolics found in an aqueous extract of strawberries.[28] These findings suggest that strawberries are a good source of phenolics for the potential management of hyperglycemia linked to type 2 diabetes. Pinto et al.[29] also observed that redcurrants have a high inhibitory activity against α-glucosidase, α-amylase, and ACE.

Apple is one of the main sources of flavonoids in the western diet, and it provides approximately 22% of the total phenols consumed per capita in the USA. An increased intake of apple has been correlated with a decreased risk of cardiovascular disease and diabetes. Adyanthaya et al.[30] observed a positive correlation between phenolic content and a high inhibitory activity against α-glucosidase in different apple cultivars; a high phenolic content also enhanced post-harvest preservation. Phenolic-enriched apples may modulate postprandial glucose levels and thus reduce the risk of development of type 2 diabetes. Barbosa et al[31] also found that phenolic-enriched apple cultivars had a high level of inhibitory activity against α-glucosidase. The main phenolic compounds found in peel extracts were quercetin derivatives, protocatechuic acid, and chlorogenic acids, whereas pulp extracts had quercetin derivatives, chlorogenic acid, and *p*-coumaric acid. A high phenolic content and inhibitory activity against α-glucosidase were also observed in peel and pulp extracts of different pear cultivars. Phenolic-enriched fruits, such as apple, pear, cherry, and berries, not only have potential to control postprandial hyperglycemia but could also maintain the cellular redox balance to prevent long-term diabetic complications. Pinto et al.[32] reported a high inhibitory activity against α-glucosidase and ACE in some Peruvian fruits. Modulation during pre- and post-harvest stages through agronomic, horticultural, biotechnologic, and bioprocessing tools could improve the quantity of bioactive compounds in these fruits and could be utilized in effective therapeutic strategies in type 2 diabetes and associated hypertension management.

ENHANCEMENT AND MOBILIZATION OF PLANT-BASED ANTIOXIDANTS, INCLUDING PHENOLICS

Phenolic phytochemicals have a crucial role in the development of functional food for current and future health- and wellness-related dietary applications.[7] Improving phenolic phytochemicals in plant-based food sources is an exciting disease-preventive strategy for combating non-communicable chronic diseases, as the cost of treatment is increasing rapidly and such drug treatments are not affordable in the less-developed countries. Novel tissue culture and bioprocessing technologies have been developed for consistent production of bioactively optimized dietary phytochemicals in food crops and also in medicinal plants.[33]

McCue et al.[34] improved the antidiabetic potential of soybean extracts by sprouting and dietary fungal bioprocessing with *Rhizopus oligosporus*. Short-term sprouting improved the inhibitory action on α-amylase, while long-term sprouting increased the inhibition of α-glucosidase. In another study with common American and Asian vegetables and spices, they observed significant antidiabetic activity (inhibition of α-amylase and α-glucosidase) in *in vitro* assays.[35] They found a strong association between antioxidant activity and α-amylase inhibition in phenolic-optimized foods. In this study, ginger showed strong inhibition of angiotensin I-converting enzyme (ACE) – so it has potential for the management of hyperglycemia-associated hypertension. Correira et al.[36] also found inhibition of α-amylase in pineapple waste bioprocessed with *Rhizopus oligosporus*; they concluded that inhibition of α-amylase is associated mainly with the particular structure of the phenolic compound in the food. In another study, a higher concentration of rosmarinic acid in oregano extracts showed a high inhibition of α-amylase.[37]

Bioprocessing of grain and legume sprouts also improves the phenolic-linked antioxidant activity and functionality in plant-based foods; this is relevant to the management of type 2 diabetes. Randhir et al.[38] found increased inhibitory activity against α-glucosidase in wheat, buckwheat, and oats after thermal processing with an autoclave. In the same study, inhibitory activity against α-amylase also increased in sprouts of buckwheat and oats after thermal processing. A higher phenolic content and high inhibitory activity against α-amylase were observed after solid-state bioconversion of mung beans with *Rhizopus oligosporus*.[39] Within a plant-system model, acid, exogenous phenolic, proline analogues, and precursor combinations, as well as microbial elicitors, can be used to stimulate the biosynthesis of phenolics and other antioxidant enzymes in order to counter oxidative stress in plants;[40] this approach may be used for enhancing the formation of bioactive phytochemicals in plant-based foods (Fig. 27.3). This model provides a scientific foundation for developing different dynamic strategies to harness the benefits of a plant's phenolic phytochemicals in designing functional foods and nutraceuticals to counter chronic diseases such as type 2 diabetes and its complications. It is important to use such knowledge for manipulating the production of an agricultural food crop when designing antioxidant-enriched foods. Phenolic biosynthesis in plants can be stimulated through different mechanisms such as seed treatment, manipulation of soil rhizosphere, external application of natural antioxidants during early growth stages (and also during the pre-harvest stage), and post-harvest treatment with natural bioprocessed elicitors (Fig. 27.4). Mobilization of antioxidant enzymes in seeds and in other plant parts is a very efficient and effective strategy for harnessing health-beneficial compounds from food crops. Legumes and other plants have been studied by using such a scientific rationale to stimulate health-relevant plant bioactives. Such innovative strategies have been developed to manipulate a plant's metabolic regulation in order to design functional food with consistent and optimized phenolic phytochemicals. Tools for improving the phenolic-linked antioxidants in plants can be used in an agricultural production system to enhance the enrichment of bioactive products – which is relevant to the management of diseases, including type 2 diabetes, and aging.

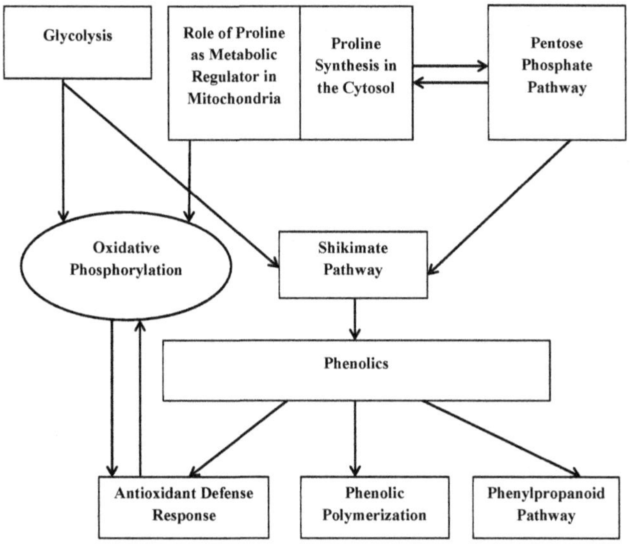

FIGURE 27.3 The role of regulation of the proline-associated pentose phosphate pathway in the biosynthesis of phenolics and stimulation of the antioxidant defense response in plants. The figure was created and adapted from the model described by Shetty & Wahlqvist.[40]

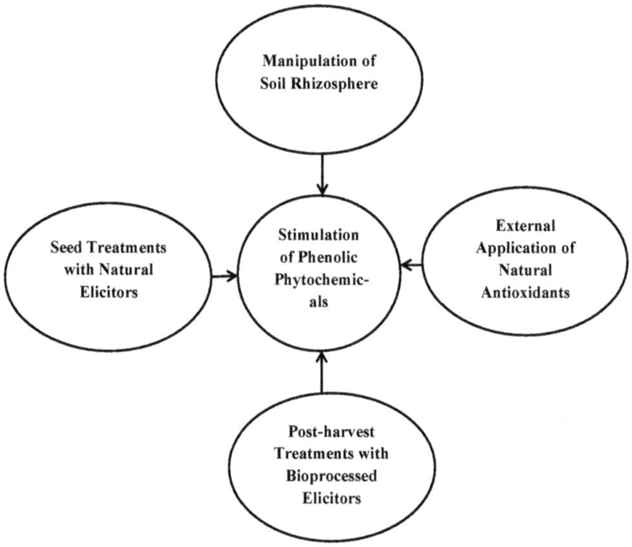

FIGURE 27.4 Different mechanisms for stimulating phenolic phytochemicals in plants. This figure was created by the authors by summarizing different agronomic, horticultural, and metabolic strategies available for stimulating the biosynthesis of phenolic phytochemicals in plants.

CONCLUSION

Maintaining cellular homeostasis by reducing oxidative pressure is an effective measure for managing obesity-associated chronic diseases such as type 2 diabetes, cardiovascular diseases, and the associated risk of aging. Plant-based antioxidants can counter oxidative stress successfully and thus help to prevent the development of many chronic diseases in humans. An intake of antioxidant-enriched food mitigates cellular oxidative stress by directly quenching free radicals or by inducing the endogenous antioxidant-enzyme defense. A balanced diet, enriched with bioactive antioxidant foods, potentially provides protection against the metabolic breakdown associated with the pathophysiology of type 2 diabetes and aging. Antioxidants not only protect cellular components but could also potentially help to repair damage caused by disease-induced oxidative stress. The management of non-communicable chronic diseases, such as type 2 diabetes, needs to include this scientific rationale for designing a cost-effective solution involving functional food.

Pharmaceutical drugs alone cannot prevent and manage the global epidemic of type 2 diabetes because of their huge costs and potentially harmful side effects, particularly for the older population. As this disease is complex and multifaceted, it requires dynamic and holistic health management strategies involving food, lifestyle, physical activity, and mental health care coupled with pharmacologic solutions. Research and translational application efforts to design and produce functional food with higher bioactive ingredients need special attention for better management of chronic diseases including type 2 diabetes and reducing health disparities globally. It is also necessary to understand the value of traditional and ethnic food systems in the management of these diseases. Many ethnic foods and traditional crop cultivars are rich sources of plant phytochemicals and can be utilized in type 2 diabetes-related aging management. These could be part of the solution for native populations, worldwide, who are more adversely affected by this disease and who could use their own traditional food systems following metabolically rationalized screening. Fruit and vegetable-enriched diets, particularly for the urban population, can also be effective in preventing and managing type 2 diabetes and associated aging. Metabolically, we cannot escape our dependency on oxygen – as oxygen is necessary to provide energy – and we also cannot avoid oxygen malfunction at the cellular level; however, we can learn better ways of managing the situation by understanding and using ecologic and biologic principles from our environment and especially from plant foods.

SUMMARY POINTS

- The pathophysiology of many human diseases, including type 2 diabetes, and aging involves the breakdown of cellular homeostasis owing to oxidative stress.
- Oxidative stress in diabetic populations is due to either (i) over-nutrition with soluble carbohydrates from refined calorie-dense foods or (ii) early exposure to malnutrition, which can cause significant cellular damage and potentially result in the development of insulin resistance, β-cell dysfunction, and the accumulation of toxic substances, leading to aging.
- Plants are a rich source of natural antioxidants, and they can be used in the diet to counter type 2 diabetes-induced oxidative pressure linked to cellular complications.
- Plant phenolics are secondary metabolites. They could have a significant role in glucose metabolism and in the control of oxidation-linked malfunction and hypertension – and thus reduce mortality in diabetics.
- Plant phytochemicals, including phenolics, can (i) directly quench free radicals generated from oxidative stress and (ii) induce the endogenous antioxidant enzyme defense system in cells.
- The benefits of phenolic phytochemicals can be harnessed by designing functional food with enriched and consistent levels of bioactive components for the effective prevention and management of type 2 diabetes and its oxidation-based complications (microvascular and macrovascular hypertension).

References

1. Bisbal C, Lambert K, Avignon A. Antioxidants and glucose metabolism disorders. *Curr Opin Clin Nutr Metabol Care* 2010;**13**:439–46.
2. Mittler R. Oxidative stress, antioxidants and stress tolerance. *Trends Plant Sci* 2002;**7**:405–10.
3. Beckman KB, Ames BN. Oxidants, antioxidants, and aging. In: Scandalios JG, editor. *Oxidative stress and the molecular biology of antioxidant defenses.* Plainview, NY: Cold Spring Harbor Laboratory Press; 1997. pp. 201–46.
4. Christie PJ, Alfenito MR, Walbot V. Impact of low-temperature stress on general phenylpropanoid and anthocyanin pathways: enhancement of transcript abundance and anthocyanin pigmentation in maize seedlings. *Planta* 1994;**194**:541–9.
5. Hanhineva K, Törröne NR, Bondia-Pons I, et al. Impact of dietary polyphenols on carbohydrate metabolism. *Int J Mol Sci* 2010;**11**:1365–402.
6. Dembinska-Kiec A, Mykkänen O, Kiec-Wilk B, Mykkänen H. Antioxidant phytochemicals against type 2 diabetes. *Br J Nutr* 2008;**99**:ES109–17.
7. Kwon Y-I. *3eases using phenolic phytochemicals.* Amherst: Doctoral thesis submitted to the University of Massachusetts; 2007. 15.

8. Caspery WF. Physiology and pathophysiology of intestinal absorption. *Am J Clin Nutr* 1992;**55**:299S–308S.
9. Song Y, Manson JA, Burring JE, et al. Association of dietary flavonoids with risk of type 2 diabetes, and markers of insulin resistance and systematic inflammation in women: a prospective study and cross-sectional analysis. *J Am Coll Nutr* 2005;**24**:376–84.
10. Hnamura T, Hagiwara T, Kawagishi H. Structural and functional characterization of polyphenols isolated from *Acerola* (*Malpighia emerginata* DC.) fruit. *Biosci Biotechnol Biochem* 2005;**69**:280–6.
11. Kim H, Peterson TG, Barnes S. Mechanisms of action of the soy isoflavone genistein: emerging role for its effects via transforming growth factor beta signaling pathways. *Am J Clin Nutr* 1998;**68**:1418S–25S.
12. Choi MS, Jung UJ, Yeo J, et al. Geneistin and daidzein prevent diabetes onset by elevating insulin level and altering hepatic gluconeogenic and lipogenic enzyme activities in non-obese diabetic (NOD) mice. *Diabetes Metab Res Rev* 2008;**24**:74–81.
13. Zaveri NT. Green tea and its polyphenolic catechins: medicinal uses in cancer and noncancer applications. *Life Sciences* 2006;**78**:2073–80.
14. Hertog MGL, Kromhout D, Aravanis C, et al. Flavonoid intake and long-term risk of coronary heart disease and cancer in the seven countries study. *Arch Intern Med* 1995;**155**:381–6.
15. Ford ES, Mokdad AH. Epidemiology of obesity in the western hemisphere. *J Clin Endocrinol Metab* 2008;**93**:s1–8.
16. Montonen J, Knekt P, Härkänen T, et al. Dietary patterns and incidence of type 2 diabetes. *Am J Epidemiol* 2005;**161**:219–27.
17. Villegas R, Shu XO, Gao Y-T, et al. Vegetable but not fruit consumption reduces the risk of type 2 diabetes in Chinese women. *J Nutr* 2008;**138**:574–80.
18. Liu S, Serdula M, Janket S-J, et al. A prospective study of fruit and vegetable intake and the risk of type 2 diabetes in women. *Diabetes Care* 2004;**27**:2993–6.
19. Kwon Y-I, Apostolidis E, Shetty K. Evaluation of pepper (*Capsicum annum*) for management of diabetes and hypertension. *J Food Biochem* 2007;**31**:370–85.
20. Kwon Y-I, Apostolidis E, Shetty K. In vitro studies of eggplant (*Solanum melongena*) phenolics as inhibitors of key enzymes relevant for type 2 diabetes and hypertension. *Bioresource Technol* 2008;**99**:2981–8.
21. Saleem F. *Anti-diabetic potentials of phenolic enriched Chilean potato and selected herbs of apiaceae and lamiaceae families*. Amherst: M.S. Thesis University of Massachusetts; 2010.
22. Jayaprakasam B, Vareed SK, Olson LK, Nair MG. Insulin secretion by bioactive anthocyanins and anthocyanidins present in fruits. *J Agric Food Chem* 2005;**53**:28–31.
23. Mulabagal V, Lang GA, DeWitt DL, et al. Anthocyanin content, lipid peroxidation and cyclooxygenase enzyme inhibitory activities of sweet and sour cherries. *J Agric Food Chem* 2009;**57**:1239–46.
24. Seymour EM, Singer AAM, Kirakosyan A, et al. Altered hyperlipidemia, hepatic steatosis, and hepatic peroxisome proliferator activated receptors in rats with intake of tart cherry. *J Medicinal Food* 2008;**11**:252–9.
25. Ankolekar A, Pinto M, Greene D, Shetty K. Phenolic bioactive modulation by *Lactobacillus acidophilus* mediated fermentation of cherry extracts for anti-diabetic functionality, *Helicobactor pylori* inhibition and probiotic *Bifidobacterium longum* stimulation. *Food Biotechnol* 2011;**25**:305–35.
26. Cheplick S, Kwon Y-I, Bhowmik P, Shetty K. Phenolic-linked variation in strawberry cultivars for potential dietary management of hyperglycemia and related complications of hypertension. *Bioresource Technol* 2010;**101**:404–13.
27. Cheplick S, Kwon Y-I, Bhowmik P, Shetty K. Clonal variation in raspberry fruit phenolics and relevance for diabetes and hypertension management. *J Food Biochem* 2010;**31**:656–79.
28. Pinto M, Kwon Y-I, et al. Functionality of bioactive compounds in Brazillian strawberry *(Fragaria x ananassa* Duch.) cultivars: evaluation of hyperglycemia and hypertension potential using *in vitro* models. *J Agric Food Chem* 2008;**56**:4386–92.
29. Pinto M, Kwon Y-I, Apostolidis E, et al. Evaluation of red currants (*Ribes nigrum* L.), black currants (*Ribes nigrum* L.), red and green gooseberries (*Ribes uva-crispa*) for potential management of type 2 diabetes and hypertension using *in vitro* models. *J Food Biochem* 2010;**34**:639–60.
30. Adayanthaya I, Kwon Y-I, Apostolidis E, Shetty K. Health benefits of apple phenolics from postharvest stages for potential type 2 diabetes management using *in vitro* models. *J Food Biochem* 2010;**34**:31–49.
31. Barbosa AC, Pinto M, Sarkar D, et al. Influence of varietal and pH variation on antihyperglycemia and antihypertension properties of long-term stored apples using *in vitro* assay models. *J Food Biochem* 2012;**36**:479–93.
32. Pinto M, Ranilla LG, Apostolidis E, et al. Evaluation of antihyperglycemia and antihypertension potential of native Peruvian fruits using *in vitro* models. *J Medicinal Food* 2009;**12**:278–91.
33. Shetty K. Biotechnology to harness the benefits of dietary phenolics; focus on Lamiaceae. *Asia Pacific J Clinical Nutr* 1997;**6**:162–71.
34. McCue P, Kwon Y-I, Shetty K. Anti-diabetic and anti-hypertensive potential of sprouted and solid-state bioprocessed soybean. *Asia Pac J Clin Nutr* 2005;**14**:145–52.
35. McCue P, Kwon Y-I, Shetty K. Anti-amylase, anti-glucosidase and anti-angiotensis I-converting enzyme potential of selected foods. *J Food Biochem* 2005;**29**:278–94.
36. Correia RTP, McCue P, Vattem DA, et al. Amylase and *Helicobacter pylori* inhibition by phenolic extracts of pineapple wastes bioprocessed by *Rhizophus oligisporus*. *J Food Biochem* 2004;**28**:419–34.
37. McCue P, Shetty K. Inhibitory effects of rosmarinic acid extracts on procine pancreatic amylase *in vitro*. *Asia Pac J Clin Nutr* 2004;**13**:101–6.
38. Randhir R, Kwon Y-I, Shetty K. Effect of thermal processing on phenolics, antioxidant activity and health-relevant functionality of select grain sprouts and seedlings. *Innovat Food Sci Emerging Technol* 2008;**9**:355–64.
39. Randhir R, Kwon Y-I, Shetty K. Mung bean processed by solid-state bioconversion improves phenolic content and functionality relevant for diabetes and ulcer management. *Innovative Food Sci Emerging Technol* 2007;**8**:197–204.
40. Shetty K, Wahlqvist M. A model for the role of proline-linked pentose phosphate pathway in phenolic phytochemical biosynthesis and mechanism of action for human health and environmental applications. *Asia Pacific J Clinical Nutr* 2004;**13**:1–24.

Index

Note: Page numbers with "*f*" denote figures; "*t*" tables.

Printed in the United States
By Bookmasters